Catfishes, a Highly Diversified Group

Volume 1

Their Outstanding Biology

Editors

Gloria Arratia

Biodiversity Institute, University of Kansas
Lawrence, KS, USA

Roberto E. Reis

Pontifícia Universidade Católica do Rio Grande do Sul
Porto Alegre, RS, Brazil

CRC Press
Taylor & Francis Group
Boca Raton London New York

CRC Press is an imprint of the
Taylor & Francis Group, an **informa** business
A SCIENCE PUBLISHERS BOOK

Cover Illustration: *Pseudoplatystoma corruscans* from Olho d'Água River, Serra da Bodoquena, Jardim, Mato Grosso do Sul, Brazil. Photograph courtesy of José Sabino.

First edition published 2025
by CRC Press
2385 NW Executive Center Drive, Suite 320, Boca Raton FL 33431

and by CRC Press
4 Park Square, Milton Park, Abingdon, Oxon, OX14 4RN

CRC Press is an imprint of Taylor & Francis Group, LLC

Library of Congress Cataloging-in-Publication Data (applied for)

ISBN: 978-1-032-42135-3 (hbk)
ISBN: 978-1-032-42136-0 (pbk)
ISBN: 978-1-003-36133-6 (ebk)

DOI: 10.1201/9781003361336

Typeset in Times New Roman
by Prime Publishing Services

Foreword

In 2003, Gloria Arratia, B.G. Kapoor, Michel Chardon, and Rui Diogo published a book with a bold title: *Catfishes*, which represented the first comprehensive treatment of the Siluriformes. The book was divided into two volumes dedicated to the biology, morphology, physiology, behavior, taxonomy, and systematics of this highly diversified group. This group, second only to Cypriniformes in number of freshwater species, and in contrast to them, is currently distributed across all continents except Antarctica. Siluriforms have an enormous representation in South America, whereas cypriniforms are absent (except for human introductions). *Catfishes* 2003 was an overview of knowledge, current research, and scientific disagreements yet to be resolved about Siluriformes, and the project required the effort of the international community studying catfishes. That community was mostly represented by European researchers, with few from other continents. *Catfishes* 2003 produced some outstanding contributions that will stand the passage of time, and the book fully reached its goals, being sold out long ago.

Twenty years later, the current knowledge on Siluriformes has reached a higher level of understanding due to the interest of a new generation of scientists dedicated to catfishes and the acquisition of new methodologies for morphological and phylogenetic studies, resulting in a significant number of molecular hypotheses of phylogenetic relationships within certain families. As a result, there have been changes in previous interpretations, but simultaneously, some major disagreements between morphological and molecular hypotheses of relationships. Thus, G. Arratia and R.E. Reis planned this new book, *Catfishes, a Highly Diversified Group*, in two volumes, subtitled *Their Outstanding Biology* and *Evolution and Phylogeny.* The first volume, *Their Outstanding Biology* aims to update the current knowledge of Siluriformes, including the external and internal anatomy of numerous catfish families, their functions, and explanation of adaptations in unique groups.

There is a major switch in the current research on Siluriformes, as recent research has emphasized South American catfishes, with a minimal representation of Africa, Asia, and North America, which points to the necessity to expand studies on these continents. It is expected that a great amount of novel information from Africa and Asia will be generated in the future, as these continents have great potential to increase the known diversity, impacting current interpretations. Thus, the vast embodiment of catfishes cannot be thoroughly covered in this book, but our main goal of updating the knowledge has been reached. Each chapter represents the current state of the art with exhaustive bibliographic verification of each subject, plus new information from the authors.

Topics and Contributors

After the introduction, this book has two sections. In *Anatomy of Catfishes*, morphology is amply covered in two chapters; Rizzato and Almeida offer a summary of some main features of the external anatomy, while Moritz brings an overview of the hard and soft internal anatomy, presented along siluriform family diversity, giving a new approach to the subject. These chapters illustrate the need for further research, as internal structures of the circulatory, digestive, urogenital, nervous, and other systems are incompletely known in Siluriforms, and this is a call for further research on these topics. A similar approach is followed in the presentation on skeletal structures in two other chapters; in addition to a general overview, Moritz, Koch, and Thieme present a specific chapter on the diversity of the pelvic girdle and fin, and Duclos and Grande explore the modularity and integration in the evolution of the Weberian apparatus, a novel subject that opens a new avenue in siluriform research. Internal "soft" anatomy is also treated by Cumplido, who introduces us to a new technique to study muscles and a new understanding of caudal myology in certain catfishes, including the discovery of new muscles. In another chapter, Egge writes about catfish venom systems, detailing anatomy, functional morphology, and venom composition and action. Anatomical studies can be enhanced and reveal unknown structures using new techniques, such as X-Ray Computed Tomography, which is briefly but accurately explained by Arce and Summer.

The last section of this book, *Biology of Catfishes*, is dedicated to selected groups with special modifications that adapt catfishes to singular lifestyles. These include parasitic (DoNascimiento and de Pinna), subterranean (Trajano and Secutti), and migratory catfishes (Zuanon and Torrente-Vilara). We also learn about size differences in catfishes and a new interpretation of miniaturization, a special characteristic affecting a large number of catfishes (Henschel, Katz and Bragança). Functional topics, such as sound production and communication (Kaatz, Lobel and Rice) and respiration (Zaccone, Icardo and Alesci), represent other expanding areas of knowledge in catfishes, and these are the chapters closing this book.

The book begins with a chapter by editors Arratia and Reis, introducing readers to the present state of catfish research—its developed areas, but also scarcely explored ones, as well as addressing challenges affecting catfishes worldwide that may not be specifically treated in the other chapters. Arratia and Reis present this invitation to explore the large catfish world and its current understanding.

Acknowledgments

A book with a wide scope, such as this one, is only possible with the assistance of many dedicated reviewers. All chapters were strictly reviewed by other specialists, who contributed their time and expertise in their respective fields, and we are very thankful to all of them. They suggested improvements where they were needed and advice on the scientific subject itself and its presentation. Thus, each chapter also reflects the contribution of each reviewer. These reviewers are Dominique Adriaens (Gent, Belgium), Gloria Arratia (Lawrence, KS, USA), Kelly Boyle (New Orleans, LA, USA), Tiago Carvalho (Bogotá, Colombia), Nicolás Cumplido (Santiago, Chile), Carlos DoNascimiento (Medellín, Colombia), Kory Evans (Houston, TX, USA), Carl Ferraris, Jr. (Portland, OR, USA), Vidal Haddad Junior (São Paulo, Brazil), Lisiane Hahn (São José dos Campos, Brazil), George Mattox (São Carlos, Brazil), Doaa Mokhtar (Assiut, Egypt). Fernando Pelicice (Porto Nacional, Brazil), Eric Parmentier (Liege, Belgium), Graham Proudlove (Manchester, England), Vinícius Reis (São Paulo, Brazil), Pedro Rizzato (São Paulo, Brazil), Lukas Rüber (Bern, Switzerland), Irene Salinas (Albuquerque, NM, USA), Brian Sidlauskas (Corvallis, OR, USA), Veronica Slobodian (Brasília, Brazil), Philipp Thieme (Montpellier, France), Richard Winterbottom (Toronto, Canada), and Wolmar Wosiacki (Belém, Brazil). Some reviewers checked more than one manuscript, and the two editors of this book reviewed every one.

Special thanks to colleagues who contributed to clarifying some specific issues raised during the review process (José Birindelli, Londrina, Brazil) and T.J. Meehan (Lawrence, KS, USA), who helped improve the English style of some manuscripts. We also thank Raju Primlani of the CRC Press for assistance in preparing the manuscripts for publication. R.E. Reis is partially supported by the National Council for Scientific and Technological Development, Brazil (CNPq processes #302760/2022-9 and #400166/2016-0). Office materials and computer facilities were supported by H.P. Schultze and G. Arratia (Lawrence, USA). To all these colleagues, we are indebted for their time and consideration.

Gloria Arratia and Roberto Reis

Contents

Section 1
Introduction

Chapter 1

An Introduction to Catfishes

A Tale of Evolutionary Success

Gloria Arratia[1,*] and *Roberto E. Reis*[2]

Catfishes, or Siluriformes, comprise a remarkable and extraordinarily diverse group of fishes that have adapted to a wide range of aquatic environments worldwide, including marine and freshwaters. From the bustling rivers of South America and Asia to the cichlid-ruled lakes of Africa, catfishes thrive in virtually all habitats. They constitute a ubiquitous faunal component in high-altitude Andean wetlands and lakes (e.g., trichomycterids in Chungará Lake at 4,517 m above sea level, the highest lake in the world containing fishes; Figure 1A), torrential rivers worldwide (Figure 1B), and Himalayan creeks (e.g., sisorid *Glyptosternum* and amblycipitid *Amblyceps* in the Tibetan Plateau) to oxygen-deprived lowland lakes (e.g., ictalurids in Lacamas Lake, USA). They live in shallow marine beaches of sunny Australia (plotosids) to the dark freshwater caves of Asia and South America (Figure 1C), or even in phreatic habitats in South America (Figure 1D; *Phreatobius goeldi*; Trajano and Secutti this volume) in the Amazon of Brazil and Bolivia and North America (*Satan eurystomus* and *Trogloglanis pattersoni*), and a few are parasites of other fishes (e.g., *Vandellia cirrhosa* in the Amazon basin; Figure 1E; see DoNascimiento and de Pinna this volume). In other words, catfishes represent a highly successful evolutionary history and inhabit different freshwater, brackish, and marine environments of the world, except Antarctica.

Such a large group is characterized by its enormous external morphological variability that goes in parallel with its internal morphological differences. Catfishes include giant species such as the European freshwater *Silurus glanis* reaching over four meters and reaching a maximum of 290 kilograms (Figure 2A); exceptionally large migratory species (e.g., *Brachyplatostoma filamentosum* migrating from the Amazon freshwater to brackish water systems with almost three meters in length and over 250 kg; see Zuanon and Torrente-Vilara this volume), as well as miniature forms (e.g., *Hara*, *Scoloplax*, *Tridens*; Figure 2B; Henschel et al. this volume) with less than 20 mm, and varying from plain dark fishes that lurk among rocks to attractive beauties that fascinate aquarists (Figure 2C). Thus, catfishes include an astonishing diversity of shapes, sizes, colors, and ecologies and currently represent the second largest freshwater fish order with over 4,200 species, only beaten by the Cypriniformes.

[1] Biodiversity Institute, University of Kansas, Dyche Hall, 1345 Jayhawk Blvd., Lawrence, KS 66045, USA.

[2] Pontifícia Universidade Católica do Rio Grande do Sul, PUCRS, Av. Ipiranga, 6681, 90619-900 Porto Alegre, RS, Brazil. Email: reis@pucrs.br

* Corresponding author: garratia@ku.edu

Figure 1. Environmental diversity and catfishes. (A) High altitude Andean lakes, e.g., Chungará Lake at about 4,500 m.a.s.l. where only one catfish is living, *Trichomycterus chungaraensis*, Trichomycteridae. (Photo from Wikipedia). (B) Section of a torrential river where catfishes inhabit the Rhithron, between and below the bottom stones; a kind of environment occupied by diverse catfishes worldwide (photograph courtesy of Cristian Correa). (C) An unusual environment, a well in the neotropics, São Gabriel da Cachoeira, Rondônia, Brazil containing *Phreatobius dracunculus* (J. Zuanon is inside the well, courtesy of Eleonora Trajano). (D) A living host, the characiform *Myloplus asterias*, parasitized by a catfish, *Vandellia cirrohsa* (Trichomycteridae), identified with arrows (photograph courtesy of Leandro Sousa); the serrasalmid is oriented to the left.

What Makes a Fish a Catfish?

The Order Siluriformes is currently composed of over 4,200 species in ca. 450 genera, classified in 40 extant and five extinct families (Table 1), but these numbers are susceptible to changes, as some of the families are not recognized in the catfish phylogeny in this book (see de Pinna et al. 2025). In addition, during the last months, there were a few major taxonomic readjustments, and the freshwater Asian Erethistidae lost its family status (Fricke et al. 2024). The largest catfish family is

Figure 2. Catfish diversity, a sample of curiosities and beauties. (A) *Silurus glanis* (Siluridae), one of the largest and heaviest known siluriform, swimming in a wildlife reserve in Eggstätt-Hemhofer Seenplatte, Germany (photograph by Andreas Harti). (B) *Pimelodus blochii* with its long barbels, swimming in Saramacca River at Uitkijk, Suriname (Pimelodidae; photograph by Vivica von Vietinghoff). (C) *Belonoglanis* cf. *tenuis* in Ngoko River at Ndongo, Cameroon (Amphiliidae; photograph by Timo Moritz). (D) The upside-down catfish, *Synodontis* sp. Juvenile, swimming in Ngoko River at Ndongo, Cameroon (Mochokidae; photograph by Timo Moritz). (E) A group of the rare *Phreatobius cisternarum* (Phreatobidae) from the Marajó Island, Pará, Brazil (photograph by Janice Muriel-Cunha).

the Neotropical Loricariidae, with one-fourth of the Order richness, followed by Trichomycteridae (423 species), Sisoridae (315 species), Heptapteridae (235 species), Mochokidae (228 species), Callichthyidae (225 species), and Bagridae (223 species), to mention those with more than two hundred species. The numbers are changing rapidly for some families (e.g., Trichomycteridae that in the last six months had one new genus and six new species described), whereas they stay unchanged year after year for others (e.g., Nematogenyidae). The Neotropical realm harbors the largest catfish richness on Earth, with over 2,500 species, followed by Asia (+600 species) and Africa (+600 species). North America and Europe have a much smaller catfish diversity, with ca. 50 and two endemic species, respectively, while ca. 200 species represent marine and brackish water catfishes of the families Ariidae and Plotosidae with a cosmopolitan distribution.

Catfishes are long known to the sciences. Although they were interpreted as a formal group of their own by Rafinesque in 1815 under the name Oplophoria, they had already been known for about one hundred years but not recognized as a group. The first anatomically based diagnosis and classification appeared in Cuvier (1817) under the name "Siluroïdes." For details on the history of catfish taxonomy and classifications throughout time, see de Pinna (1998), and for a historical

Table 1. Catfish families are ordered alphabetically, listing valid taxa, their habitats, and geographical distribution. Most genera and species content of extant families, according to Fricke et al. (2024); fossil species not included in these counts. Fossil (†) families are documented by Arratia (2025). Freshwater** – mostly freshwater with some estuarine species. Abbreviations: CA (Central America), NA (North America), SA (South America). ***Name described in Arratia (2025).

Families	Nr. of genera	Nr. of species	Habitat	Geographical Distribution
†Andinichthyidae	--	--	Freshwater	SA
†Astephidae	1	1	Freshwater	NA
†Bachmanniidae	1	1	Freshwater	SA
†Hypsidoridae	1	2	Freshwater	NA
†Kooiichthyidae***	1	1	Marine	SA
Aiilidae	7	25	Freshwater	Asia
Akysidae	5	59	Freshwater	Asia
Amblicipitidae	4	49	Freshwater	Asia
Amphiliidae	13	106	Freshwater	Africa
Ancharidae	2	6	Freshwater	Africa
Ariidae	42	157	Brackish and Marine	Cosmopolitan
Aspredinidae	13	49	Freshwater**	SA
Astroblepidae	1	82	Freshwater	Panama and SA
Auchenipteridae	25	128	Freshwater**	Panama and SA
Auchenoglanididae	3	21	Freshwater	Africa
Austroglanididae	1	3	Freshwater	Africa
Bagridae	17	223	Freshwater	Africa and Asia
Callichthyidae	8	225	Freshwater	SA
Cetopsidae	5	44	Freshwater	SA
Chacidae	1	4	Freshwater	Asia
Clariidae	16	120	Freshwater	Africa and Asia
Claroteidae	9	65	Freshwater	Africa
Cranoglanididae	1	3	Freshwater	Asia
Diplomystidae	2	9	Freshwater	SA
Doradidae	31	96	Freshwater	SA
Heptapteridae	23	235	Freshwater	Mexico and SA
Heteropneustidae	1	5	Freshwater	Asia
Horobagridae	4	10	Freshwater	Asia
Ictaluridae	7	51	Freshwater	CA and NA
Kryptoglanidae	1	1	Freshwater	Asia
Lacantuniidae	1	1	Freshwater	NA
Loricariidae	115	1,056	Freshwater	CA and SA
Malapteruridae	2	19	Freshwater	Africa
Mochokidae	9	228	Freshwater	Africa
Nematogenyidae	1	1	Freshwater	SA
Pangasiidae	4	30	Freshwater**	Asia
Pimelodidae	30	116	Freshwater	Panama and SA
Plotosidae	10	42	Brackish and Marine	Indian and W. Pacific oceans
Pseudopimelodidae	6	56	Freshwater	SA
Ritiidae	1	7	Freshwater	Asia
Schilbeidae	5	35	Freshwater	Africa
Scoloplacidae	1	6	Freshwater	SA
Siluridae	13	106	Freshwater	Europe and Asia
Sisoridae	26	315	Freshwater	Asia
Trichomycteridae	43	423	Freshwater	CA and SA
Total	449	4,217		

account of high-level phylogeny, see Diogo (2003, 2004) and Diogo and Peng (2010). In the time span of 200 years, the taxonomic knowledge of the group increased enormously, especially after the influence of Hennig (1966), which gave a better understanding of the Order Siluriformes by searching for its phylogenetic position among the Teleostei. This included specific catfish families, extending to the phylogenetic relationships among families, as well as those of the Siluriformes with other ostariophysans and other clupeocephalans (e.g., Fink and Fink 1981, 1996; Betancur-R. et al. 2017; Arratia 2018).

Siluriform Identity

A group so diversified as the Siluriformes could be difficult to diagnose; however, despite their enormous variability, one can empirically recognize a catfish from other teleosts. Catfishes have no scales, although some species possess different arrangements of bony plates covering most or parts of the body; typical teleostean cycloid scales are absent in all catfishes. In addition, although other teleostean groups may have barbels outside siluriforms, the catfish head may possess from one to four pairs (see Rizzato and Almeida this volume). Their skull is generally depressed dorsoventrally (platybasic condition; Figure 3); the cranial roof is generally broad and has one bone aligned with the frontals posteriad and a characteristically-shaped mesethmoid anteriad to which the flat premaxillae (occasionally one premaxilla) attach. Additionally, the maxillae are commonly atrophied (Arratia

Figure 3. A few examples of threatened species. (A) Historical specimen of *Diplomystes chilensis* collected in Maipo Basin (Chile) over 100 years ago; the species is now considered extinct (courtesy of Andrew Williston) and is only represented by a few museum specimens. (B) *Diplomystes incognitus*, Endangered, another member of the primitive family Diplomystidae from Chile (photograph by Claudio Quezada-Romegialli). (C) The beauty of *Hypancistrus zebra* (Loricariidae), Critically Endangered, heavily demanded by the aquarium trade, and threatened by the Belo Monte hydropower plant, the main reason for its decline (photograph by Leandro Sousa).

2003a) and lacking teeth. Moreover, the position and orientation of the hyomandibula and the palatoquadrate make the identification of a catfish easy. In Siluriformes, the hyomandibula occupies the position of the metapterygoid of other teleosts, whereas the metapterygoid occupies the position of both the ectopterygoid and entopterygoid in other teleosts.

In contrast to other teleostean orders, Siluriformes are unique in several anatomical features. For instance:

1) Absence of scales or scaleless skin and highly developed sensory capabilities. The so-called naked skin possesses different kinds of receptors and alarm cells among catfish groups, being a highly sophisticated sensory organ unlikely other teleosts (e.g., Arratia and Huaquín 1995; Arratia 2003b).
2) Loss of certain bones, e.g., basisphenoid, intercalar, parietal bone as a separate ossification, wall of the trigemino-facialis chamber, myodomes, sclerotic bones, a dermal supraorbital, supramaxilla, dermopalatine, symplectic, posteroventral process of quadrate, and subopercle. Some of these losses, which are not exclusive to catfishes, may have different phylogenetic interpretations: (a) the absences of basisphenoid, dermopalatine, and supramaxilla have been interpreted as ostariophysan synapomorphies (Fink and Fink 1981); (b) the absence of an independent parietal bone, symplectic and subopercle as synapomorphies of Siluriformes (e.g., Wiley and Johnson 2010); and (c) the absences of the intercalar bone in the braincase, of a dermal supraorbital, and of sclerotic bones as synapomorphies of Siluriformes and Gymnotiformes (Fink and Fink 1981; Arratia 2003a).
3) Reduction in size of number of bones, e.g., the ectopterygoid and entopterygoid may be rudimentary or even absent in some catfishes (e.g., Arratia 1992, 2003a; Wiley and Johnson 2010); the infraorbital series is usually reduced in most catfishes to the canal bearing bones or ossicles in contrast to the flat, laminar series of infraorbital bones in other teleosts.
4) Possible fusion between bones, a conflictive subject based on possible theoretical phylogenetic fusions of some of them, such as parietal and supraoccipital bones (= parieto-supraoccipital; Arratia and Gayet 1995) or the posttemporal and supracleithrum at the upper part of the shoulder girdle that includes only one bone in siluriforms in contrast to two independent bones in other teleosts. A recent study based on ontogenetic series of *Ictalurus punctatus* (7.7 mm notochordal length to 44.9 mm SL) and *Noturus gyrinus* (5.4 mm SL to 26.4 mm SL) showed the presence of only one ossification center for the bone at the posteromedian region of the skull. Consequently, the author proposed that there is only one bone—supraoccipital—in siluriforms, which originates from one ossification center and that the paired parietals of other teleosts are lost. Furthermore, the so-called postemporo-supracleithrum would originate from a single ossification center that represents the supracleithrum (Kubicek 2022).

   However, a main question remains: Can ictalurids, that phylogenetically are recognized as advanced catfishes (e.g., Mo 1991; de Pinna 1998; Diogo 2004; Hardman 2005; Sullivan et al. 2006; de Pinna et al. 2025) be interpreted as representatives of all Siluriformes? Ontogenetic series from embryo to adults of cyprinids, catostomids, and cobitids show differences in the appearance and formation of certain bones, e.g., of the vertebral column and caudal endoskeleton revealing that there is variability in the formation of bones in the three families (G. Arratia pers. obser.) and the same could be expected from certain siluriform families.
5) Possession of dorsal- and pectoral-fin spines, a siluriform synapomorphy (e.g., Wiley and Johnson 2010), which may be associated with venom glands in different catfishes (Egge this volume).
6) Parurohyal present (Arratia and Schultze 1990; Wiley 2008; Wiley and Johnson 2010).

The platybasic skull of catfishes involves some transformations that may be related to the benthic and nocturnal habits of most of them (Arratia 2003a). The platybasic skull can be interpreted as an adaptation for stability when lying on the bottom, as well as to reduce the current drag of torrential rivers where many of them live. Also, associated to the platybasic condition, there are changes linked with the development of the chondrocranium and the reduction of the eye size that characterize many catfishes. Additionally, the small eyes are covered by the skin in some catfishes, e.g., Nematogenyidae (G. Arratia pers. obser.), in contrast to the larger eye not covered by the skin in diplomystids and other catfishes (e.g., Arratia 1987; Moritz this volume). The loss or fusion of several bones, the structure and orientation of the suspensorium, and the unique development and morphology of the palatoquadrate may also be associated with the platybasic condition of the skull (e.g., Daget 1964; Alexander 1965; Arratia and Schultze 1991; Arratia 1992, 2003a; Adriaens and Verraes 1997, 1998; Adriaens and Vandewalle 2003).

Last 30 Years

Without question, some special conferences and publications have played a major role in the enrichment of knowledge of Siluriformes, sometimes from a regional perspective (e.g., Malabarba et al. 1998 on Neotropical catfishes either fossil and extant) or from a broader perspective (Arratia et al. 2003). It is clear to us that the 1997 *International Symposium on Phylogeny and Classification of Neotropical Fishes* was a key element in the development of ichthyological studies that followed this major initiative. The volume published as a result of the symposium (Malabarba et al. 1998) had ca. 100 pages dedicated to phylogenetic relationships of Neotropical Siluriformes (de Pinna 1998) and specific chapters on three families, the Auchenipteridae (Soares-Porto 1998), Callichthyidae (Reis 1998), and Loricariidae (Montoya-Burgos et al. 1998; Schaefer 1998). The study of three families was a small representation of the South American catfish richness, but it was a stimulant beginning to developing interest in the particular Neotropical siluriform biodiversity among the South American scientific community. Additionally, one chapter (Gayet and Meunier 1998) was dedicated to some Bolivian fossils, including Late Cretaceous/Paleocene catfishes, an issue that remains controversial until the present time (see Arratia 2025; Otero et al. 2025).

One year after the publication of *Phylogeny and Classification of Neotropical Fishes* (Malabarba et al. 1998), Ferrari and de Pinna (1999) recognized 35 catfish families, whereas Teugels (2003) recognized only 31, with 438 genera and 2,753 species being the Loricariidae the most speciose with 653 species in 95 genera and the Trichomycteridae, the second largest one. The analysis of the taxa indicates different study trends because in the last 10 years, 15 new loriicariid genera have been described (e.g., *Rhinotocinclus*; Reis and Lehmann 2022), but only two for trichomycterids *Potamoglanis*; (Henschel et al. 2018) and *Rhinotridens* (Datovo et al. 2023). A significant contribution was a detailed checklist and type catalog for siluriforms by Ferraris (2007), which in last years has been overseen by Fricke and collaborators at the California Academy of Sciences. During this period, two new families were discovered, the Lacantuniidae, based on a single species from Mexico (Rodiles-Hernández et al. 2005), and the Kryptoglanidae, a subterranean catfish from Peninsular India (Britz et al. 2014). Additionally, the Ritidae, which was previously interpreted as a subfamily of Bagridae, was erected to family rank by Nelson et al. (2016). A major change was a new interpretation of the superfamily Pimelodoidea and the recognition of its three subfamilies (Heptapterinae, Pimelodinae, and Pseudopimelodinae) (Diogo 2004) as valid families (Hardman 2005; Sullivan et al. 2006) based on molecular studies.

Five years after the publication of *Phylogeny and Classification of Neotropical Fishes,* the book *Catfishes* (Arratia et al. 2003) was published, and its content (26 chapters) and the list of authors (30) showed marked tendencies in catfish research. The majority of the authors (ca. 57%) were European in contrast to few authors from other continents as Africa (ca. 3%), Asia (ca. 3%), Australia (ca. 3%), North America (ca. 17%), and South America (ca. 10%), revealing that Europe was at that time an active center of study of siluriforms, including well-known researches who covered mainly anatomy, function, behavior, ecology, and little of catfish systematics (ca. 27%). *Catfishes, a Highly Diversified Group, Volume 1: Their Outstanding Biology* (2025) reveals a major change in the nationality of its

55 authors, with the majority from South America (ca. 52%), followed by ca. 20% from North America, ca. 14% from Europe, ca. 6% from Africa, and a similar percent for Asia. These counts do not represent a biased selection because South America, with Brazil at the top, is leading catfish studies worldwide. There has not been only a switch in the composition of authors but also in their goals, being the phylogenetic understanding of families (either morphological, molecular, or both), a major goal, as shown in *Catfishes, a Highly Diversified Group. Volume 2: Evolution and Phylogeny* (Arratia and Reis 2025).

Catfishes Today

These two books dedicated to catfishes are the result of the research efforts and professional experience of 28 ichthyologists in book 1 and 42 in book 2. Book 1—*Catfishes and their Outstanding Biology*—as suggested by the title, deals with the external and internal morphology of catfishes and their different patterns among families. While the characteristic and peculiar external structures and their functions and variability have called the attention of many researchers for a long time, a compilation are clearly and succinctly presented in at least four chapters by P. Rizatto and M. Almeida, J. Egge, and T. Moritz. Different to some previous works, the authors deliver their results and observations in a comparative way among families, illustrating the high variability characterizing external structures in siluriforms and their distribution among families.

As explained above, a significant amount of information is already available concerning catfish skeletal structures, many of which are diagnostic. Consequently, the current knowledge of the skeletal anatomy of many catfish families also reaches high levels, and their diversity is beautifully illustrated in the chapters by Moriz et al. (this volume). In contrast, the knowledge of muscles and associated structures like tendons and ligaments is not as extensive as that of bones, but novel information on the musculature and myomeres of the caudal region of some catfishes is presented and beautifully illustrated by Cumplido (this volume). A novel approach to one of the most characteristic skeletal systems of catfishes—the Weberian apparatus—is addressed by Duclos and Grande, who are opening a new line of research on this subject. In contrast to the information on external structures and bones, internal soft structures such as those belonging to the gas bladder (Kaatz et al. this volume), special respiratory organs (Zaccone et al. this volume), digestive, circulatory, urinary, nervous systems, and others (see Moritz this volume) appears to be incompletely known and in need of further research.

Reproductive Mechanisms, Eggs, and Ontogenetic Growth Series

Throughout this survey, it becomes evident that there is scarce information on catfish reproduction, including reproductive seasons for different species, reproductive organs, behavior, development of eggs, care of eggs and larvae, and early ontogenetic development of most catfishes worldwide. Most of the current knowledge on one or other aspects mentioned above concerns catfishes used in aquaculture and aquarium trades, so that part of the information is published in technical journals associated with these trades.

An important issue in this kind of research is to study the sex proportions in different taxa, because this could be critical for some species and their survival, especially in the current times of climate change. For instance, in two of the most basal, early branches of Siluriformes, Diplomystidae, and Nematogenyidae, there is a marked disproportion between sexes, with males being more numerous than females (Arratia 1983). Similar results also were obtained for some species of Trichomycteridae (Arratia 1983; Manríquez et al. 1988). The significance of such differences is unknown, and there is no data on the proportion of sexes for most catfishes.

The research concerning reproductive annual periods involves intensive, at least monthly, fieldwork, collecting information on the fishes that may extend to 24 months (e.g., Manríquez et al. 1988), and parallel lab studies concerning gonadic changes, but also the periodic collection (usually monthly) of environmental conditions, such as chemical characteristics of the water, climatological conditions of the region, and associated aquatic flora and fauna. Consequently, this type of research is time-consuming and requires significant funding to support field and laboratory work. Again, most of the current knowledge on these subjects is on aquarium and aquaculture species, with very little known about wild species.

Among the two early branches of catfishes, information on the gonadal changes—macromorphological and histological—is only known for *Nematogenys* (Huaquín et al. 2002). This is not the result of a neglected attitude from researchers but of the difficulties in doing research with fishes that are considered Critically Endangered (IUCN; see below Endangered Species and Climate Change), as well as incomplete knowledge of their habitats in different locations across their geographical distribution. These are not isolated constraints for diplomystids and nematogenyids, but they can also be applied to many wild catfish species worldwide. Consequently, the early ontogenetic development of most structures for many catfishes is unknown, and this is a subject in need of attention so that the origin and variability in the development of different structures can be understood for families and be included as potential characters in morphological phylogenies so that homology questions can be addressed.

Catfishes and the Business World

In addition to their biological and ecological importance in freshwater and marine environments, catfishes are substantial components of the aquarium trade and the aquaculture industry of certain countries. In other words, some species are economically important, some because of their beautiful aspects, others because of their meat.

The catfish aquarium trade includes any action in which species are taken from their natural habitats and sold to aquarists. The aquarium trade has grown considerably, not only because there are hundreds of types of catfish for aquariums but also because they have fascinating appearances. Some are particularly beautiful, others are fun to watch, are generally peaceful, and some even help keep the fish tanks clean. Among the most popular species are a few from Africa, as the "up-side down catfish" (*Synodontis nigriventris* from the Democratic Republic of Congo) and from Asia, as the "glass catfishes" (*Kryptopterus vitreolus*). Most aquarium species are from Central and South America; for instance, the "tiger shovelnose catfish" (*Pseudoplatystoma fasciatum*, Pimelodidae), the "pictus catfish" (*Pimelodus pictus*, Pimelodidae), the "cory catfishes" (many species of *Corydoras*, Callichthyidae), the "bumblebee catfishes" (*Microglanis iherengi* and *Pseudopimelodus bufonius*, Pseudopimelodidae), and many "sucker-mouth catfish" (Loricariidae of the genera *Otocinclus*, *Hypancistrus*, *Peckoltia*, among many others). Although the aquarium trade is currently a well-established, blooming business, especially in Europe and North America, it is not always clear how some of the fish have been obtained, especially considering that currently, many countries have well-known rules and laws concerning collection permits of specific species and number of specimens that could be collected.

On the other hand, the aquaculture industry in certain countries is also a booming business. For instance, ictalurids (e.g., *Ictalurus punctatus* or "channel catfish") represent the largest industry of aquaculture in North America and are also substantial in the aquaculture industry in Asia, especially China; the North African clariid, *Clarias gariepinus*, is popularly farmed in South Africa, Kenya, Mali, and Nigeria, whereas the pangasiid catfishes are a closely related group of species found in Vietnam and other Southeast Asian countries, the most widely cultured of which is *Pangasianodon hyphophthalmus*. The silver catfish, *Rhamdia quelen*, has been the focus of research and development in farming in Argentina, Uruguay, and southern Brazil. It is adaptable to a wide range of temperatures, but it does best in mild, sub-tropical regions. Based on a recent report by the FAO's Cultured Aquatic Species Information Program, catfishes have been cultured in almost 90 countries globally. Eventually, they could compete with the Tilapia and Carp trades for global dominance.

Two problems related to catfish aquaculture, however, are the worldwide introduction of alien species like the ones mentioned above and the hybridization of related species with subsequent accidental release in the natural habitat. The introduction of *Ictalurus punctatus*, *Clarias gariepinus*, and *Pangasianodon hypophthalmus* is well known, and these species are currently established in many river basins of eastern Brazil. On the other hand, aquaculture of the more robust hybrid between *Pseudoplatystoma fasciatum* and *P. corruscans* in Brazil resulted in the introduction of hybrid individuals in the natural habitat that sometimes interbreeds, causing a genetic deterioration in the native population.

Endangered Species and Climate Change

The geological history of Earth has revealed, throughout time, some major crises when some major climatic changes accompanied by volcanism, earthquakes, meteorites, desertification, glaciation, etc., affected the planet or some specific geographic regions. Many of these crises marked some major events, such as the Permo/Triassic or the Cretaceous/Paleocene extinction boundaries, while others had less dramatic and lethal effects. Still others were just locally restricted to small areas. Despite these major geological/climatic events, more recently, there are others playing roles in species diversity and their fate, such as the monumental impact that the increasing human population and (anthropogenic) effects have imposed on Nature. Thus, the anthropogenic impact is finally bringing the planet into a major crisis that also affects all oceanic and freshwater ecosystems and consequently is impacting the "normal" cycle of different species. Others are a direct result of climate changes that lead to the desertification of landmasses, causing aquatic environments to slowly shrink due to drought, desertification, and the damming of rivers that supply them. Examples include the large Chad and Volta Lakes in Africa, Poopo Lake in South America, and Aral Lake in Asia.

The anthropogenic actions can be of different kinds, either overfishing or contaminating waters with chemical (e.g., mining and forestry residues) or simple garbage, changing the course of rivers, modifying lakes, or installing hydroelectric plants at the origin of a river or along its course. Consequently, many aquatic species have been declared "threatened" all over the world, and catfish are among them. Thus, many countries have established an Endangered Species Act, and there is an International Union for Conservation of Nature Red List of Threatened Species, also known as the IUCN Red List, which provides an inventory of the global conservation status and extinction risk of biological species with different strategies to protect the existing ones.

We believe that it is a major responsibility to create awareness of this problem in the ichthyological community as well as in the general public. Among our catfishes, there are many that are considered as threatened species at this moment. This is a time that is particularly risky for many catfishes, especially those without a bony armor and possessing sensible skins characterized by the presence of taste buds, chemosensory organs, electroreceptors or ampullary organs, and superficial neuromasts all over the body. Among these considered as Critically Endangered are at least seven species of Diplomystidae from Chile and the only species of Nematogenyidae, both families with critical positions in the Tree of Life of Siluriformes. Among diplomystids, *Diplomystes chilensis* (Figure 3A) represents one of the oldest fishes ever described for Chile and the first diplomystid for South America (Molina 1782). The last collections of this catfish were done by R. Philippi in the Maipo Basin and published in 1866, whereas those reported by Eigenmann (1927) from Santiago Market were transported from Valdivia (southern Chile) so that they would correspond to *D. camposensis*. Considering that there is not a single record of *Diplomystes chilensis* around Valparaiso and Santiago (with highly polluted basins – Aconcagua and Maipo) in more than 100 years, Arratia and Quezada-Romeguialli (2017) proposed to consider *D. chilensis* as extinct, which is a dramatic decision but not surprising after the high degree of anthropogenic impact over these basins. Conservation of rare, ancient species (such as those of Diplomystidae) is fundamental because they contribute disproportionately to biodiversity (Faith 1992), and their extinction would cause a severe loss of important branches of the Tree of Life. The strictly freshwater family Diplomystidae is one of such groups of species. With nine species, it represents the earliest or one of the earliest branching groups of the Siluriformes, with extensive morphological and molecular evidence placing them as the sister group of all other catfish worldwide (e.g., Lundberg and Baskin 1969; Arratia 1987; Grande 1987; Mo 1991; de Pinna 1998; Diogo 2004; de Pinna et al. 2025). The family Nematogenyidae, with its only species (*Nematogenys inermis*), is another catfish endemic to Chile that is currently interpreted as Critically Endangered (IUCN). Nematogenyidae is another phylogenetically important taxon because it also represents one of the earliest branchings of Siluriformes, and it is considered a sister to all other loricarioids (e.g., Sullivan et al. 2006; de Pinna et al. 2025). Endangered species are also found in other continents; for instance, in North America, the ictalurids *Trogloglanis pattersoni* ("toothless blindcat"), *Satan eurystomus* ("widemouth blindcat"), *Ictalurus pricei* ("yaqui catfish"), and *Prietella phreatophila* ("Mexican blindcat") are notable. At the same time, the "cave catfish" (*Clarias cavernicola*), a Critically Endangered species of airbreathing catfish, is only known to live in the Aigamas Cave and the nearby Dragon's Breath Cave, Otjozondjupa region, Namibia. From Southeast Asia, the "Mekong catfish" (*Pangasianodon gigas*) is also considered Critically Endangered.

Catfishes that are harvested for the aquarium trade can also be driven to extinction. Some species of sucker-mouth catfishes that are already threatened by the construction of the Belo Monte hydroelectric powerplant in the Xingu River of Brazil are heavily collected for the aquarium trade. Species belonging to the genera *Hypancistrus*, *Peckoltia*, and *Panaqolus* are among the most popular, but the Critically Endangered *Hypancistrus zebra* (Figure 3C) is a flagship among them.

But the question remains: What are we doing as ichthyologists, educators, and citizens of Earth to help Nature and its biological species face an uncertain future provoked principally by economic interests and development?

Catfishes and the Near Future

The study of catfishes has advanced enormously in the last decades, but there is room for additional extensive studies on anatomy and morphology, many aspects of biology (e.g., migration, reproduction, and development), diversity, phylogenetics, and biogeography. Our knowledge of the diversity of catfishes has vastly increased in the last 50 years. Not very long ago, Nelson (1976), in the first edition of his seminal book *Fishes of the World*, listed 2,000 species of catfishes, of which about 1,200 are in South America. Some 20 years ago, Reis et al. (2003) listed 1,648 valid catfish species for South and Central America and estimated an additional 630 species yet to be discovered and described. That projection, nonetheless, was already surpassed years ago, as we now have over 2,500 catfish species in South and Central America (Table 1). Despite this enormous advance in the discovery of diversity, the situation has not improved much. Less than a decade ago Reis et al. (2016) estimated that the entire Neotropical fish fauna had 5,617 species and projected that the final number should be somewhere between 8,000 and 9,000, with roughly one-third of the fauna still to be discovered, the same proportion applying to catfishes. The situation in Africa is worse than in the neotropics, with more species to be discovered. Pavanelli et al. (2018) estimated that the final number of fish species in the African continent is similar to that estimated for the neotropics, but two-thirds of the fauna are still to be described. Catfish diversity in North America and Europe is well-documented, with likely no or few species left to be described. However, extensive study is still needed in Asia and Africa, where the kind of comprehensive discovery efforts currently underway in South America have yet to be initiated.

Acknowledgments

Our sincerest thanks go to those who enthusiastically contributed with photographs, enriching the presentation and meaning of this chapter. We are grateful to our colleagues Cristian Correa (Valdivia, Chile), Claudio Quezada-Romegialli (Arica, Chile), Timo Moritz, and Vivica von Vietinghoff (Jena, Germany), Janice Muriel-Cunha (Bragança, Brazil), Leandro Sousa (Altamira, Brazil), Eleonora Trajano (São Paulo, Brazil), and Andrew Williston (Cambridge, MA, USA) and the professional photographer Andreas Harti (Germany). RER is partially funded by the Conselho Nacional de Desenvolvimento Científico e Tecnológico – CNPq (process #302760/2022-9 and #400166/2016-0).

References

Adriaens, D. and W. Verraes. 1997. The ontogeny of the chondrocranium in *Clarias gariepinus*; trends in siluroids. J. Fish Biol. 50: 1221–1257.

Adriaens, D. and W. Verraes. 1997. Ontogeny of the osteocranium in the African catfish, *Clarias gariepinus* (Burchel, 1822) (Siluriformes: Clariidae): Ossification sequence as a response to functional demands. J. Morph. 235: 138–237.

Adriaens, D. and P. Vandewalle. 2003. Embryonic and larval development in catfishes. pp. 639–666. *In*: Arratia, G., B.C. Kapoor, M. Chardon and R. Diogo (eds.). Catfishes. Science Publishers, Inc., Enfield, Plymouth.

Alexander, R. McN. 1965. Structure and function in catfish. J. Zool. 148: 88–152.

Arratia, G. 1983. Preferencias de habitat de peces siluriformes de aguas continentales de Chile (Fam. Diplomystidae y Trichomycteridae). Stud. Neotrop. Fauna E. 18(4): 217–237.

Arratia, G. 1987. Description of the primitive family Diplomystidae (Siluriformes, Teleostei): Morphology, taxonomy and phylogenetic implications. Bonner Zool. Monogr. 24: 1–103.

Arratia, G. 1990. Development and diversity of the suspensorium of trichomycterids and comparison with loricarioids (Telostei: Siluriformes). J. Morph. 205: 193–218.

Arratia, G. and H.-P. Schultze. 1991. The palatoquadrate and its ossifications: Development and homology within osteichthyans. J. Morph. 208: 1–81.

Arratia, G. 1992. Development and variation of the suspensorium of primitive catfishes (Teleostei; Ostariophysi) and their phylogenetic relationships. Bonn. Zool. Monogr. 42: 1–148.

Arratia, G. and M. Gayet. 1995. Sensory canals and related Tertiary siluriforms crania from Bolivia and North America and comparison with recent forms. J. Vert. Paleont. 15: 482–505.

Arratia, G. and L. Huaquín. 1995. Morphology of the lateral line system and of the skin of diplomystid and certain primitive loicarioid catfishes and systematics and ecological considerations. Bonn. Zool. Monogr. 36: 1–110.

Arratia, G. 1998. *Silvinichthys*, a new genus of trichomycterid catfishes from the Argentinian Andes, with redescription of *Trichomycterus nigricans*. Ichthyol. Explor. Freshw. 9: 347–370.

Arratia, G. 2003a. Catfish head skeleton. pp. 3–46. *In*: Arratia, G., B.C. Kapoor, M. Chardon and R. Diogo (eds.). Catfishes. Science Publishers, Inc., Enfield, Plymouth.

Arratia, G. 2003b. The skin of catfishes. pp. 177–199. *In*: Arratia, G., B.C. Kapoor, M. Chardon and R. Diogo (eds.). Catfishes. Science Publishers, Inc., Enfield, Jersey, Plymouth.

Arratia, G., B.C. Kapoor, M. Chardon and R. Diogo (eds.). 2003. Catfishes. Science Publishers, Inc., Enfield, Jersey, Plymouth.

Arratia, G. and C. Quezada Romigialli. 2017. Understanding morphological variability in a taxonomic context in Chilean diplomystids (Teleostei: Siluriformes), including the description of a new species. PeerJ 5(4): e2991. DOI: 10.7717/peerj.2991.

Arratia, G. 2018. Otomorphs (= otocephalans or ostarioclupeomorphs) revisited. Neotrop. Ichthyol. 16(3): e180079. DOI: 10.1590/1982-0224-20180079.

Arratia, G. 2025. Siluriforms and the South American fossil record: their significance and the origin and evolutionary radiation of the group. *In*: Arratia, G. and R.E. Reis (eds.). Catfishes, a Highly Diversified Group. Volume 2: Evolution and Phylogeny. CRC Press, Boca Raton, FL.

Arratia, G. and R.E. Reis (eds.). 2025. Catfishes, a Highly Diversified Group. Volume 2. Evolution and Phylogeny. CRC Press,

Bardach, J.E. and J. Atema. 1971. The sense of taste in fishes. pp. 293–336. *In*: Beidler, L. (ed.). Handbook of Sensory Biology. Berlin, Heidelberg, New York: Springer.

Betancur-R., R., E.O. Wiley, G. Arratia, A. Acero, N. Bailly, M. Miya et al. 2017. Phylogenetic classification of bony fishes. BMC Evolutionary Biology. 42 pgs. DOI 10.1186/s12862-017-0958-3.

Britz, R., F. Kakkassery and R. Raghavan. 2014. Osteology of *Kryptoglanis shajii,* a stygobitic catfish (Teleostei: Siluriformes) from Peninsular India with a diagnosis of the new family Kryptoglanidae. Ichthyol. Explor. Freshwaters 24: 193–207.

Cumplido, N. This volume. The Caudal Fin musculature of catfishes and the homology of the posterior myomeres. pp. 135–161. *In*: G. Arratia and R.E. Reis (eds.). Catfishes, a Highly Diversified Group. Vol. 1: Their Outstanding Biology. Science Publishers, Inc., Enfield, Jersey, Plymouth.

Cuvier, G. 1817. Le Règne Animal, distribué d'aprés son organization. Tome 2. Les Reptiles, Les Poissons, les Mollusques et les Annélides. Deterville, Paris.

Daget, J. 1964. Le crâne des Téléostéens. Mem. Mus. Nat. Hist. Natur., Sér. A 31: 163–341.

Datovo, A., L. Ochoa, G. Vita, P. Presti, W.M. Ohara and M. de Pinna. 2023. A new genus and species of miniature tridentine catfish from the Amazon basin (Siluriformes: Trichomycteridae). Neotr. Ichthyol. 21: e230076.

Diogo, R. 2003. Higher-level phylogeny of Siluriformes. pp. 317–352. *In*: Arratia, G., B.C. Kapoor, M. Chardon and R. Diogo (eds.). Catfishes. Science Publishers, Enfield (NH), Jersey and Plymouth.

Diogo, R. 2004. Morphological Evolution, Adaptations, Homoplasies, Constraints and Evolutionary Trends: Catfishes as a Case Study on General Phylogeny and Macroevolution. Enfield, Science Publisher, Inc.

Diogo, R. and Z. Peng. 2010. State of the art of Siluriformes higher-level phylogeny. pp. 464–516. *In*: Grande, T., F. Poyato-Ariza and R. Diogo (eds.). Gonorynchiformes and Ostariophysan Relationships. Science Publishers, Enfield (NH), Jersey and Plymouth.

DoNascimiento, C. and M.C.C. de Pinna. 2025. Parasitic catfishes. pp. 232–248. *In*: Arratia, G. and R.E. Reis (eds.). Catfishes, a Highly Diversified Group. Volume 1: Their Outstanding Biology. CRC Press, Boca Raton, FL..

Egge, J. 2025. Siluriform venom systems. pp. 160–181. *In*: Arratia, G. and R.E. Reis (eds.). Catfishes, a Highly Diversified Group. Volume 1: Their Outstanding Biology. CRC Press, Boca Raton, FL.

Faith, D.P. 1992. Conservation evaluation and phylogenetic diversity. Biol Conserv. 61(1): 1–10. https://doi.org/10.1016/0006-3207(92)91201-3.

Ferraris, C.J. and M.C.C. de Pinna. 1999. Higher-level names for catfishes (Actinopterygii: Ostariophysi, Siluriformes). Proc. Calif. Acad. Sci. 51: 1–17.

Ferraris, C.J. 2007. Checklist of catfishes, recent and fossil (Osteichthyes: Siluriformes), and catalogue of siluriform primary types. Zootaxa 1418: 1–628.

Fink, S.V. and W.L. Fink. 1981. Interrelationships of ostariophysan fishes (Teleostei). Zool. Jour. Linnean Soc. 72: 297–353.

Fink, S.V. and W.L. Fink. 1996. Interrelationships of ostariophysan fishes (Teleostei). pp. 209–249. *In*: Stiassny, M., L.R. Parenti and G.D. Johnson (eds.). Interrelationships of Fishes. Academic Press, New York, NY.

Fricke, R., W.N. Eschmeyer and J.D. Fong. 2024. Eschmeyer's catalog of fishes: genera/species by family/subfamily. (http://researcharchive.calacademy.org/research/ichthyology/catalog/SpeciesByFamily.asp). Electronic version accessed 12 January 2024.

Gayet, M. and F.J. Meunier. 1998. Maastrichtian to early late Paleocene freshwater Osteichthyes of Bolivia: additions and comments. pp. 85–110. *In*: Malabarba, L.R., R.E. Reis, R.P. Vari, Z.M. Lucena and C.A.S. Lucena (eds.). Phylogeny and Classification of Neotropical Fishes. EDIPUCRS, Porto Alegre, Brazil.

Gayet, M. and F.J. Meunier. 2003. Paleontology and paleobiogeography of catfishes. pp. 491–519. *In*: Arratia, G., B.G. Kapoor, M. Chardon and R. Diogo (eds.). Catfishes, vol. 2. Enfield, NH, USA.

Grande, L. 1987. Redescription of †Hypsidoris farsonensis (Teleostei: Siluriformes), with a reassessment of its phylogenetic relationships. J. Vert. Paleont. 7: 24–54.

Grizzle, J.M. and A. Rogers 1979. Anatomy and Histology of the Channel Catfish. Agricultural Xxperimental Station, Auburn University, Alabama, 2nd Ed.

Hardman, M. 2005. The phylogenetic relationships among non-diplomystid catfishes as inferred from mitochondrial cytochrome *b* sequences: the search for the ictalurid sister taxon (Otophysi: Siluriformes). Mol. Phylogenet. Evol. 37: 700–720.

Henschel, E., J.L.O. Mattos, A. Katz and W.J.E.M. Costa. 2018. Position of enigmatic miniature trichomycterid catfishes inferred from molecular data (Siluriformes). Zool. Scr. 47(1): 44–53.

Henschel, E., A.M. Katz and P.H.N. Bragança. 2025. Miniature catfishes: Diversity, ecology and evolution. pp. 201–231. *In*: Arratia, G. and R.E. Reis (eds.). Catfishes, a Highly Diversified Group. Volume 1: Their Outstanding Biology. CRC Press, Boca Raton, FL.

Hennig, W. 1996. Phylogenetic Systematics. University of Illinois Press, Urbana.

Huaquín, L., D. Veliz and G. Arratia. 2002. Estudio comparativo de ovarios y cubiertas ovocitarias en peces siluriformes de aguas continentals de Chile. Gayana 66(2): 269–274.

Kaatz, I., P. Loebel and A. Rice. 2025. Sound production and communication in catfishes. pp. 301–322. *In*: G. Arratia and R.E. Reis (eds.). Catfishes, a Highly Diversified Group. Vol. 1: Their Outstanding Biology. Science Publishers, Inc., Enfield, Jersey, Plymouth.

Kubicek, K. 2022. Developmental osteology of *Ictalurus punctatus* and *Noturus gyrinus* (Siluriformes: Ictaluridae) with a discussion of siluriform bone homologies. Vertebrate Zool. 72: 661–727.

Lundberg, J.G. and J.N. Baskin. 1969. The caudal skeleton of the catfishes, order Siluriformes. Am. Mus. Novit. 2398: 1–49.

Malabarba, R.L., R.E. Reis, R.P. Vari, Z.M. Lucena and C.A.S. Lucena (eds.). 1998. Phylogeny and Classification of Neotropical Fishes. EDIPUCRS, Porto Alegre, Brazil.

Manríquez, A., L. Huaquín, M. Arellano and G. Arratia. 1988. Aspectos reproductivos de *Trichomycterus areolatus* Valenciennes, 1846 (Pisces: Teleostei: Siluriformes) en Rio Angostura, Chile. Stud. Neotrop. Fauna E. 23(2): 89–102.

Mo, T. 1991. Anatomy, relationships and systematics of the Bagridae (Teleostei: Siluroidei), with a hypothesis of siluroid phylogeny. Theses Zoologicae 17. Königstein: Koeltz Scientific Books.

Molina, J.I. 1782. Sagio sulla storia naturale del Chile. Lib. 4. The Geographical, Natural and Civil History of Chili [Irvig's Englis translation) 1809: 181–191.

Montoya-Burgos, J.I., S. Muller, C. Weber and J. Pawlowski. 1998. Phylogenetic relationships of the Loricariidae (Siluriformes) based on mitochondrial rRNA gene sequences. pp. 363–374. *In*: Malabarba, L.R., R.E. Reis, R.P. Vari, Z.M. Lucena and C.A.S. Lucena (eds.). Phylogeny and Classification of Neotropical Fishes. EDIPUCRS, Porto Alegre, Brazil.

Moritz, T. 2025. An overview on the anatomy of catfishes and its diversity. pp. 48–80. *In*: Arratia, G. and R.E. Reis (eds.). Catfishes, a Highly Diversified Group. Volume 1: Their Outstanding Biology. CRC Press, Boca Raton, FL.

Nelson, J.S. 1976. Fishes of the World. John Wiley and Son, New York.

Nelson, J.S., T. Grande and M.V.H. Wilson. 2016. Fishes of the World. 5th edition. John Wiley and Sons, Inc., Hoboken, New Jersey.

Ono, R.D. 1980. Fine structure and distribution of epidermal projections associated with taste of some loricariids. J. Morph. 164: 139–159.

Otero, O., L. Cavin and R. Céspedes. 2025. Siluriformes from the Upper Cretaceous – Paleocene of Bolivia. *In*: Arratia, G. and R.E. Reis (eds.). Catfishes, a Highly Diversified Group. Volume 2. Evolution and Phylogeny. CRC Press, Boca Raton, FL.

Pavanelli, C.S., G.C. Deprá, H.J. Message, B.F. Morales, R.P. Ota, R.R. Ota et al. 2018. Predictive model for descriptions of freshwater African fish species in a historical, political and biogeographic context. Malawi: Presentation in the VI International Conference of the Pan-African Fish and Fisheries Association.

Philippi, R.A. 1866. Bemerkungen über die chilenischen Flufische. Mber. Kgl. Preuss Akad. Wiss., Berlin: 708–717.

de Pinna, M.C.C. 1998. Phylogenetic relationships of Neotropical Siluriformes (Teleostei Ostariophysi): Historical overview and synthesis of hypotheses. pp. 279–330. *In*: Malabarba, L.R., R.E. Reis, R.P. Vari, Z.M. Lucena and C.A.S. Lucena (eds.). Phylogeny and Classification of Neotropical Fishes. EDIPUCRS, Porto Alegre, Brazil.

de Pinna, M.C.C., L. Peixoto, V. Tagliacollo and M. Britto. 2025. Phylogenetic relationships and evolution of the major groups of Siluriformes. *In*: Arratia, G. and R.E. Reis (eds.). Catfishes, a Highly Diversified Group. Volume 2: Evolution and Phylogeny. CRC Press, Boca Raton, FL.

Rafinesque, C.S. 1815. Analyse de la Nature, ou Tableau de l'Univers et des Corps Organisés. Jean Barravecchia, Palerme.

Reis, R.E. 1998. Systematics, biogeography and the fossil record of the Callichthyidae: a review of the available data. pp. 351–362. *In*: Malabarba, L.R., R.E. Reis, R.P. Vari, Z.M. Lucena and C.A.S. Lucena (eds.). Phylogeny and Classification of Neotropical Fishes. EDIPUCRS, Porto Alegre, Brazil.

Reis, R.E., J.S. Albert, F. Di Dario, M.M. Mincarone, P. Petry and L.A. Rocha. 2016. Fish biodiversity and conservation in South America. J. Fish Biol. 89: 12–47.

Reis, R.E. and P. Lehmann A. 2022. A new genus of armored catfish (Siluriformes: Loricariidae) from the Greater Amazon, with a review of the species and description of five new species. Neotr. Ichthyol. 20(2): 1–100.

Reis, R.E., S.O. Kullander and C.J. Ferraris, Jr. 2003. Check List of the Freshwater Fishes of South and Central America. EDIPUCRS, Porto Alegre, Brazil.

Reutter, M. 1978. Tate buds in the bullhead (Teleostei). Adv. Anat., Embryol. Cell Biol. 55: 1–98.

Rizzato, P.P. and M.A. Almeida. 2024. An overview on some of the main features of the external anatomy of catfishes. pp. xx. *In*: Arratia, G. and R.E. Reis (eds.). Catfishes, a Highly Diversified Group. Science Publishers, Inc., Enfield, Jersey, Plymouth.

Rodiles-Hernández, R., D.A. Hendricks, J.G. Lundberg and J.H. Humphries. 2005. *Lacantunia enigmatica* (Teleostei: Siluriformes) a new phylogenetically puzzling freshwater fish from Mesoamerica. Zootaxa 1000: 1–24.

Schaefer, S. 1998. Conflict and Resolution: Impact of new taxa on phylogenetic studies of the Neotropical cascudinhos (Siluroidei: Loricariidae). pp. 375–400. *In*: Malabarba, L.R., R.E. Reis, R.P. Vari, Z.M. Lucena and C.A.S. Lucena (eds.). Phylogeny and Classification of Neotropical Fishes. EDIPUCRS, Porto Alegre, Brazil.

Soares-Porto, M.L. 1998. Monophyly and interrelationships of the Centromochlinae (Silurofoemes: Auchenipteridae). pp. 331–350. *In*: Malabarba, L.R., R.E. Reis, R.P. Vari, Z.M. Lucena and C.A.S. Lucena (eds.). Phylogeny and Classification of Neotropical Fishes. EDIPUCRS, Porto Alegre, Brazil.

Sullivan, J.P., J.G. Lundberg and M. Hardman. 2006. A phylogenetic analysis of the major groups of catfishes (Teleostei: Siluriformes) using *mg1* and *rag2* nuclear gene sequences. Mol. Phylogenetics, Evol. 41: 636–662.

Teugels, G.G. 2003. State of the art of recent Siluriformes systematics. pp. 317–352. *In*: Arratia, G., B.G. Kapoor, M. Chardon and R. Diogo (eds.). Catfishes, vol. 2. Enfield, NH, USA.

Trajano, L. and S. Secutti. 2025. Diversity and life cycles of subterranean catfishes. pp. 280–298. *In*: Arratia, G. and R.E. Reis (eds.). Catfishes, a Highly Diversified Group. Volume 1: Their Outstanding Biology. CRC Press, Boca Raton, FL.

Wiley, E.O. 2008. Homology, identity, and transformation. pp. 308–337. *In*: Arratia, G., H.-P. Schultze and M.V.H. Wilson (eds.). Mesozoic Fishes 4 – Homology and Phylogeny. Verlag Dr. Friedrich Pfeil, Munich.

Wiley, E.O. and G.D. Johnson. 2010. A teleost classification based on monophyletic groups. pp. 123–182. *In*: Nelson, J.S., H.-P. Schultze and M.V.H. Wilson (eds.). Origin and Phylogenetic Interrelationships of Teleosts. Verlag Dr. Friedrich Pfeil, Munich.

Zuanon, J. and G. Torrente-Vilara. 2025. Whiskers ahead – Catfish migrations in freshwater and marine environments. pp. 249–279. *In*: Arratia, G. and R.E. Reis (eds.). Catfishes, a Highly Diversified Group. Volume 1: Their Outstanding Biology. CRC Press, Boca Raton, FL.

Section 2
Anatomy of Catfishes

Chapter 2

An Overview of Some of the Main Features of the External Anatomy of Catfishes

Pedro P. Rizzato[1,*] and *Malu A. Almeida*[2]

Siluriformes is an astonishingly diverse order. According to Fricke et al. (2023), this monophyletic group currently comprises about 4,000 species. As taxonomists from all over the world continue to explore and describe its diversity, the numbers keep growing larger and more impressive: between 2014 and 2023, approximately 600 new catfish species were described, Loricariidae and Trichomycteridae alone contributing with 203 and 140 new species each, respectively.

Such staggering taxonomic diversity is partly explained by their worldwide distribution: when fossils are included, catfishes are reported from all continents (Grande and Eastman 1986; Nelson et al. 2016; Arratia 2025). This cosmopolitanism allowed them to colonize a variety of aquatic environments, mostly freshwater but also marine (notably members of Ariidae and Plotosidae). Catfishes may be found in high elevations (e.g., members of Trichomycteridae, Astroblepidae, and Sisoridae) as well as in subterranean habitats (see Trajano and Secutti this volume), including epikarstic (the South American trichomycterid *Ituglanis epikarsticus*; Bichuette and Trajano 2004) and phreatic (e.g., the South American genus *Phreatobius* and the Asiatic *Horaglanis*; Muriel-Cunha and de Pinna 2005; Fernández et al. 2007; Shibatta et al. 2007; Raghavan et al. 2023).

The colonization of such a vast array of habitats improved diversification and resulted in the development of an equally astonishing morphological diversity, which includes adaptations, in some cases extreme, to several different types of environments. As a group, catfishes exhibit an impressive diversity in terms of ecology, behavior, diet, feeding, locomotion, and so forth, hardly rivaled by other fish groups, which is reflected in their morphology. The combination of all these factors makes any effort to describe the morphological diversity of catfishes a huge challenge. On the other hand, it makes the exercise of exploring any aspect of their morphology an informative and fruitful (not to mention pleasant!) enterprise in terms of investigating the possible causes and consequences of the morphological diversification of living organisms.

[1] Departamento de Zoologia, Instituto de Biociências, Universidade de São Paulo, Rua do Matão, Travessa 14, no. 101, Cidade Universitária, 05508-090 São Paulo, SP, Brazil,

[2] Laboratório de Ictiologia de Ribeirão Preto, Departamento de Biologia, Faculdade de Filosofia, Ciências e Letras de Ribeirão Preto, Universidade de São Paulo, Avenida dos Bandeirantes, no. 3900, 14040-901 Ribeirão Preto, São Paulo, Brazil.

Email: maluaalmeida@usp.br

* Corresponding author: rizzatopp@usp.br

The purpose of this chapter is to provide a general overview of some of the main features of the external anatomy of catfishes, trying to illustrate how much morphological diversity can be found among representatives of the group. We discuss the general body shape, body size, skin, barbels, and fins. We selected these aspects and structures because they nicely illustrate the extent of morphological variation that can be found among catfishes. Therefore, the chapter is not intended to be exhaustive but illustrative: it is intended to provide the readers with a sense of how diverse the external anatomy of catfishes can be. In addition, many of the features we address are useful for recognizing and diagnosing catfish taxa, and we also intend to demonstrate how these features may be related to several different aspects of their biology. Throughout the text, we provide the readers with general references where more detailed information on different aspects of the anatomy of catfishes can be found. Our major goal is to demonstrate how catfishes, with their astonishing morphological diversity, can, at the same time, amaze us and incite us to ask and explore questions about the diversity and diversification of life on our planet.

General Body Shape

Catfishes exhibit a vast array of body shapes (Figures 1 and 2). Several members of different, not necessarily closely-related catfish families exhibit a generalized, "typical" body shape often associated with catfishes (Figures 1A and B): long, more or less deep, deeper and roughly round in cross-section in the trunk region, gradually becoming more compressed toward the caudal peduncle with a wide and depressed head, well-developed paired and median fins (often including an adipose fin), a well-developed, usually forked caudal fin, and pelvic fins positioned around the middle of the body, the dorsal and pectoral fins bearing strong spines. This general description can be applied to all, most, or at least some members of the following families: Akysidae, Amblycipitidae, Amphiliidae, Anchariidae, Ariidae, Auchenipteridae, Austroglanididae, Bagridae, Claroteidae, Cranoglanididae, Diplomystidae (Figure 1A), Doradidae, Heptapteridae, Horabagridae, Ictaluridae, Lacantuniidae, Mochokidae (Figure 1B), Pangasiidae, Pimelodidae, Pseudopimelodidae, and Sisoridae.

Members of Cetopsidae, and especially of the loricarioid families (Nematogenyidae, Trichomycteridae, Callichthyidae, Scoloplacidae, Astroblepidae, and Loricariidae), exhibit body shapes that depart considerably from the "typical" catfish body shape described above. Cetopsids usually have a distinctively round, robust anterior body profile, especially due to the head shape that is generally less depressed and round in lateral profile, giving some of them the aspect of a whale (the genus name *Cetopsis*. Figure 1C, means "similar to a whale," and cetopsids are popularly known as Whale Catfishes). Several auchenipterids (especially the centromochlines) resemble cetopsids in having a similarly round, blunt head, and some of them are also compared to whales (e.g., *Centromochlus orca*).

Nematogenyids (Figure 1D) and most trichomycterids (Figure 1E) have a long and roughly cylindrical body that is more homogeneously deep in the trunk and caudal peduncle, and the posterior margin of the caudal fin is usually truncated or round. These characteristics are similar to some amblycipitids (e.g., *Liobagrus* and *Nahangbagrus*) and amphiliids (e.g., *Paramphilius*). However, some trichomycterids have more divergent body shapes, including more elongate forms in the subfamilies Glanapteryginae, Microcambevinae, Tridentinae, Stegophilinae, and Vandellinae, as well as short, deep-bodied forms in the subfamily Sarcoglanidinae (*Malacoglanis gelatinosus* and *Sarcoglanis simplex*). Other loricarioids include: callichthyids, which may have relatively short bodies (e.g., some species of *Aspidoras*, *Corydoras*: Figure 1F, and *Scleromystax*); astroblepids, which may have more or less depressed bodies; scoloplacids, which have strongly depressed bodies with a relatively wide head region; and loricariids, in which body shape diversity is considerably high and hardly rivaled by other catfish families, in part as a consequence of its enormous taxonomic diversity. Body shape diversity in Loricariidae spans from relatively short and deep forms, such as some hypostomines (Figure 1G), to very long and strongly depressed forms, such as species of *Farlowella* (Figure 2A), *Planiloricaria*, and *Pseudohemiodon* (Figure 2B). The latter genera,

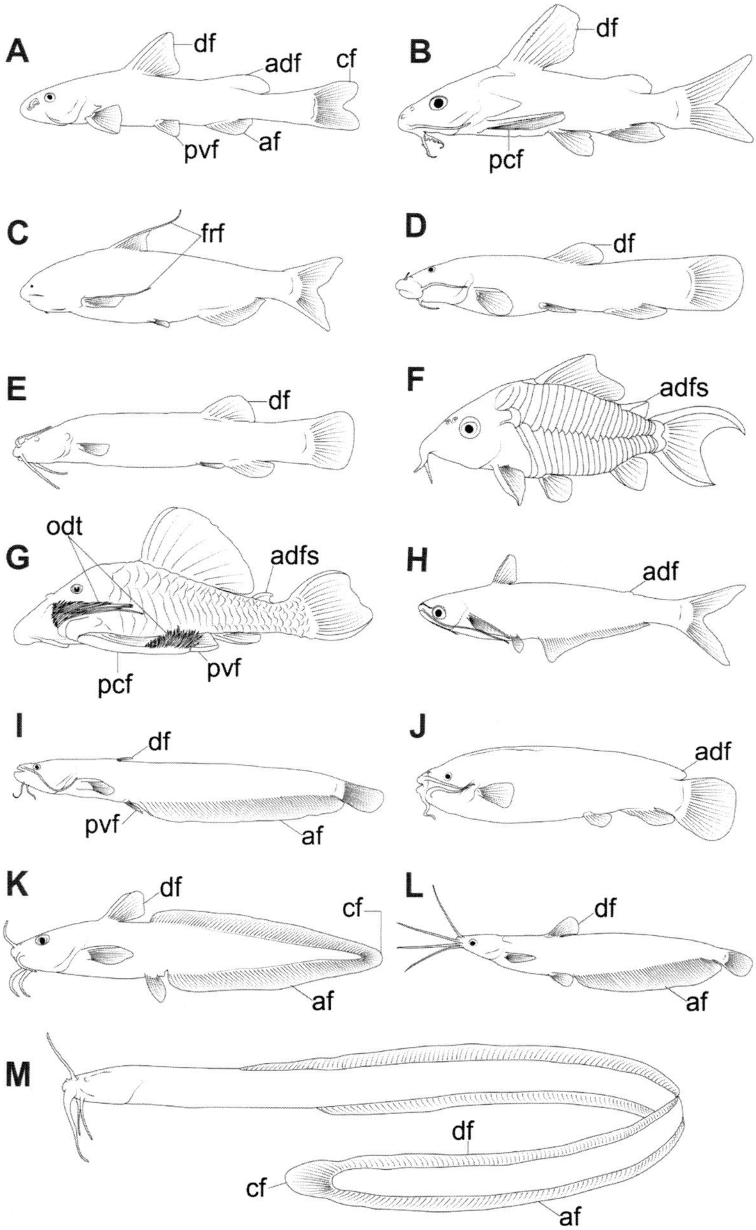

Figure 1. Body shape variation (in left lateral view) among selected catfishes. (A) *Diplomystes incognitus* (Diplomystidae). (B) *Synodontis ornatipinnis* (Mochokidae). (C) *Cetopsis coecutiens* (Cetopsidae). (D) *Nematogenys inermis* (Nematogenyidae). (E) *Trichomycterus maracaya* (Trichomycteridae). (F) *Corydoras splendens* (Callichthyidae). (G) *Panaque armbrusteri* (Loricariidae). (H) *Clupisoma sinense* (Ailiidae). (I) *Silurus burmanensis* (Siluridae). (J) *Malapterurus electricus* (Malapteruridae). (K) *Plotosus lineatus* (Plotosidae). (L) *Heteropneustes fossilis* (Heteropneustidae). (M) *Channallabes apus* (Clariidae). Abbreviations: adf, adipose fin; adfs, adipose fin spine; af, anal fin; cf, caudal fin; df, dorsal fin; frf, fin-ray filament; odt, odontodes; pcf, pectoral fin; pvf, pelvic fin.

members of the subfamily Loricariinae, further differ in members of *Farlowella* having thin bodies (Figure 2A) while members of *Planiloricaria* and *Pseudohemiodon* have wide bodies, especially the head and trunk (Figure 2B). Some members of Amphiliidae (*Andersonia*, *Belonoglanis*, *Phractura*, and *Trachyglanis*) and Sisoridae (e.g., *Sisor*) roughly resemble loricariines, with a strongly depressed body and a long and thin caudal peduncle. Some mochokids, such as species of *Chiloglanis*, resemble astroblepids or might be even, at first look, confused with some loricariids, apart from the absence

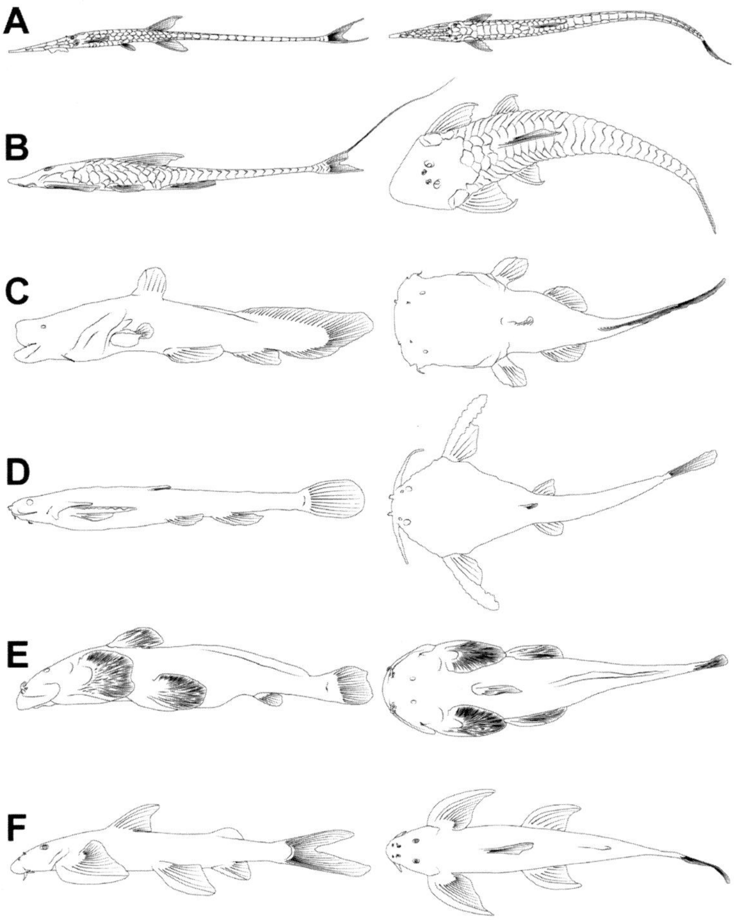

Figure 2. Body shape variation among selected catfishes with a depressed body (left column: left lateral view; right column: dorsal view). (A) *Farlowella wuyjugu* (Loricariidae). (B) *Pseudohemiodon unillano* (Loricariidae). (C) *Chaca chaca* (Chacidae). (D) *Acanthobunocephalus scruggsi* (Aspredinidae). (E) *Pseudexostoma brachysoma* (Sisoridae). (F) *Congoglanis alula* (Amphiliidae).

of osseous plates, and, of course, from their geographical distribution (loricariids, exclusively Neotropical, are known as the Armored Suckermouth Catfishes, while species of *Chiloglanis* are known as the African Suckermouth Catfishes).

Some other catfishes exhibit body shapes that depart even more strongly from the generalized, "typical" catfish body shape described above. Several catfishes have strongly depressed bodies, especially species with more benthic-oriented habits (this is the case of scoloplacids and loricariines, mentioned above). For example, members of Chacidae have a moderately long but strongly depressed body, especially in the head region, and as such, they roughly resemble Goosefishes (Lophiidae), which explains one of their popular names, Angler Catfishes (Figure 2C). Members of the pseudopimelodid genus *Lophiosilurus*, especially *L. alexandri*, also have a body shape similar to Goosefishes (as suggested by the name of the genus). Members of Aspredinidae usually have a dorsoventrally flat body, especially in the head and anterior portion of the trunk that, in some (e.g., *Acanthobunocephalus*: Figure 2D, *Bunocephalus* and *Pseudobunocephalus*), are roughly circular or diamond-shaped in dorsal profile, followed posteriorly by a long and roughly cylindrical caudal peduncle, which explain their popular names Banjo Catfishes (Nelson et al. 2016) or even Frying Pan Catfishes (Burgess et al. 1989). Some species of the akysid genus *Acrochordonichthys* resemble those aspredinids in body shape. Aspredinidae also includes elongate forms such as

Aspredo, *Aspredinichthys*, and *Platystacus*, and in an interesting parallel, the highly modified akysid genus *Breitensteinia* roughly resemble these elongated aspredinids in body shape because of the higher number of vertebrae (especially in the caudal peduncle) in comparison with other akysids.

On the other hand, some catfishes have more compressed than depressed bodies. Members of Ailiidae have a long, laterally flat body, in some cases with a keeled trunk and ventral body margin, similar to members of Clupeidae (herrings), and the name of one genus in this family, *Clupisoma*, can be literally translated as "body of a herring" (or "herring-bodied": Figure 1H). Schilbeids, the horabagrid genera *Pachypterus* and *Platytropius,* the plotosid genera *Neosilurus* and *Porochilus*, the pangasiid genera *Helicophagus* and *Pseudolais*, and some auchenipterids (e.g., *Auchenipterus*, *Epapterus*, and *Pseudepapterus*) also have a long and more laterally flat body, especially in the caudal peduncle. *Kryptoglanis sajii* and most members of Siluridae (e.g., *Silurus*: Figure 1I) also have similarly compressed bodies, and some of them (e.g., *Phalacronotus*) exhibit a long caudal peduncle and anal fin and a small caudal fin, giving them a body shape like some Neotropical Knifefishes (Gymnotiformes).

Other more specific body shapes characterize members of some families. Malapterurids have a more round, cylindrical, somehow globose body, with the adipose fin displaced posteriorly, the caudal fin rounded or squared, and without dorsal fin (Figure 1J). Their unique appearance ["In every respect a revolting fish" (Copley 1941)] inspired the name of the malapterurid genus *Paradoxoglanis*, meaning "strange catfish" (Norris 2002). Plotosids have a long caudal peduncle and usually a more laterally compressed body, with the pelvic fin anteriorly displaced, combined with a caudal fin that, in most species of the family, extends far anteriorly dorsally and is continuous ventrally with the anal fin, resembling a tadpole tail (Figure 1K, see Fins below). Some species of the bagrid genus *Bagrichthys* (e.g., *B. hypselopterus*) and the mochokid genus *Synodontis* (e.g., *S. batensoda*, *S. eupterus*, *S. membranaceus*, and *S. resupinatus*) have a distinctively deep trunk. In *Bagrichthys* and *Bagroides*, the dorsal profile of the body anterior to the dorsal fin is strongly arched, giving these fishes a characteristic body shape.

Sisorids are extremely variable in shape, ranging from short-bodied to elongate forms, although they usually have the head and trunk characteristically flat ventrally, the latter sometimes bearing an adhesive apparatus (see Skin below). Some sisorids such as *Chimarrichthys* (= *Euchiloglanis*), *Myersglanis*, *Oreoglanis*, *Parachiloglanis*, *Pareuchiloglanis* and *Pseudexostoma* (Figure 2E) have a moderately to strongly depressed head, the anterior margin of which is round and continuous laterally with thick, wide-based maxillary barbels (see Barbels below). These catfishes have, thus, a characteristic, conspicuously round head in dorsal and ventral profile and well-developed and round pectoral fins that give them a discoid shape anteriorly (Figure 2E), roughly resembling the gobioid *Rhyacichthys aspro* (Rhyacichthyidae) or even some hillstream loaches of the family Gastromyzontidae. The amphillid genus *Congoglanis* resembles those sisorids in having well-developed pectoral and pelvic fins, in addition to a round anterior head margin and a body roughly triangular with round vertices in cross-section (Figure 2F), well adapted to withstand strong currents in fast-flowing rivers.

A common trend among catfishes is body elongation, which occurs in lower or higher degrees in members of Akysidae (*Breintensteinia*), Amblycipitidae (some *Amblyceps*), Amphiliidae (several genera), Aspredinidae (e.g., *Aspredinichthys*, *Aspredo*, and *Platystacus*), Auchenipteridae (e.g., *Trachelyopterichthys*), Bagridae (e.g., *Hemileiocassis, Olyra*, and *Sundolyra*), Claroteidae (*Leptoglanis*), Doradidae (e.g., some *Leptodoras* such as *L. juruensis*), Heptapteridae (e.g., *Acentronichthys*, *Chasmocranus*, *Nemuroglanis*, *Phenacorhamdia*, and *Pseudochasmocranus*), Kryptoglanidae, Phreatobiidae, Plotosidae (e.g., *Cnidoglanis* and *Euristhmus*), Sisoridae (*Conta* and *Sisor*) and Trichomycteridae (several genera). However, in terms of body elongation, the more extreme examples are found among members of Clariidae and Heteropneustidae. All members of these two families, commonly regarded as sister taxa, have a long body and usually a long and strongly depressed head, but in *Heteropneustes* (single genus of Heteropneustidae), the body is more compressed posteriorly, and the dorsal fin is short (up to seven rays) (Rema et al. 1999; Hossain et al. 2013; Figure 1L) while in clariids the dorsal fin is long (usually with more than 30 rays, see Fins below). However, while some clariids (e.g., *Clarias*, *Dinotopterus*, and *Heterobranchus*) have relatively shorter bodies, comparable to those found in *Heteropneustes*, others have considerably more elongated, eel-like bodies (e.g., *Channalabes*: Figure 1M, *Dolichallabes*, *Gymnallabes*, *Platyallabes*, *Platyclarias*, and some *Clariallabes*). As discussed by several authors (e.g., Boulenger 1908; Pellegrin 1927; Cabuy et al. 1999; Teugels and Adriaens 2003; Agnèse and Teugels 2005), clariids may be ordered in an artificial, graded series from species (such as those in *Heterobranchus*) with a more fusiform body, a large head, and well developed paired and unpaired fins (including the adipose fin), to species (e.g., *Channallabes apus*: Figure 1M) with a long, cylindrical, eel-like body with similarly long dorsal and anal fins confluent with the caudal fin, a small head, and lacking adipose and paired fins, featuring several "intermediate" body shapes between these two extremes (e.g., *Clariallabes*, *Tanganikallabes*, and *Pseudotanganikallabes*; see Teugels and Adriaens 2003, figs. 16.3 and 16.4). According to some authors (e.g., Poll 1977; Teugels and Adriaens 2003; Agnèse and Teugels 2005), these elongated, eel-like body shapes evolved more than once independently along the clariid phylogeny, possibly associated with an adaptation to living in the mud (Adriaens et al. 2002; Agnèse and Teugels 2005). In addition, some clariids are found in caves (*Clarias cavernicola* and *Uegitglanis zammaranoi*) and other subterranean habitats (*Horaglanis* spp.) and exhibit typical characteristics of subterranean organisms such as reduced eyes and body pigmentation. Clariids and Heteropneustids, therefore, represent how much body shape variation can be found even among closely-related catfishes.

Body Size

Body size can vary enormously even among closely related taxa, such as members of the same family. Auchenipterids vary from miniature species of about 30 mm SL (*Gelanoglanis*) to large-sized catfishes of about 600 mm SL, such as some species of *Ageneiosus* that are used as food sources (Calegari et al. 2019). Loricariids range from about 20 mm (e.g., *Nannoplecostomus eleonorae*, Ribeiro et al. 2012, tab. 3) to about 600 mm (*Panaque schaeferi*, Lujan et al. 2010; *Acanthicus hystrix*, Chamon 2016; *Pseudacanthicus major*, Chamon and Costa e Silva 2018); doradids range from about 30 mm (*Physopyxis*) to about one meter (*Oxydoras*; Sousa and Rapp Py-Daniel 2005; Birindelli 2014); and sisorids range from about 20 mm (*Erethistes*) to about two meters in length (*Bagarius*).

Several medium to large-sized catfishes represent important human food resources, especially pimelodids and bagrids, with many species surpassing one meter in length. Some catfishes, however, can grow much bigger, reaching truly gigantic sizes. Stone (2007), for example, lists the world's 20 largest freshwater fishes, six of which are catfishes. According to de Pinna (1996), the sisorid *Bagarius yarrelli* can reach almost two meters in length, making it one of the largest catfishes in Asia. The pimelodid *Brachyplatystoma filamentosum* can surpass two meters (Stone (2007) reports 3.6 m for this species), making it the largest catfish in South America. The pangasiids *Pangasianodon gigas* and *Pangasius sanitwongsei* can reach three meters in length and 300 kg, being some of the largest catfishes in the world. The largest catfishes known are the silurids *Silurus soldatovi* and *S. glanis*, which are reported to reach about four and five meters, respectively, the latter surpassing 300 kg (Stone 2007).

On the other extreme, several catfishes are small to minute-sized, in some cases being considered miniatures (see Henschel et al. this volume). Most aspredinids are less than 15 cm long, and miniaturized species occur in the tribe Hoplomyzontini (de Pinna 1996). The loricariid subfamily Hypoptopomatinae includes species that are usually 20 to 40 mm, although some species may surpass 100 mm (de Queiroz et al. 2013). Several catfish genera are composed exclusively of small-sized species, such as the akysid *Pseudobagarius* (usually less than 70 mm in length when adults), the sisorid *Pseudolaguvia* and *Caelatoglanis* (the largest species not surpassing 50 mm in length; Tamang and Sinha 2014), and the bagrid *Nanobagrus* (not surpassing 50 mm in length when adults) and *Hyalobagrus* (species less than 40 mm in length and with a translucent body). There is even an entire catfish family composed exclusively of miniatures, Scoloplacidae, which includes a few species of small catfishes that do not reach more than 20 mm in length as adults, making them some of the smallest living vertebrates (Schaefer et al. 1989). This, combined with the fact that several parts of their body are covered by odontodes (see Odontodes below), justifies their popular name, Spiny Dwarf Catfishes. Scoloplacids were neglected for more than a century by taxonomists due in part to their small size, even though they are distributed over a wide region in South America and are often abundant in certain areas (Schaefer et al. 1989).

According to de Pinna (1996), both gigantism and miniaturization in catfishes do not occur abruptly in phylogenetic terms but gradually across lineages. This is suggested by the fact that miniaturized species tend to occur in clades composed mostly of small-sized species (e.g., Aspredinidae, Trichomycteridae, and Hypoptopomatinae), while giant species tend to occur in groups composed of medium to large-sized species (Pangasiidae, Pimelodidae, and Sisoridae; see also Ng 2015). As highlighted by de Pinna (1996), data on gigantism is far less extensive than on miniaturization (see Henschel et al. this volume). However, the fact that some of the world's smallest and largest freshwater fish species can be found within the same fish order beautifully illustrates the extent of morphological variation among catfishes.

Skin

Catfishes do not have scales. The absence of scales is one of the most characteristic features of catfishes and is considered a siluriform synapomorphy (e.g., Fink and Fink 1981; Arratia and Huaquín 1995; Wiley and Johnson 2010). However, a considerable number of catfishes have different types and arrangements of bony plates or scutes covering the entire body or some of its parts (see below), forming a complete or incomplete armor-like couverture (e.g., Figures 1F and 1G; Figures 2A and 2B). In these cases, the morphological variation in the bony armor may be useful for differentiating taxa. General information on the histological composition of the skin of catfishes is provided by Bhatti (1938) and Arratia (2003b). Here, we provide a general overview trying to encompass the range of morphological variation that can be found on the skin of different catfishes.

Bony plates or scutes are present in most or all members of the following catfish families: Callichthyidae, Doradidae, Loricariidae, and Scoloplacidae. Of these, only Doradidae is not a member of Loricarioidea, and among the latter, plates are absent in nematogenyids, trichomycterids, and astroblepids. However, as highlighted by Schaefer (2003), the distribution and arrangement of bony plates vary among the armored loricarioids, and may be useful to differentiate families (Schaefer 2003, fig. 25). Callichthyids have their body covered by an extensive armor formed by two rows of overlapping bony plates on each side, one dorsal and one ventral (Figure 1F). These catfishes can breathe air through their hindgut and thus survive for short periods out of the water, and it is possible that their thick, protective armor also prevents them from drying out too quickly when outside the water (Burgess 1989). Scoloplacids also have two bilateral series of body plates, one dorsal and one ventral, in addition to an unpaired midventral series of plates between the anus and anal fin and a unique, movable, shield-shaped rostral plate covered by a variable number of recurved odontodes on the snout (Schaefer et al. 1989; Rocha et al. 2012). These two families share with Loricariidae the presence of dermal plates arranged in at least two paired longitudinal series, one dorsal at least between the dorsal and caudal fin, and one ventral at least between the pelvic and anal fin (Schaefer 2003). Odontodes can also be found in several other parts of their body, including over the plates of the bilateral series and the dorsal and pectoral-fin spines (see Odontodes below).

Loricariids have at least one, but frequently more than one, additional plate series between the bilateral dorsal and ventral series of plates. At least one mid-lateral plate series associated with the trunk lateral-line canal is present in all loricariids, but most have a total of five plate series on the trunk, with two additional series between the dorsal, mid-lateral, and ventral series (Schaefer 2003, fig. 25C). These plates are usually extensive, encasing the entire dorsum and frequently also the abdomen and anterior region of the head, although, according to Schaefer (2003), the series between the dorsal, mid-lateral, and ventral series may be truncated to various degrees in several loricariid taxa (e.g., most loricariines and hypopotopomatines). In addition, loricariids are unique among loricarioids, including the armored ones (i.e., callichthyids and scoloplacids), in having plates on the lateral cheek region below the infraorbital series. Loricariids, have, therefore, with a few exceptions (Lithogeneinae, see below), their entire body, including the head, covered by a conspicuous, extensive armor of bony plates (Figures 1G, 2A and 2B), which explains their popular name, Armored Suckermout Catfishes, and the scientific name of the family, Loricariidae (from the type-genus *Loricaria*; the Latin word *Lorica* means "cuirass" or "armor").

Members of the subfamily Lithogeneinae are unusual among loricariids in having the body almost completely naked, the plates being extremely reduced and restricted to up to three paired series on the posterior portion of the trunk and the caudal peduncle, these representing the dorsal, mid-lateral, and ventral series present in all loricariids, plus a few plates on the dorsal and ventral midline of the body and on the head (Schaefer 2003; Schaefer and Provenzano 2008). According to Schaefer (2003), the condition of lithogenines may be regarded as pedomorphic in relation to the remaining loricariids. By having an almost naked body, as well as a modified, movable pelvic fin (see Fins, below), lithogenines resemble astroblepids, a resemblance that caused controversies regarding the phylogenetic relationships between these groups (see discussion in Schaefer 2003).

The arrangement of bony plates in doradids is distinct from those in the armored loricarioids. Popularly known as Thorny Catfishes, they have, with a few exceptions (*Liosomadoras morrowi*), a mid-lateral row of bony plates or scutes in the body, most of which with a central, posteriorly-directed spine or thorn (e.g., Sabaj and Arce 2021, fig. 1). The scutes may be oblique or approximately perpendicular to the long axis of the body (e.g., Birindelli 2014), and their number is variable (from 15 to 46 according to Birindelli 2014) but quite conservative within each species, being therefore useful for distinguishing them. In some species, however, they are weakly developed. In *Wertheimeria maculata*, for example, the mid-lateral scutes lack spines (except for those scutes in the caudal peduncle of smaller specimens) and are covered by thick skin (Birindelli 2014, fig. 39a). Doradids also possess a unique infranuchal scute (Birindelli 2014; Sabaj and Arce 2021), and several of them have caudal procurrent rays flattened into plates (named by Sabaj and Arce (2021) as "procumbent plates"), in some cases (e.g., *Agamyxis*, *Hypodoras*, and some *Platydoras*) so robust that they contact the lateral margins of the mid-lateral scutes, completely encasing the caudal peduncle in bony armor. The doradid *Agamyxis*, "the thorniest of Thorny Catfishes" (Sabaj and Arce 2021), additionally has the external bony surfaces of the body as well as the fin rays (including the anal and pelvic ones) ornamented with numerous small to large accessory spines. According to Sabaj and Arce (2021), similar accessory spines are found in other doradids (e.g., *Acanthodoras*) but never in the anal and pelvic fins.

Besides the four families mentioned above, bony plates may be found in some members of other families. *Sisor*, for example, is the only sisorid and the only Asian catfish genus with bony plates on the surface of the body, and in some, the posteriormost mid-dorsal plate bears a prominent ridge that forms a spine-like structure anterior to the adipose fin (de Pinna 1996; Ng 2003, 2015). According to these authors, the bony plates of *Sisor* are arranged in three series: one along the lateral line on each side of the body (formed by expanded lateral-line ossicles), one along the mid-dorsal line formed by autogenous plates, and one on each side of the dorsal-fin base. Some aspredinids (e.g., *Dupouyichthys*, *Ernstichthyis*, *Hoplomyzon*, and *Micromyzon*) may also have different arrangements of dorsal, ventral, and lateral series of bony plates on the body (Friel and Carvalho 2017).

The skin of catfishes that do not have bony plates is usually smooth and may be thin or thick depending on the group and the part of the body. The thickness of the skin in the head, for example, is associated with the exposition and degree of ornamentation of the underlying skull bones. For instance, pseudopimelodids usually do not have their skull bones exposed on the surface of the head due to the thickness of their skin. On the other hand, bagrids such as *Hemibagrus* and *Sperata* have the bones of the skull, with their ornaments of ridges and grooves, readily visible externally due to the thin skin cover. In sisorids (e.g., *Conta*, *Sisor*, and *Erethistes*) and some aspredinids (e.g., *Bunocephalus*), the head is usually bony and often conspicuously ornamented. In Heteropneustidae, the head is extensively covered by osseous plates. Among members of Clariidae, considered sister taxa to Heteropneustidae, some also have the head extensively covered by osseous plates (e.g., *Clarias*), while others have the head covered by powerful jaw muscles readily noticeable below the skin (e.g., *Channallabes*, *Dolichallabes*, *Gymnallabes*, *Horaglanis*, *Platyallabes*, *Platyclarias*, *Uegitglanis*; Cabuy et al. 1999; Devaere et al. 2005). Similarly, phreatobiids and some heptapterids (*Brachyglanis*, *Gladioglanis*, *Leptorhamdia*, and *Myoglanis*) have their head covered dorsally by hyperdeveloped jaw muscles. Other parts of the body may also be covered by thick skin, such as the dorsal fin of amblycipitids.

Skin is an important character for diplomystids (Arratia 1987; Arratia and Huaquín 1995; Arratia and Quezada-Romegialli 2017), in which it is characterized by a high number of taste buds distributed over the entire body (including the fins and inside the mouth) as well as large superficial neuromasts organized in lines ("pit-lines" of Arratia 1987; Arratia and Huaquín 1995; Arratia and Quezada-Romegialli 2017), and is covered by a variable number of papillae. In *Diplomystes chilensis*, these papillae are long (see Arratia and Quezada-Romegialli 2017, figs. 12A and 13), and these catfishes seem to have a "hairy" skin (diplomystids are popularly known as Velvet Catfishes; Nelson et al. 2016). Some other catfishes also have peculiar characteristics associated with the skin that allow, in some cases, their prompt recognition. Malapterurids, for example, have a distinctively "blubbery" skin that gives them a fleshy, globose aspect (Figure 1J; Norris 2002). Another peculiar catfish is the so-called Greasefish, the trichomycterid *Rhizosomichthys totae*, which can be readily recognized by the presence of thick "rings" of adipose tissue encircling the trunk, forming conspicuous fat rolls under the skin (Arratia 1990; Schaefer and Fernández 2009, fig. 1). Finally, in the auchenipterid genus *Spinipterus*, conspicuous, conical spines easily visible through the skin are present on the dorsal surface of skull-roofing bones, on the preopercle, and on the posterior cleithral process, being particularly large in the latter (Akama and Ferraris 2011, fig. 3; Rocha et al. 2019, figs. 1, 3, 4, 6 and 7).

Other catfishes have very particular structures in different parts of their skin, associated with different functions. Some plotosids, especially those living in saltwater, have well-developed dendritic organs on the ventral region of the body between the anus and the anal fin. These structures are highly vascularized and comprised of several papillae, and their function is probably osmoregulation. In addition, adult male *Euristhmus* have a patch of rugose epidermis dorsal to the base of the pectoral fin (Murdy and Ferraris 2006, fig. 1), the function of which is unknown. Some amblycipitids have a cup-like depression between the opercular flap and the dorsal base of the pectoral fin, first described by Hora (1933), that is unique to them among catfishes and appears to have a respiratory function (Chen and Lundberg 1995, fig. 15).

Keratinized Epidermal Structures

In chacids, the skin is granulous, covered by keratinized structures (Mistri et al. 2018), and several cutaneous flaps or cirri are present on the skin, including on the margin of the mouth, reinforcing their similarity with anglerfishes (Lophiidae). Some pseudopimelodids (e.g., *Lophiosilurus*, *Rhyacoglanis*) have thick, rough skin covered by several unculiferous epidermal structures (Shibatta and Vari 2017; Silva et al. 2021). Unculiferous epidermal structures are also present in representatives of several different families (e.g., Amphiliidae, Aspredinidae, Sisoridae, Doradidae, and the mochokid *Chiloglanis*) (Birindelli 2014; Shibatta and Vari 2017). They are especially prominent in some akysids, in which they are usually arranged in longitudinal rows along the body, except for *Parakysis* in which they are evenly distributed over the entire body surface. Aspredinids also have large keratinized tubercles arranged in parallel rows along the body, and these catfishes shed the entire outer layer of their skin (Friel 1994, 2003; Friel and Carvalho 2017). Periodic molting of the skin has also been observed in some akysids kept in aquaria (e.g., *Acrochordonichthys* and

Breitensteinia) (Ng and Ng 2001; Ng and Siebert 1998). In addition, some aspredinids carry their developing embryos attached to the ventral surface of their body, in some cases (e.g., *Platystacus*, *Aspredo*, and *Aspredinichthys*) by means of fleshy stalks called cotylephores (Friel 1994, 2003; Friel and Carvalho 2017).

The auchenipterid *Ageneiosus pardalis* and species of *Pseudauchenipterus* have well-developed, papilla-like epidermal tubercles on the tympanic area (Birindelli 2014; Calegari et al. 2019). Some other auchenipterids have, as a sexually dimorphic character, unculiferous tubercles on the skin of nuptial males, referred to as breeding tubercles by Birindelli (2014). According to Birindelli (2014) and Calegari et al. (2019), these breeding tubercles may be present on the dorsal surface of the head (*Auchenipterus*, *Entomocorus*, *Epapterus*, and *Pseudepapterus*), on the maxillary barbels (the same taxa mentioned above in addition to *Tetranematichthys*), and on the tip of the anteriormost anal-fin rays (*Auchenipterus*, *Epapterus*, and *Pseudepapterus*). In *Entomocorus*, these tubercles are also present on the dorsal-fin spinous ray, pectoral-fin spinous ray, pelvic fin, ventral region, and isthmus. According to Sarmento-Soares and Martins-Pinheiro (2020, fig. 4A and B), papilla-like tubercles are present on the ventral surface of the head of nuptial males of *Gelanoglanis nanonocticolus*, *G. pan* and *G. varii* as a sexually dimorphic character, as well as a filament on the first pectoral ray (see Fins, below) and a falcate pelvic-fin margin.

Adhesive Organs

Perhaps the most specialized modifications of the skin among catfishes are those found in rheophilic sisorids. The skin of sisorids may be smooth (e.g., *Ayarnangra*, *Gagata*, *Pseudecheneis*, some *Erethistoides*, and *Glyptothorax*) or rough, in the latter case, granular or tuberculated (e.g., *Caelatoglanis*, *Conta*, *Erethistes*, some *Erethistoides*, and *Glyptothorax*). Some sisorids (e.g., *Caelatoglanis*, *Conta, Pseudolaguvia*, *Glyptothorax* and, most notably, *Pseudecheneis*, Figure 3), however, have an adhesive apparatus ventrally in the body, regarded as an adaptation to their fast-flowing habitats since they use it to attach themselves to rocks in order to avoid being washed away by the current (although, according to Roberts 2001, some species that live in slow-flowing habitats may also exhibit adhesive organs).

The adhesive apparatus consists of a region of corrugated skin forming patterns of ridges and grooves ventrally on the "abdominal area," which is usually flat, bearing unculiferous structures usually arranged in regular rows and, in some cases, provided with hooks (e.g., *Pseudecheneis*). The area covered by the adhesive apparatus may be restricted to an arc just posterior to the branchiostegal membrane (e.g., some *Glypthotorax*) or extend almost the entire length of the ventral body region between the pectoral and pelvic fins (e.g., *Conta*) (de Pinna 1996). In *Conta*, the adhesive apparatus is thin and long and consists of unculi-studded integumentary ridges oriented obliquely or parallel to the longitudinal axis of the body (Ng 2015). In *Pseudolaguvia* and *Glyptothorax*, the adhesive apparatus consists of longitudinal pleats of skin arranged in an elliptical or rhomboidal field (Tamang and Sinha 2014; Ng and Kottelat 2016; Lokeshwor and Marak 2022, fig. 2). In *Pseudolaguvia* and *Glyptothorax*, as in *Conta*, the grooves on the adhesive apparatus are parallel or oblique to the longitudinal body axis, while in *Pseudecheneis* these grooves are thick, flat, and arranged transversely to the longitudinal body axis (Figure 3) (de Pinna 1996; Thomson and Page 2006; Ng 2015, figs. 14 and 16). De Pinna (1996), however, remarked that in young *Pseudecheneis sulcatus* the integumentary folds are also oblique and gradually become transversely oriented. In addition, in some of these sisorids (e.g., *Conta*, *Pseudolaguvia*, and some *Glyptothorax*), the adhesive organ also bears a median depression that is believed to assist on adhesion by creating an area of lower pressure for suction (Ng 2015). The morphology of the apparatus is thus highly variable across sisorid taxa and may be used to separate even species, for example, in the genera *Glyptothorax* (Ng and Kottelat 2016) and *Pseudolaguvia* (e.g., Lokeshwor and Marak 2022).

Even in sisorids that lack an adhesive apparatus ventrally in the body, other structures that help support the strong water currents of their fast-flowing habitats may be present. For

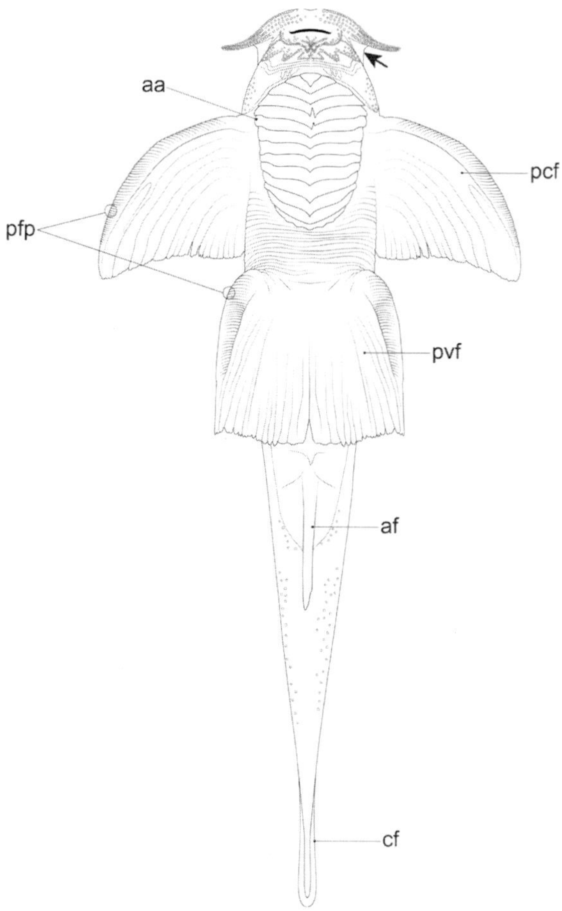

Figure 3. *Pseudecheneis sympelvica* (Sisoridae) in ventral view. Abbreviations: aa, adhesive organ; af, anal fin; cf, caudal fin; pcf, pectoral fin; pfp, paired-fin pads; pvf, pelvic fin. The arrow points to the wing-like skin flap at the base of the maxillary barbel.

example, in *Coraglanis*, *Glaridoglanis* and *Glyptosternon*, the fin rays may help adhesion, and in *Chimarrichthys* (= *Euchiloglanis*), *Glaridoglanis*, *Glyptosteron*, *Exostoma*, *Oreoglanis*, *Myersglanis*, *Pareuchiloglanis*, and *Pseudexostoma* also the lips (e.g., Conway et al. 2012, fig. 1E–H). In *Glyptosternum* and *Chimarrichthys*, unculiferous laminae may be present on the ventral surfaces of the head (including lips, barbels, and, in some cases, the isthmus) (Ng 2015) and fins (usually the outermost rays of the pectoral and pelvic fins) (Roberts 2001), in the latter case forming the so-called paired-fin pads (Conway et al. 2012; Lujan and Conway 2015). Paired-fin pads are present on the paired fins of all genera of Glyptosternina (*sensu* Thomson and Page 2006, table 3), in *Pseudecheneis* (Figure 3), and in some species of *Glyptothorax* (e.g., Thomson and Page 2006, figs. 14 and 15). According to Conway et al. (2012), the largest unculiferous paired-fin pads are those found in amphiline and doumeine amphiliids and glyptosternine sisorids. Roberts (2001) commented and described briefly the different types of adhesive organs found in sisorids and related to the type of habitat. Thomson and Page (2006), Lujan and Conway (2015, fig. 4f), and Ng and Kottelat (2016) provided pictures of some of these structures, and de Pinna (1996) and Ng (2015) commented on their evolution. Information on the microanatomy of some of these structures is available for a few sisorid taxa (e.g., de Pinna 1996; Das and Nag 2004; Conway et al. 2012; Ng 2015 and references therein).

Odontodes

Another structure found in the skin of catfishes, which deserves special attention, is odontodes. These dermal denticles, whose structure is identical to teeth but which are located outside of the oral cavity, are exclusive of loricarioids among catfishes (Isbrücker and Nijssen 1992; de Pinna 1998; Schaefer and Buitrago-Suárez 2002) and have been lost very few times within the group (e.g., in the glanapterygine trichomycterids *Pygidianops eigenmanni* and *Typhlobelus ternetzi*) (de Pinna 1998). These structures may be present at least in some part of the life cycle in different parts of the body, depending on the group. In *Nematogenys*, they are present in the pectoral-fin spinous ray (Arratia 1987, fig. 46C; Isbrücker and Njissen 1992). In scoloplacids, they are present on the head, bony plates (including the rostral plate; see Skin, above), pectoral girdle, and spinous fin rays, but are absent in the branched fin rays (Figure 4A). In callichthyids, odontodes are present all over the body (e.g., *Corydoras aeneus*) (Sire and Huysseune 1996, fig. 31), including on the surface and margin of the bony plates (see Skin, above), on the external surface of the skull bones and the fin elements (Figure 4B; Bhatti 1938; Isbrücker and Njissen 1992; Sire and Huysseune 1996). In astroblepids, large odontodes are present on the first unbranched ray of the dorsal, anal, pectoral, and pelvic fins. Furthermore, two types of odontodes are present in astroblepids, which differ in attachment and association with other epidermal structures (Schaefer and Buitrago-Suárez 2002). So-called type I odontodes are like those found in other loricarioids, while type II are apparently exclusive of astroblepids and are described as forming mechanosensory organs (Schaefer and Buitrago-Suárez 2002; Haspel et al. 2012).

In trichomycterids, odontodes are restricted to the head (Figure 5A) and involved in an interesting specialization unique to the group among catfishes and teleosts in general: the operculo-interopercular apparatus. Both the opercle and interopercle of trichomycterids are robust and compact bones, highly modified into odontode-bearing structures. They are connected to the bones of the suspensorium by a series of well-developed, strong ligaments, forming a tightly interconnected functional complex used for anchoring and moving on substrates (de Pinna 1998; Adriaens et al. 2010; Datovo and Bockmann 2010). This apparatus gives them the ability to resist and ascend strong currents by means of "elbowing" movements of the head (some are believed to be capable even of climbing vertical surfaces of waterfalls) or, in the case of the parasitic species of the family, to penetrate and anchor themselves into their hosts (de Pinna 1998; Adriaens et al. 2010; Datovo and Bockmann 2010). There is a high degree of morphological variation in this apparatus among trichomycterids associated with the different ecological characteristics of each taxon, beautifully investigated, described, and illustrated by Adriaens et al. (2010) and Datovo and Bockmann (2010).

In loricariids, similarly to callichthyids, odontodes are present all over the body, including the fin elements (Figure 4C; Isbrücker and Njissen 1992). They may be attached to bony plates or directly to other skeletal elements, including bones of the skull and pectoral girdle. For example, hypoptopomatines, which have the pectoral girdle exposed ventrally, have odontodes supported directly by the elements of the pectoral girdle, while in other loricariids, when bony plates cover the pectoral girdle ventrally, the odontodes are supported by the plates instead of the pectoral girdle (Reis et al. 2006). Several modifications of the odontodes may be present in different groups of Loricariidae, associated with different aspects of their biology. Most hypoptopomatines, for example, have medially directed, often larger odontodes on the ventral surface of the unbranched ray of the pelvic fins that help them attach themselves to underwater plants (Pereira et al. 2014). Other loricariids (several hypostomines) have evertible cheek plates, usually with more or less developed odontodes in different numbers, sizes, and shapes (Isbrücker and Njissen 1992; Armbruster 2004; Geerinckx and Adriaens 2006), reported to be used in agonistic displays and combats (Burgess 1989).

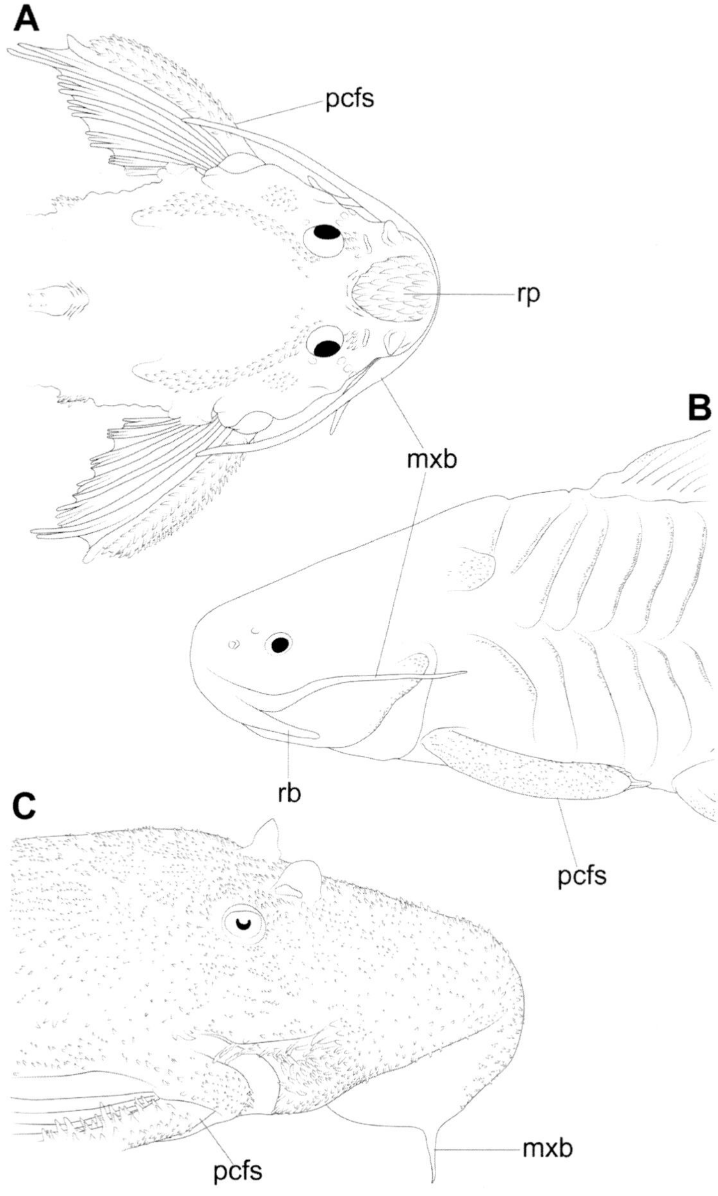

Figure 4. Diversity of odontode distribution among representative loricarioids. (A) *Scoloplax dicra* (Scoloplacidae). (B) *Callichthys callichthys* (Callichthyidae). (C) *Acanthicus* sp. (Loricariidae). Abbreviations: mxb, maxillary barbel; pcfs, pectoral-fin spine; rb, rictal barbel; rp, rostral plate.

Odontodes and Sexual Dimorphism

Several loricariids (as well as some callichthyids, Rapp Py-Daniel and Cox-Fernandes 2005) have a high degree of sexual dimorphism in the size of the odontodes, and according to Reis et al. (2006), sexual selection for male weaponry and display may have been an important influence in the diversification of the family. Some nuptial male loricariids have hypertrophied odontodes on the snout, such as delturines, several hypostomines, some loricariines (*Ixinandria*, *Rineloricaria*, *Sturisoma* and *Sturisomatichthys*), the rhinelepine *Pseudorinelepis*, and the hypoptopomatines *Isbrueckerichthys epakmos* and *Pareiorhaphis* (Reis et al. 2006). The condition reaches an extreme in nuptial males of the hypostomine *Neblinichthys pilosus*, popularly known as the Punk Catfish, which bears several extremely long hypertrophied odontodes on the snout (Isbrücker and Njissen 1992, figs. 10 and 11; Armbruster 2004; Taphorn et al. 2010; Armbruster and Taphorn 2013). Some of the above also have hypertrophied odontodes on the cheek (delturines, some loricariines, and the hypoptopomatines *Isbrueckerichthys* and *Pareioraphis*). In *Pareiorhaphis hystrix*, in addition to the hyperdeveloped odontodes that vary in size, thickness, density, and direction among individuals, nuptial males may also exhibit hypertrophied fleshy lobes on the cheek that vary in thickness and shape (Fagundes et al. 2020, fig. 4). Hypertrophied odontodes on the cheek may also be present in both males and females and, regardless of season, in some loricariid species (e.g., several hypostomines).

Other male loricariids may have hypertrophied odontodes in other parts of the body, such as on the pectoral fin (e.g., delturines, *Panaque*, Figure 1G, *Pareiorhaphis*) (Pereira and Reis 2002; Lujan et al. 2010; Reis et al. 2006; Pereira et al. 2014). In nuptial males of the hypoptopomatine *Hirtella carinata*, a unique pattern of hypertrophied, bristle-like odontodes on the head and body is present, in the latter forming five longitudinal rows on the lateral plates (the name *Hyrtella* comes from the Latin *hyrtus*, "hairy") (Pereira et al. 2014). In nuptial males of *Hypostomus unicolor* and *H. ammophilus*, hypertrophied, bristle-like odontodes cover the entire dorsal and lateral surfaces of the caudal peduncle (Pereira et al. 2014). Similar odontodes are also present on the lateral surface of the body of *Harttia leiopleura* (Pereira et al. 2014). In *Microplecostomus forestii*, long hypertrophied odontodes are present on the lateral portion of the body of mature males, and, interestingly, they are bicuspid (Silva et al. 2016, fig. 2), somehow resembling the oral teeth of loricariids and most other loricarioids, which are frequently bifid (de Pinna 1998).

Sexual dimorphism in loricariids goes beyond the aspect of their odontodes, and may include differences in body size, lip shape, labial papillae, oral teeth, and shape or size of the pectoral and anal fins (see Isbrücker and Njissen (1992) for a beautifully illustrated review). In loricariines, it is often pronounced, including in some cases modifications on the lips associated with parental care (e.g., *Furcodontichthys*, *Hemiodontichthys*, *Limatulichthys*, *Loricariichthys*, and *Pseudoloricaria*) (Isbrücker and Njissen 1992, fig. 27; Rapp Py-Daniel and Cox-Fernandes 2005; Rapp Py-Daniel and Ohara 2013; Covain and Van der Sleen 2017). One of the most famous and interesting sexually dimorphic characteristics of loricariids, however, occurs in members of the hypostomine genus *Ancistrus*, popularly known as Bristle-Nosed Catfishes. Large, long, conspicuous fleshy tentacles develop on the snout of these catfishes, being more numerous and well-developed in males, possibly to mimic larvae, as suggested by Sabaj et al. (1999). Male *Ancistrus* guard their eggs and larvae for a few days in a cavity nest, and it is possible that the tentacles on the snout of male *Ancistrus* mimic their larvae, increasing the odds of females choosing the tentacled male for reproduction. Other possible functions of these tentacles in *Ancistrus*, discussed by Sabaj et al. (1999), have been hypothesized, although, as remarked by the authors, they are not mutually exclusive with the hypothesis of larval mimicry. This is only one of the several interesting reproductive behaviors and strategies that can be found among loricariids involving sexually dimorphic modifications and adaptations in different parts of their bodies, the significance of most of which, however, remains unclear.

Barbels

Barbels represent one of the most typical features of catfishes, justifying their popular name, which refers to cat whiskers. There is enormous morphological variation in the barbels among catfishes related to characteristics of their biology and behavior, and this variation may be useful for recognizing and distinguishing taxa.

Catfish barbels vary in position: they may be found on the corners of the mouth (maxillary and rictal barbels), on the chin (mental or mandibular barbels, commonly in two pairs), and on the nasal openings: on the anterior nasal opening in *Nematogenys* (Arratia and Huaquín 1995, fig. 7B) and trichomycterids (Figure 5A), on the posterior nasal opening in several families

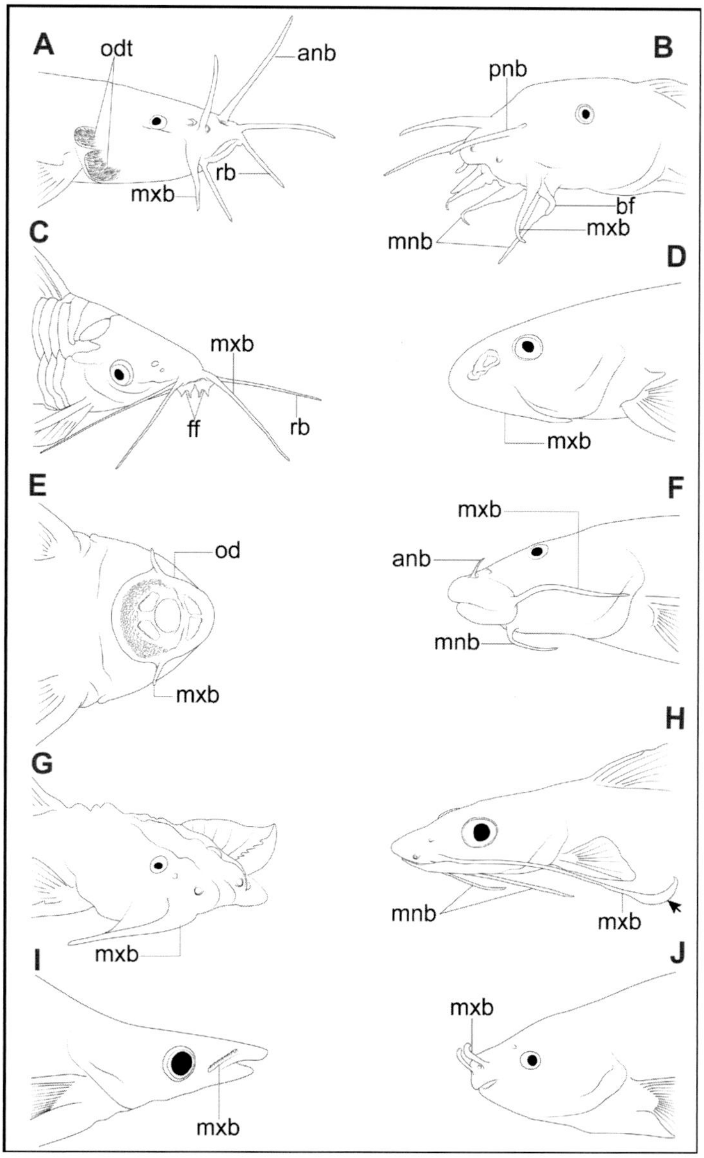

Figure 5. Barbel variation among selected catfishes. (A) *Trichomycterus* sp. (Trichomycteridae). (B) *Cnidoglanis macrocephalus* (Plotosidae). (C) *Dianema urostriatum* (Callichthyidae). (D) *Diplomystes incognitus* (Diplomystidae). (E) *Hypostomus rhantos* (Loricariidae). (F) *Nematogenys inermis* (Nematogenyidae). (G) *Hoplomyzon sexpapilostoma* (Aspredinidae). (H) *Phyllonemus typus* (Claroteidae, arrow point to the leaf-like structure on the distal portion of the maxillary barbel). (I) *Ageneiosus ucayalensis* (Auchenipteridae). (J) *Ceratoglanis pachynema* (Siluridae). Abbreviations: anb, anterior nasal barbel; bf, barbel-like filament; ff, fleshy flaps; mnb, mental or mandibular barbels; mxb, maxillary barbel; od, oral disk; odt, odontodes; pnb, posterior nasal barbel; rb, rictal barbel.

such as Clariidae, Ictaluridae and Plotosidae (Figure 5B). Some catfishes also bear variably developed fleshy flaps similar to barbels on the lips, such as some callichthyids (e.g., *Dianema*, corydoradines; Figure 5C) and some sisorids (*Gagata* and *Sisor*) (Ng 2015), and in some plotosids (e.g., *Cnidoglanis*; Figure 5B), there is a barbel-like filament at the corner of the mouth formed from a labial fold. So-called accessory barbels associated with the mental barbels are found in the akysids *Parakysis* and *Breitensteinia* (Ng and Siebert 1998; Ng 2009, fig. 6), and some aspredinids (e.g., *Aspredinichthys*, *Ernstichthys intonsus*, and *Hoplomyzon papillatus*) have several additional pairs of barbels or barbel-like papillae on the ventral surface of the body (Stewart 1985, fig. 3; Friel and Carvalho 2017). Some trichomycterids (the sarcoglanidine *Malacoglanis gelatinosus* and some species of the microcambevine genus *Microcambeva*) have a paired barbel-like structure on the branchiostegal region (e.g., Costa and Katz 2021, fig. 9C).

Barbels vary in number: diplomystids (Figure 5D), astroblepids, and loricariids (Figure 5E) have only one pair of maxillary barbels, while ictalurids, akysids, sisorids, clariids, and plotosids (Figure 5B) usually have four pairs of barbels—the maxillary, two mental, and one pair of nasal barbels. Barbel number may vary even among members of the same family. For example, most ariids have three pairs of barbels (the maxillary and two mental ones). However, *Batrachocephalus* only has two pairs of mental barbels (the maxillary barbel is absent), and *Osteogeneiosus* only has one pair of maxillary barbels (which, in this species, is bony throughout its extent; see below) and lack the mental ones. Mental barbels are also absent in astroblepids, loricariids (Figure 5E), trichomycterids (Figure 5A), and the auchenipterids *Ageneiosus* and *Tympanopleura* (Birindelli 2014; Calegari et al. 2019). Pangasiids usually have two pairs of barbels, the maxillary and one pair of mental ones, but in adult *Pangasianodon gigas*, only the maxillary pair is present (Nelson et al. 2016). In *Nematogenys*, there is also a single pair of mental barbels, in addition to the maxillary and one pair of short nasal barbels on the anterior nasal openings (Figure 5F). A single pair of mental barbels is also found in the auchenipterids *Gelanoglanis* and *Tetranematichthys*, in the silurids *Silurichthys* and *Micronema*, and in the ariid *Bagre*. Scoloplacids are described as having biramous maxillary barbels and uniramous or biramous mental barbels, in addition to mandibular barbels that may or may not be present between the maxillary and mental ones (Schaefer et al. 1989, figs. 4, 6, 8 and 10).

Barbels vary also in length. Pimelodids are known as Long-Whiskered Catfishes (e.g., Nelson et al. 2016) due to the presence of long barbels, especially the maxillary, which in some species surpasses the body length (e.g., *Platystomatichthys sturio*). Other catfishes also have similarly long barbels, such as several heptapterids and some species of the bagrid genera *Bagrus*, *Hemibagrus*, *Mystus*, and *Sperata*. Subterranean catfishes commonly have longer barbels than their epigean counterparts (e.g., subterranean species of *Trichomycterus*) (Rizzato et al. 2011; Bichuette and Rizzato 2012). In other catfishes, such as *Chaca* (Figure 2C), some loricariids (Figure 5E), and some pseudopimelodids (e.g., *Lophiosilurus alexandri*), the barbels are short or almost unnoticeable. In the auchenipterids, *Ageneiosus*, *Tetranematichthys*, and *Tympanopleura*, the maxillary barbel is short and thin, and it is lodged in a sulcus below the posterior portion of the upper lip (Calegari et al. 2019).

Barbels vary in shape. Catfish barbels are usually cylindrical, and teretiform, but some catfishes have ramified, fimbriate, or fringed barbels, such as some doradids (the so-called fimbriate-barbel doradines) (Birindelli 2014, figs. 5 and 6; Sabaj and Arce 2021), mochokids (e.g., some *Mochokus*, *Mochokiella*, *Microsynodontis*, and *Synodontis*, Figure 1B), and loricariids (e.g., some loricariines; Covain and Fisch-Muller 2007). Among doradids, fimbriae or papillae may be present on the maxillary barbel, either on its ventral face (*Anduzedoras*, *Doras*, *Hassar*, *Hemidoras*, *Leptodoras*, *Nemadoras*, *Ossancora*, *Tenellus*, and *Trachydoras*) or its dorsal face (some *Ossancora*), and in the mental barbels (*Hemidoras*, most *Ossancora*), in this case forming one (*Hemidoras*, *Nemadoras humeralis*, and *Ossancora fimbriata*) or two rows (*N. cristinae*, *O. asterophysa*, *O. eigenmanni*, and *Trachydoras nattereri*) (Sabaj et al. 2014, fig. 1). In the doradid *Rhinodoras*, transverse striae are present on the maxillary and mental barbels (Birindelli 2014, fig. 6a), and in the aspredinid *Ernstichthys intonsus*, the maxillary barbel has threadlike secondary barbels (Stewart 1985, fig. 3).

Sisorids often have wide, broad-based barbels, especially the maxillary pair located at the borders of the mouth (Ng 2015). These barbels are formed by a wing-like skin flap connecting the base of the maxillary barbel to the head, which in some sisorids (e.g., *Bagarius*, *Pseudecheneis*, Figure 3, *Sisor*) extend to at least half the barbel length (Ng 2015). Similar barbels are also found in some aspredinids (e.g., *Aspredo*, *Ernstichthys*, *Hoplomyzon*, Figure 5G, *Platystacus*). Other catfishes also have broad-based barbels, such as the akysids *Acrochordonichthys* and *Akysis*, and some trichomycterids (*Scleronema*, some microcambevines, and sarcoglanidines). In sisorids such as *Chimarrichthys* (= *Euchiloglanis*), *Myersglanis*, *Oreoglanis*, *Parachiloglanis*, *Pareuchiloglanis*, and *Pseudexostoma* (Figure 2E) the thick, wide-based maxillary barbels are continuous medially with the round anterior margin of the head giving these catfishes a characteristic, conspicuously round head in dorsal and ventral profile (e.g., Conway et al. 2012, fig. 1E). In addition, the tip of the maxillary barbel in *Oreoglanis* and *Pseudexostoma* (Figure 2E) is round instead of tapering to a point (Ng 2015).

The shape of the barbel varies considerably among ariids, as remarked by Kailola (2004). It may be round in cross-section (e.g., *Doiichthys novaeguineae*), flattened and strap-like (*Hexanematichthys sagor*), flattened and fleshy (*Arius venosus*, *A. leptonotacanthus*), wisp-like (*A. acutirostris*), compressed, tape-like or ribbon-like (*Bagre*), stiff and rod-like (*Osteogeneiosus*, some *Sciades*), or with a low basal membrane (*Cochlefelis spatula*, *Potamarius*). Anchariids are described as having "fringed" barbels (e.g., Ng and Sparks 2005); their maxillary barbels are thick and flattened, formed by a cylindrical core from which a flattened skin membrane extends posteriorly, either uniformly (in *Ancharius*) or in a flared fashion (in *Gogo*) (see Ng and Sparks 2005, fig. 1). A skin flap along the posterior edge of the maxillary barbels is also present in some members of Claroteidae, Mochokidae and Sisoridae, and strap-like maxillary barbels are present in some ariids (e.g., *Bagre*), but in this case, according to Ng and Sparks (2005), the cartilage supporting the barbel is flattened, instead of a skin extension along the anterior or posterior margin of the barbel. Some catfishes have membranous, ribbon-like barbels, such as the mochokid *Synodontis membranaceus*, the ailiid *Laides*, and the pimelodid *Hypophthalmus* (notably, *Hypophthalmus fimbriatus*; Littman et al. 2021, figs. 1D, 14 and 15). In the claroteid *Phyllonemus typus*, only the distal portion of the maxillary barbel is enlarged, forming a leaf-like structure (Figure 5H; Boulenger 1906; Risch 1987; Burgess 1989).

Barbels are movable, and their movement is possible due to modifications to their internal musculoskeletal support. The maxillary barbel, for example, is internally supported by the maxillary bone, which is usually movable due to modifications on the musculoskeletal system of the jaws and suspensorium (forming the so-called palatine-maxillary system; Gosline 1975; Diogo et al. 2003). Similarly, musculoskeletal modifications associated with barbel movement are present in the mental barbels when these are present (Diogo and Chardon 2000) and in catfishes with a well-developed nasal barbel (e.g., trichomycterids). In some catfishes, the movable maxillary barbel fits into a more or less developed groove below the eye (suborbital groove; Birindelli 2014, fig. 7; Calegari et al. 2019) when retracted, such as in auchenipterids, *Cetopsis*, some ailiids (e.g., *Laides*), and some bagrids (e.g., *Olyra*, *Sundolyra*). In the sisorid *Gagata*, groves are present on the ventral surface of the lower jaw where the mental barbels fit when adducted (Ng 2015).

While in most catfishes, the maxillary barbel moves horizontally, in auchenipterids, it moves vertically instead (Birindelli 2014; Calegari et al. 2019). This is related to the sexual behavior of these fishes, which usually involves the so-called amplexus (Calegari et al. 2019), in which the male uses the stiffened, anteriorly-rotated barbels together with the anteriorly rotated dorsal-spine (see Fins, below) to "hold" the female in position during spawning. Barbel movement may also serve other, more specific functions (Jayaram 1978). In some *Tetranematichthys* (e.g., *Tetranematichthys wallacei*, and *T. quadrifilis*), for example, fleshy, digitiform structures are present distally on their single pair of mental barbels (Vari and Ferraris 2006, fig. 4) and, as described by Vari and Ferraris (2006), a well-developed and conspicuous muscle mass from the *protractor hyoideus* attaches to the base of the barbel, which is probably used by this catfish to lure prey, in a manner comparable to the escae of ceratioids (Lophiiformes). Similarly, members of *Chaca* are described as using movements of their maxillary barbels, which are in some cases unpigmented distally, to "angle" fishes by luring them closer to their mouths (Burgess 1989; Nelson et al. 2016).

Barbels and Sexual Dimorphism

Several barbel modifications are found among auchenipterids associated with sexual dimorphism. Nuptial males of several auchenipterine genera (*Ageneiosus* (Figure 5I), *Auchenipterus*, *Entomocorus*, *Epapterus*, *Pseudepapterus*, *Tympanopleura*, *Tetranematichthys*, and some *Trachelyopterus*) have the maxillary barbels distinctively elongated in comparison to non-nuptial males and females. In some (e.g., *Tetranematichthys*), the maxillary barbel is distinctly thickened and well-ossified for the entire or almost entire length. The maxillary barbel in the silurid *Ceratoglanis* is also inflexible, stiffened by a cartilaginous core that forms a distal hook (Figure 5J), and in the ariid *Osteogeneiosus*, as mentioned above, the entire maxillary barbel is ossified (Marceniuk et al. 2012, fig. 13C). In addition, nuptial males of the genera *Auchenipterus*, *Entomocorus*, *Epapterus*, *Pseudepapterus*, and *Tetranematichthys* have well-developed unculiferous tubercles on the maxillary barbels, called breeding tubercles by Birindelli (2014), and in *Ageneiosus* (Figure 5I), *Tympanopleura*, *Tetranematichthys* and *Trachelyopterus insignis*, strongly ossified tubercles are present on the maxilla (Calegari et al. 2019). In some species of the centromochline genus *Gelanoglanis*, soft papillae are present along the maxillary barbel arranged in a single row of regular "bumps," but the association of those structures with sexual dimorphism is yet to be investigated since it may be present in both juveniles and adults (Calegari et al. 2019; Sarmento-Soares and Martins-Pinheiro 2020).

Fins

Catfishes may have the full "basic" set of fins usually present in teleosts: the median dorsal, caudal, and anal fins and the paired pectoral and pelvic fins (Figure 1A). A more or less developed adipose fin is also commonly present (see Adipose Fin below). However, in several catfishes, some of these fins may be reduced or entirely absent, and different combinations of the presence and absence of fins may be found (Figures 1 and 2). The absence of fins can be associated with characteristics of the biology of a given species. For example, burrowing clariids (e.g., *Channallabes*; Figure 1M; *Dolichallabes* and *Gymnallabes*) usually have the pectoral and pelvic fins reduced or entirely absent, as well as an elongated body (see Body Shape, above). Similar characteristics are also found in other burrowing catfishes, such as some psamophilic trichomycterids, in which fin reduction may be even more extreme. For example, in the microcambevine genus *Listrura*, the body is elongated, and the fins are reduced or absent: the pelvic fin is always absent, and in some cases also the dorsal fin (e.g., *L. boticario*, *L. depinnai*, *L. menezesi*; Costa and Katz 2021; Villa-Verde et al. 2022). In glanapterygines, the body is also elongate, and while members of *Glanapteryx* are nearly finless, members of *Pygidianops* and *Typhlobelus*, described as "some of the most aberrant Siluriformes known" (de Pinna and Zuanon 2013), may entirely lack the dorsal, pelvic, anal and, in some cases (*P. magoi* and *T. auriculatus*), the pectoral fins, in the most extreme cases of fin losses among ostariophysans (Schaefer et al. 2005; de Pinna and Kirovsky 2011; de Pinna and Zuanon 2013).

When fins are present, they may vary in shape, length, size, position, degree of development, number and types of rays, orientation, and so forth, variations that correlate with different aspects of the biology, ecology, and behavior of a given catfish species and may be extremely useful for distinguishing and diagnosing taxa. Species of the above-mentioned microcambevine genus *Listrura*, for example, may be differentiated by characteristics of their pectoral fins: *L. tetraradiata* has a relatively well-developed pectoral fin with four rays, some (e.g., *L. boticario* and *L. camposi*) have the pectoral fin formed by a well-developed first ray followed by one or two rudimentary rays, and others (e.g., *L. costai* and *L. nematopteryx*) have a filamentous pectoral fin formed by a single, elongate fin ray (Costa and Katz 2021). Costa and Katz (2021) correlate this morphological variation on the pectoral fin with different degrees of specialization to fossorial life.

Fin Spines

Spine-like (i.e., spinous) rays are often present on the dorsal and pectoral fins of catfishes. These structures are formed from modified fin rays and may include a locking mechanism involving other elements of the fin skeleton: adjacent lepidotrichia and pterygiophores in the case of the dorsal fin; elements of the pectoral girdle in the case of the pectoral fin (Arratia 2003a). In the dorsal fin, two spinous rays are usually present, one very short, often named "spinelet," that forms a locking mechanism for the second one, which is usually long and robust. Similarly, the first or anteriormost pectoral-fin ray is modified into a robust, lockable spine-like structure in several catfishes (see Kubicek et al. 2019 for a review). The presence of spinous rays on the dorsal and pectoral fins represents a defensive strategy for these fishes that may prevent them from being ingested (as dramatically demonstrated by one incident described by Benoist et al. (2010) involving a drinking party and one very unlucky individual of *Corydoras aeneus*).

However, there is high variation among catfish taxa in terms of the presence or absence of a spinous ray and its locking mechanism on the dorsal and pectoral fins, as well as in the shape of the spine-like structure when present (e.g., smooth, with serrae on the anterior margin, on the posterior margin, on both; see Kubicek et al. 2019 and Ballen and de Pinna 2021 for reviews on catfish pectoral spine diversity). In several catfish groups, instead of a well-developed, robust spine, the pectoral and dorsal fins bear only a weak, reduced spine-like structure or are entirely spineless. In addition, even when a robust spine is present, the locking mechanism may be absent. For example, the dorsal fin of aspredinids is short and often spineless or with a weak, flexible spine-like structure, although in *Acanthobunocephalus* (Figure 2D), a rigid spine is present and with a functional locking mechanism (Carvalho and Reis 2020). Loricarioids nicely illustrate the variation among closely related taxa on the presence of a spine-like ray and spine-locking mechanism on the dorsal fin. A locking mechanism is present in scoloplacids and callichthyids, but it is absent in astroblepids. Among loricariids, some have a functional lock mechanism (e.g., hypostomines), while in others it is not functional. The dorsal fin of nematogenyids and trichomycterids is entirely spineless, therefore, without a locking mechanism, and it is also displaced posteriorly (Figure 1D and E).

Fin Filaments

Filaments may be present on the fins of catfishes, more commonly on the dorsal, caudal (either on the dorsal, ventral, or both lobes), and pectoral fins. For example, the pangasiid *Pangasius sanitwongsei* has filamentous extensions in the dorsal, pectoral, pelvic, and anal fins (Roberts and Vidthayanon 1991). Similar filamentous extensions are present in other pangasiids but only in males (Roberts and Vidthayanon 1991). In some aspredinids (e.g., *Aspredo*, *Platystacus*), the first dorsal-fin ray is present as a long filament, which is longer in males than females (Friel and Carvalho 2017). Similarly, mature males of some cetopsids have elongated filaments on the dorsal fin, while in other members of the family (e.g., *C. coecutiens* (Figure 1C), *C. oliveirai*) elongate filaments in the dorsal and pectoral fins are present in all specimens (Vari et al. 2005). Rapp Py-Daniel and Cox-Fernandes (2005, fig. 1) and Sabaj et al. (2014) report the presence of a long filamentous extension in the dorsal fin of mature males of *Nemadoras* (*N. cristianae*, *N. humeralis*, and *N. trimaculatus*), a sexually dimorphic characteristic unusual for doradids. Among loricariines, several species have long filamentous extensions on the pectoral and dorsal fins and on one or both lobes of the caudal fin (Covain and Van der Sleen 2017). According to Marceniuk et al. (2012), a filament is present as part of the dorsal-fin spine in all species of the ariid *Bagre* except for *B. panamensis* and in some species of *Arius*. In members of the heptapterid genus *Mastiglanis*, long filaments are present on the dorsal and pectoral fins, which are used, in combination with their also long barbels, in a remarkable sit-and-wait foraging tactic (described and illustrated by Zuanon et al. 2006, fig. 2).

In *Sisor*, the uppermost principal caudal-fin ray is enlarged and elongated as a filament in some cases longer than the body, and these fishes are sometimes called "Whiptail Catfishes" (e.g., Javed and Mirza 2011). De Pinna (1996) remarked that this ray in *Sisor* is also uniquely large, its diameter about five times larger than the diameter of the remaining rays. The name "Whiptail Catfishes" is also commonly applied to loricariines, where a similarly long filament on the upper lobe of the caudal fin is commonly present (Figure 2B; Covain and Van der Sleen 2017). In some bagrids, such as *Bagrus*, *Bagrichthys*, and *Sperata*, the most external caudal-fin rays on both the dorsal and ventral lobes may extend as long filaments, a condition commonly found in catfishes with deeply forked caudal fins according to de Pinna (1996).

Dorsal and Anal Fins

A well-developed dorsal fin is usually present in catfishes (Figures 1A and B), although some reductions in the fin occur in specific taxa. For example, the dorsal fin is considerably short in chacids, with only three or four rays in addition to a strongly serrated spine-like ray (Figure 2C), and in the aspredinid *Amaralia*, the dorsal fin is formed by only two or three rays (Friel and Carvalho 2017). In silurids, the dorsal fin is either short, with two to five fin rays, or absent (e.g., *Ceratoglanis*, *Micronema*, and *Phalacronotus*). The dorsal fin also lacks in malapterurids (Figure 1J), *Kryptoglanis*, and some ailiids.

In the bagrid *Mystus bocourti*, the anterior portion of the dorsal fin is extremely large. In *Bagrichthys*, the dorsal fin is also large, especially in *B. hypselopterus*, giving these fishes their popular name, "Lancer." This fin is also considerably large in some plotosids such as *Euristhmus* (e.g., *E. sandrae*) and *Paraplotosus* (e.g., *P. butler*), in some corydoradines (e.g., male *Corydoras longipinnis*), and in some species of *Synodontis* (*S. alberti*, *S. eupterus*, and *S. velifer*).

The anal fin is also usually present and well-developed in catfishes, and there is high variation in its length among taxa (Figure 1). In heteropneustids, the anal fin may have about 50–75 rays, and in silurids, about 40–110 rays (Rema Devi and Raghunathan 1999; Hossain et al. 2013). Variation in the length of the anal fin may be high even among closely-related taxa. For example, among aspredinids, the anal fin base may be short, with less than 20 rays (e.g., *Amaralia*, *Bunocephalus*, *Hoplomyzon*, *Mircomyzon*, and *Xyliphius*) or long, with more than 50 rays (e.g., *Aspredo*, *Aspredinichthys*, and *Platystacus*). Among auchenipterids, centromochlines have a short anal-fin base, while some auchenipterines, such as *Auchenipterus*, *Epapterus*, *Pseudepapterus*, and *Trachelyopterichthys*, have a long anal-fin base (Calegari et al. 2019).

Adipose Fin

A more or less developed adipose fin is often present in catfishes (Figures 1A and B; Arratia 2003a; Birindelli 2014; Nelson et al. 2016). The adipose fin in catfishes is usually formed from a median fin fold present in the larva that extends from the head to the dorsal margin of the caudal fin and tends to disappear with ontogeny (e.g., de Pinna et al. 2018, fig. 5). When retained in adults (e.g., *Trichomycterus dali*), the dorsal-fin fold is considered a paedomorphic feature (Rizzato et al. 2011). As the larval fin fold disappears during ontogeny, a more or less developed adipose fin may be formed between the dorsal and caudal fins, and in some cases, it remains confluent with the caudal fin. An adipose fin confluent with the caudal fin occurs in members of several families, such as Amblycipitidae (e.g., *Amblyceps accari*, *Liobagrus*, and *Xiurebagrus dorsalis*), Heptapteridae (several taxa; Bockmann and Slobodian 2017), and Ictaluridae (e.g., some species of *Noturus* such as the Tadpole Madtom, *Noturus gyrinus*). In some cases, the adipose fin is vestigial, present merely as a low adipose ridge, such as in some heteropneustids and trichomycterids (e.g., some copionodontines, *Scleronema*, *Cambeva variegata*, *Trichomycterus astromycterus*; Bockmann et al. 2023). In others, the adipose fin is entirely lacking, such as in cetopsines (Figure 1C), plotosids (Figure 1K), silurids (Figure 1I), aspredinids (Figure 2D), *Kryptoglanis*, *Nematogenys* (Figure 1D), and most trichomycterids (Figure 1E). In most catfishes, where an adipose fin is present, it appears as a soft structure. However, in some species, a spine-like structure is present associated with its anterior margin (e.g., callicthyids, some loricariids, *Sisor*; Arratia 2003a).

The presence or absence of an adipose fin is, therefore, a variable character even among closely-related taxa, as well as its shape, length, and degree of development and support. For example, among akysids, the adipose fin is commonly short and deep in *Akysis* and *Pseudobagarius*, long and low in *Acrochordonichthys*, and present as a low adipose ridge in *Parakysis* and *Breitensteinia* (Ng and Siebert 1998; Ng 2009). Among species of *Hemibagrus*, it may have a long (e.g., *H. sabanus*, *H. guttatus*, *H. macropterus*, and *H. vietnamicus*) or short base (e.g., *H. punctatus*, *H. hoevenii*, and *H. spilopterus*). The adipose fin is very short in pangasiids, auchenipterids, and ailiids (Figure 1H), but some species of the latter two families lack an adipose fin.

Among loricariids, an adipose fin is absent in all loricariines (Figures 2A and B). When present in the remaining members of the family, it usually bears an ossified spine-like structure along its anterior margin (e.g., Figure 1G), but among lithogenines, such spine is present in *Lithogenes villosus* and *L. valencia* but absent in *L. wahari* (Schaefer and Provenzano 2008). A similar spine is present in callichthyids (Figure 1F; Arratia 2003a) and in the astroblepids with an adipose fin, but some members of the latter family lack an adipose fin. The adipose fin of catfishes, even when present and well developed, is not usually supported by endoskeletal elements, but at least in the clariid *Dinotopterus* and *Heterobranchus*, neural spines penetrate on the adipose fin (Tan et al. 2023, fig. 1). According to Arratia (2003a) and Buckland-Nicks (2016), the adipose fin of other catfishes may also have support from endoskeleton (horabagrids) or rays (bagrids, pimelodids, and mochokids).

Caudal Fin

The shape of the caudal fin may be highly variable among catfishes, even among closely related taxa (e.g., members of the same family). For example, in amblycipitids, the caudal-fin margin can be furcated (several *Amblyceps*), emarginated (e.g., some *Amblyceps*, *Xiurenbagrus dorsalis*, and *X. gigas*), truncated (e.g., some *Amblyceps*, *Xiurenbagrus xiurenensis*, and some *Liobagrus*), subtruncate (some *Liobagrus*) or round (some *Liobagrus*). Characteristically round caudal-fin margins are also found in aspredinids (Figure 2D), chacids (Figure 2C), clariids (Figure 1M), malapterurids (Figure 1J), and *Nematogenys* (Figure 1D). In some *Amblyceps* (e.g., *A. protentum*), the caudal fin is deeply forked and asymmetrical, and the upper lobe lanceolate and longer than the ventral lobe (e.g., Ng and Wright 2009, fig. 1). Similarly, asymmetrical caudal-fin margins are common among members of *Olyra* (Ng and Ferraris 2016), *Sundolyra* (Ng et al. 2015), and *Hemibagrus*, and in some of the latter, a filamentous extension on the upper lobe may also be present. In *Hemibagrus olyroides*, all rays in the upper lobe are considerably elongated, giving the caudal fin a strikingly asymmetric aspect (Ng and Kottelat 2013, fig. 29). Among heptapterids, species with asymmetrical caudal-fin lobes are also found, the longer lobe being in some cases the upper (e.g., *Imparfinis*), in others the lower (e.g., *Gladioglanis*, Bockmann and Slobodian 2017).

An interesting example of intrafamilial variation in the caudal fin is found among doradids. As remarked by Sabaj and Arce (2021), the caudal fin of doradids occupies the entire spectrum from evenly rounded (only in *Acanthodoras*) to truncate or emarginate with gently rounded lobes (*Agamyxis*) to shallowly forked with rounded or bluntly pointed lobes (e.g., *Centrochir*, *Platydoras*, and *Oxydoras*) to deeply forked with pointed lobes (*Centrodoras*, *Doraops*, *Hassar*, *Hemidoras*, *Leptodoras*, and *Rhinodoras*). In addition, some doradids (e.g., *Hypodoras*, *Physopyxis*) usually have the branched rays of the upper caudal-fin lobe longer, and in *Astrodoras*, the upper lobe is relatively narrow and pointed, and the lower lobe is broad and rounded, in such a way that the distal margin of the caudal fin has an uneven S-shape (Sabaj and Arce 2021). Sabaj and Arce (2021) discuss the possible relationships of caudal fin shapes with habitat and behavior in doradids.

The caudal fin is confluent with the adipose fin in chacids and extends far anteriorly both dorsally and ventrally by means of several procurrent caudal fin rays (Figure 2C). The caudal fin also extends far anteriorly in the dorsum of some plotosids (*Oloplotosus*, *Paraplotosus*, and *Plotosus*, Figure 1K, *Tandanus*), in some cases originating anterior to the vertical through the pelvic fins (e.g., *Paraplotosus*). In *Phreatobius*, the caudal fin extends dorsally and ventrally by means of several well-developed procurrent rays, and ventrally it is continuous with the anal fin (Muriel-Cunha and de Pinna 2005; Fernández et al. 2007; Shibatta et al. 2007). In some clariids, the dorsal, caudal, and anal fins are also confluent, but in this case, the dorsal and anal fins are very long (e.g., *Channallabes*, Figure 1M; *Gymnallabes*). Other clariids have the dorsal and caudal fins separated, in some cases with a clearly defined adipose fin between them (e.g., *Clarias lamottei*, *C. ngamensis*, *Dinotopterus*, and *Heterobranchus*), but even in these, the dorsal fin is always long, with more than 30 rays. Interestingly, in *Heteropneustes*, regarded as a sister taxon of Clariidae, the dorsal fin is short (Figure 1L) and usually has up to seven rays.

Pectoral and Pelvic Fins

Well-developed pectoral and pelvic fins are usually present in catfishes, although, as mentioned earlier, several groups have lost either one or both of them, especially elongate taxa. Several modifications may occur in the paired fins of catfishes, some of which were already mentioned: the presence of spinous or filamentous rays, adhesive pads (Figure 3), odontodes (Figure 4), or even modification in the shape and position of these fins (e.g., Figure 2E and F). Perhaps the most interesting modifications of the paired fins among catfishes are those found in the pelvic fin and girdle of astroblepids and lithogenine loricariids. Astroblepids have thickened pelvic fins with specialized musculature that allow them to grasp surfaces and, combined with their sucker-like mouth (see Mouth below), enable them to climb even vertical surfaces by moving their pelvic girdle relative to the pectoral girdle (one popular name for astroblepids is "Climbing Catfishes" (Shelden 1937; Schaefer 2003; De Crop et al. 2013). A similar condition is found in the very distinctive loricariids of the subfamily Lithogeneinae. These peculiar, "naked" loricariids (see Skin below) also have a specialized, movable pelvic fin that, in combination with the ventral mouth modified into a sucker disk and with enlarged odontodes on the fin rays also present in other loricariids, helps them adhere to and even climb surfaces. As highlighted by Schaefer and Provenzano (2008), these peculiar morphological adaptations occur in astroblepids and lithogenine loricariids but not in the remaining members of Loricariidae. These and other similarities between lithogenines and astroblepids partly explain some of the past controversies over the relationship between Astroblepidae and Loricariidae (discussed in Schaefer 2003), some authors considering that the former should be regarded as a subfamily of Loricariidae (Astroblepinae; e.g., Bailey and Baskin 1976), others placing *Lithogenes* in Astroblepidae (e.g., Nijssen and Isbrücker 1987).

Fins and Sexual Dimorphism

The most interesting modifications on the fins associated with sexual dimorphism among catfishes are found in auchenipterids. These catfishes exhibit marked sexual dimorphism that may involve several morphological changes in mature males: shape, size and position of anal fin, development and fusion of anal-fin pterygiophores, enlargement of anal-fin rays, enlargement, elongation and ossification of the dorsal-fin spinous ray (and maxillary barbel; see barbels, above), development of serrae on the anterior surface of the dorsal-fin spinous ray, development of a gonopodium associated with the anal fin, and development of nuptial tubercles (see Skin, above) (Calegari et al. 2019). These modifications are more or less directly related to reproduction by insemination, which is a characteristic feature of the family (Birindelli 2014; Calegari et al. 2019). In adult male auchenipterids, an intromittent organ (gonopodium) is present on the anal fin and is used to transfer spermatozoids to females, and modifications on the anal-fin shape of adult males are probably involved with sustaining the intromittent organ during the copula (Calegari et al. 2019). The position of the urogenital opening in male auchenipterids varies: it may be near the anus (as in most catfishes), at the base of the first anal-fin ray in centromochlines, or the tip of the anterior anal-fin rays in auchenipterines (Birindelli 2014, fig. 18; Sarmento-Soares and Martins-Pinheiro 2020). As remarked by Calegari et al. (2019), in centromochlines, the anal fin of both nuptial and non-nuptial males differs in shape from the anal fin of females, while in other auchenipterids the anal-fin shape is modified only in nuptial males. In mature male centromochlines, the anal fin is also oblique relative to the body axis and posteriorly oriented. As remarked by Calegari et al. (2019), this different orientation is associated with modifications on the anal-fin proximal radials, including the partial fusion of these elements.

More specific modifications are found in some members of the family. In *Epapterus*, there is a conspicuous gap distally between the anterior anal-fin rays, which are modified for insemination (at the tip of which the urogenital opening is located), and the remaining anal-fin rays of nuptial males (Birindelli 2014, fig. 18b; Calegari et al. 2019). In nuptial males of *Auchenipterus* there is a blunt, posteriorly-directed hook on the last unbranched anal-fin ray (Birindelli 2014; Calegari et al. 2019). As expected, modifications associated with reproduction by insemination are also present in female auchenipterids: the urogenital pore is internally enlarged, ending in a cavity where the spermatozoids transferred by males are retained (Birindelli 2014, fig. 19; Calegari et al. 2019), and a flap associated with the genital papilla is present in females of most centromochlines, being noticeably thick in some (e.g., *Glanidium albescens*, Sarmento-Soares and Martins-Pinheiro 2020, fig. 5D).

Regarding the dorsal fin, according to Birindelli (2014) and Calegari et al. (2019), the dorsal spine is elongated in the auchenipterids *Centromochlus existimatus*, *C. heckelii*, and *Tatia intermedia*, while in *Asterophysus*, *Epapterus*, *Pseudepapterus*, and *Spinipterus*, it is short. In nuptial males of some auchenipterines (e.g., *Ageneiosus*, *Tetranematichthys*, *Tympanopleura*, and some *Trachelyopterus*), the dorsal spine is distinctly longer, noticeably arched, and capable of rotating anteriorly, forming an angle of more than 90° relative to its base when fully rotated. In addition, in some (e.g., *Ageneiosus*, *Auchenipterichthys*, *Trachelyopterichthys*, and some *Trachelyopterus*), modifications are present in the serrae of the dorsal spine (see Birindelli 2014). These modifications are associated with the amplexus behavior during spawning, which is also characteristic of these catfishes (Calegari et al. 2019).

Contrasting with the dorsal and anal fins, modifications on the pelvic fin are not common in auchenipterids, except for *Entomocorus*, in which the first ray of the pelvic fin is modified, enlarged, and elongate in mature males (Birindelli 2014; Calegari et al. 2019). Among loricariids, modifications may be present in the pelvic fins associated with sexual dimorphism, such as the presence of a skin fold at the edge of the first ray (Isbrücker and Nijssen 1992; Rapp Py-Daniel and Cox-Fernandes 2005; Pereira et al. 2014).

Conclusion

As highlighted in the title of this chapter, only selected features of the external anatomy of catfishes are presented and discussed. Other aspects and structures of their external anatomy could have been explored, and in every one of them, an enormous variation would be found among taxa; however, due to constraints in space, a selection of some more illustrative features was made. A few other teleostean orders are comparably so morphologically diverse, not to mention taxonomically, and with their successful colonization of many different types of habitats around the globe, catfishes provide seemingly endless examples of morphological diversification and adaptation, several of which are unique to the Siluriformes or even to specific taxa within the group. We hope that, with the examples provided, we were able to demonstrate how catfish external morphological diversity can be at the same time challenging and fascinating, how it can be useful for recognizing, differentiating, and diagnosing catfish taxa, and how it can be used to investigate and demonstrate the different ways by which the anatomy of a given species can be associated with many different aspects of its biology. There is still a lot to be known about the external anatomy of catfishes, and as new species are described and more detailed morphological studies about the already-known taxa are provided, our knowledge and understanding of catfish morphology, biology, and phylogeny tend to increase, as well as the number of interesting questions that can be asked about their evolution and diversification. We foresee a long future of morphological studies on Siluriformes, and we hope that the examples presented here will inspire future researchers interested in contributing to the knowledge of such a challenging and fascinating group.

Acknowledgments

We sincerely thank Mariangeles Arce, Mark Sabaj, Kyle Luckenbill and John Lundberg (ANSP); Cristina Cox-Fernandes and Jeff Podos (UMass); Lynne Parenti and Kate Bemis (USNM); Scott Schaefer, Radford Arrindell and Bruno Melo (ANSP); Jacqueline Webb (URI); Alessio Datovo, Mario de Pinna and Murilo Pastana (MZUSP) for providing access to specimens housed in their respective institutions during the preparation of this chapter. We are grateful to Gabriel L. Gomes for the illustration in Figure 3. PPR is funded by "Programa de Apoio a Novos Docentes" of Pró-Reitoria de Pesquisa e Inovação (PRPI), University of São Paulo (#2022.1.9345.1.2). MAA is funded by the Coordenação de Aperfeiçoamento de Pessoal de Nível Superior – Brasil (CAPES) – Finance Code 88887.480948/2020-00 at Programa de Pós-Graduação em Biologia Comparada, Faculdade de Filosofia, Ciências e Letras de Ribeirão Preto, University of São Paulo, including a CAPES-PRINT (Programa Institucional de Internacionalização) fund – Finance Code 88887.695435/2020-00; and by the Böhlke Memorial Endowment Fund, The Academy of Natural Sciences, Drexel University. We are grateful to Flávio A. Bockmann (LIRP) for sharing with us its passion for fish morphology in general and of catfishes in particular.

References

Adriaens, D., S. Devaere, G.G. Teugels and W. Verraes. 2002. Intraspecific variation in limblessness in catfishes, related to a specialized fossorial habit. Biol. J. Linn. Soc. 75: 367–377.

Adriaens, D., J.N. Baskin and H. Coppens. 2010. Evolutionary morphology of trichomycterid catfishes: about hanging on and digging in. pp. 337–362. *In*: Nelson, J.S., H.-P. Schultze and M.V.H. Wilson (eds.). Origin and Phylogenetic Interrelationships of Teleosts. Verlag Dr. F. Pfeil, München, Germany.

Agnèse, J.F. and G.G. Teugels. 2005. Molecular phylogeny of Clariidae (Teleostei, Siluriformes): implications for their body shape evolution, biogeography and taxonomy. Mol. Phyl. Evol. 36: 546–553.

Akama, A. and C.J. Ferraris Jr. 2011. *Spinipterus*, a new genus of small, spiny catfish (Siluriformes: Auchenipteridae) from the Peruvian Amazon. Zootaxa 2992: 51–60.

Armbruster, J.W. 2004. Phylogenetic relationships of the suckermouth armoured catfishes (Loricariidae) with emphasis on the Hypostominae and the Ancistrinae. Zool. J. Linn. Soc. 141: 1–80.

Armbruster, J.W. and D.C. Taphorn. 2013. Description of *Neblinichthys peniculatus*, a new species of loricariid catfish from the río Paragua drainage of Venezuela. Neotrop. Ichthyol. 11: 65–72.

Arratia, G. 1987. Description of the primitive family Diplomystidae (Siluriformes, Teleosteri, Pisces): morphology, taxonomy and phylogenetic implications. Bonner Zool. Monog. 24: 1–120.

Arratia, G. 1900. The South American Trichomycterinae (Teleostei: Siluriformes), a problematic group. pp. 395–403. *In*: Peters, G. and H. Hutterer (eds.). Vertebrates in the Tropics. Museum A. Koenig, Bonn.

Arratia, G. and L. Huaquín. 1995. Morphology of the lateral line system and of the skin of diplomystid and certain primitive loricarioid catfishes and systematics and ecological considerations. Bonner Zool. Monog. 36: 1–110.

Arratia, G. 2003a. The Siluriformes pstcranial skeleton—an overview. pp. 121–158. *In*: Arratia, G., B.G. Kapoor, M. Chardon and R. Diogo (eds.). Catfishes. Science Publishers, Enfield, New Hampshire, USA.

Arratia, G. 2003b. The skin of catfishes—a review. pp. 177–199. *In*: Arratia, G., B.G. Kapoor, M. Chardon and R. Diogo (eds.). Catfishes. Science Publishers, Enfield, New Hampshire, USA.

Arratia, G., B.G. Kapoor, M. Chardon and R. Diogo. 2003. Catfishes. Science Publishers, Enfield, New Hampshire, USA.

Arratia, G. and C. Quezada-Romegialli. 2017. Understanding morphological variability in a taxonomic context in Chilean diplomystids (Teleostei: Siluriformes), including the description of a new species. PeerJ. 5: 1–40.

Arratia, G. 2025. Siluriforms and the South American fossil record: their significance and the origin and evolutionary radiation of the group. *In*: Arratia, G. and R.E. Reis (eds.). Catfishes, a Highly Diversified Group. Volume 2: Evolution and Phylogeny. CRC Press, Boca Raton, FL.

Bailey, R.M. and J.N. Baskin. 1976. *Scoloplax dicra*, a new armored catfish from the Bolivian Amazon. Occas. pap. Mus. Zool. Univ. Mich. 674: 1–14.

Ballen, G.A. and M.C.C. de Pinna. 2021. A standardized terminology of spines in the order Siluriformes (Actinopterygii: Ostariophysi). Zool. J. Linn. Soc. 194: 1–25.

Benoist, L.B.L., B. van der Hoven, A.C. de Vries, B. Pullens, E.J.O. Kompanje and C.W. Moeliker. 2019. A jackass and a fish: A case of life-threatening intentional ingestion of a live pet catfish (*Corydoras aeneus*). Acta Oto-Laryngologica CaseReports 4: 1–4.

Bhatti, H.K. 1938. The integument and dermal skeleton of Siluroidea. Zool. Soc. Lond. 24: 1–102.

Bichuette, M.E. and E. Trajano. 2004. Three new subterranean species of *Ituglanis* from central Brazil (Siluriformes: Trichomycteridae). Ichthyol. Explor. Freshw. 15: 243–256.

Bichuette, M.E. and P.P. Rizzato. 2012. A new species of cave catfish from Brazil, *Trichomycterus rubbioli* sp.n., from Serra do Ramalho karstic area, São Francisco River basin, Bahia State (Silurifomes: Trichomycteridae). Zootaxa 3480: 48–66.

Birindelli, J.L.O. 2014. Phylogenetic relationships of the South American Doradoidea (Ostariophysi: Siluriformes). Neotrop. Ichthyol. 12: 451–564.

Bockmann, F.A. and V. Slobodian. 2017. Family Heptapteridae – Three-barbeled catfishes. pp. 233–252. *In*: van der Sleen, P. and J.S. Albert (eds.). Field Guide to the Fishes of the Amazon, Orinoco & Guianas. Princeton, New Jersey, USA.

Bockmann, F.A., J. Ferrer, P.P. Rizzato, A.L.H. Esguícero, L.F. Duboc and L.F.S. Ingenito. 2023. Anatomy, ecology, and behavior of a new species of *Scleronema* Eigenmann, 1917 (Siluriformes: Trichomycteridae) from coastal drainages in the southern Brazilian Atlantic Rainforest, with comments on the monophyly and phylogeny of the genus. Zootaxa 5297(1): 1–47.

Boulenger, G.A. 1906. Fourth contribution to the ichthyology of Lake Tanganyika.-Report on the collection of fishes made by Dr. W. A. Cunnington during the Third Tanganyika Expedition, 1904–1905. Trans. Zool. Soc. London. 17: 537–601.

Boulenger, G.A. 1908. A revision of the African Silurid fishes of the subfamily Clariinae. Proceedings of the Zoological Society of London 1907: 1062–1097.

Buckland-Nicks, J.A. 2016. New details of the neural architecture of the salmonid adipose fin. J. Fish Biol. 89(4): 1991–2003.

Burgess, W.E. 1989. An Atlas of Freshwater and Marine Catfishes: A Preliminary Survey of the Siluriformes. T. F. H. Publications, USA.

Cabuy, E., D. Adriaens, W. Verraes and G.G. Teugels. 1999. Comparative study on the cranial morphology of *Gymnallabes typus* (Siluriformes: Clariidae) and their less anguilliform relatives, *Clariallabes melas* and *Clarias gariepinus*. J. Morphol. 240: 169–194.

Calegari, B.B., R.P. Vari and R.E. Reis. 2019. Phylogenetic systematics of the driftwood catfishes (Siluriformes: Auchenipteridae): a combined morphological and molecular analysis. Zool. J. Linn. Soc. 187: 661–773.

Carvalho, T.P. and R.E. Reis. 2020. A new miniature species of *Acanthobunocephalus* (Silurifomes: Aspredinidae) from the lower Purus River basin, Amazon Basin, Brazil. Copeia 108: 347–357.

Chamon, C.C. 2016. Redescription of *Acanthicus hystrix* Agassiz, 1829 (Siluriformes: Loricariidae), with comments on the systematics and distribution of the genus. Zootaxa-4088: 395–408.

Chamon, C.C. and T. Costa e Silva. 2018. *Pseudacanthicus major*: description of one of the largest known Loricariidae (Hypostominae: Ancistrini), a species from rio Tocantins basin, Brazil. Zootaxa 4387: 499–510.

Chen, X.P. and J.G. Lundberg. 1995. *Xiurenbagrus*, a new genus of amblycipitid catfishes (Teleostei: Siluriformes), and phylogenetic relationships among the genera of Amblycipitidae. Copeia 1995(4): 780–800.

Conway, K.W., N.K. Lujan, J.G. Lundberg, R.L. Mayden and D.S. Siegle. 2012. Microanatomy of the paired-fin pads of ostariophysan fishes (Teleostei: Ostariophysi). J. Morphol. 273: 1127–1149.

Copley, H. 1941. A short account of the freshwater fishes of Kenya. J. East Afr. Nat. Hist. 16: 1–24.

Costa, W.J.E.M., E. Henschel and A.M. Katz. 2020. Multigene phylogeny reveals convergent evolution in small interstitial catfishes from the Amazon and Atlantic forests (Siluriformes: Trichomycteridae). Zool. Scr. 49: 159–173.

Costa, W.J.E.M. and A.M. Katz. 2021. Comparative morphology, phylogeny, classification and evolution of interstitial habits in microcambevine catfishes (Siluriformes: Trichomycteridae). Taxonomy 1(4): 313–344.

Covain, R. and S. Fisch-Muller. 2007. The genera of the neotropical catfish subfamily Loricariinae (Siluriformes: Loricariidae): a practical key and synopsis. Zootaxa 1462: 1–40.

Covain, R. and P. Van der Sleen. 2017. Subfamily Loricariinae—Loricariine Armored Catfishes. pp. 287–296. *In*: van der Sleen, P. and J.S. Albert (eds.). Field Guide to the Fishes of the Amazon, Orinoco & Guianas. Princeton, New Jersey, USA.

Das, D. and T.C. Nag. 2004. Adhesion by paired pectoral and pelvic fins in a mountain-stream catfish, *Pseudocheneis sulcatus* (Sisoridae). Environ. Biol. Fishes 71: 1–5.

Datovo, A. and F.A. Bockmann. 2010. Dorsolateral head muscles of the catfish families Nematogenyidae and Trichomycteridae (Siluriformes: Loricarioidei): comparative anatomy and phylogenetic analysis. Neotrop. Ichthyol. 8: 193–246.

De Crop, W., E. Pauwels, L. Van Hoorebeke and T. Geerinckx. 2013. Functional morphology of the Andean climbing catfishes (Astroblepidae, Siluriformes): Alternative ways of respiration, adhesion, and locomotion using the mouth. J. Morphol. 274: 1164–1179.

Devaere, S., D. Adriaens, G.G. Teugels, N.M. De Clerck and A.A. Postnov. 2005. Skeletal morphology of the holotype of *Gymnallabes nops* Roberts & Stewart, 1976, using micro CT-scanning. Cybium 29: 281–293.

Diogo, R. and M. Chardon, 2000. The structures associated with catfish (Teleostei: Siluriformes) mandibular barbels: origin, anatomy, function and synonymy. Neth. J. Zool. 50: 455–478.

Diogo, R., M. Chardon and P. Vandewalle. 2003. Functional morphology of catfishes: movements of barbels. pp. 203–220. *In*: Arratia, G., A.S. Kapoor, M. Chardon and R. Diogo (eds.). Catfishes. Science Publishers, Enfield, New Hampshire, USA.

Fagundes, P., E.H.L. Pereira and R.E. Reis. 2020. Iterative taxonomic study of *Pareiorhaphis hystrix* (Siluriformes, Loricariidae) suggests a single, yet phenotypically variable, species in South Brazil. PLoS ONE 15: 1–26.

Fernández, L., L.J. Saucedo, F.M. Carvajal-Vallejos and S.A. Schaefer. 2007. A new phreatic catfish of the genus *Phreatobius* Goeldi 1905 from groundwaters of the Iténez River, Bolivia (Siluriformes: Heptapteridae). Zootaxa 1626: 51–58.

Fink, S.V. and W.L. Fink. 1981. Interrelationships of ostariophysian fishes (Teleostei). Zool. J. Linn. Soc. 72: 297–353.

Fricke, R., W.N. Eschmeyer and J.D. Fong. 2023. Species by Family/subfamily. (http://researcharchive.calacademy.org/research/ichthyology/catalog/SpeciesByFamily.asp).Electronic version accessed 28 jun 2023.

Friel, J.P. 1994. A Phylogenetic Study of the Neotropical Banjo Catfishes (Teleostei: Siluriformes: Aspredinidae. Ph.D Dissertation, Duke University, Durham, North Carolina, US.

Friel, J.P. 2003. Family Aspredinidae (Banjo catfishes). pp. 261–267. *In*: Reis, R.E., S. Kullander and C.J. Ferraris, Jr. (eds.). Check List of the Freshwater Fishes of South and Central America. Edipucrs, Porto Alegre, Brazil.

Friel, J.P. and T.P. Carvalho. 2017. Family Aspredinidae - Banjo Catfishes. pp. 202–207. *In*: van der Sleen, P. and J.S. Albert (eds.). Field Guide to the Fishes of the Amazon, Orinoco & Guianas. Pinceton, New Jersey, USA.

Geerinckx, T. and D. Adriaens. 2006. Development of the Osteocranium in the Suckermouth armored catfish *Ancistrus* cf. *triradiatus* (Loricariidae, Siluriformes). J. Morphol. 268: 254–274.

Gosline, W.A. 1975. The palatine-maxilary mechanism in catfishes with comments on the evolution and zoogeography of modern siluroids. Occas. Pap. Calif. Acad. Sci. 120: 1–31.

Grande, L. and J.T. Eastman. 1986. A review of Antarctic ichthyofaunas in the light of new fóssil discoveries. Palaeontology 29: 113–137.

Haspel, G., A. Schwartz, A. Streets, D.E. Camacho and D. Soares. 2012. By the teeth of their skin, cavefish find their way. Curr. Biol. 22: 1–2.

Henschel, E., A.M. Katz and P.H.N. Braganca. 2025. Miniature catfishes: diversity, ecology and evolution. pp. 201–231. *In*: Arratia, G. and R.E. Reis (eds.). Catfishes, a Highly Diversified Group. Volume 1: Their Outstanding Biology. CRC Press, Boca Raton, FL.

Hora, S.L. 1933. Siluroid fishes of India, Burma and Ceylon, I. loach-like fishes of the genus *Amblyceps* Blyth. Rec. Ind. Mus. Culcutta 34: 607–621.

Hossain, M.S., S.S. Sarker, S.M. Sharifuzzaman and Chowdhury, sayedur. 2013. New species of stinging catfish *Heteropneustes nani* (Siluriformes: Heteropneustidae) from Noakhali, Bangladesh. Vertebr. Zool. 63(3): 259–267.

Isbrücker, I.J.H. and H. Njissen. 1992. Sexualdimorphismus bei Harnischwelsen (Loricariidae). Odontoden, Zahne, Lippen, Tentakel, Genitalpapilen und Folssen. Aquarien- und Terrarien Zeitschrift (DATZ) Sonderheft Harnischwelse 19–33.

Javed, M.N. and M.R. Mirza. 2011. *Sisor pakistanicus* (Teleostei, Sisoridae), a new catfish from the river Chenab, Pakistan. Biologia (Pakistan) 57: 15–21.

Jayaram, K.C. 1978. Functional responses of catfish barbels. Bull. Zool. Surv. India 1: 77–80.

Kailola, P.J. 2004. A phylogenetic exploration of the catfish Family Ariidae (Otophysi: Siluriformes). The Beagle, Records of the Museums and art Galleries of the occasional papers of the Northern Territory 20: 87–166.

Kubicek, K.M., R. Britz and K.W. Conway. 2019. Ontogeny of the catfish pectoral-fin spine (Teleostei: Siluriformes). J. Morphol. 280: 339–359.

Littmann, M.W., J.G. Lundberg and M.S. Rocha. 2021. Revision of the South American catfish genus *Hypophthalmus* (Siluriformes, Pimelodidae) with descriptions of two new species from the Amazon and Orinoco Basins. Proc. Acad. Nat. Sci. 167: 171–223.

Lokeshwor, Y. and P.D. Marak. 2022. *Pseudolaguvia meghalayaensis*, a new diminutive sisorid catfish (Teleostei: Siluriformes) from northeastern India. Zootaxa 5175: 367–376.

Lujan, N.K., M.H. Hidalgo and D.J. Stewart. 2010. Revision of *Panaque* (Panaque), with descriptions of three new species from the Amazon basin (Siluriformes, Loricariidae). Copeia. 2010(4): 676–704.

Lujan, N.L. and K.W. Conway. 2015. Life in the fast lane: a review of reophily in freshwater fishes. pp. 107–136. *In*: Riesch, R., M. Tobler and M. Plath (eds.). Extremophile Fishes: Ecology, Evolution, and Physiology of Teleosts in Extreme Environments. Springer International Publishing, Switzerland.

Marceniuk, A.P., N.A. Menezes and M.R. Britto. 2012. Phylogenetic analysis of the family Ariidae (Ostariophysi: Siluriformes), with a hypothesis on the monophyly and relationships of the genera. Zool. J. Linn. Soc. 165: 534–669.

Mistri, A., U. Kumari, S. Mittal and A.K. Mittal. 2018. Morphological specializations of the epidermis of an angler catfish *Chaca chaca* (Siluriformes, Chacidae) in relation to its ecological niche: A scanning electron microscopic investigation. Microsc. Res. Tech. 81: 439–448.

Murdy, E.O. and C.J. Ferraris, Jr. 2006. A revision of the marine eel-tailed catfish genus *Euristhmus* (Teleostei: Siluriformes: Plotosidae). The Beagle, Records of the Museums and Art Galleries of the Northern Territory 22: 77–90.

Muriel-Cunha, J. and M.C.C. de Pinna. 2005. New data on cistern catfish, *Phreatobius cisternarum*, from subterranean waters at the mouth of the Amazon River (Siluriformes, *incertae sedis*). Pap. Avulsos de Zool. 45: 327–339.

Nelson, J.S., T.C. Grande and M.V.H. Wilson. 2016. Fishes of the World. Hoboken: John Wiley & Sons Inc., USA.

Ng, H.H. and D.J. Siebert. 1998. A revision of the akysid catfish genus *Breitensteinia* Steindachner (Teleostei: Siluriformes) with descriptions of two new species. J. Fish Biol. 53: 645–657.

Ng, H.H. and P.K.L. Ng. 2001. A revision of the akysid catfish genus *Acrochordonichthys* Bleeker. J. Fish Biol. 58: 386–418.

Ng, H.H. 2003. A revision of the south Asian sisorid catfish genus *Sisor* (Teleostei: Siluriformes). J. Nat. Hist. 37: 2871–2883.

Ng, H.H. and J.S. Sparks. 2005. Revision of the endemic Malagasy catfish Family Anchariidae (teleostei: Siluriformes), with descriptions of a new genus and three new species. Ichthyol. Explor. Freshw. 16: 303–323.

Ng, H.H. 2009. Parakysis hystriculus, a new species of catfish (Siluriformes: Akysidae) from Sumatra. Zootaxa 1977: 39–48.

Ng, H.H. and J.J. Wright. 2009. A new torrent catfish from western Thailand (Siluriformes: Amblycipitidae). Copeia 2009(2): 369–377.

Ng, H.H. 2015. Phylogenetic systematics of the Asian catfish family Sisoridae (Actinopterygii: Siluriformes). Ichthyol. Explor. Freshw. 26(2): 97–157.

Ng, H.H., R.K. Hadiaty, J.G. Lundberg and K.R. Luckenbill. 2015. A new genus and species of bagrid catfish from northern Sumatra (Siluriformes: Bagridae). Proc. Acad. Nat. Sci. 164: 149–157.

Ng, H.H. and C.J. Ferraris Jr. 2016. A new species of anguilliform catfish (Actinopterygii: Siluriformes: Bagridae) from Bangladesh and northeastern India. Zootaxa 4079: 381–387.

Ng, H.H. and M. Kottelat. 2016. Revision of the Asian catfish genus *Hemibagrus* Bleeker, 1862 (Teleostei: Siluriformes: Bagridae). Raffles Bull. Zool. 61(1): 205–291.

Ng, H.H. and M. Kottelat. 2016. The *Glyptothorax* of Sundaland: a revisionary study (Teleostei: Sisoridae). Zootaxa 4188: 1–92.

Nijssen, H. and I.J.H. Isbrücker. 1987. *Spectracanthicus murinus*, nouveaux genre et espèce de Poisson-Chat cuirassé du Rio Tapajós, Est. Pará, Brésil, avec des remarques sur d'autres genres de Loricariidés (Pisces, Siluriformes, Loricariidae). Rev. Française d'Aquar. Herpét. 13: 93–98.

Norris, S.M. 2002. A revision of the African electric catfishes, family Malapteruridae (Teleostei, Siluriformes), with erection of a new genus and descriptlions of fourteen new species, and an annotated bibliography. Ann. Mus. Roy. Afrique Centrale, Sci. Zool. 289: 1–155.

Pellegrin, J. 1927. La disparition des nageoires paires chez les poissons africains du groupe des clariinés. Ann. Sci. Nat. (Zool.) Paris 10: 209–222.

Pereira, E.H.L. and R.E. Reis. 2002. Revision of the loricariid genera Hemipsilichthys and Isbrueckerichthys (Teleostei: Siluriformes), with descriptions of five new species of Hemipsilichthys. Ichthyol. Explor. Freshw. 13: 97–146.

Pereira, E.H.L., A.M. Zanata, M. Cetra and R.E. Reis. 2014. A remarkable sexually dimorphic new genus and species of neoplecostomine catfish (Siluriformes, Loricariidae) from a coastal drainage of eastern Brazil. Copeia 2014(4): 673–681.

de Pinna, M.C.C. 1996. A phylogenetic analysis of the Asian catfish families Sisoridae, Akysidae, and Amblycipitidae, with a hypothesis on the relationships of the Neotropical Aspredinidae (Teleostei, Ostariophysi). Fieldiana Zool. 84: 1–83.

de Pinna, M.C.C. 1998. Phylogenetic relationships of Neotropical Siluriformes (Teleostei: Ostariophysi): Historical overview and synthesis of hypotheses. pp. 279–330. *In*: Malabarba, L.R., R.E. Reis, R.P. Vari, Z.M.S. Lucena and C.A.S. Lucena (eds.). Phylogeny and Classification of Neotropical Fishes. Edipucrs, Porto Alegre, Brazil.

de Pinna, M.C.C. and A.L. Kirovsky. 2011. A new species of sand-dwelling catfish, with a phylogenetic diagnosis of *Pygidianops* Myers (Siluriformes: Trichomycteridae: Glanapteryginae). Neotrop. Ichthyol. 9(3): 493–504.

de Pinna, M.C.C. and J. Zuanon. 2013. The genus *Typhlobelus*: monophyly and taxonomy, with description of a new species with a unique pseudotympanic structure (Teleostei: Trichomycteridae). Copeia 2013(3): 441–453.

de Pinna, M.C.C., R. Burger and A.M. Zanata. 2018. A new species of *Copionodon* lacking a free orbital rim (Siluriformes: Trichomycteridae). Neotrop. Ichthyol. 16(2): e170146 [1–9].

Poll, M. 1977. Les genres nouveaux *Platyallabes* et *Platysclarias* comparés au genre *Gymnallabes* GTHR. Synopsis nouveau des genres de Clariidae. Bull. Acad. Roy. Belg. Classe Sci. 63: 122–149.

de Queiroz, L.J., R.E. Reis and P. Lehmann. 2013. Chapter 32: Hypoptopomatinae. pp. 204–223. *In*: Queiroz, L.J, G. Torrente-Vilara, W.M. Ohara, T.H.S. Pires, J. Zuanon and C.R.C. Doria (orgs.). Peixes do Rio Madeira. Santo Antonio Energia, São Paulo, SP, BR.

Raghavan, R., R.L. Sundar, C.P. Arjun, R. Britz and N. Dahanukar. 2023. Evolution in the dark: unexpected genetic diversity and morphological stasis in the blind, aquifer-dwelling catfish *Horaglanis*. Vertebr. Zool. 73: 57–74.

Rapp Py-Daniel, L. and C. Cox-Fernandes. 2005. Dimorfismo sexual em Siluriformes e Gymnotiformes (Ostariophysi) da Amazônia. Acta Amazonica 35: 97–110.

Rapp Py-Daniel, L. and W.M. Ohara. 2013. Loricariinae. pp. 224–301. *In*: Queiroz, L.J, G. Torrente-Vilara, W.M. Ohara, T.H.S. Pires, J. Zuanon and C.R.C. Doria (orgs.). Peixes do Rio Madeira. Santo Antonio Energia, São Paulo, SP, BR.

Reis, R.E, E.H.L. Pereira and J.W. Armbruster. 2006. Delturinae, a new loricariid catfish subfamily (Teleostei, Siluriformes), with revisions of *Delturus* and *Hemipsilichthys*. Zool. J. Linn. Soc. 147: 277–299.

Rema Devi, K. and M.B. Raghunathan. 1999. *Heteropneustes longipectoralis* (Siluriformes: Heteropneustidae) a new species from the Anamalai Hills, in the western Ghats. Rec. Zool. Surv. India 97(3): 109–115.

Ribeiro, A.C., F.C.T. Lima and E.H.L. Pereira. 2012. A new genus and species of a minute suckermouth armored catfish (Siluriformes: Loricariidae) from the Rio Tocantins drainage, central Brazil: the smallest known loricariid catfish. Copeia-2012(4): 637–647.

Risch, L. 1987. Description of four new bagrid catfishes from Africa (Siluriformes: Bagridae). Cybium 11: 21–38.

Rizzato, P.P., E.P.D. Costa-Jr, E. Trajano and M.E. Bichuette. 2011. *Trichomycterus dali*: a new highly troglomorphic catfish (Silurifomes: Trichomycteridae) from Serra da Bodoquena, Mato Grosso do Sul State, Central Brazil. Neotrop. Ichthyol. 9: 477–491.

Roberts, T.R. and C. Vidthayanon. 1991. Systematic revision of the Asian catfish family Pangasiidae, with biological observations and descriptions of three new species. Proc. Acad. Nat. Sci. 143: 97–143.

Roberts, T.R. 2001. *Ayarnangra estuarius*, a new genus and species of sisorid catfish from the Ayeyarwaddy basin, Myanmar. Nat. Hist. Bull. Siam Soc. 49: 81–87.

Rocha, M., H. Lazzarotto and L.R. Py-Daniel. 2012. A New Species of *Scoloplax* with a Remarkable New Tooth Morphology within Loricarioidea (Siluriformes: Scoloplacidae). Copeia 4: 670–677.

Rocha, M.S., F. Rossoni, A. Akama and J. Zuanon. 2019. A new species of spiny driftwood catfish *Spinipterus* (Siluriformes: Auchenipteridae) from the Amazon basin. J. Fish Biol. 96: 243–250.

Sabaj, M.H., J.W. Armbruster and L.M. Page. 1999. Spawning in *Ancistrus* (Siluriformes: Loricariidae) with comments on the evolution of snout tentacles as a novel reproductive strategy: larval mimicry. Ichthyol. Explor. Freshw. 10: 217–229.

Sabaj, M.H., M. Arce, L.M. de Sousa and J.L.O. Birindelli. 2014. *Nemadoras cristinae*, new species of thorny catfish (Siluriformes: Doradidae) with redescriptions of its congeners. Proc. Acad. Nat. Sci. 163: 133–178.

Sabaj, M.H. and M. Arce, 2021. Towards a complete classification of the Neotropical thorny catfishes (Siluriformes: Doradidae). Neotrop. Ichthyol. 19: 01–51.

Sarmento-Soares, L. and R.F. Martins-Pinheiro. 2020. A reappraisal of phylogenetic relationships among auchenipterid catfishes of the subfamily Centromochlinae and diagnosis of its genera (Teleostei: Siluriformes). Proc. Acad. Nat. Sci. 167: 85–146.

Schaefer, S.A., S.H. Weitzman and H.A. Britski. 1989. Review of the Neotropical Catfish Genus *Scoloplax* (Pisces: Loricarioidea: Scoloplacidae) with comments on reductive characters in phylogenetic analysis. Proc. Acad. Nat. Sci. 141: 181–211.

Schaefer, S.A. and U.A. Buitrago-Suárez. 2002. Odontode morphology and skin surface features of Andean astroblepid catfishes (Siluriformes, Astroblepidae). J. Morphol. 254: 139–48.

Schaefer, S.A. 2003. Relationships of *Lithogenes villosus* (Siluriformes, Loricariidae): Evidence from high-resolution computed microtomography. Am. Mus. Novit. 3401: 1–55.

Schaefer, S.A., F. Provenzano, M.C.C. de Pinna and J.N. Baskin. 2005. New and noteworthy Venezuelan Glanapterygine catfishes (Siluriformes, Trichomycteridae), with discussion of their biogeography and Psammophily. Am. Mus. Novit. 3496: 1–27.

Schaefer, S.A. and F. Provenzano. 2008. The Lithogeninae (Siluriformes, Loricariidae): Anatomy, interrelationships, and description of a new species. Am. Mus. Novit. 3637: 1–49.

Schaefer, S.A. and L. Fernández. 2009. Redescription of the Pez Graso, *Rhizosomichthys totae* (Trichomycteridae), of Lago de Tota, Colombia, and aspects of cranial osteology revealed by microtomography. Copeia 2009(3): 510–522.

Shelden, F.F. 1937. Osteology, myology, and probable evolution of the nematognath pelvic girdle. Ann. N. Y. Acad. Sci. 37: 1–96.

Shibatta, O.A., J. Muriel-Cunha and M.C.C. de Pinna. 2007. A new subterranean species of *Phreatobius* Goeldi, 1905 (Siluriformes, Incertae sedis) from the Southwestern Amazon basin. Pap. Avulsos de Zool. 47: 191–201.

Shibatta, O.A. and R.P. Vari. 2017. A new genus of Neotropical rheophilic catfishes, with four new species (Teleostei: Siluriformes: Pseudopimelodidae). Neotrop. Ichthyol. 15: 1–30.

Silva, G.S.C., B.F. Melo, F.F. Roxo, L.E. Ochoa, O.A. Shibatta, M.H. Sabaj et al. 2021. Phylogenomics of the bumblebee catfishes (Siluriformes: Pseudopimelodidae) using ultraconserved elements. J. Zoolog. Syst. Evol. Res. 59: 1662–1672.

Sire, J.Y. and A. Husseyne. 1996. Structure and development of the Odontodes in an Armoured Catfish, *Corydoras* aeneus (Siluriformes, Callichthyidae). Acta Zool. 77: 51–72.

Sousa, L.M. and L.H. Rapp Py-Daniel. 2005. Description of two new speices of *Physopyxis* and redescription of *P. lyra* (Siluriformes: Doradidae). Neotrop. Ichthyol. 3: 625–636.

Stewart, D.J. 1985. A review of the South American catfish tribe Hoplomyzontini (Pisces, Aspredinidae), with descriptions of new species from Ecuador. Fieldiana Zool. 25: 1–19.

Stone, R. 2007. The Last of the Leviathans. Science 316(5832): 1684–1688.

Tamang, L. and B. Sinha. 2014. Two new species of the South Asian catfish genus *Pseudolaguvia* from northeastern India (Teleostei: Sisoridae). Zootaxa 3887: 37–54.

Tan, Z.W., E. Toh, Y. Cai and H.H. Tan. 2023. Discovery of *Encheloclarias* (Teleostei: Clariidae) from Singapore, with notes on morphology and distribution. Raffles Bull. Zool. 71: 196–206.

Taphorn, D.C., J.W. Armbruster, H. López-Fernández and C.R. Bernard. 2010. Description of *Neblinichthys brevibracchium* and *N. echinasus* from the upper Mazaruni River, Guyana (Siluriformes: Loricariidae), and recognition of *N. roraima* and *N. yaravi* as distinct species. Neotrop. Ichthyol. 8: 615–624.

Teugels, G.G. and D. Adriaens. 2003. Taxonomy and phylogeny of Clariidae. An overview. pp. 465–487. *In*: Arratia, G., A.S. Kapoor, M. Chardon and R. Diogo (eds.). Catfishes. Science Publishers, Enfield, New Hampshire, USA.

Thomson, A.W. and L.M. Page. 2006. Genera of the Asian Catfish Families Sisoridae and Erethistidae (Teleostei: Siluriformes). Zootaxa-1345: 1–96.

Trajano, E. and S. Secutti. 2025. Diversity and life cycles of Subterranean Catfishes. pp. 289–298. *In*: Arratia, G. and R.E. Reis (eds.). Catfishes, a Highly Diversified Group. Volume 1: Their Outstanding Biology. CRC Press, Boca Raton, FL.

Vari, R.P. and C.J. Ferraris, Jr. 2006. The catfish genus *Tetranematichthys* (Auchenipteridae). Copeia 2006: 168–180.

Villa-Verde, L., M.C.C. de Pinna, V.J.C. Reis and O.T. Oyakawa. 2022. Secretive fish diversity: a new species of *Listrura* (Siluriformes, Trichomycteridae) from a supposedly well-known river in south-eastern Brazil. J. Fish Biol. 100: 1299–1310.

Wiley, E.O. and G.D. Johnson. 2010. A teleost classification based on monophyletic groups. pp. 123–182. *In*: Nelson, J.S., H.P. Schultze and M.V.H. Wilson (eds.). Origin and Phylogenetic Interrelationships of Teleosts. Verlag Dr. F. Pfeil, München, Germany.

Zuanon, J., F.A. Bockmann and I. Sazima. 2006. A remarkable sand-dwelling fish assemblage from central Amazonia, with comments on the evolution of psammophily in South American freshwater fishes. Neotrop. Ichthyol. 4: 107–118.

Chapter 3

An Overview of the Anatomy of Catfishes and its Diversity

Timo Moritz

The huge diversity of catfishes, from bottom dwellers to pelagic swimmers, from miniature species to more than three-metre giants, including voracious predators, algae scrapers, and even blood-drinking parasites, is reflected in adaptations on their external (Rizzato and Almeida this volume) and internal anatomy (Figure 1). Looking at the body shape and behaviour of other otomorphs, it becomes evident that the ancestors of catfishes developed from free swimming, laterally compressed, scaled, and day-active fishes (Alexander 1965). When transforming into bottom-dwelling and nocturnal predators, the progenitors of catfishes underwent many transformations and adaptations which were decisive for their evolutionary success. The mere number of catfish species and their morphological diversity (Teugels 1996; Armbruster 2011; Fricke et al. 2024) do not allow a comprehensive guide through all their anatomical details and variations. An overview of general anatomy is given here, with a focus on siluriform apomorphies and characters of evolutionary importance for certain groups. Topics treated in other chapters of this book series are omitted.

The skeleton is probably the best-studied anatomic system of fishes because it exhibits clear differences between species but is usually conservative enough to be reasonably uniform within a species. This allows its use not only for taxonomy (e.g., delimitation and identification of species and other taxa) but also for phylogenetic studies. Furthermore, the technique of clearing and double staining offers a superb approach to studying bones and cartilages in their original positions and relations to each other (Figure 1) and has been intensively used, in combination with other methodologies, like dry skeletons or CT-scanning, to compile comparative data. Therefore, the skeleton will cover much of the information in this overview.

For an essay, the skeleton of vertebrates can be organised in various ways. The more common approaches are by (1) ontogenetic origin, (2) position, and/or (3) functional units. For a simplified approach, a mixture of anatomical position and functional units is chosen here. For a general overview of the teleostean skeletal anatomy, see Stiassny (2000) and Hilton (2011). The skeletal anatomy of catfishes has been studied in detail for certain taxa, like the basal diplomystids (Arratia 1987, 1992), some of the species-rich loricariids (Schaefer 1987, 1991, 1997; Schaefer and Provenzano 2008), callichthyids (Reis 1998), cetopsids (de Pinna and Vari 1995), doradids and auchenipterids (Birindelli 2014), mochokids (Vigliotta 2008), schilbeids (de Vos 1995), bagrids (Mo 1991), the clariid *Clarias* (Teugels 1986), the trichomycterid *Silvinichthys* (Arratia 1998), or also the enigmatic

Ocean Museum Germany, Katharinenberg 14-20, 18439 Stralsund, Germany.
Institut für Biowissenschaften, Universität Rostock, Albert-Einstein-Straße 3, 18059 Rostock, Germany.
Email: timo.moritz@meeresmuseum.de

Figure 1. Diversity of catfish skeletons, as illustrated by cleared and double-stained specimens (in left lateral view, except when noted) of selected catfish species. (A) *Diplomystes mesembrinus* (Diplomystidae). (B) *Tridentopsis cahuali* (Trichomycteridae). (C) *Astroblepus* sp. (Astroblepidae). (D) *Paralithoxus surinamensis* (Loricariidae), dorsal view. (E) *Loricariichthys maculatus* (Loricariidae). (F) *Ancistrus* cf. *dolichopterus* (Loricariidae). (G) *Helogenes marmoratus* (Cetopsidae). (H) *Corydoras brevirostris* (Callichthyidae). (I) *Aspredo aspredo* (Aspredinidae). (J) *Cetopsis orinoco* (Cetopsidae). (K) *Pseudauchenipterus nodosus* (Auchenipteridae). (L) *Chasmocranus longior* (Heptapteridae). (M) *Silurus glanis* (Siluridae). (N) *Sperata aor* (Bagridae). (O) *Phractura clauseni* (Amphiliidae). (P) *Malapterurus electricus* (Malapteruridae). (Q) *Chiloglanis voltae* (Mochokidae). (R) *Siluranodon auritus* (Schilbeidae). (S–T) *Synodontis sorex* (Mochokidae), frontal (S) and lateral view (T). (U) *Mochokus niloticus* (Mochokidae). (V) *Pangasianodon hypophthalmus* (Panagasidae).

Kryptoglanis (Britz et al. 2014; Lundberg et al. 2014). Furthermore, certain anatomical regions have been at the centre of studies, such as the pectoral girdle (Diogo et al. 2001), the caudal skeleton (Lundberg and Baskin 1969), the pelvic girdle (Shelden 1937; Tilak 1967; Moritz et al. this volume), and the Weberian apparatus (Duclos and Grande this volume). However, for many body regions and taxa, much information has yet to be discovered.

By comparing a catfish with a typical ostariophysean, e.g., *Brycon* (Weitzman 1962), it becomes obvious that the skeleton of catfishes is formed by a smaller number of bones (Table 1). In several cases, it remains unclear whether a given bone was lost or if it is fused to another bone. Ontogenetic studies often can help to clarify such questions and provide information for phylogenetic studies, but the amount of developmental information for catfishes is still small to cover the huge variety of taxa and anatomical diversity (Figure 1; Rizzato and Almeida this volume). Some ontogenetic information is available, for example, for *Corydoras* (Huysentruyt et al. 2011), *Clarias* (Adriaens and Verraes 1998), *Diplomystes* (Arratia 1987, 1992, 2003b), and ictalurids (Arratia 2003b; Kubicek 2022).

Some catfishes exhibit sexual dimorphism in their skeleton. It is most expressed in auchenipterids, whose males may have a modified anal fin, an enlarged dorsal-fin spine, or even enlarged maxillae; these modifications are sometimes restricted to the mating season (Ferraris 1988; Calegari et al. 2019; Sarmento-Soares and Martins-Pinheiro 2020; Rizzato and Almeida this volume). In mochokids, the male anal and caudal fins may elongate, and the humeral process may enlarge (Friel and Vigliotta 2006). *Scoloplax* seems to represent a special case since it exhibits sexual dimorphism in many skeletal parts (Schaefer 1990). In some loricariids and most callichthyids, males usually have stronger pectoral-fin spines; in the former, they may carry large odontodes, while in the latter, the difference in size may be extreme but limited to reproductive phases (Reis 1998). Nevertheless, the following descriptions are generally valid for males and females of most catfish species. A certain variability in the skeleton due to age, developmental state, or habitat may occur. The cranium, an intricate, multipart, and variable character complex, will be treated first, followed by the postcranial skeleton, and finally, a short overview of the soft anatomy of catfishes.

Head Skeleton

Fish skulls are highly mobile with many movable elements, especially in teleosts, where suction feeding is optimised (Westneat 2006). This also holds true for catfishes in general, many of which are predators that engulf their prey with a single snap. Yet, in some evolutionary lineages, other feeding strategies have developed, such as algae scraping or biting off pieces of tissue, which require special adaptations. Being primarily bottom dwellers, most catfishes have dorso-ventrally flattened bodies and skulls, showing a so-called platybasic condition (Arratia 2003a). Compared to other Ostariophysi, several skeletal elements are much reduced or lacking as distinct elements either by fusion with other bones or complete reduction (Table 1). Many of these reductions are putative synapomorphies with Gymnotiformes (Table 1). A thorough overview of the head skeleton of catfishes was provided by Arratia (2003a).

Neurocranium

The skull of catfishes is usually flat and broad. Therefore, the small eyes are more distant from each other (Figure 2A–D) than in other teleosts, and an interorbital septum is absent (Allis 1919; Alexander 1965). Small eyes require less muscle mass; therefore, the so-called myodome—a cavity in the neurocranium ventral to the orbit that houses the eye muscles in other fishes—is generally absent in catfishes (Alexander 1965; Arratia 2003a). The neurocranium forms a robust roof, which may be smooth on its external surface, have ornamentations with ridges, or carry odontodes (Figure 2A–D; Arratia 2003a; Rizzato and Almeida this volume). The bones may be thick and massive, but some taxa have more delicate or even reduced bones in the skull roof. Contrasting with other teleosts, catfishes lack a separate parietal bone; it is commonly regarded as being fused

Table 1. Bones in the head skeleton of a generalised teleost and their respective characteristics in catfishes in general.

Bone	Remarks for Catfishes
Ethmoidal Region	
Mesethmoid	• Variable in catfishes, but often rather flat with anterolateral expansions usually supporting the premaxillae • Processes not common, e.g., *Trachydoras*, Doradidae (Birindelli 2014) or some chiloglanidinae (Vigliotta 2008)
Lateral ethmoid	• Forming the complete nasal capsule in Callichthyidae and Loricariidae (Schaefer 1987; Reis 1998) • Sometimes contributing to the dorsal region of the neurocranium, e.g., in some mochokids, pimelodids and auchenipterids (Calegari et al. 2019)
Nasal	• Tube-like in catfishes; rarely plate-like (e.g., in some doradids and auchenipterids) (Birindelli 2014; Calegari et al. 2019)
Vomer	• Missing in Aspredinidae, *Scoloplax*; *Chaca*, *Phreatobius*, *Gelanoglanis* and *Kryptoglanis* (Schaefer 1990; de Pinna 1993; Birindelli 2014; Britz et al. 2014; Calegari et al. 2019) • Often without teeth, e.g., in Amphiliidae, Aspredinidae, Sisoridae, Loricariidae, and Cranoglanididae (Arratia 2003a; Calegari et al. 2019) • Accessory vomerine tooth plates in Pangasiidae, Ariidae, Schilbeidae, Claroteidae, and *Helogenes* (de Pinna 1993; Calegari et al. 2019)
Orbital region	
Frontal	• Often separated from its contralateral by one or two fontanels
Orbitosphenoid	• Present
Pterosphenoid	• Absent as a separate element in some catfishes, e.g., *Auchenipterus*, *Scoloplax* (Schaefer 1990), may be fused to frontal, e.g., *Trachelyoperus* (Ferraris 1988), or fused with prootic and sphenotic, e.g., some trichomycterines (Arratia 2003a)
Basisphenoid	• Absent; this is considered an apomorphy of Ostariophysi (Fink and Fink 1981, 1996), but a "suprasphenoid" is present in the respective position in several catfish taxa, e.g., *Hypophthalmus*, Doradidae, and Auchenipteridae (Howes 1983a)
Parasphenoid	• Present
(Auto-)sphenotic	• Present, but in catfishes without postorbital spine • fused to the prootic and pterosphenoid in some trichomycterines (Arratia 2003a)
Antorbital	• "Antorbital" in catfishes connected via ligaments to various bones, questionable homology with antorbital of other teleosts
Circumorbitals	• Circumorbitals or infraorbitals are usually tube-like and represent ossifications of infraorbital lateral-line canal segments; present in variable numbers (from none to nine), exact homologies questionable.
Supraorbital	• Absent: The absence of supraorbital is considered a synapomorphy of Siluriformes and Gymnotiformes (Fink and Fink 1981); trichomycterids with non-homologous "supraorbital" tendon-bone (Adriaens et al. 2010; de Pinna et al. 2020; Costa 2021)
Sclerotic bones	• Absent: This absence is considered a synapomorphy of Siluriformes and Gymnotiformes (Fink and Fink 1981, 1996)
Otic region	
Parietal	• Absent as a separate element: either fused with supraoccipital = parieto-supraoccipital (Arratia and Gayet 1995) or absent (Kubicek 2022); absence of a separately ossifying parietal considered an apomorphy of Siluriformes (Fink and Fink 1981, 1996)
Pterotic	• Present as a single element representing a fusion between autopterotic and dermopterotic (as in most teleosts) • Fused to posttemporo-supracleithrum in Loricarioidea except Nematogenyidae and Loricariidae (Reis 1998)
Intercalar	• Absent: This absence is considered a synapomorphy of Siluriformes and Gymnotiformes (Fink and Fink 1981, 1996)

Table 1 contd. ...

...Table 1 contd.

Bone	**Remarks for Catfishes**
Occipital area	
Supraoccipital	• Large; for possible fusion with parietals, see parietal • Posteriorly-directed crest is absent in a few taxa, e.g., Auchenipteridae, Mochokidae, Doradidae and some aspredinids (Calegari et al. 2019)
Exoccipital	• Fused with basioccipital in Callichthyidae and *Scoloplax* (Reis 1998)
Epiotic/ Epioccipital	• Missing in some taxa, e.g., some trichomycterids, *Nematogenys*, *Hypostomus*, *Scoloplax*, and *Synodontis* • Forming part of the dorsal cephalic shield in Doradidae and Auchenipteridae (de Pinna 1998; Birindelli 2014)
Basioccipital	• Lateral expansion is present in some taxa, e.g., Ariidae, Auchenipteridae, Doradidae, and Mochokidae (Vigliotta 2008)
Extrascapular	• Usually small • Absent (as a separate element) in Aspredinidae, Auchenipteridae, Amphiliidae, Erethistidae, Mochokidae, *Nematogenys*, and Siluridae (Mo 1991; Vigliotta 2008; Birindelli 2014); in Doradidae, may be fused to epiotic (Birindelli 2014)
Hyopalatine arch	
Palatine	• Corresponds only to autopalatine; dermopalatine absent in Ostariophysi (Fink and Fink 1981) • Cartilage of autopalatine is not connected to the cartilage of the quadrate and metapterygoid; this condition is considered an apomorphy of Siluriformes (de Pinna 1993; Wiley and Johnson 2010) • Reduced in Siluridae (Bornbusch 1991) • Anterior part of the palatine presents as separate ossification in some trichomycterids (de Pinna 1989)
Endo- and ectopterygoid	• True endopterygoid (= entopterygoid and mesopterygoid) and ectopterygoid absent (Arratia 1992, 2003a) • A sesamoid bone may be present in a similar position of the endopterygoid; the presence of such sesamoid bone is considered an apomorphy of Siluriformes; respective bone absent in loricarioids (Schaefer 1987; Reis 1998) a sesamoid bone may be present in similar position of the ectopterygoid in a few taxa, e.g., Pimelodidae, Ariidae (Calegari et al. 2019)
Metapterygoid	• Often with an axe-shaped enchondral portion in Ostariophysi, only the anterior half of the axe is present in Siluriformes plus Gymnotiformes (Fink and Fink 1996) • Placed anterodorsal to quadrate, forming part of the ventral border of the suspensorium; this condition is considered an apomorphy of Siluriformes (Fink and Fink 1981; de Pinna 1993) • Variable in size in catfishes; absent in *Kryptoglanis* (Britz et al. 2014)
Quadrate	• Ventral process, associated with symplectic in other teleosts, absent in catfishes: this condition is considered an apomorphy of Siluriformes (Wiley and Johnson 2010)
Symplectic	• Absent; this absence is considered a synapomorphy of Siluriformes and Gymnotiformes (Fink and Fink 1981)
Hyomandibula	• Articulating bones in neurocranium variable (for details, see text) • Often sutured with metapterygoid (see Birindelli 2014 for exceptions)
Jaws	
Premaxilla	• Forming the upper half of the mouth cleft • Teeth are absent in some taxa, e.g., Callichthyidae, Aspredinidae, some loricariids, some auchenipterids and some doradids (Reis 1998; Birindelli 2014)
Maxilla	• Usually smaller than the premaxilla, not associated with the mouth cleft, supporting the maxillary barbel instead • Teeth absent, except for *Diplomystes* and some fossils (Arratia 2025)
Supramaxilla	• Absent: This absence is considered an apomorphy of Ostariophysi (Fink and Fink 1996; Wiley and Johson 2010)

Table 1 contd. ...

...Table 1 contd.

Bone	**Remarks for Catfishes**
Articular, angular and retroarticular	• Fused into a single bone, the anguloarticulo-retroarticular
Dentary	• Only loosely connected to its contralateral in Loricariidae and *Astroblepus* • Without teeth in some callichthyids (Reis 1998)
Coronomeckelian	• Often small, much reduced or absent in mochokids, hoplomyzontines, malapterurids, callichthyids, loricariids, *Astroblepus* and *Scoloplax* (Vigliotta 2008; Birindelli 2014)
Lower hyoid arch	
Interhyal	• Absent in Amplycipitidae, Amphiliidae, Akysidae, Erethistidae, Mochokidae, Sisorids, *Astroblepus*, *Scoloplax*, adult clariids, and adult trichomycterids (Arratia and Schultze 1990; Arratia 2003a; Birindelli 2014)
Anterior and posterior ceratohyal	• Both present
Branchiostegal rays	• Variable number, from 3 to 20; usually around 8 to 10 (McAllister 1968; Reis 1998; Arratia 2003a; Birindelli 2014)
Dorsal and ventral hypohyal	• Dorsal hypohyal is often small or absent, e.g., in most loricarioids (Reis 1998), Amphiliidae, Akysidae, Aspredinidae, Pseudopimelodidae, Sisoridae and *Chiloglanis* (Birindelli 2014).
Basihyal (= glossohyal)	• Absent: The absence of a basihyal is considered an apomorphy of Siluriformes (Arratia and Schultze 1990; de Pinna 1993) • Occasionally atavistically present in *Diplomystes* (Arratia and Quezada-Romegialli 2017)
Urohyal	• Absent: A non-homologous "parurohyal" is present instead, with a unique origin that includes a cartilaginous ossification in addition to a tendon one (Arratia and Schultze 1990)
Opercular area	
Preopercle	• Without horizontal limb: This absence is considered an apomorphy of Siluriformes • Suprapreopercle(s): additional tubular bone(s) dorsal to preoperculum, widely distributed among catfishes, e.g., in Amphiliidae, most auchenipterids, Bagridae, Diplomystidae, Erethistidae, Malapteruridae, Ictaluridae, *Nematogenys*, Siluridae and Sisoridae (Birindelli 2014) • Subpreopercle: additional tubular bone anteroventral to preoperculum, present in Amphiliidae, Auchenipteridae, Bagridae, Cetopsidae, Diplomystidae, Doradidae, Erethistidae, Heptapteridae, Ictaluridae, Malapteruridae, *Nematogenys*, Pangasiidae, Pimelodidae and Sisoridae (Birindelli 2014)
Opercle	• Rather triangular instead of rectangular in shape; this condition is considered a synapomorphy of Siluriformes and Gymnotiformes (Fink and Fink 1981)
Subopercle	• Absent; this absence is considered an apomorphy of Siluriformes (Fink and Fink 1981)
Interopercle	• Small and triangular (shortened in longitudinal axis); this condition is considered an apomorphy of Siluriformes • Deep and short in cetopsids (de Pinna and Vari 1995)
Gill arches	
Pharyngobranchials	• Pharyngobranchial 1 often absent (de Pinna 1993; Reis 1998) • Pharyngobranchial 2 is often absent, except for Plotosidae, Pangasiidae, *Diplomystes* and *Horabagrus* (Arratia 1987; de Pinna 1993) • Pharyngobranchial 4 ossified in most catfishes (de Pinna 1993) • Only a single pharyngobranchial tooth plate is present: this condition is considered a synapomorphy of Siluriformes and Gymnotiformes (Fink and Fink 1981) • Structure located at tips of epibranchial 1 in *Lacatunia* ossified and not regarded as homologous to pharyngobranchial, therefore called "pseudo-pharyngobranchial bone" (Rodiles-Hernández et al. 2005)
Epibranchials	• 1–4 present (as typical for teleosts)
Ceratobranchials	• 1–5 present (as typical for teleosts)
Hypobranchials	• Hypobranchial 3 is usually unossified (de Pinna 1993) • Hypobranchial 2 is sometimes unossified (de Pinna 1993; Birindelli 2014)
Basibranchials	• Absent in Hoplomyzontini and *Scoloplax* (de Pinna 1993)

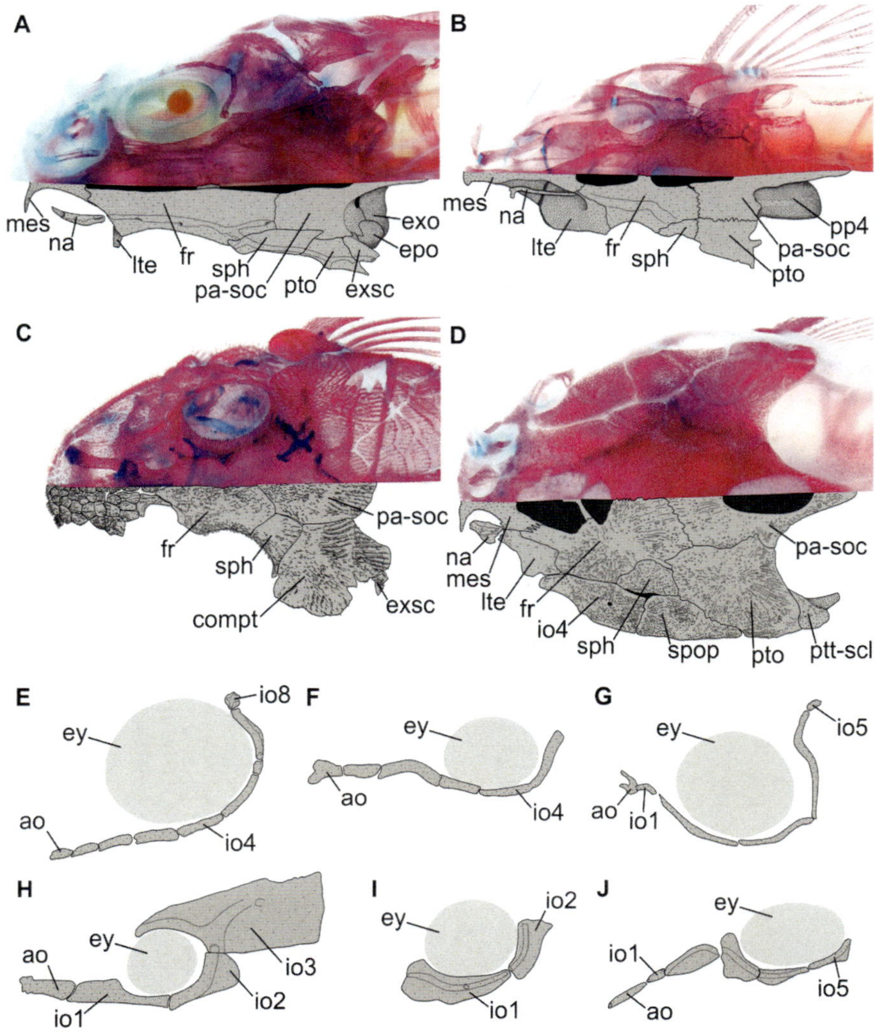

Figure 2. Neurocranium in dorsal view (A–D; cleared and double-stained specimen on the upper half, interpretative drawing on the lower half) and infraorbital series in left lateral view (E–J) of selected catfishes. (A, E) *Diplomystes mesembrinus* (Diplomystidae). (B) *Phractura clauseni* (Amphiliidae). (C, J) *Ancistrus* cf. *dolichopterus* (Loricariidae). (D) *Clarias agboyensis* (Clariidae). (F) *Sperata aor* (Bagridae). (G) *Pangasianodon hypophthalmus* (Pangasidae). (H) *Clarias agboyensis* (Clariidae). (I) *Corydoras brevirostris* (Callichthyidae). Abbreviations: ao: 'antorbital'; compt: compound-pterotic bone; epo: epioccipital; exo: exoccipital; exsc: extrascapular; ey: eyeball; fr: frontal; io: infraorbital; lte: lateral ethmoid; mes: mesethmoid; na: nasal; pa-soc: parieto-supraoccipital; pp4: parapophysis of 4th vertebra; pto: pterotic; ptt-scl: posttemporo-supracleithrum; sph: sphenotic; spop: suprapreopercle.

to the supraoccipital (Arratia and Gayet 1995; Fink and Fink 1996), but a recent study indicates its absence in catfishes (Kubicek 2022). Posterior to the neurocranium, additional flat bones, known as the nuchal plates, may be present, covering the dorsum between the head and dorsal fin (see below). Also, a small extrascapular is often present (Fink and Fink 1981; Arratia and Gayet 1995). Teleosts typically exhibit a supraoccipital crest, which in catfishes may be moderate, small or absent (Figure 2A–D). When present, it often sutures with a nuchal plate or a neural spine; in some catfishes, the crest contacts the first pterygiophore of the dorsal fin.

Dorsally, the neurocranium has a large median opening, the fontanel, for most of its part located between the frontal bones (Figure 2A, B, and D). It is often divided below the surface into two portions by an epiphyseal bar and may also be completely separated by superficial bones into anterior and posterior fontanels with distinct, taxon-specific shapes and sizes (Vigliotta 2008; Birindelli 2014; Calegari et al. 2019). *Helogenes* uniquely has additional paired fontanels laterally in the frontals (de Pinna and Vari 1995). The fontanel is rarely absent and is believed, for instance, to have been lost once in the common ancestor of *Scoloplax*, *Astroblepus* and loricariids (Figure 2C) (Schaefer 1990; de Pinna 1993).

The mesethmoid of catfishes is rather flat and carries processes only in a few cases. With few exceptions, the premaxilla rests on the ventral face of the mesethmoid. Typically, in teleosts, the lateral ethmoid forms the anterior border of the orbita, but in Callichthyidae and Loricariidae it also shapes the complete nasal capsule (Schaefer 1987; Reis 1998). A vomer is usually present and bears teeth in several catfishes (Table 1). Additional vomerine tooth plates have been reported in some taxa. The sphenotic, at the posterior margin of the orbit, possesses in other teleosts a characteristic lateral spine as an attachment site for the *M. levator hyoideus*, but such a spine is absent in catfishes (Alexander 1965). Only in mochokids has a comparable spine developed anew (Vigliotta 2008). Posterior to the orbita, one would expect a *trigemino-fascialis* chamber in fishes (Allis 1919), but it is missing in catfishes (Howes 1983a). An intercalar is absent in siluriforms and gymnotiforms. The exoccipitals of catfishes are connected via cartilage to the neural arch of the vertebral complex (see below; Duclos and Grande this volume); in auchenipterids and doradids, this connection is via a suture (Birindelli 2014). In loricarioids, the anterior part of the bony swim-bladder capsule fuses to the neurocranium (Aquino and Schaefer 2002). In some other taxa, the often highly reduced swim bladders are also encapsulated in bone (see below). In some cases, it is unclear how many bones may participate in the final ossification: basi- and exoccipitals, epiotics, pterotic, posttemporo-supracleithrum, parts of vertebrae and/or ossified myoseptae.

Orbital Area

The eyeball of teleosts is usually supported by sclerotic bones, but these are absent in siluriforms and gymnotiforms (Fink and Fink 1981). Primitively and in many recent teleosts, the eye is surrounded by flat circumorbital bones, consisting of the supraorbital, antorbital, and usually, six infraorbital, the most posterior of which is called dermosphenotic (Nelson 1969). All but the supraorbital may carry the infraorbital canal. The supraorbital is absent in catfishes and gymnotiforms, but a sesamoid or "tendon-bone supraorbital" is present in the respective position in some trichomycterids, and furthermore is possibly not even homologous within this taxon (Costa 2021). The posterior-most tubular bone surrounding the eye does likely not correspond to the dermosphenotic in catfishes, where it is apparently reduced or fused to the (auto-)sphenotic.

In most catfishes, circumorbitals largely lost their flat, plate-like appearance and were reduced to tube-like canal ossifications (Figure 2E–J; de Pinna 1993). Few catfishes secondarily exhibit flat circumorbitals, such as doradids, callichthyids (Figure 2I), loricariids (Figure 2J), most clariids (Figure 2H), and some auchenipterids. Siluriforms have a variable number (none to nine) of circumorbitals (Figure 2E–J; Arratia and Huaquín 1995; Arratia 1998; Reis 1998). The exact homologies of the circumorbitals among catfishes, and with those of other ostariophysans, remain largely unresolved. Numbering them should, therefore, be seen as descriptive without implying homology. Usually, in teleosts, the most anterior bone carrying the infraorbital canal is the antorbital, but in characiforms, the antorbital does not bear a canal. Whether the first element in the circumorbital series of catfishes represents a true antorbital, the original first infraorbital (i.e., the lacrimal), a fusion of both, or a different element remains questionable. In some taxa, such as callichthyids, a corresponding bone is missing (Figure 2I) (Reis 1998). In catfishes in which the infraorbital series stretches far anteriorly on the snout (e.g., Figure 2E and F), its first element exhibits a peculiar condition, being connected via ligaments to other bones, e.g., palatine, maxilla, lateroethmoid, and sometimes the premaxilla (Arratia 1992, 2003a). Uniquely within teleosts, this bone helps support the manoeuvrability of the maxillary barbel in catfishes.

Jaws

Ancestrally in teleosts, the maxilla is the largest tooth-bearing bone in the upper jaw, often dorsally accompanied by one or two supramaxillae. In several teleost lineages, the premaxilla overtook the maxilla's task, bordering the mouth cleft. This also happened in catfishes, where the maxilla is usually small and toothless, except for diplomystids (Figure 3A) (Arratia 1987) and some fossil

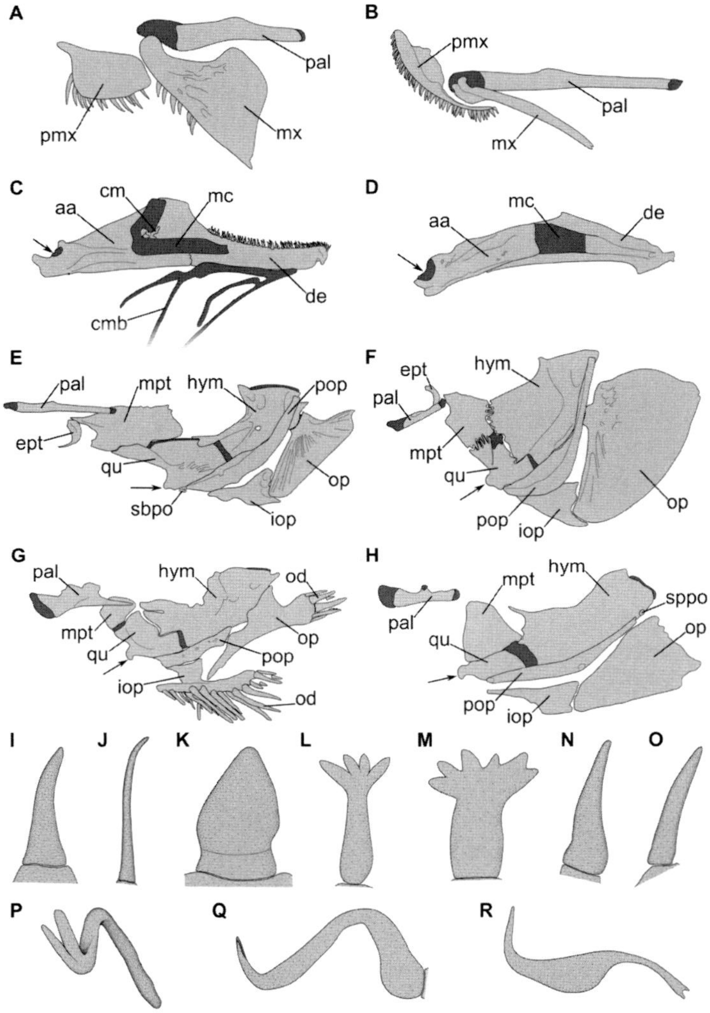

Figure 3. Upper jaw plus palatine in left lateral view (A–B), lower jaw in medial view (C–D, anterior to the right), suspensorium with opercular area in left lateral view (E–H), and tooth diversity (I–R) of selected catfishes. (A) *Diplomystes mesembrinus* (Diplomystidae). (B, C, E) *Sperata aor* (Bagridae). (D) *Corydoras paleatus* (Callichthyidae). (F) *Siluranodon auratus* (Schilbeidae). (G) *Hatcheria macraei* (Trichomycteridae). (H) *Synodontis petricola* (Mochokidae). The arrow points to the articulatory head or grove, respectively, of the jaw joint. (I) *Sperata aor* (Bagridae), premaxillary tooth, lateral view. (J) *Tridentopsis simils* (Trichomycteridae), dentary tooth, lateral view. (K) *Plotosus lineatus* (Plotosidae), vomerine tooth, lateral view, rotated. (L) *Scoloplax baileyi* (Scoloplacidae), premaxillary tooth, frontal view, rotated (after Rocha et al. 2012). (M) *Phenacorhamdia taphorni* (Heptapteridae), premaxillary tooth, frontal view, rotated (after Donascimiento and Milani 2008). (N) *Ancistrus* cf. *dolichopterus* (Loricariidae), odontode of the dorsal-fin spine, lateral view. (O) *Corydoras paleatus* (Callichthyidae), odontode of adipose-fin spine, lateral view. (P, Q) *Ancistrus* cf. *dolichopterus* (Loricariidae), dentary tooth, lateral (P) and anterolateral (Q) view. (R) *Synodontis petricola* (Mochokidae), dentary tooth, lateral view. Abbreviations: aa: anguloarticulo-retroarticular; cm: coronomeckelian; cmb: cartilages of mandibular barbels; de: dentary; ept: endopterygoid; hym: hyomandibular; iop: interopercle; mc: Meckel's cartilage; mpt: metapterygoid; mx: maxilla; od: odontode; op: opercle; pal: palatine; pmx: premaxilla; pop: preopercle; qu: quadrate; sbpo: subpreopercle; sppo: suprapreopercle.

representatives (Grande 1987; Arratia 2025). In many other derived teleosts, the maxilla works as a lever pushing the premaxilla anteriorly when opening the mouth. In catfishes, however, the maxilla, together with the autopalatine, supports the movement of the maxillary barbel (Figure 3A–B; Adriaens and Verraes 1997).

The usually plate-like and teeth-carrying premaxilla (Figure 3A–B) is firmly attached to the mesethmoid and works as an actual upper jaw in most catfishes. It may be connected to the palatine in different ways (Arratia 1992). In a clade including callichthyids, scoloplacids, astroblepids, and loricariids, of which the latter two developed a sucker mouth, the premaxilla became highly mobile and is directly controlled by parts of the *M. adductor mandibulae* (Howes 1983b; Schaefer 1990; de Pinna 1993). The suckermouth in loricariids is highly specialised, allowing scraping and breathing while sucked on the substrate (Geerinckx et al. 2007). Its possible evolution was presented by Schaefer and Lauder (1986). Further, sucker mouth-bearing catfishes originated independently in Africa (chiloglanidine mochokids) and Asia (glyptosternoid sisorids). Other mouth shapes that deviate significantly from a large gape are, for example, those of callichthyids (Figure 1H), which have small mouths and, as adults, no teeth on their reduced premaxilla (Reis 1998); mochokids, with jaws specialised for scraping; cetopsids which can bite junks out of bigger prey; and some trichomycterids, which feed on blood, mucus, or scales.

The lower jaw of teleosts may include four major bones, the dentary, retroarticular, angular and articular, the latter two being fused to an angulo-articular in clupeocephalans. Apomorphically in catfishes, the retroarticular is also fused into this complex, forming an anguloarticulo-retroarticular bone (Figure 3C–D; de Pinna 1993), commonly called simply angulo-articular in literature. An additional tendon-bone, the coronomeckelian, may be present but often reduced in size (Table 1; Figure 3C). The dentary is the largest bone in the lower jaw of catfishes and the only one bearing teeth. The contralateral lower jaws are firmly connected to each other, except in loricariids and *Astroblepus*, where they are linked only by cartilage (Schaefer 1990). The dentary also usually carries a portion of the preoperculomandibular canal, with much variation in the number of pores and sometimes absent, e.g., in *Sisor*, aspredinids or chiloglanidins (Birindelli 2014). Medioventral to the lower jaw, there are accessory cartilages of different sizes supporting the mandibular barbels (Figure 3C). The barbels themselves comprise cartilaginous cells and dense elastin fibres in their bases and central rods (Ghiot and Bouchez 1980; Benjamin 1990). Few taxa, such as *Hypophthalmus*, have an accessory cartilage supporting the lip fold (Howes 1983a).

Teeth and Odontodes

In teleosts, oral teeth may occur on the dermal bones of the mouth margin and cavity, as well as on additional tooth plates commonly associated with chondral bones. In catfishes, the ectopterygoid and endopterygoid are much reduced and tooth-less. A dermopalatine is absent in Ostariophysi, but non-homologous palatine tooth plates are sometimes present. Accessory vomerine tooth plates may also occur (Table 1). Additional tooth plates are located on the branchial arches. The maxilla is edentulous except in diplomystids (Figure 3A; Arratia 1987) among living siluriforms.

Teeth in catfishes are usually simple and cone-shaped (Figure 3I–J) and used to grasp and hold prey. But many specialised teeth have evolved on the premaxilla and dentary (Figure 3). Loricarioids apomorphically have bifurcated teeth (Figure 3P–Q; Schaefer 1990; de Pinna 1993), and wood-scraping loricariids have short, wide, spoon-shaped teeth (Armbruster 2011). Multicuspid teeth are rare among catfishes (Figure 3L–M). Mochokids have S-shaped teeth on the premaxilla and dentary (Figure 3R; Vigliotta 2008). Usually, teeth become gradually smaller towards the margin of a bone, but in mochokids, they are often grouped in separate patches of different sizes (Vigliotta 2008). The pharyngeal teeth of the branchial arches may be used to grind food items such as grain in loricariines (Armbruster 2004) and doradids (Goulding 1980). In *Plotosus*, some vomerine teeth are blunt, almost molariform (Figure 3K; Kapoor et al. 2003).

Within Loricarioidei, teeth are also found on other parts of the body (i.e., other than the mouth); these are called odontodes (Figure 3N–O; Reis 1998; Rivera-Rivera and Montoya-Burgos 2017; Mori and Nakamura 2022; Rizzato and Almeida this volume). While in trichomycterids, they are restricted to opercular bones (Figure 3G), in *Nematogenys*, they occur only on the pectoral fin. In callichthyids, loricariids and *Scoloplax* odontodes are associated with dermal bony plates or fin rays. In some male loricariids, odontodes may reach enormous sizes (Rizzato and Almeida this volume).

Opercular Area

The lateral side of the head posterior to the circumorbitals and jaws is usually covered by the opercular series in bony fishes, including opercle, preopercle, subopercle and interopercle. The preopercle carries part of the preoperculomandibular lateral-line canal and usually consists of a vertical and a horizontal limb, giving it an L-shaped appearance. In catfishes, the horizontal limb of the preopercle is absent, a suborpercle is missing, and the interopercle is rather small in several taxa and absent in loricariids and *Scoloplax* (Schaefer 1990). In several catfish taxa, additional canal bones are present dorsal and/or ventral of the preopercle, the so-called suprapreopercle(s) and subpreopercle(s), respectively (Figure 3E and H; Table 1) (Birindelli 2014). In clariids, the suprapreopercle is flat and large and covers a significant portion of the cheek region (Figure 2D; Bernt and Stiassny 2022).

In trichomycterids, the opercle and interopercle carry odontodes (Figure 3G). Especially in parasitic forms, these odontodes play an important role in the attachment on (or in) the host; in other trichomycterids, they can be used to anchor in the substrate and even for climbing waterfalls (Eigenmann 1918; Adriaens et al. 2010; Rizzato and Almeida this volume).

Hyopalatine Arch

The main articulation of the hyopalatine arch with the neurocranium is via the hyomandibula. There is remarkable variability in catfishes regarding how the articulatory groove on the skull is formed; one to four bones may be involved: (1) the sphenotic only; (2) the sphenotic and pterotic; (3) sphenotic, pterotic, and pterosphenoid; (4) sphenotic, pterotic, and prootic; or (5) all of these bones (Arratia 2003a). Catfishes are unique in having the *pars autopalatina* separated from the rest of the palatoquadrate cartilage throughout ontogeny (Arratia 1987, 1992; Arratia and Schultze 1991). Thus, in later stages, the palatine generally has no direct contact with the remaining part of the arch (Fink and Fink 1981, 1996). In Ostariophysi, a dermopalatine is absent, and the palatine bone corresponds only to the autopalatine of other teleosts. In ostariophysans, it articulates with the maxilla, lateral ethmoid and vomer (Arratia 1992), but the latter articulation was lost in catfishes except for trichomycterids (de Pinna 1993). In catfishes, the palatine is linked in various ways to the premaxilla (Arratia 1992); it is usually rod-like and elongated (Figure 3B and E–H), extending posterior to its articulation with the lateral ethmoid and serving as a base for the maxilla and maxillary barbel (Figure 3A–B).

True ecto- and endopterygoids seem to be missing in catfishes, but small bones with sesamoid origin and unclear homology are often present in their respective position (Figure 3E–F) (Arratia 1992; Arratia 2003a; Calegari et al. 2019). The presence of a tendon-bone "entopterygoid," or sesamoid 1, seems to be a siluriform apomorphy, but the involved ligaments may differ among taxa (Arratia 1992). An additional pterygoid bone, called "ectopterygoid," pterygoid type 1, or sesamoid 2, is reported for some taxa (Arratia 1992; Calegari et al. 2019).

The metapterygoid is a chondral bone, but in catfishes, it often enlarges as membrane bone and occupies much of the area occupied by the endo- and ectopterygoid in other teleosts (Arratia 1992). Compared to other ostariophysans, the metapterygoid lies much more anterior in comparison with the quadrate (Figure 3E–H). Tooth-bearing dermal plates on or close to the metapterygoid have been reported from *Parapimelodus* (Pimelodidae), *Tachysurus* (Bagridae) and *Chrysichthys* (Claroteidae) (Tilak 1965; Arratia 1992).

The quadrate may be quite round on its dorsal and posterior margins, or it may possess several processes (Arratia 1992). However, a ventroposterior process commonly present in other teleosts is absent in catfishes (Figure 3E–H). The absence of a symplectic is regarded as a synapomorphy for siluriforms plus gymnotiforms. However, symplectic cartilages in the respective position, interpreted as the remnant of a symplectic, have been reported for *Hypophthalmus* (Howes 1983a) and *Diplomystes* (Arratia 1992). Many catfishes have sutured connections between the quadrate and hyomandibula (Arratia 1992), and in some taxa, the preopercle also ankyloses to the hyomandibula (Arratia 2003a).

Lower Hyoid Arch

The lower hyoid arch in other teleosts is connected to the suspensorium via a small interhyal (Figure 4A–D), which in catfishes attaches to the area where the hyomandibula and quadrate meet (Howes 1983a). Yet, some catfishes have lost the interhyal (Table 1). On the ventral side of the posterior and anterior ceratohyals, the branchiostegal rays span the gill membrane (Figure 4A–D). Catfishes usually have eight to ten branchiostegal rays, but the range extends from three in advanced loricarioids (e.g., *Farlowella*) to 20 in *Wallago* (McAllister 1968; Schaefer 1990; Reis 1998; Arratia 2003a; Birindelli 2014). Callichthyids are unique in possessing a branchiostegal cartilage (Figure 4B; Reis 1998). However, high numbers of branchiostegal rays seem to be primitive for teleosts (McAllister 1968), and when present in catfishes, it is considered a derived condition (Birindelli 2014). Ventral and dorsal hypohyals are ancestrally present in catfishes (Arratia 1987). The dorsal hypohyal, however, becomes smaller or is even completely reduced independently in various groups (Table 1; Figure 4A–D; Reis 1998). The absence of a basihyal is an apomorphy of catfishes (Arratia and Schultze 1990; de Pinna 1993). In addition, a tooth plate that could be assigned to the basihyal is missing (Fink and Fink 1981).

Within the *Musculus sternohyoides*, which retracts the hyoid arch, teleosts have a sesamoid ossification, the urohyal. In catfishes, the bone in the respective position also has cartilaginous portions on its dorsomedian part and originates from paired sesamoid ossifications, in contrast to other teleosts, where it forms directly as an unpaired sesamoid ossification. To emphasise its non-homology with the urohyal of other teleosts, researchers refer to the structure of catfishes as "parurohyal" (Arratia and Schultze 1990). Few taxa (Auchenipteridae, Doradidae, and Mochokidae) have lost the parurohyal (Vigliotta 2008).

Gill Arches

The branchial apparatus of catfishes usually includes almost the whole set of bones typically found in other teleosts. These are (infra-)pharyngobranchials 1 to 4, epibranchials 1 to 4, ceratobranchials 1 to 5, hypobranchials 1 to 3, and basibranchials 2 to 3. Only the latter are unpaired, and while primitive teleosts and still other ostariophysans also have a basibranchial 1, this element is missing in catfishes (de Pinna 1993). Pharyngobranchials 1 and 2 are absent in most siluriforms (de Pinna 1993; Calegari et al. 2019).

While generally in teleosts, the basibranchial and pharyngobranchial elements may carry separate tooth plates, in siluriforms and gymnotiforms there is a single pharyngobranchial tooth plate, and tooth plates corresponding to the basibranchials 1 to 3 are never present (Fink and Fink 1996). Most catfishes also have a large tooth plate on the ceratobranchial 5.

Catfishes ancestrally have a double row of gill rakers on their first ceratobranchial, as other ostariophysans, but they are often reduced to a single row (Birindelli 2014). The size and number of gill rakers are closely related to diet (Figure 4E–G): most species have a few rigid gill rakers, usually fewer than 15 (Figure 4E). They are completely missing in Parakysinae, *Sisor* and *Chaca* (de Pinna 1993). The few filter-feeding taxa, e.g., *Siluranodon* (Schilbeidae) (Figure 4G), *Synodontis membranacea* and *S. batensoda* (Mochokidae), have high numbers of long and slender

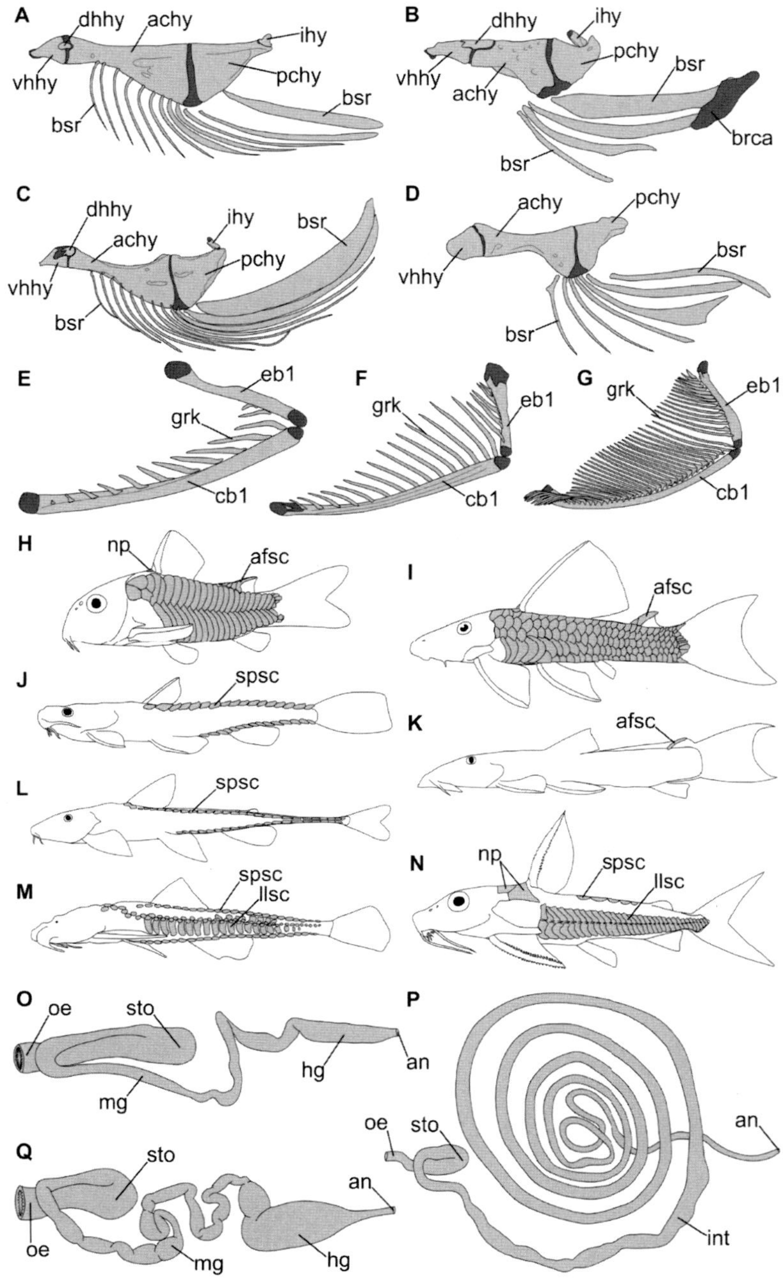

Figure 4. Right lower hyal arch, medial view (A–D), first left gill arch, lateral view (E–G), postcranial body armour (H–N), and alimentary tract (O–Q) pulled apart of selected catfishes. (A, F) *Sperata aor* (Bagridae). (B) *Corydoras paleatus* (Callichthyidae). (C, G) *Siluranodon auritus* (Schilbeidae), secondary ramifications on gill rakers omitted. (D) *Hatcheria macraei* (Trichomycteridae). (E) *Silurus glanis* (Siluridae). (H) *Corydoras brevis* (Callichthyidae). (I) *Ancistrus* cf. *dolichopterus* (Loricariidae). (J) *Scoloplax baskini* (Scoloplacidae) (based on Rocha et al. 2008). (K) *Astroblepus* sp. (Astroplepidae). (L) *Phractura clauseni* (Amphiliidae). (M) *Hoplomyzon papillatus* (Aspredinidae) (based on Stewart 1985). (N) *Hemidoras* (Doradidae). (O) Predatory *Bagrus docmak* (Bagridae). (P) Aufwuchs-feeder *Ancistrus* cf. *dolichopterus* (Loricariidae). (Q) Omnivorous *Sciades herzbergii* (Ariidae). Shading: dark grey: cartilage; light grey: bone (A–G), superficial bone (H–N), or alimentary tract (O–Q). Abbreviations: achy: anterior ceratohyal; afsc: adipose-fin scute; an: anus; brc: branchiostegal cartilage; brca: branchial cartilage; brs: branchiostegal ray; cb1: ceratobranchial 1; dhhy: dorsal hypohyal; eb1: epibranchial 1; grk: gillraker; hg: hind gut; ihy: interhyal; int: intestine; llsc: lateral-line scute; mg: middle gut; np: nuchal plate; oe: oesophagus; pchy: posterior spsc: scute derived from vertebral spine; sto: stomach; ceratohyal; vhhy: ventral hypohyal.

gill rakers which may carry secondary ramifications. In *Hypophthalmus*, up to 200 gill rakers per ceratobranchial were reported (Howes 1983a).

Postcranial Skeleton

The postcranial skeleton comprises the vertebral column and the fins, including their supporting structures. Intermuscular bones are common among teleosts, such as in otomorphs; however, catfishes do not possess them. Similarly, catfishes lack a long series of supraneural bones, although it is discussed whether a structure at the beginning of the dorsal fin represents a pterygiophore or a last supraneural (Arratia 2003b). Furthermore, parts of supraneurals may have become part of the vertebral complex (see below).

Body Armour

The absence of scales is a siluriform apomorphy (e.g., Fink and Fink 1981). Protective bony armour, however, has developed several times independently and on various parts of the body (Figure 4H–N). Posterior to the neurocranium and in front of the dorsal fin, a nuchal shield is often present (Figure 4N). Usually, it consists of three paired nuchal plates associated with pterygiophores of the dorsal fin (for interpretation of the first pterygiophore as a supraneural; see next paragraph). Nuchal shields can be large and well-formed, like those in Auchenipteridae, Ariidae, Doumeinae, and most Mochokidae. In doradids and some mochokids, the nuchal plates get in contact with the posttemporo-supracleithrum, forming an even stronger framework of head, pectoral girdle, and dorsal fin. Sometimes, the dermal bones of the neurocranium or pectoral girdle are enlarged, covering more body area than usual, like the humeral process in *Synodontis* (Figures 1T and 5P).

Bony plates covering at least partially the body have developed not less than five times independently in siluriforms. Rivera-Rivera and Montoya-Burgos (2017) listed the presence of dermal plates for Callichthyidae, Loricariidae, Scoloplacidae, Doradidae, Amphiliidae, Aspredinidae, and Sisoridae. In the last three taxa, however, the bony armour is not of dermal origin but formed from outgrowths of vertebrae reaching the body surface and enlarging there (Figure 4H–N) (Mahajan 1967; Friel 1994; Ferraris and Vari 2012). Dermal plates of callichthyids, loricariids, and *Scoloplax* carry odontodes (Figure 3N–O). Size and amount of dermal armour vary significantly between higher taxa (Figure 4H–N) but also within families like loricariids and doradids (e.g., Sabaj and Arce H. 2021, fig. 1). Loricariids like *Farlowella* or *Loricariichthys* are almost completely covered in body armour, principally leaving only openings for the mouth, eyes, nostrils, fins, and anus. For astroblepids, it is generally accepted that they lost the body armour; however, they possess a spine in the adipose fin (Figures 1C and 4K). The shape, position, and structure of this spine are equal in Callichthyidae, Loricariidae and *Astroblepus*. For *Corydoras*, it was shown that these spines fin are actually dermal scutes (Stewart et al. 2019). Therefore, the adipose-fin spine in *Astroblepus* should be regarded as the last remainder of a formerly more extensive body armour (Figure 4K).

The lateral-line canal of the body is often stiffened by small tube-like ossicles (Alexander 1965), which are so far poorly studied. In some teleosts, lateral-line scales are also reduced to bony tubes only, so future studies should clarify if the catfishes' bony tubes are apomorphic or indeed represent remnants of scales.

Vertebral Column

Many catfishes have between 38 and 45 vertebrae (Birindelli 2014), although the total range is from as few as 22 in some pangasiids to more than 100 in some anguilliform clariids. One portion of the vertebral column of Ostariophysi is particularly striking: the so-called Weberian apparatus. It comprises parts of the first vertebrae modified to transmit sound from the swim bladder to the inner ear. The Weberian apparatus has undergone many changes during evolution, and several apomorphies on various taxonomic levels can be found here. While in other Ostariophysi, the first four vertebrae are incorporated into the Weberian apparatus, the number in catfish is more variable. For details and further references on the Weberian apparatus, see Duclos and Grande (this volume). Sometimes, parapophyses of vertebrae from the Weberian apparatus participate in forming an osseous capsule for the (anterior part of the) swim bladder (see below).

The shape of "normal" vertebrae in catfishes usually resembles that of a typical ostariophysan, comprising a simple centrum with arches and quite slender neural and haemal spines. At least three times independently, in some amphiliids, bunocephalids, and loricarioids, dorsal and ventral appendages of vertebrae broadened, partly projecting dorso- and ventrolateral. Vertebrae reinforcing the base of the dorsal fin have often modified and enlarged neural spines embracing anterior pterygiophores.

The presence and absence of supraneurals in siluriforms are confusing and need more detailed studies. The first supraneural, i.e., the one corresponding to the first vertebra, is always absent in Ostariophysi, and the second is missing in Characiphysi (Fink and Fink 1996). Supraneural 3 is part of the neural complex in some siluriforms, e.g., diplomystids, some ictalurids, pimelodids, and schilbids (Hoffmann and Britz 2006). Most catfishes, however, seem to lack supraneural 3 completely throughout development (Britz and Hoffmann 2006; Hoffmann and Britz 2006). Free supraneurals could be expected between the occiput and the beginning of the dorsal fin. Thus, the first element supporting the dorsal fin is sometimes interpreted as supreneural (Alexander 1965; Birindelli 2014) or, contrasting this view, as pterygiophore (Arratia 2003b). I tentatively follow the latter view regarding the first element of the dorsal-fin support as a true pterygiophore associated with the spinelet (see below). Even more confusing, in a few taxa, there are small supplementary ossifications between the occiput and dorsal fin-origin, like in *Rhamdia* (Heptapteridae) (Arratia 2003b), some callichthyids, loricariids, and *Astroblepus* (Reis 1998). Yet, due to their phylogenetic distribution, it seems likely that these are new formations. Furthermore, the respective bone in *Astroblepus* has no sign of cartilaginous origin but appears to be a direct ossification in the vertical septum (pers. obs.); therefore, it is likely not homologous to true supraneurals of chondral origin.

Compared to other Ostariophysi, catfishes have fewer ribs, usually only 6 to 11, rarely more like some *Diplomystes* have 13 (Arratia 1987), and sometimes as few as four (Walsh 1990; Birindelli 2014). Furthermore, the orientation of the ribs is horizontal in catfishes and gymnotiforms. This situation is "reversed" in some free-swimming taxa, with ribs approaching the ventral midline (Alexander 1965).

Fin Spines

Characteristic and apomorphic for catfishes are the possession of a strong spine in the dorsal and pectoral fins (Arratia 2003b; Wiley and Johnson 2010). Ontogenetic studies showed that catfish spines develop like lepidotrichs from two hemitrichs, usually with several segments (Reed 1924; Kubicek et al. 2019). The prevailing use of these spines is defence against predators by increasing the catfish's diameter (Figure 1S) and harming the predator. In order to make them more effective and harmful, spines often possess anterior and/or posterior serrations (Figure 5A–G; Vanscoy et al. 2015). The type, size, and location of the serrae are often phylogenetically informative and helpful in species identification (Vigliotta 2008). Taxa-bearing odontodes often have numerous and enlarged teeth on their spines, much more pronounced in some loricariids. Few species have combined pointed spines with venomous glands, like plotosids (Figure 5D), *Heteropneustes* and *Noturus* (Wright 2009; Egge this volume).

Most catfishes possess a locking mechanism, preventing the dorsal-fin spine from being depressed by exterior force. The large defensive spine is actually the second spine (Figure 5H). In front of it and connected via a strong ligament generally lies a small, roughly triangular spine called a spinelet. When the defensive spine is erected, it is locked in place between the pterygiophores and secured by the spinelet (Figure 5H–I). In order to unlock, the spine has to be moved slightly dorsally; other movements will block the spine even more. To enforce spine stability, the anterior part of

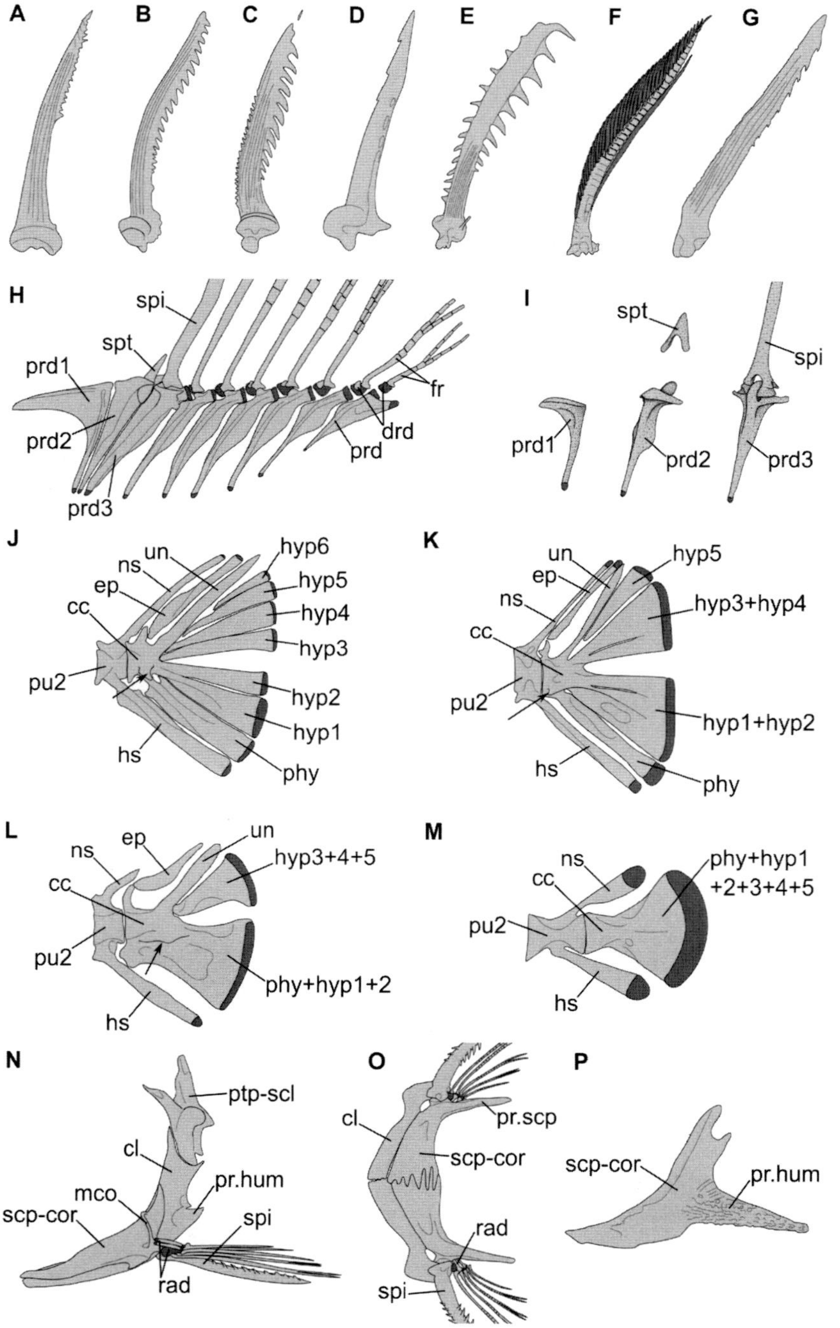

Figure 5. Spines of catfishes, right pectoral fin spine in dorsal view (A–F), dorsal fin spine in lateral view (G), dorsal fin of *Sperata* (H–I), caudal-fin skeleton in lateral view (J–M, rays omitted), and pectoral girdle (N–O) of selected catfishes. (A) *Silurus glanis* (Siluridae). (B) *Chrysichthys auratus* (Claroteidae). (C) *Synodontis petricola* (Mochokidae). (D, M) *Plotosus lineatus* (Plotosidae). (F) *Amphilius atesuensis* (Amphiliidae), non-osseous pinnate appendages in dark grey. (G) *Synodontis petricola* (Mochokidae). (H–I) *Sperata aor* (Bagridae), supporting structures of the dorsal fin, lateral view (H), and first elements separated in oblique, fronto-lateral view. (J) *Diplomystes mesembrinus* (Diplomystidae). (K) *Bagre marinus* (Ariidae). (L) *Chiloglanis voltae* (Mochokidae). (N) *Sperata aor* (Bagridae), lateral view. (O) *Bunocephalus coracoideus* (Aspredinidae), ventral view. (P) Scapulo-coracoid of *Synodontis petricola* (Mochokidae), lateral view. Abbreviations: cc: compound centrum; cl: cleithrum; drd: distal radial; ep: epural; fr: fin ray; hs: hemal spine; hyp: hypural; mso: mesocoracoid; ns: neural spine; phy: parhypural; pr.hum: humeral process; pr.scp: scapulo-coracoid process; prd: proximal radial; rad: proximal radial; scp-cor: scapulo-coracoid; spi: spine; spt: spinelet; un: uroneural.

the fin supports is in many taxa connected or strongly associated with the skull via connections with vertebrae, pectoral girdle, extensions of the neurocranium and/or body armour like nuchal plates. Some groups lost the locking mechanisms, e.g., some loricariids and sisorids, as well as most aspredinids and heptapterids. Some taxa secondarily lost the dorsal-fin spine (Fink and Fink 1996), like Siluridae, Clariidae plus Heteropneustidae, Cetopsidae, Trichomycteridae, some Pimelodidae, *Siluranodon* (Schilbeidae) and *Olyra* (Bagridae).

Less common, like in *Chaca*, the pectoral spine became reduced or convergently lost its spiny character, e.g., in *Kryptoglanis*, Trichomycteridae, Amphiliidae (Figure 5F), Sisoridae, and Malapteruridae (Kubiceck et al. 2019). Dorsal and pectoral spines look much alike but differ completely in their supporting structures. The pectoral spine articulates on the scapulocoracoid via three articular facets (Alexander 1965; Fine et al. 1997). In several taxa, a locking mechanism is present, which may be supported by the cleithrum forming an exact fitting groove for the anterior articulatory process so that the spine can only be removed when turning it at a certain angle (Fine et al. 1997). Several taxa produce sounds using a rough surface on the spine base which is rubbed against the cleithrum (Pfeiffer and Eisenberg 1965; Fine et al. 1997; Kaatz et al. 2010). The dorsal spine can produce respective sounds only in some sisorids (Tilak 1963a). In a few cases, spines are also used in terrestrial locomotion (Pace and Gibb 2014). Adipose-fin spines of callichthyids, loricariids and *Astroblepus* are plates of the dermal body armour (Figure 1H–I, K). The adipose-fin spines in *Sisor* and some amphiliids are regarded by some authors as transformed scutes (e.g., Stewart et al. 2014) but apparently are outgrowths of neural spines. The anal fin in catfishes has no spine, but usually, the first two rays are unbranched.

Dorsal and Anal Fins

A complete set of fins in catfishes may comprise unpaired (dorsal, adipose, caudal, and anal) and paired (pelvic and pectoral) fins. The rays of the dorsal and anal fin are supported by a roundish distal radial. Towards the body axis lies a single, long rod-like element, where more basal teleosts have two: the proximal and middle radials. The absence of a separate middle radial ossification represents an apomorphy for siluriforms plus gymnotiforms (Fink and Fink 1981). In Siluridae, the distal radials are also absent in the dorsal fin (Bornbusch 1991). The first three pterygiophores of the dorsal fin, especially the second and third ones, are more massive than the following ones and carry lateral projections to support the defensive spine and spinelet (Figure 5H–I).

In general, the dorsal fin in catfishes is rather short, often with only seven or fewer rays. An exception is clariids with more than 30 rays and even above 100 in the anguilliform genera *Channallabes* and *Gymnallabes*. The dorsal fin is missing in malapterurids, ailiids, the schilbeid *Parailia* (de Vos 1995), the silurids *Hemisilurus*, *Ceratoglanis* and some *Kryptopterus* species (Bornbusch 1991), as well as in the trichomycterids *Glanapteryx*, *Pygidianops,* and *Typhlobelus* (de Pinna 1989; Schaefer et al. 2005).

Anal fins are variable in length in catfishes, usually with nine to 13 branched rays, but with a total range of five to over 60 (Birindelli 2014). A total absence of an anal fin seems unique to the trichomycterids *Glanapteryx anguilla* (de Pinna 1989) and *Pygidianops magoi* (Schaefer et al. 2005). In male auchenipterids, the first rays are transformed into a copulatory organ (Loir et al. 1989; Meisner et al. 2000). Bornbusch (1991) listed a posterodorsal shift of fin-ray articulation as apomorphic for silurids; this gradual character seems to be present also in other taxa, especially those with long anal fins, and may need further studies.

An adipose fin apparently originated at the common origin of Characyphysi. In catfishes, it is usually long and thick and sometimes large, e.g., *Synodontis* (Mochokidae) and *Bagrus* (Bagridae). It is rather small, or even absent, as in Aspredinidae, Cetopsinae, Siluridae, *Chaca*, *Heteropneustes*, *Nematogenys*, some schilbeids, and few auchenipterids. Usually, there is no skeletal element connected to the adipose fin, but some taxa have small cartilages at their base. Loricarioids (except trichomycterids) carry a spine-like scute on their anterior margin (Figure 5H–I and K). Sometimes, neural spines are prolonged into the adipose fin, e.g., in clariids (Teugels 1983), or to form a spine on the anterior margin of the adipose fin, e.g., some amphiliids (Figure 1L) and *Sisor.* In *Mochokus* (Mochokidae), there are even rays but without pterygiophores in the adipose fin. Plotosids have a long, continuous fin from the back of the dorsal fin continuous with the caudal and anal fin. It remains open whether the dorsal part of this fin represents just procurrent rays of the caudal fin or rays without pterygiophores in an adipose fin similar to what is found in *Mochokus*.

Caudal Fin

As members of the Ostariophysi, catfish evolution already started from a derived condition of the caudal fin skeleton, as they have several elements fused and others reduced in comparison to stem teleosts. Both ural centra (in diural terminology) form a compound centrum to which hypural 2 and the most anterior uroneural (so-called pleurostyle) are fused; formerly, it was supposed that this fusion also includes preural centrum 1 (e.g., Fink and Fink 1996), but recent studies did not find a preural centrum 1 during any time of the ontogeny of ostariophysans (e.g., Wiley et al. 2015; Cumplido et al. 2020). Yet, a separate ural centrum 2, to which hypural 3, 4 and 5 sometimes fuse, is visible during development and persists in adults in certain catfish taxa (Lundberg and Baskin 1969; de Pinna and Ng 2004). Although the presence of a distinct ural centrum 2 is a primitive teleost character, it is regarded as a derived novelty within siluriforms (de Pinna and Ng 2004). In Characiphysi, also the parhypural and hypural 1 are fused to the compound centrum (Fink and Fink 1996). Above the compound centrum, an almost complete neural arch and never more than a single epural are present in catfishes (Figure 5J–M) (Lundberg and Baskin 1969; see Arratia 1983 reporting intraspecific variation for these structures in some catfishes).

Catfishes have up to six hypurals (Figure 5J), but there is a trend towards the fusion of hypural plates (Figure 5J–M). The resulting combinations are too diverse to be listed here but may lead to a condition with only two plates in adults: a lower hypural plate consisting of parhypural, hypural 1 and 2, and an upper hypural plate consisting of hypural 3 to 5. Such a condition can be found in some mochokids (Figure 5L) and some aspredinids. Even fusion of the parhypural and all hypurals, forming a single plate, occurs in Plotosidae (Figure 5M), Chacidae, some loricariids, and *Kryptoglanis* (Lundberg and Baskin 1969; Britz et al. 2014).

A lateral outgrowth of the parhypural, the hypurapophysis, is important for muscle attachment. Catfishes may possess additional outgrowths at the base of the first and second hypural. Lundberg and Baskin (1969) distinguished five types of hypurapophysis configurations in catfishes, ranging from their total absence to an extended shelf formed by three sequential hypurapophyses. The primitive condition, as found in diplomystids and several other taxa, seems to be a "normal" hypurapophysis on the parhypural plus a small separate one on hypural 2 (Lundberg and Baskin 1969).

Primitively, catfishes possess a forked caudal fin with 9+9 (= i8+8i) principal caudal rays, like in *Diplomystes* (Lundberg and Baskin 1969; Arratia 1987; de Pinna 1993). Few taxa have increased numbers of principal caudal rays, while various reductions are present, with the lowest count being 5+5 in most aspredinids (Lundberg and Baskin 1969). Procurrent rays count from zero to two in aspredinids to high numbers in plotosids (depending on homology interpretation; see above), usually around 12 to 25 (Lundberg and Baskin 1969). Rounded or truncated fin shapes have evolved several times.

Paired Fins

Typical teleosts have two paired fins: pectoral and pelvic. As the pelvic girdle is discussed by Moritz et al. (this volume), only the pectoral girdle is regarded here. Catfishes have only three larger skeletal elements in their pectoral girdle (Figure 5N) (Diogo et al. 2001). The dorsal part, made of posttemporal and supracleithrum in other teleosts, consists of a single element in catfishes. It is interpreted as a fused product, the posttemporo-supracleithrum (e.g., Alexander 1965; Howes 1983a; Arratia 2003b; Birindelli 2014). Ontogenetic studies so far have not clearly shown its compound nature (Arratia 2003b). Where scapula and coracoid are in other teleosts, there is only one bone in catfishes, likely a fusion, called scapulocoracoid (Figure 5N). The cleithrum represents the largest bone in the pectoral girdle, like most teleosts. Plesiomorphically, the mesocoracoid forms a thin arch in catfishes (Figure 5N), but it is small, missing, or fused with the scapulocoracoid (Alexander 1965; Reis 1998; Arratia 2003b; Vigliotta 2008). Postcleithra, of which basal teleosts have up to three, are never present in catfishes.

Another striking difference to non-siluriform teleosts is the strong connection of the pectoral girdle to the skull. In many catfishes, this connection becomes immobile via a suture of the posttemporo-supracleithrum to the neurocranium (Alexander 1965; Vigliotta 2008; Birindelli 2014; Calegari et al. 2019), and in a few cases to the parapophysis of the fourth vertebra (de Pinna 1996). In callichthyids, loricariids, *Scoloplax*, and *Astroblepus*, the posttemporo-supracleithrum fuses with the pterotic (Schaefer 1990). Due to their depressed body, the vertical arm of the cleithrum is shortened, while the horizontal arm is enlarged. A posterior cleithral (= humeral) process above the pectoral spine is often present (Figure 5N and P) (Vigliotta 2008; Birindelli 2014) and can be helpful in species identification (e.g., Pinton et al. 2016). Scapulo-coracoid and cleithrum are more or less sutured to the respective bone of the other side (Figure 5O) (Tilak 1963b; Alexander 1965; Birindelli 2014). The connection is weaker in some pelagic taxa and weak in *Asterophysus*, allowing it to swallow extremely large prey (Zuanon and Sazima 2005). The scapulo-coracoid supports the pectoral radials and forms an articulatory area for the pectoral spine, usually equipped with a locking mechanism (see above).

Usually, catfishes have six to eight pectoral-fin rays in a total range of four to 15 (Birindelli 2014). The rays articulate on small, roundish, cartilaginous distal radials, yet there are fewer distal radials than rays; in a few cases, distal radials are completely reduced, e.g., in clariids, heteropneustids, and some trichomycterids (Tilak 1963b; de Pinna 1989). More anteriorly are the proximal radials, primitively four in teleosts but only three (rarely two) in catfishes (Arratia 1992, 2003b). Few trichomycterids have completely reduced their pectoral fin (Schaefer et al. 2005).

Muscles and Ligaments

The seminal work of Winterbottom (1973) gave a comprehensive overview of the striated muscles of teleosts and is still highly valuable for identifying muscles in catfishes. Subsequent studies focusing on certain taxonomic groups have refined or altered this nomenclature. Most important for catfishes are the changes in the interpretation of the *adductor mandibulae* complex by Diogo and Chardon (2000a) and Datovo and Vari (2013, 2014); the latter authors introduced a new nomenclature. Similarly, Cumplido (this volume) investigated the caudal fin musculature.

Muscles of the Jaws, Suspensorium and Opercular Area

The flat, broad skulls with small eyes significantly influence the size and position of the jaw musculature in catfishes. Typically, catfishes have a massive *adductor mandibulae* complex, which, in teleosts, mostly originates on the lateral part of the hyopalatinal arch. In several catfish taxa, it is markedly enlarged and extends onto the skull roof, much pronounced in, e.g., cetopsids, malapterurids, or *Horaglanis* (Clariidae), where dorsally, the posterior neurocranium is completely covered by muscle (de Pinna and Vari 1995). Apomorphic for Siluriphysi, the ligament connecting the *adductor mandibulae* to the lower jaw, the ligamentum primordium, attaches to the cartilage of the dorsal tip of the lower jaw instead of close to the jaw articulation (Fink and Fink 1981, 1996; Arratia 1992). In teleosts, the interoperculo-mandibular ligament couples the opercular apparatus with the jaws to promote suction feeding. Jaw mechanics in scraping catfishes changed significantly so that the respective ligament got lost in loricariids, *Scoloplax*, and *Astroblepus* (Schaefer and Lauder 1986; Schaefer 1990; Geerinckx and Adrians 2006). This modification also enables breathing while the mouth holds onto a substrate (Geerinckx et al. 2007).

Catfishes have two specialised muscles to move the highly mobile maxillary barbel. Their evolutionary origin was unclear for a long time (Alexander 1965; Howes 1983b), but new evidence suggests that the *retractor tentaculi* derived from the *stegalis* (formerly A_3) portion of the *adductor mandibulae*, and the *extensor tentaculi* are derived fibres of the *adductor arcus palatini* (Adriaens and Verraes 1997; Diogo and Chardon 2000a; Datovo and Vari 2014). It was suggested that the *retractor tentaculi* is not homologous throughout siluriforms, despite its presence in all taxa (Howes 1983b; Geerinckx et al. 2007; Datovo and Vari 2014). External and internal mandibular barbels are supported by bases (or roots) of cartilages and dense elastin (Figure 3C), to which the controlling muscles are attached (Benjamin 1990). Generally, these are six different muscles or muscle portions in catfishes, plus sometimes additional muscles in certain taxa (Diogo and Chardon 2000b).

The opercle has the typical set of muscles for teleosts, i.e., *dilatator*, *adductor*, and *levator operculi*. In trichomycterids, which use their toothed opercular apparatus (Figure 3G) for fixation on structures, an additional muscle of uncertain homology, the *protractor operculi*, is of major importance and comprises a substantial portion of the cheek musculature (de Pinna 1998; Adriaens et al. 2010). Also, hypostomine loricariids can erect odontodes anterior of the operculum, but they use the hypertrophied *dilatator operculi* for that (Geerinckx and Adriaens 2006).

Baudelot's Ligament

The pectoral girdle of catfishes is directly connected to the cranium via the dorsal process of the posttemporo-supracleithrum. In teleosts, further anchorage generally comes from (1) the ventral process of the posttemporal connecting directly or via a short ligament to the neurocranium and (2) via Baudelot's ligament, connecting the supracleithrum and/or cleithrum to the basioccipital or the first vertebrae. A ventral process of the posttemporal part of the posttemporo-supracleitrum is missing in siluriforms (Fink and Fink 1981). Yet, Baudelot's ligament, also called transscapular ligament, is much derived. Synapomorphic with gymnotiforms, it is robust and distally bifurcated (Fink and Fink 1981). In catfishes, it is further ossified; its proximal part attaches to the neurocranium, the lateral distal part fuses to the posttemporo-supracleithrum, and the medial distal part connects to the parapophysis of the fourth vertebral centrum (Arratia 2003b). In loricariids, it fuses with the compound-pterotic (Schaefer 1987).

Body Musculature

The body musculature of catfishes looks much like that of a "typical" teleost but without intermuscular bones. Furthermore, behind the pectoral girdle, the hypaxial musculature leaves a gap where the swim bladder projects laterally. Anteriorly, the body musculature originates on the skull, and pectoral girdle, and its force is transmitted via myosepta and follows myotomes towards the posterior body portion and caudal fin.

Muscles of the Fins

The derived condition of the pectoral girdle in catfishes makes the homologisation of muscles difficult (Diogo et al. 2001). The pectoral fin rays have muscles likely corresponding to the *abductor* and *adductor* muscles of other teleosts, each with *superficialis* and *profundus* portions (Miano et al. 2013). The pectoral spine usually has four larger muscles: the *arrector dorsalis* and *ventralis*, as well as the spine *abductor* and *adductor*, with the latter two causing most of the movements (Miano et al. 2013). In species using the pectoral spine for stridulation, like doradids and mochokids, the respective muscles can be significantly enlarged (Alexander 1965).

Muscularisation of the fin rays in the pelvic girdle follows a similar setting. Four muscles are usually involved: *abductor superficialis* and *profundus*, as well as *adductor superficialis* and *profundus*; the first ray additionally has an *arrector ventralis* and *dorsalis* (Winterbottom 1973). Diverse pelvic girdles in catfishes (Moritz et al. this volume) indicate more complex uses of pelvic fin rays than in generalised teleosts. Respectively, muscle arrangement and course are more variable among taxa. Lateral processes on the basipterygium may deviate the *arrector dorsalis* (often via a tendon), giving the first ray more power and mobility (Shelden 1937). This is especially important for species moving on a substrate in water as well as on land, e.g., *Astroblepus* and loricariids. The pelvic girdle is kept in position only by the hypaxial musculature, with some portions branching to directly attach to processes and sometimes flanges of the basipterygium.

Dorsal and anal rays are moved by paired erector and depressor muscles. These muscles are well-developed on the dorsal fin spine. Inclinator muscles are rare in catfishes. Most taxa have only a single pair of inclinators on the dorsal fin, but up to four pairs are also reported (Birindelli 2014). Adipose fins are generally passive structures, but for *Horabagrus*, a musculoskeletal linkage was reported, allowing some controlled movement (Stewart and Hale 2013). The musculature of the caudal fin is presented in its own chapter (Cumplido this volume).

Electric Organs

Electric catfishes (Malapteruridae) are capable of producing strong discharges for stunning prey and deterring predators, but they may use discharges for social interaction, too (Rankin and Moller 1992). Their electric organ surrounds the postcranial body like a tube, directly below the skin and above the normal body musculature. It seems to be derived from pectoral muscles (Crampton 2019). In contrast to most other electric organs, the electrolytes are more irregularly organised (Galant 2019). Its discharges reach 300 V and last two to three milliseconds (Welzel and Schuster 2021). Electric catfishes are well-protected against electric shocks of their own and from external sources, but the protective mechanism is still unknown (Welzel and Schuster 2021).

Synodontis (Mochokidae) species have a weakly electric organ that evolved from the sonic swim-bladder muscles (Hagedorn et al. 1990; Kéver et al. 2020). The function of these discharges remains obscure but likely is related to communication (Orlov and Baron 2005; Kéver et al. 2020). For some other not related catfish taxa, i.e., *Clarias* (Clariidae), *Ompok* (Siluridae), *Auchenoglanis* (Auchenoglanididae), and *Ictalurus* (Ictaluridae), weak electric discharges have been reported (Albert and Crampton 2005). Yet, no electric organs have been discovered in those taxa, and it seems likely that the discharges constitute trans-membrane potentials of the skin and other organs (Crampton 2019). Nevertheless, they seem to be used in communication during aggression or mating behaviour (Baron et al. 1994, 2005; Olshanskiy et al. 2009).

Organs of the Peritoneal Cavity

Several studies have been conducted on various soft tissues of catfishes, but they scarcely cover a significant portion of diversity. Rarely are certain organ systems studied in a significant number of species, like the swim bladder in doradids and auchenipterids (Brinidelli et al. 2009, 2012). Often, soft tissue information is given only for very few or even single species, which makes it difficult to draw general conclusions. For organs like the spleen, liver, kidneys, bladder, and heart, as well as the immune and circulatory system, the little information available so far usually does not significantly deviate from other teleosts.

Alimentary Canal

Typically, for teleosts, the catfishes' digestive tract can be divided into oesophagus, stomach, intestine, and rectum (Figure 4O–P; Wilson and Castro 2011). Yet, there is substantial diversity in shape, histology, glands, supplementary structures, and length. Although diet is not the only factor influencing the morphology of the alimentary tract, it is apparently crucial. Predatory catfishes differ a lot (Figure 4O) from algivorous and detritivorous taxa (Figure 4P), while omnivorous species and other feeding types are somewhere between those two extremes (Figure 4Q).

While the oesophagus in detritivorous catfishes is simple tube-like, it has many folds in predatory fishes, allowing stretching for bigger food items (Mello et al. 2017). In either case, it usually has many mucous glands. A stomach is poorly developed, with only a thin wall in detritivores but distinct and thick-walled in predators with three different regions typical for teleosts: cardia, fundus, and pylorus. The stomach in catfish is usually U-, J- or Y-shaped, but other forms have also been reported. Species specialised in larger prey have much more extendable stomachs. In the anterior part of the gut, many teleosts possess pyloric caeca to increase the surface area of the intestine (Buddington and Diamond 1986), but they are absent in catfishes (Mello et al. 2017).

While the intestines are short in predators, down to only 20% of the standard length (Figure 4O; Mello et al. 2017), they may become extremely long, for example, in detritivores loricariids (Figure 4P), from which intestines of more than 17-fold the standard length have been reported (German 2009). The long loricariid intestines are coiled in loops more or less in a horizontal plane (Delariva and Agostinho 2001). In such taxa, a distinct hindgut is not developed (Figure 4P). Surprisingly, *Astroblepus* has a relatively short intestine despite its feeding habits and relatedness to loricariids.

Swim Bladder

In non-Ostariophysi, the swim bladder mainly regulates buoyancy and is, therefore, often reduced or absent in bottom-living taxa. As the swim bladder in Ostariophysi, however, is also of importance for hearing, it is generally present in catfishes, even though sometimes only small remnants are left, like in some chiloglanidines (Vigliotta 2008). Few taxa, like *Trogloglanis* (Ictaluridae), *Rhynchodoras*, and most *Leptodoras* species (Doradidae), lost their swim bladder (Birindelli et al. 2009). A "simple" and likely plesiomorph catfish swim bladder is heart-shaped, with the tip of the heart pointing posteriorly. Externally, it looks like a single bladder but likely corresponds to the anterior and posterior bladder, usually present in Ostariophysi (Fink and Fink 1996). Inside, a T-shaped septum separates a larger anterior chamber from two posterior chambers. The Weberian apparatus mechanically connects the swim bladder to the inner ear, and the pneumatic duct directly connects the swim bladder with the oesophagus. Yet, in loricariids, the pneumatic duct gets lost (Bridge and Haddon 1892). Secondarily, pelagic catfishes also increase buoyancy by a high proportion of fat in the body (Howes 1983a), swallowing air (Reis 1998) or constant fin movement.

The wall of the swim bladder is formed by a thin internal and a thicker external layer. The lateral body musculature usually leaves a window so that the anterior part of the swim bladder gets in close contact with the environment. Catfishes show more variability in their swim-bladder morphology than other ostariophysans, including shape, size, appendages, internal division and surface structure, and connection to other anatomical structures (Bridge and Haddon 1892; Howes 1983a; Birindelli et al. 2009, 2012). In some taxa like Cetopsinae, Clariidae, Sisoridae, Pseudopimelodidae, and loricarioids, the highly reduced swim bladder is enclosed in a bony capsule originating from various skeletal structures. In contrast, some other catfishes have large posterior portions of the swim bladder. Several taxa, like doradids, auchenipterids, ariids and mochokids, have the ability to produce sounds by bringing their swim bladder to vibration. An elastic spring apparatus for sound production likely developed several times independently (Kaatz et al. this volume).

Accessory Breathing Organs

Typically, for fish, catfishes take up oxygen from the surrounding water via their gills (Zaccone et al. this volume). Few taxa have developed air-breathing mechanisms. Clariids possess a suprabranchial organ derived from their branchial arches, where bush-like appendages take oxygen from engulfed air (Teugels and Adriaens 2003). *Heteropneustes*, although sister-group to clariids shows another adaptation: long lung-like appendages extend from the pharynx in the posterior direction (Olson et al. 1990). Callichthyidae and some Loricariidae have highly vascularised areas in their intestine, allowing them to breath swallowed air. Used air is released via the anus. Air-breathing behaviour in catfishes is usually elicited by hypoxia, but in callichthyids, gulping air is independent of the water's oxygen content. In this taxon and some loricariids, the swallowed air also greatly contributes to gaining buoyancy (Armbruster 1998; Reis 1998). Some trichomycterids and *Scoloplax* likely possess a respiratory stomach (Armbruster 1998; Reis 1998); the used air is released via mouth and gill openings in these cases.

Reproductive Organs

Sex determination in catfishes principally depends on sex chromosomes. Heteromorphic sex chromosomes have been reported for males, e.g., in *Ictalurus* (Patiño et al. 1996) and *Silurus* (Liu et al. 2010), or for females, e.g., in *Hypostomus* (Artoni et al. 1998) and *Imparfinis* (Vissotto et al. 1997). In *Clarias gariepinus*, even both systems, male and female heteromorphic sex chromosomes, exist at the same time (Nguyen et al. 2021), and for *Bunocephalus coracoideus* (Aspredinidae), even a multiple-sex chromosome system was confirmed (Ferreira et al. 2016). Yet, it is not surprising that environmental factors also affect sex differentiation. For example, temperature influences sex determination in *Ictalurus punctatus* (Patiño et al. 1996), and aquaculture-bred *Silurus meridionalis* are always females despite a natural sex ratio 1:1 (Liu et al. 2010). Once differentiated, catfish seem unable to change their sex.

Differing reproductive strategies in catfishes affect shape, size, and presence/absence of supporting structures in reproductive organs (Melo et al. 2011). Testes can consist of many digitiform lobes of various numbers and sizes or be compact without lobes; some testes have areas without sperm production for secretory and/or storage function (Melo et al. 2011; Birindelli 2014). Female reproductive organs seem less variable. Adaptations of oocyte morphology to the reproductive strategy have been reported in loricariids: the zona radiata is thickest in species carrying their eggs, less thick in eggs being scattered, and thinnest in cavity-breeding species (Suzuki et al. 2000).

Auchenipterids have internal insemination (Birindelli 2014). The urogenital opening, which lies usually close to the anus, is shifted in males towards the first anal fin ray via a long and thick urogenital tube. Females possess structures for sperm storage. Internal insemination was also verified for the loricarioids *Scoloplax* (Spandella et al. 2008) and *Astroblepus* (Spadella et al. 2012).

Skin

Most catfishes and supposedly their last common ancestor possess naked skin (Fink and Fink 1981). The skin may carry papillae (Arratia 2003b), which are large and expressed in *Diplomystes* (Arratia 1987). Some taxa bear bony armour on their skin (see above, Figures 1, 4H–N), while in others, parts of the skin may keratinise in adults, e.g., labial tubercles on the pectoral fin or parts of the flanks (Roberts 1982; Arratia 2003b). In some sisorids and akysids, most of the body is covered by keratinous unculi (Roberts 1982). Breeding tubercles, an apomorphy of Ostariophysi, are regarded to be absent from catfishes (Roberts 1982); however, Birindelli (2014) suggests calling the large tubercles in male auchenipterids breeding tubercles.

Besides delimitating the body from the environment and protecting against pathogens, the skin serves in sensing, metabolism and behaviour. Therefore, several specialised tissues can be found here: (1) neuromasts, open and/or in lateral-line canals, superficial or in pits; (2) electrosensitive ampullary organs; (3) chemosensory organs (taste buds concentrated on barbels and head, but also distributed on the rest of the body); (4) various secretory glands, e.g., for production of mucous or toxins. An ostariophysan peculiarity is club cells, which carry alarm substances (Pfeiffer 1963); some catfish taxa lost them (Pfeiffer 1967; Arratia and Huaquín 1995). Furthermore, catfishes have an axillary gland above the base of the pectoral spine; its function(s) is poorly studied (Wright 2015). A review on the skin of catfishes is given by Arratia (2003c).

Nervous System and Sensory Organs

The transition from a diurnal to a nocturnal lifestyle resulted in many changes in the sensory organs during early catfish evolution. Eyesight lost importance, while mechanical sensing, hearing, tasting, and smelling became more significant. Increased mechanosensory abilities are well-illustrated by the catfishes' typical barbels: short barbels seem optimised for food searching, while long barbels are especially important for orientation (Alexander 1965). Furthermore, with electrosensitivity, synapomorphically shared with gymnotiforms, a new sensory system arose. The changes in sensory organs were accompanied by adaptations in the peripheral and central nervous system.

Brain

Brain morphology is largely influenced by sensory systems and phylogenetic origin. During the last decades, much information on catfish brain morphology has been gathered (e.g., Bass 1981; Trajano 1994; Abrahão and Shibatta 2015; Angulo and Langeani 2017; Abrahão et al. 2018). Most studies on the catfish brain refer to that of ictalurids (e.g., Atema 1971; Bass 1981; Tong and Finger 1983; Meek and Nieuwenhuys 1998). Available data is still somewhat randomly distributed, covering only a minor portion of the catfish diversity, so possible apomorphies and evolutionary trends have to be seen as provisional ideas.

The catfish olfactory lobes and the telencephalon are enlarged compared to other Ostariophysi, reflecting the importance of smelling in catfishes. Accordingly, the optic lobes are smaller but not very small, likely also because other information, such as hearing and somatosensation, is processed here. In several catfishes, the tectum opticum still is the second largest brain section (Angulo and Langeani 2017; Abrahão et al. 2018). Yet, the cerebellum is much enlarged in most catfishes, extending partly or completely anterior over the optic lobes (Albert 2001; Albert et al. 1998; Angulo and Langeani 2017). Directly posterior to the cerebellum, there are paired electrosensory lateral-line lobes in Silurophysi, which are highly laminated (Tong and Finger 1983; Albert 2001; Albert et al. 1998). In comparison with other teleosts, these lobes, where mechanosensory and electrosensory information is collected, are quite large (Ito et al. 2007). Behind the cerebellum and between the lateral-line lobes are the facial lobes, where information on the taste buds is collected via the facial nerve. As expected, the facial lobes are also large in catfishes compared to many teleosts (Ito et al. 2007), but they are also not small in other Ostariophysi, where tasting plays an important role.

Many more remarkable characters of the siluriform brains would need detailed descriptions of nervous structures and pathways that are beyond the scope of this review. For a detailed introduction to teleost brains, see Meek and Nieuwenhuys (1998), and for the autonomic nervous system, see Nilsson (2011).

Eye

Eyesight is of less importance in catfishes. The size reduction of eyes is an apomorphy for Silurophysi (Fink and Fink 1981). In several lineages, eyes have been mostly, as in Parakysinae, Cetopsidae and Aspredinidae, or even completely been reduced, as in the aspredinids *Micromyzon* (Friel and Lundberg 1996) and *Xyliphius* (Carvalho et al. 2017), or in the cetopsids *Cetopsis oliveirai* (Lundberg and Rapp Py-Daniel 1994). Furthermore, eye reduction or complete loss is common in about 50 species of cave-dwelling catfishes (Proudlove 2010; Trajano and Secutti this volume). Nevertheless, to increase light yield, most catfishes possess a tapetum lucidum (Nicol et al. 1973).

The larger a fish's eye, the longer the controlling muscles have to be. Therefore, catfishes require relatively small muscle portions, and a myodom for housing them is absent. Large eyes of pelagic catfish taxa are secondary, as in Schilbeidae, Pangasiidae, Auchenipteridae, *Kryptopterus* (Siluridae) or *Hypophthalmus* (Pimelodidae). In these cases, the corresponding muscles are also larger, and the eyes have a lateral position, allowing them to cover a field of vision above and below the animal. While most fish are unable to regulate pupil opening, all but a few loricariids regulate it by a dorsal flap of skin called omega operculum, which can expand or shrink (Armbruster 1998; Douglas et al. 2002). *Nematogenys* seems to regulate light incidence by varying pigmentation in the skin above the eye (G. Arratia pers. comm.).

Lateral-Line System, Hearing and Equilibrioception

Hearing, reception of water motion, and equilibrioception are highly connected in fishes and are all perceived by neuromasts. Partly, there is an overlap in these sensory systems, as otoliths for orientation in space may also vibrate on sound oscillations, and acoustic waves may also trigger the lateral-line system. For enhanced hearing abilities of catfishes, see Ladich (2023).

For teleosts, catfishes typically have three semicircular canals and three otoliths per side, i.e., sagitta, lapillus, and astericus. While in non-Ostariophysi, the sagitta is the largest otolith, in Ostariophysi, it is the lapillus (Long and Stewart 2010).

The lateral line of catfishes usually runs straight along the flanks. In auchenipterids, it is pronounced sinusoidal (Ferraris 1988; Birindelli 2014), and in silurids, pangasiids, and *Hypophthalmus* ramified (Howes 1983a). Several taxa, such as callichthyids, amblycipitids, or *Scoloplax*, have a much-reduced lateral line (Reis 1998; Schaefer et al. 1989). Loricariids are unique in that they have direct contact with the lateral-line canal with the swim-bladder membrane (Aquino and Schaefer 2002).

Catfishes, in general, have reduced the supratemporal canal (Lekander 1949) but gained a pterotic branch on the postotic canal (Schaefer and Aquino 2000); the latter got secondarily lost in *Scoloplax* and *Astroblepus* (Schaefer and Aquino 2000). Loricarioids, sisorids, chiloglanidines, and most aspredinids lost the latero-sensory canal on the lower jaw (Schaefer 1990; de Pinna 1996; Brinidelli 2014; Calegari et al. 2019), and some taxa lost the complete preopercular-mandibular canal, like trichomycterids and *Scoloplax* (Adriaens et al. 2010). Besides sensory cells enclosed in the lateral-line canals, catfishes possess variable numbers and patterns of superficial neuromasts, which can be isolated or arranged in lines or fields (Arratia and Huaquín 1995). They are innervated via the same nerves as the lateral-line system (Gibbs 2004).

Electroreception

While electroreception in lower vertebrates is common, it evolved only twice among teleosts: within Osteoglossomorpha and Siluriphysi (Gibbs 2004; Baker et al. 2013). The ampullary electroreceptors of Siluriphysi are optimised for receiving low frequencies below 30 Hz and are used for passive electrolocation and detecting biopotentials of prey organisms (Zakon 1986, 1988). They have evolved from the lateral line system and are, therefore, innervated by the lateral line nerves (Gibbs 2004). Electroreceptors are present on the body and more densely on the head, but their exact distribution is known only in a few catfishes. The ampullary receptors are formed by a variable number of electrosensitive cells which are connected via a jelly-filled canal to a pore in the skin. In the marine *Plotosus*, these canals are much longer than in other catfishes (Obara 1976). Ariids, which alter between freshwater and marine environments, possess two different types of electroreceptors (Whitehead et al. 2015). Within catfishes, tuberous receptors have been only found in a blind *Cetopsis* species (Andres et al. 1988).

Tasting and Smelling

Smelling and tasting are well-developed in catfishes. The olfactory lamellae are large and numerous. In cetopsids, the olfactory organ is well-developed and surrounded by a large cartilaginous capsule (de Pinna and Vari 1995). Pheromones may play an important role in catfish behaviour but are hardly studied. Axillary glands in all types of catfishes (Wright 2015) and tissue on the pectoral fins of sexually active male callichthyids may be involved in pheromone production (Reis 1988).

Taste buds are distributed all over the body, including the mouth cavity, but are dense on the barbels (Ovalle and Shinn 1977; Arratia and Huaquín 1995).

Conclusion

Catfishes apparently evolved some 100 to 130 million years ago in the early Cretaceous (Lavoué 2020; Arratia 2025). Independent of various ostariophysan phylogenetic hypotheses, they descended from fishes looking similar to today's generalised characids. Changing to a bottom-oriented, nocturnal lifestyle highly affected their morphology and apparently was key to successful radiation. Striking is the development of a depressed, scale-less body, barbels, electrosensitivity, and also the possession of defensive fin spines (Alexander 1965).

Today, catfishes comprise more than 4,200 species, almost twelve per cent of all teleosts. Various catfish lineages underwent diverse adaptations. Much deviating from the typical catfishes with a predatory lifestyle, the food-scraping, armoured loricariids and the rather small, loach-like trichomycterids represent successful lineages comprising one-quarter and one-eighth of all siluriform species, respectively. Quite some catfishes secondarily returned to a free-swimming lifestyle (Howes 1983a), regaining some typical pelagic characters, like lateral compression, small pelvic girdles, or large eyes, and keeping the advantageous siluriform innovations, like electrosensitivity, long barbels, and defensive spines.

The morphology of catfishes is highly diverse, and most character complexes have been superficially treated in this review. Many anatomical systems remain understudied, especially soft tissues. Despite the numerous, multifaceted studies on siluriforms over the last decades, catfish morphology still leaves numerous open questions regarding functional morphology and character diversity, ontogeny, and evolution.

References

Abrahão, V.P. and O.A. Shibatta. 2015. Gross morphology of the brain of *Pseudopimelodus bufonius* (Valenciennes, 1840) (Siluriformes: Pseudopimelodidae). Neotrop. Ichthyol. 13: 255–264.

Abrahão, V.P., F. Müller Pupo and O.A. Shibatta. 2018. Comparative brain gross morphology of the Neotropical catfish family Pseudopimelodidae (Osteichthyes, Ostariophysi, Siluriformes), with phylogenetic implications. Zool. J. Linn. Soc. 184: 750–772.

Adriaens, D. and W. Verraes. 1997. Ontogeny of the maxillary barbell muscle in *Clarias gariepinus* (Siluroidei: Clariidae), with some notes on the palatine-maxillary mechanism. J. Zool., Lond. 241: 117–133.

Adriaens, D. and W. Verraes. 1998. Ontogeny of the osteocranium in the African catfish, *Clarias gariepinus* Burchell (1822) (Siluriformes: Clariidae): ossification sequence as a response to functional demands. J. Morph. 235: 183–237.

Adriaens, D., J.N. Baskin and H. Coppens. 2010. Evolutionary morphology of trichomycterid catfishes: about hanging on and digging in. pp. 337–362. *In*: Nelson, J.S., H.-P. Schultze and M.V.H. Wilson (eds.). Origin and Phylogenetic Interrelationships of Teleosts. Verlag Dr. Friedrich Pfeil, München, Germany.

Albert, J.S., M.J. Lannoo and T. Yuri. 1998. Testing hypotheses of neural evolution in gymnotiform electric fishes using phylogenetic character data. Evol. 52: 1760–1780.

Albert, J.S. 2001. Species Diversity and Phylogenetic Systematics of American Knifefishes (Gymnotiformes, Teleostei). University of Michigan Museum of Zoology, Michigan, USA.

Albert, J.S. and W.G.R. Crampton. 2005. Electroreception and electrogenesis. pp. 431–472. *In*: Evans, D.H. and J.B. Claiborne (eds.). The Physiology of Fishes. Taylor & Francis, Boca Raton, USA.

Allis, E.P. Jr. 1919. The myodome and trigenimo-facialis chamber of fishes and the corresponding cavities in higher vertebrates. J. Morph. 32: 207–326.

Alexander, R.M. 1965. Structure and function in the catfish. J. Zool. 148: 88–152.

Andres, K.H., M. von Düring and E. Petrasch. 1988. The fine structure of ampullary and tuberous electroreceptors in the South American blind catfish *Pseudocetopsis* spec. Anatomy and Embryology 177: 523–535.

Angulo, A. and F. Langeani. 2017. Gross brain morphology of the armoured catfish *Rineloricaria heteroptera*, Isbrucker and Nijssen (1976), (Siluriformes: Loricariidae: Loricariinae): A descriptive and quantitative approach. J. Morph. 278: 1689–1705.

Aquino, A.E. and S.A. Schaefer. 2002. The temporal region of the cranium of loricarioid catfishes (Teleostei: Siluriformes): morphological diversity and phylogenetic significance. Zool. Anz. 241: 223–244.

Armbruster, J.W. 1998. Modifications of the digestive tract for holding air in loricariid and scoloplacid catfishes. Copeia 1998: 663–675.

Armbruster, J.W. 2004. Phylogenetic relationships of the suckermouth armoured catfishes (Loricariidae) with emphasis on the Hypostominae and the Ancistrinae. Zool. J. Linnean Soc. 141: 1–80.

Armbruster, J.W. 2011. Global catfish biodiversity. Am. Fish. Soc. Symp. 77: 15–37.

Arratia, G. 1987. Description of the primitive family Diplomystidae (Siluriformes, Teleostei, Pisces): morphology, taxonomy and phylogenetic implications. Bonner Zool. Monogr. 24: 1–120.

Arratia, G. and H.-P. Schultze. 1990. The urohyal: development and homology within osteichthyans. J. Morph. 203: 247–282.

Arratia, G. and H.-P. Schultze. 1991. Palatoquadrate and its ossifications: development and homoloy within osteichthyans. J. Morph. 208: 1–81.

Arratia, G. 1992. Development and variation of the suspensorium of primitive catfishes (Teleostei: Ostariophysi) and their phylogenetic relationships. Bonner Zool. Monogr. 32: 1–149.

Arratia, G. and M. Gayet. 1995. Sensory canals and related bones of tertiary siluriform crania from Bolivia and North America and comparison with recent forms. J. Vert. Paleontol. 15: 482–505.

Arratia, G. and L. Huaquín. 1995. Morphology of the lateral, line system and of the skin of diplomystid and systematic and ecological considerations. Bonn Zool. Monogr. 36: 1–109.

Arratia, G. 1998. *Silvinichthys*, a new genus of trichomycterid catfishes from the Argentinian Andes, with redescription of *Trichomycterus nigricans*. Ichthyol. Explor. Freshwaters 9: 347–370.

Arratia, G. 2003a. Catfish head skeleton—an overview. pp. 3–46. *In*: Arratia, G., B.G. Kapoor, M. Chardon and R. Diogo (eds.). Catfishes. Science Publishers, Enfield, USA.

Arratia, G. 2003b. The siluriform postcranial skeleton—an overview. pp. 121–157. *In*: Arratia, G., B.G. Kapoor, M. Chardon and R. Diogo (eds.). Catfishes. Science Publishers, Enfield, USA.

Arratia, G. 2003c. The skin of catfishes—a review. pp. 177–199. *In*: Arratia, G., B.G. Kapoor, M. Chardon and R. Diogo (eds.). Catfishes. Science Publishers, Enfield, USA.

Arratia, G. and C. Quezada-Romegialli. 2017. Understanding morphological variability in a taxonomic context in Chilean diplomystids (Teleostei: Siluriformes), including the description of a new species. PeerJ 5: e2991.

Arratia, G. 2025. Siluriforms and the South American fossil record: their significance and the origin and evolutionary radiation of the group. *In*: Arratia, G. and R.E. Reis (eds.). Catfishes, a Highly Diversified Group. Volume 2: Evolution and Phylogeny. CRC Press, Boca Raton, FL.

Artoni, R.F., P.C. Venere and L.A.C. Bertollo. 1998. A heteromorphic ZZ/ZW sex chromosome system in fish, genus *Hypostomus* (Loricariidae). Cytologia 63: 421–425.

Atema, J. 1971. Structures and functions of the sense of taste in the catfish (*Ictalurus natalis*). Brain, Behav. Evol. 4: 273–294.

Baker, C.V.H., M.S. Modrell and J.A. Gillis. 2013. The evolution and development of vertebrate lateral line electroreceptors. J. Exp. Biol. 216: 2515–2522.

Baron, V.D., A.A. Orlov and A.S. Golubtsov. 1994. African *Clarias* catfish elicits long-lasting weak electric pulses. Experientia 50: 644–647.

Baron, V.D., A.A. Orlov and D.E. El'yashev. 2005. Investigations of electric and acoustic activity of some representatives of the ichthyofauna of South Vietnam. J. Ichthyol. 45: S271–S279.

Bass, A.H. 1981. Organization of the telencephalon in the channel catfish, *Ictalurus puncatus*. J. Morph. 169: 71–90.

Benjamin, M. 1990. The cranial cartilages of teleosts and their classification. J. Anat. 169: 153–172.

Bernt, M.J. and M.L.J. Stiassny. 2022. A new species of air-breathing catfish (Clariidae: *Clarias*) from Salonga National Park, Democratic Republic of the Congo. Am. Mus. Nov. 3990: 1–20.

Birindelli, J.L.O., L.M. Sousa and M.H. Sabaj Pérez. 2009. Morphology of the gas bladder in thorny catfishes (Siluriformes: Doradidae). Proc. Acad. Nat. Sci. 158: 261–296.

Birindelli, J.L.O., A. Akama and H.A. Britski. 2012. Comparative morphology of the gas bladder in driftwood catfishes (Siluriformes: Auchenipteridae). J. Morph. 273: 651–660.

Birindelli, J.L.O. 2014. Phylogenetic relationships of the South American Doradoidea (Ostariophysi: Siluriformes). Neotrop. Ichthyol. 12: 451–564.

Bornbusch, A.H. 1991. Monophyly of the catfish family Siluridae (Teleostei: Siluriformes), with a critique of previous hypotheses of the family's relationships. Zool. J. Linn. Soc. 101: 105–120.

Bridge, T.W. and A.C. Haddon. 1892. Contributions to the anatomy of fishes. II. The air-bladder and Weberian ossicles in siluroid fishes. Proc. R. Soc. Lond. 52: 315–320.

Britz, R. and M. Hoffmann. 2006. Ontogeny and homology of the claustra in ostophysan Ostariophysi (Teleostei). J. Morph. 267: 909–923.

Britz, R., F. Kakkassery and R. Raghavan. 2014. Osteology of *Kryptoglanis shajii*, a stygobitic catfish (Teleostei: Siluriformes) from peninsular India, with diagnosis of a new family Kryptoglanidae. Ichthyol. Explor. Freshwaters 24: 193–207.

Buddington, R.K. and J.M. Diamond. 1986. Aristotle revisited: The function of pyloric caeca in fish. PNAS 83: 8012–8014.

Calegari, B.B., R.P. Vari and R.E. Reis. 2019. Phylogenetic systematics of the driftwood catfishes (Siluriformes: Auchenipteridae): a combined morphological and molecular analysis. Zool. J. Linn. Soc. 187: 661–773.

Carvalho, T.P., R.E. Reis and M.H. Sabaj. 2017. Description of a new blind and rare species of *Xyliphius* (Siluriformes: Aspredinidae) from the Amazon basin using high-resolution computed tomography. Copeia 2017: 14–28.

Costa, W.J. 2021. Comparative osteology, phylogeny and classification of the eastern South American catfish genus *Trichomycterus* (Siluriformes: Trichomycteridae). Taxonomy 1: 160–191.

Crampton, W.G.R. 2019. Electroreception, electrogenesis and electric signal evolution. J. Fish Biol. 95: 92–134.

Cumplido, N., M.L. Allende and G. Arratia. 2020. From devo to evo: patterning, fusion and evolution of the zebrafish terminal vertebra. Front. Zool. 17: 18.

Cumplido, N. 2025. The caudal fin musculature of catfishes and the homology of the posterior myomeres. pp. 133–159. *In*: Arratia, G. and R.E. Reis (eds.). Catfishes, a Highly Diversified Group. Volume 1: Their Outstanding Biology. CRC Press, Boca Raton, FL.

Datovo, A. and F.A. Bockmann. 2010. Dorsolateral head muscles of the catfish families Nematogenyidae and Trichomycteridae (Siluriformes: Loricarioidei): comparative anatomy and phylogenetic analysis. Neotrop. Ichthyol. 8: 193–246.

Datovo, A. and R.P. Vari. 2013. The jaw adductor muscle complex in teleostean fishes: evolution, homologies and revised nomenclature (Osteichthyes: Actinopterygii). PLoS ONE 8: e60846.

Datovo, A. and R.P. Vari. 2014. The adductor mandibulae muscle complex in lower teleostean fishes (Osteichthyes: Actinopterygii): comparative anatomy, synonymy, and phylogenetic implications. Zool. J. Linn. Soc. 171: 554–622.

Delariva, R.L. and A.A. Agostinho. 2001. Relationship between morphology and diets of six neotropical loricariids. J. Fish Biol. 58: 832–847.

Diogo, R. and M. Chardon. 2000a. Homologies among different adductor mandibulae sections of teleostean fishes, with special regard to catfishes (Teleostei: Siluriformes). J. Morph. 243: 193–208.

Diogo, R. and M. Chardon. 2000b. The structures associated with catfish (Teleostei: Siluriformes) mandibular barbels: origin, anatomy, function, taxonomic distribution, nomenclature and synonymy. Neth. J. Zool. 50: 455–478.

Diogo, R., C. Oliveira and M. Chardon. 2001. On the osteology and myology of catfish pectoral girdle, with a reflection on catfish (Teleostei: Siluriformes) plesiomorphies. J. Morph. 249: 100–125.

DoNacimiento, C. and N. Milani. 2008. The Venezuelan species of *Phenacorhamdia* (Siluriformes: Heptapteridae), with the description of two new species and a remarkable new tooth morphology for siluriforms. Proc. Acad. Nat. Sci. 157: 16–33.

Douglas, R.H., S.P. Collin and J. Corrigan. 2002. The eyes of suckermouth armoured catfish (Loricariidae, subfamily *Hypostomus*): pupil response, lenticular longitudinal spherical aberration and retinal topography. J. Exp. Biol. 205: 3425–3433.

Duclos, K. and T. Grande. 2025. Siluriform Weberian apparatus: modularity and integration. pp. 81–102. *In*: Arratia, G. and R.E. Reis (eds.). Catfishes, a Highly Diversified Group. Volume 1: Their Outstanding Biology. CRC Press, Boca Raton, FL.

Egge, J. 2025. Siluriform venom systems. pp. 160–181. *In*: Arratia, G. and R.E. Reis (eds.). Catfishes, a Highly Diversified Group. Volume 1: Their Outstanding Biology. CRC Press, Boca Raton, FL.

Eigenmann, C.H. 1918. The Pygidiidae, a family of South American catfishes. Mem. Carnegie Mus. 7: 259–398.

Ferraris, C.J. Jr. 1988. The Auchenipteridae: Putative Monophyly and Systematics with a Classification of the Neotropical Doradoid Catfishes (Ostariophysi: Siluriformes). Ph.D. Thesis, The City University of New York, New York, USA.

Ferraris, C.J. and R.P. Vari. 2012. Review of the African catfish genus *Andersonia* (Teleostei: Siluriformes). Zootaxa 3210: 39–49.

Ferreira, M., C. Garcia, D.A. Matoso, I.S. de Jesus and E. Feldberg. 2016. A new multiple sex choromsome system $X_1X_1X_2X_2/X_1Y_1X_2Y_2$ in Siluriformes: cytogenetic characterization of *Bunocephalus coracoideus* (Aspredinidae). Genetica 144: 591–599.

Fine, M.L., J.P. Friel, D. McElroy, C.B. King, K.E. Loesser and S. Newton. 1997. Pectoral spine locking and sound production in the channel catfish *Ictalurus punctatus*. Copeia 1997: 777–790.

Fink, S.V. and W.L. Fink. 1981. Interrelationships of the ostariophysan fishes (Teleostei). Zool. J. Linn. Soc. 72: 297–353.

Fink, S.V. and W.L. Fink. 1996. Interrelationships of the ostariophysan fishes (Teleostei). pp. 209–249. *In*: Stiassny, M.L.J., L.R. Parenti and G.D. Johnson (eds.). Interrelationships of Fishes. Academic Press, San Diego, USA.

Fricke, R., W.N. Eschmeyer and J.D. Fong. 2024. Eschmeyer's catalog of fishes: genera/species by family/subfamily. (http://researcharchive.calacademy.org/research/ichthyology/catalog/SpeciesByFamily.asp) (accessed Jan/2024).

Friel, J.P. 1994. A Phylogenetic Study of the Neotropical Banjo Catfishes (Teleostei: Siluriformes: Aspredinidae). Ph.D. Thesis, Duke University, Durham, USA.

Friel, J.P. and J.G. Lundberg. 1996. *Micromyzon akamai*, gen. et sp. nov., a small and eyeless banjo catfish (Siluriformes: Aspredinidae) from the river channels of the lower Amazon basin. Copeia 1996: 641–648.

Friel, J.P. and T.R. Vigliotta. 2006. *Synodontis acanthoperca*, a new species from the Ogôoué River system, Gabon with comments on spiny ornamentation and sexual dimorphism in mochokid catfishes (Siluriformes: Mochokidae). Zootaxa 1125: 45–56.

Galant, J.R. 2019. The evolution and development of electric organs. pp. 91–123. *In*: Carlson, B.A., J.A. Sisneros, A.N. Popper and R.R. Fay (eds.). Electroreception: Fundamental Insights from Comparative Approaches. Springer, New York, USA.

Geerinckx, T. and D. Adriaens. 2006. The erectile cheek-spine apparatus in the bristlenose catfish *Ancistrus* (Loricariidae, Siluriformes), and its relation to the formation of a secondary skull roof. Zool. 109: 287–299.

Geerinckx, T., M. Brunain, A. Herrel, P. Aerts and D. Adriaens. 2007. A head with a suckermouth: a functional-morphological study of the head of the suckermouth armoured catfish *Ancistrus* cf. *triradiatus* (Loricariidae, Siluriformes). Belg. J. Zool. 137: 47–66.

German, D.P. 2009. Inside the guts of wood-eating catfishes: can they digest wood? J. Comp. Physiol. B 179: 1011–1023.

Ghiot, F. and N. Bouchez. 1980. The Central Rod of the Barbels of a South American Catfish, *Pimelodus clarias*. Copeia 1980: 908–909.

Gibbs, M.A. 2004. Lateral line receptors: where do they come from developmentally and where is our research going? Brain Behav. Evol. 64: 163–181.

Goulding, M. 1980. The Fishes and the Forest: Explorations in Amazonian Natural History. University of California Press, Berkeley, USA.

Grande, L. 1987. Redescription of *Hypsidoris farsonensis* (Teleostei: Siluriformes), with a reassessment of its phylogenetic relationships. J. Vert. Paleontol. 7: 24–54.

Hagedorn, M., M. Womble and T.E. Finger. 1990. Synodontid catfish: a new group of weakly electric fish. Brain Behav. Evol. 35: 268–277.

Hilton, E.J. 2011. Bony fish skeleton. pp. 434–448. *In*: Farrell, A.P. (ed.). Encyclopedia of Fish Physiology: From Genome to Environment. Volume 1. Academic Press, San Diego, USA.

Hoffmann, M. and R. Britz. 2006. Ontogeny and homology of the neural complex of otophysan Ostariophysi. Zool. J. Linn. Soc. 147: 301–330.

Howes, G.J. 1983a. Problems in catfish anatomy and phylogeny exemplified by the Neotropical Hypophthalmidae (Teleostei: Siluroidei). Bull. Br. Mus. nat. Hist. (Zool.) 45: 1–39.

Howes, G.J. 1983b. The cranial muscles of loricarioid catfishes, their homologies and value as taxonomic characters (Teleostei: Siluroidei). Bull. Br. Mus. nat. Hist. (Zool.) 45: 309–345.

Huysentruyt, F., T. Geerinckx, M. Brunain and D. Adriaens. 2011. Development of the osteocranium in *Corydoras aeneus* (Gill, 1858) Callichthyidae, Siluriformes. J. Morph. 272: 573–582.

Ito, H., Y. Ishikawa, M. Yoshimoto and N. Yamamoto. 2007. Diversity of brain morphology in teleosts: brain and ecological niche. Brain Behav. Evol. 69: 76–86.

Johnson, G.D. and C. Patterson. 1993. Percomorph phylogeny: A survey of acanthomorphs and a new proposal. Bull. Mar. Sci. 52: 554–626.

Kaatz, I.M., D.J. Stewart, A.N. Rice and P.S. Lobel. 2010. Differences in pectoral fin spine morphology between vocal and silent clades of catfishes (Order Siluriformes): ecomorphological implications. Curr. Zool. 56: 73–89.

Kaatz, I., P. Lobel and A. Rice. 2025. Sound production and communication in catfishes. pp. 299–320. *In*: Arratia, G. and R.E. Reis (eds.). Catfishes, a Highly Diversified Group. Volume 1: Their Outstanding Biology. CRC Press, Boca Raton, FL.

Kapoor, B.G., B. Khanna, R. Diogo, E. Parmentier and M. Chardon. 2003. Internal anatomy of catfishes. pp. 159–176. *In*: Arratia, G., B.G. Kapoor, M. Chardon and R. Diogo (eds.). Catfishes. Science Publishers, Enfield, USA.

Kéver, L., E. Parmentier, A.H. Bass and B.P. Chagnaud. 2020. Morphological diversity of acoustic and electric communication systems of mochokid catfish. J. Comp. Neurol. 529: 1787–1809.

Kubicek, K.M., R. Britz and K.W. Conway. 2019. Ontogeny of the catfish pectoral-fin spine (Teleostei: Siluriformes). J. Morph. 280: 339–359.

Kubicek, K.M. 2022. Developmental osteology of *Ictalurus punctatus* and *Noturus gyrinus* (Siluriformes: Ictaluridae) with a discussion of siluriform bone homologies. Vertebr. Zool. 72: 661–727.

Ladich, F. 2023. Hearing in catfishes: 200 years of research. Fish Fish. 24: 618–634.

Lavoué, S. 2020. Origins of Afrotropical freshwater fishes. Zool. J. Linn. Soc. 188: 345–411.

Lekander, B. 1949. The sensory line system and the canal bones in the head of some Ostariophysi. Acta Zool. 30: 1–131.

Liu, Z.H., Y.G. Zhang and D.S. Wang. 2010. Studies on feminization, sex determination, and differentiation of the Southern catfish, *Silurus meridionalis*—a review. Fish Physio. Biochem. 36: 223–235.

Loir, M., C. Cauty, P. Planquette and P.Y. le Bail. 1989. Comparative study of the male reproductive tract in seven families of South-American catfishes. Aquat. Living Resour. 2: 45–56.

Long, J.M. and D.R. Stewart. 2010. Verification of otolith identity used by fisheries scientists for aging channel catfish. Trans. Am. Fish. Soc. 139: 1775–1779.

Lundberg, J.G. and J.N. Baskin. 1969. The caudal skeleton of the catfishes, order Siluriformes. Am. Mus. Novit. 2398: 1–49.

Lundberg, J.G. and L. Rapp Py-Daniel. 1994. *Bathycetopsis oliveirai*, gen. et sp. nov., a blind and depigmented catfish (Siluriformes: Cetopsidae) from the Brazilian Amazon. Copeia 1994: 381–390.

Lundberg, J.G., K.R. Luckenbill, K.K. Subhash Babu and H.H. Ng. 2014. A tomographic osteology of the taxonomically puzzling catfish *Kryptoglanis shaji* (Siluriformes, Siluroidei, incertae sedis): description and a first phylogenetic interpretation. Proc. Acad. Nat. Sci. Phil. 163: 1–42.

Mahajan, C.L. 1967. *Sisor rabdophorus*—a study in adaptation and natural relationship. III. The vertebral column, median fins and their musculature. J. Zool., Lond. 152: 297–318.

Meek, J. and R. Nieuwenhuys. 1998. Holosteans and teleosts. pp. 759–937. *In*: Nieuwenhuys, R., H.J. Donkelaar and C. Nicholson (eds.). The Central Nervous System of Vertebrates. Springer, Berlin, Germany.

McAllister, D.E. 1968. Evolution of branchiostegals and classification of teleostome fishes. Bull. Natl. Mus. Can. 221: 1–239.

Melo. R.M.C., F.P. Arantes, Y. Sato, J.E. dos Santos, E. Rizzo and N. Bazzoli. 2011. Comparative morphology of the gonadal structure related to reproductive strategies in six species of Neotropical catfishes (Teleostei: Siluriformes). J. Morph. 272: 525–535.

Mello, G.C.G., M.L. Santos, F.P. Arantes, T.C. Pessali, M.F.G. Brito and J.E. Santos. 2017. Morphological characterisation of the digestive tract of the catfish *Lophiosilurus alexandri* Steindachner, 1876 (Siluriformes, Pseudopimelodidae). Acta Zool. 100: 14–23.

Meisner, A.D., J.R. Burns, S.H. Weitzman and L.R. Malabarba. 2000. Morphology and histology of the male reproductive system in two species of internally inseminating South American catfishes, *Trachelyopterus lucenai* and *T. galeatus* (Teleostei: Auchenipteridae). J. Morph. 246: 131–141.

Miano, J.P., K.E. Loesser-Casey and M.L. Fine. 2013. Description and scaling of pectoral muscles in ictalurid catfishes. J. Morph. 274: 467–477.

Mo, T.-P. 1991. Anatomy, Relationships and Systematics of the Bagridae (Teleostei: Siluroidei) with a Hypothesis of Silurid Phylogeny. Koeltz Scientific Books, Koenigstein, Germany.

Mori, S. and T. Nakamura. 2022. Redeployment of odontode gene regulatory network underlies dermal denticle formation and evolution in suckermouth armored catfish. Sci. Rep. 12: 6172.

Moritz, T., A.-K. Koch and P. Thieme. 2025. Diversity and evolution of the pelvic girdle and fin in catfishes. pp. 103–132. *In*: Arratia, G. and R.E. Reis (eds.). Catfishes, a Highly Diversified Group. Volume 1: Their Outstanding Biology. CRC Press, Boca Raton, FL.

Nelson, G.J. 1969. Infraorbital bones and their bearing on the phylogeny and geography of osteoglossomorph fishes. Am. Mus. Novit. 2394: 1–37.

Nguyen, D.H.M., T. Panthum, J. Ponjarat, N. Laopichienpong, E. Kraichak, W. Singchat et al. 2021. An investigation of ZZ/ZW and XX/XY sex determination system in North African catfish (*Clarias gariepinus*, Burchell, 1822). Front. Genet. 11: 562856.

Nicol, J.A.C., H.J. Arnott and A.C.G. Best. 1973. Tapeda lucida in bony fishes (Actinopterygii): a survey. Can. J. Zool. 51: 69–81.

Nilsson, S. 2011. Autonomic nervous system of fishes. pp. 80–88. *In*: Farrell, A.P. (ed.). Encyclopedia of Fish Physiology. Elsevier, London, UK.

Obara, S. 1976. Mechanisms of electroreception in ampullae of Lorenzini of the marine catfish *Plotosus*. pp. 128–147. *In*: Reuben, J.P. and D.P. Purpura (eds.). Electrobiology of Nerve, Synapse and Muscle. Raven Press, New York, USA.

Olshanskiy, V.M., O.A. Soldatova, K.S. Morshnev and N.T. Nga. 2009. Electrical activity of Asian catfish *Clarias macrocephalus* (Claridae, Siluriformes) during spawning behavior. Dokl. Biol. Sci. 429: 554–558.

Olson, K.R., J.S.D. Munshi, T.K. Ghosh and J. Ojha. 1990. Vascular organization of the head and respiratory organs of the air-breathing catfish, *Heteropneustes fossilis*. J. Morph. 203: 165–179.

Orlov, A.A. and V.D. Baron. 2005. Responses of the electrogeneration system of *Synodontis* (Mochokidae, Siluriformes) to weak electric fields. Dokl. Biol. Sci. 403: 566–569.

Ovalle, W.K. and S.L. Shinn. 1977. Surface morphology of taste buds in catfish barbels. Cell Tis. Res. 178: 375–384.

Pace, C.M. and A.C. Gibbs. 2014. Sustained periodic terrestrial locomotion in air-breathing fishes. J. Fish Biol. 84: 639–660.

Patiño, R., K.B. Davis, J.E. Schoore, C. Uguz, C.A. Strüssmann, N.C. Parker et al. 1996. Sex differentiation of channel catfish gonads: normal development and effects of temperature. J. Exp. Zool. 276: 209–218.

Pfeiffer, W. 1963. Alarm substances. Experientia 19: 113–123.

Pfeiffer, W. and J.F. Eisenberg. 1965. Die Lauterzeugung der Dornwelse (Doradidae) und der Fiederbartwelse (Mochokidae). Z. Morph. Ökol. Tiere 54: 669–679.

Pfeiffer, W. 1967. Schreckreaktion und Schreckstoffzellen bei Ostariophysi und Gonorhynchiformes. Z. vergl. Physiol. 56: 380–396.

de Pinna, M.C. 1989. A new sarcoglanidine catfish, phylogeny of its subfamily, and an appraisal of the phyletic status of the Trichomycterinae (Teleostei, Trichomycteridae). Am. Mus. Nov. 2950: 1–39.

de Pinna, M.C. 1993. Higher-level Phylogeny of Siluriformes, with a New Classification of the Order (Teleostei, Ostariophysi). Ph.D. Thesis, University of New York, New York, USA.

de Pinna, M.C. and R.P. Vari. 1995. Monophyly and phylogenetic diagnosis of the family Cetopsidae, with synonymization of the Helogenidae (Teleostei: Siluriformes). Smithson. Contrib. Zool. 571: 1–26.

de Pinna, M.C. 1996. A phylogenetic analysis of the Asian catfish families Sisoridae, Akysidae, and Amblycipitidae, with a hypothesis on the relationships of the Neotropical Aspredinidae (Teleostei, Ostariophysi). Fieldiana Zool. 84: 1–83.

de Pinna, M.C. 1998. Phylogenetic relationships of Neotropical Siluriformes (Teleostei: Ostariophysi): historical overview and synthesis of hypotheses. pp. 279–330. *In*: Malabarba, L.R., R.E. Reis, R.P. Vari, Z.M.S. Lucena and C.A.S. Lucena (eds.). Phylogeny and Classification of Neotropical Fishes. EDIPUCRS, Porto Alegre, Brazil.

de Pinna, M.C. and H.H. Ng. 2004. The second ural centrum in Siluriformes and its implication for the monophyly of superfamily Sisoridae (Teleostei, Ostariophysi). Am. Mus. Nov. 3437: 1–23.

de Pinna, M.C., V. Reis and H. Britski. 2020. A new species of *Trichogenes* (Siluriformes, Trichomycteridae), with a discussion on the homologies on the anterior orbital bone in trichomycteridae and other loricarioids. Am. Mus. Nov. 3951: 1–27.

Pinton, A., S. Le Fur and O. Otero. 2016. Evaluation of the fossil fish-specific diversity in a Chadian continental assemblage: exploration of morphological continuous variation in *Synodontis* (Ostariophysi, Siluriformes). J. Morph. 277: 1486–1496.

Proudlove, G.S. 2010. Biodiversity and distribution of the subterranean fishes of the world. pp. 41–64. *In*: Trajano, E., M.E. Bichuette and B.G. Kapoor (eds.). Biology of Subterranean Fishes. CRC Press, Boca Raton, USA.

Rankin, C.H. and P. Moller. 1992. Temporal patterning of electric organ discharges in the African electric catfish, *Malapterurus electricus* (Gmelin). J. Fish Biol. 40: 49–58.

Reed, H.D. 1924. The morphology and growth of spines of siluroid fishes. J. Morph. 38: 431–451.

Reis, R.E. 1998. Anatomy and phylogenetic analysis of the neotropical callichthyid catfishes (Ostariophysi, Siluriformes). Zool. J. Linn. Soc. 124: 105–168.

Rivera-Rivera, C.J. and J.I. Montoya-Burgos. 2017. Trunk dental tissue evolved independently from underlying dermal bony plates but is associated with surface bones in living odontode-bearing catfish. Proc. R. Soc. B. 284: 20171831.

Rizzato, P. and M.A. Almeida. 2025. An overview of some of the main features of the external anatomy of catfishes. pp. 19–47. *In*: Arratia, G. and R.E. Reis (eds.). Catfishes, a Highly Diversified Group. Volume 1: Their Outstanding Biology. CRC Press, Boca Raton, FL.

Roberts, T.R. 1982. Unculi (horny projections arising from single cells), an adaptive feature of the epidermis of ostariophysan fishes. Zool. Scr. 11: 55–76.

Rocha, M.S., R. Ribeiro de Oliveira and L.H. Rapp Py-Daniel. 2008. *Scoloplax baskini*: a new spiny dwarf catfish from rio Aripuanã, Amazonas, Brazil (Loricarioidei: Scoloplacidae). Neotrop. Ichthyol. 6: 323–328.

Rocha, M., H. Lazzarotto and L. Rapp Py-Daniel. 2012. A new species of *Scoloplax* with a remarkable new tooth morphology within Loricarioidea (Siluriformes: Scoloplacidae). Copeia 2012: 670–677.

Rodiles-Hernández, R., D.A. Hendrickson, J.G. Lundberg and J.M. Humphries. 2005. *Lacantunia enigmatica* (Teleostei: Siluriformes) a new and phylogenetically puzzling freshwater fish from Mesoamerica. Zootaxa 1000: 1–24.

Sabaj, M.H. and M. Arce H. 2021. Towards a complete classification of the Neotropical thorny catfishes (Siluriformes: Doradidae). Neotrop. Ichthyol. 19: e210064.

Sarmento-Soares, L.M. and R.F. Martins-Pinheiro. 2020. A reappraisal of phylogenetic relationships among auchenipterid catfishes of the subfamily Centromochlinae and diagnosis of its genera (Teleostei: Siluriformes). Proc. Acad. Nat. Sci. 167: 85–146.

Schaefer, S.A. and G.V. Lauder. 1986. Historical transformation of functional design: evolutionary morphology of feeding mechanisms in loricarioid catfishes. Syst. Zool. 35: 489–508.

Schaefer, S.A. 1987. Osteology of *Hypostomus plecostomus* (Linnaeus): with a phylogenetic analysis of the loricariid subfamilies (Pisces, Siluroidei). Contrib. Sci. 394: 1–31.

Schaefer, S.A., S.H. Weitzman and H.A. Britski. 1989. Review of the Neotropical catfish genus *Scoloplax* (Pisces: Loricarioidea: Scoloplacidae) with comments on reductive characters in phylogenetic analysis. Proc. Acad. Nat. Sci. Philadelphia 141: 181211.

Schaefer, S.A. 1990. Anatomy and relationships of the scoloplacid catfishes. Proc. Acad. Nat. Sci. 142: 167–210.

Schaefer, S.A. 1991. Phylogenetic analysis of the loricariid subfamily Hypoptopomatinae (Pisces: Siluroidei: Loricariidae), with comments on generic diagnoses and geographic distribution. Zool. J. Linn. Soc. 102: 1–41.

Schaefer, S.A. 1997. The Neotropical cascudinhos: systematics and biogeography of the *Otocinclus* catfishes (Siluriformes: Loricariidae). Proc. Acad. Nat. Sci. Philadelphia 148: 1–120.

Schaefer, S.A. and A.E. Aquino. 2000. Postotic laterosensory canal and pterotic branch homology in catfishes. J. Morph. 246: 212–227.

Schaefer, S.A., F. Provenzano, M. de Pinna and J.N. Baskin. 2005. New and noteworthy Venezuelan glanapterygine catfishes (Siluriformes, Trichomycteridae), with discussion of their biogeography and psammophily. Am. Mus. Nov. 3496: 1–27.

Schaefer, S.A., and F. Provenzano. 2008. The Lithogeninae (Siluriformes, Loricariidae): anatomy, interrelationships, and description of a new species. Am. Mus. Nov. 3637: 1–49.

Shelden, F.F. 1937. Osteology, myology, and probable evolution of the nematognath pelvic girdle. Ann. N. Y. Acad. Sci. 37: 1–95.

Spadella, M.A., C. Oliveira and I. Quagio-Grassiotto. 2008. Morphology and histology of male and female reproductive systems in the inseminating species *Scoloplax distolothrix* (Ostariophysi: Siluriformes: Scoloplacidae). J. Morph. 269: 1114–1121.

Spadella, M.A., C. Oliveira, H. Ortega, I. Quagio-Grassiotto and J.R. Burns. 2012. Male and female reproductive morphology in the inseminating genus *Astroblepus* (Ostariophysi: Siluriformes: Astroblepidae). Zool. Anz. 251: 38–48.

Stewart, D.J. 1985. A review of the South American catfish tribe Hoplomyzontini (Pisces, Aspredinidae), with descriptions of new species from Ecuador. Fieldiana Zool. 1360: 1–19.

Stewart, T.A. and M.E. Hale. 2013. First description of a musculoskeletal linkage in an adipose fin: innovations for active control in a primitively passive appendage. Proc. R. Soc. B 280: 20122159.

Stewart, T.A., W.L. Smith and M.I. Coates. 2014. The origins of adipose fins: an analysis of homoplasy and the serial homology of vertebrate appendages. Proc. r. Soc. B: Biol. Sci. 281(1781): 20133120.

Stewart, T.A., M.M. Bonilla, R.K. Ho and M.E. Hale. 2019. Adipose fin development and its relation to the evolutionary origins of median fins. Sci. Rep. 9: 512.

Stiassny, M.L.J. 2000. Skeletal system. pp. 109–118. *In*: Ostrander, G.K. (ed.). The Laboratory Fish. Academic Press, San Diego, USA.

Suzuki, H.I., A.A. Agostinho and K.O. Winemiller. 2000. Relationship between oocyte morphology and reproductive strategy in loricariid catfishes of the Paraná River, Brazil. J. Fish Biol. 57: 791–807.

Teugels, G.G. 1983. La structure de la nageoire adipeuse dans les genres *Dinotopterus*, *Heterobranchus* et *Clarias* (Pisces; Siluriformes; Clariidae). Cybium 7: 11–14.

Teugels, G.G. 1986. A systematic revision of the African species of the genus *Clarias* (Pisces; Clariidae). Ann. Mus. R. Afr. Cent. 247: 1–199.

Teugels, G.G. 1996. Taxonomy, phylogeny and biogeography of catfishes (Ostariophysi, Siluroidei): an overview. Aquat. Living Resour. 9: 9–34.

Teugels, G.G. and D. Adriaens. 2003. Taxonomy and phylogeny of Clariidae—an overview. pp. 467–487. *In*: Arratia, G., B.G. Kapoor, M. Chardon and R. Diogo (eds.). Catfishes. Science Publishers, Enfield, USA.

Tilak, R. 1963a. The osteocranium and Weberian apparatus of fishes of the family Sisoridae (Siluroidea): a study in adaptation and taxonomy. Z. wiss. Zool. 168: 281–320.

Tilak, R. 1963b. Studies on the nematognathine pectoral girdle in relation to taxonomy. Ann. Mag. Nat. Hist. 63: 145–155.

Tilak, R. 1965. The comparative morphology of the osteocranium and the Weberian apparatus of Tachysuridae (Pisces: Siluroidei). J. Zool. 146: 150–174.

Tilak, R. 1967. Studies on the osteology of the nematognathine girdle in relation to taxonomy. J. zool. Soc. India 19: 101–110.

Tong, S.-L. and T.E. Finger. 1983. Central organization of the electrosensory lateral line system in bullhead catfish *Ictalurus nebulosus*. J. comp. Neurol. 217: 1–16.

Trajano, E. 1994. Comparative study of the brain and olfactory organ of the troglobitic catfish, *Pimelodella kronei* (Ribeiro 1907), and its putative ancestor, *P. transitoria* (Ribeiro 1912) (Siluriformes Pimelodidae). Trop. Zool. 7: 145–160.

Trajano, E. and S. Secutti. 2025. Diversity and life cycles of subterranean catfishes. pp. 280–298. *In*: Arratia, G. and R.E. Reis (eds.). Catfishes, a Highly Diversified Group. Volume 1: Their Outstanding Biology. CRC Press, Boca Raton, FL.

Vanscoy, T., J.G. Lundberg and K.R. Luckenbill. 2015. Bony ornamentation of the catfish pectoral-fin spine: comparative and developmental anatomy, with an example of fin-spine diversity using the Tribe Brachyplatystomini (Siluriformes, Pimelodidae). Proc. Acad. Nat. Sci. 164: 177–212.

Vigliotta, T.R. 2008. A phylogenetic study of the African catfish family Mochokidae (Osteichthyes, Ostariophysi, Siluriformes), with a key to genera. Proc. Acad. Nat. Sci. 157: 73–136.

Vissotto, P.C., F. Foresti and C. Oliveira. 1997. A ZZ/ZW sex chromosome system in *Imparfinis mirini* (Pisces, Siluriformes). Cytologia 62: 61–66.

de Vos, L. 1995. A systematic revision of the African Schilbeidae (Teleostei, Siluriformes): with an annotated bibliography. Ann. Mus. Roy. Afrique Centr., Sci. Zool. 271: 1–450.

Walsh, S.J. 1990. A Systematic Revision of the Neotropical Catfish Family Ageneiosidae (Teleostei: Ostariophysi: Siluriformes). Ph.D. Thesis, University of Florida, Gainesville, USA.

Welzel, G. and S. Schuster. 2021. Efficient high-voltage protection in the electric catfish. J. Exp. Biol. 224: jeb239855.

Weitzman, S.H. 1962. The osteology of *Brycon meeki*, a generalized characid fish, with an osteological definition of the family. Stanford Ichth. Bull. 7: 114–123.

Westneat, M.W. 2006. Skull biomechanics and suction feeding in fishes. pp. 29–75. *In*: Shadwick, R.E. and G.V. Lauder (eds.). Fish Biomechanics. Academic Press, Amsterdam, Netherlands.

Whitehead, D.L., A.R.G. Gauthier, R.M.E. Cameron, M. Perutz and I.R. Tibbetts. 2015. Ultrastructure of the ampullary organs of *Plicocollis argyropleuron* (Siluriformes: Ariidae). J. Morph. 276: 1405–1411.

Wiley, E.O. and G.D. Johnson. 2010. A teleost classification based on monophyletic groups. pp. 123–182. *In*: Nelson, J.S., H.-P. Schultze and M.V.H. Wilson (eds.). Origin and Phylogenetic Interrelationships of Teleosts. Verlag Dr. Friedrich Pfeil, München, Germany.

Wiley, E.O., A.M. Fuiten, M.H. Doosey, B.K. Lohman, C. Merkes and M. Azuma. 2015. The caudal skeleton of the zebrafish, *Danio rerio*, from a phylogenetic perspective: a polyural interpretation of homologous structures. Copeia 2015: 740–750.

Wilson, J.M. and L.F.C. Castro. 2011. Morphological diversity of the gastrointestinal tract in fishes. pp. 1–55. *In*: Grosell, M., A.P. Farrell and C.J. Brauner (eds.). The Multifunctional Gut of Fish. Elsevier, Amsterdam, Netherlands.

Winterbottom, R. 1973. A descriptive synonymy of the striated muscles of the Teleostei. Proc. Acad. Nat. Sci. 125: 225–317.

Wright, J.J. 2009. Diversity, phylogenetic distribution, and origins of venomous catfishes. BMC Evol. Biol. 9: 282.

Wright, J.J. 2015. Evolutionary history of venom glands in the Siluriformes. pp. 279–301. *In*: Gopalakrishnakone, P. and A. Malhotra (eds.). Evolution of Venomous Animals and their Toxins. Springer, Dodrecht, Netherlands.

Zaccone, G., J.M. Icardo and A. Alesci 2025. Insights into catfish respiratory mechanisms. pp. 321–335. *In*: Arratia, G. and R.E. Reis (eds.). Catfishes, a Highly Diversified Group. Volume 1: Their Outstanding Biology. CRC Press, Boca Raton, FL.

Zakon, H.H. 1986. The electroreceptive periphery. pp 103–156. *In*: Bullock, T.H. and W. Heiligenberg (eds.). Electroreception. John Wiley and Sons, New York, USA.

Zakon, H.H. 1988. The electroreceptors: diversity in structure and function. pp. 813–850. *In*: Atema, J., R.R. Fay, A.N. Popper and W.N. Tavolga (eds.). Sensory Biology of Aquatic Animals. Springer, New York, USA.

Zuanon, J. and I. Sazima. 2005. The ogre catfish: prey scooping by the auchenipterid *Asterophysus batrachus*. Aqua J. Ichthyol. Aquat. Biol. 10: 15–22.

Chapter 4

Siluriform Weberian Apparatus

Modularity and Integration

Kevin K. Duclos[1] and *Terry C. Grande*[2,*]

The Weberian apparatus, originally described by Weber (1820), is a synapomorphy that diagnoses the ostariophysan subgroup Otophysi (Rosen and Greenwood 1970). This complex structure, found in these "hearing specialists" (e.g., carps, minnows, characins, catfishes, and knifefishes) consists of a series of modified anterior neural arches, supraneurals, and pleural ribs that link the gas bladder and the sinus impar of the inner ear (Alexander 1964, 1965). This mechanical linkage enables otophysan fishes to hear high-frequency sounds up to several thousand Hertz (Braun and Grande 2008; Duclos et al. 2021). As a fish swims through a high-frequency sound field, displacement waves cause the gas bladder to pulsate. These pulsations, in turn, vibrate the Weberian ossicles and cause them to pivot, ultimately transmitting sound into the inner ear (Grande and Young 2004). The Weberian apparatus thus forms a direct bony pathway between the sound resonator (the gas bladder) and the inner ear.

The assembly of the Weberian apparatus might singularly be the most important evolutionary event leading to the ecological success and diversification of freshwater fishes, the great majority of which (more than ten thousand species) are otophysans (Nelson et al. 2016). However, the evolutionary history of the Weberian apparatus in otophysans has not been fully elucidated. While myriad hypotheses about the evolution of the Weberian apparatus have been advanced (e.g., Rosen and Greenwood 1970; Fink and Fink 1981; Chardon and Vandewalle 1997; Ladich 2000; Grande and Braun 2002; de Pinna and Grande 2003; Chardon et al. 2003; Grande and de Pinna 2004), one ubiquitous notion is that the Weberian apparatus must have evolved over long periods, with individual vertebrae and Weberian ossicles undergoing mosaic evolution. Although the mechanisms, both evolutionary and developmental, at the root of the morphological diversity of the otophysan Weberian apparatus are poorly understood, the property necessary for mosaic evolution to occur is modularity (Von Dassow and Munro 1999; Evans et al. 2019; Raff and Raff 2000; Hansen 2003; Zelditch and Goswami 2021). Modularity is broadly defined as the capacity for systems to behave as quasi-independent units. Modularity allows for the independent variation of traits throughout ontogeny and along evolutionary trajectories, increasing the relative variational freedom among traits and enabling more exploration of morphospace throughout evolution (Rasskin-Gutman 2005; Zelditch and Goswami 2021). Conversely, morphological integration relates to the coordinated

[1] The University of Calgary, 2500 University Drive, Calgary, Alberta, T2N 1N4 Canada.
[2] Loyola University Chicago, 1032 W. Sheridan Rd., Chicago, Illinois 60660, USA.
Email: kevin.duclos@ucalgary.ca
* Corresponding author: tgrande@luc.edu

variation among traits during development (Klingenberg 2008, 2014; Hallgrímsson et al. 2009). Thus, modularity and integration are interrelated properties constraining the directionality and lability of phenotypic variation throughout ontogeny and evolutionary history (Klingenberg 2008, 2014; Hallgrímsson et al. 2009; Evans et al. 2023).

Although modularity might seem to be a prerequisite property facilitating the evolution of the diversity of complex systems such as the Weberian apparatus, few studies have formally assessed it (Bird and Hernandez 2009; Bird et al. 2020; Duclos et al. 2021). Duclos et al. (2021) demonstrated the existence of modularity in the Weberian apparatus of the cypriniform *Danio rerio*, suggesting that certain elements can vary coordinately or independently from one another. This, in turn, supports the hypothesis that the Weberian apparatus of otophysans likely evolved in subunits that ultimately came to function together as a sound transducer (e.g., Fink and Fink 1981; Patterson 1984; Grande and de Pinna 2004; Chardon et al. 2003).

Despite growing acknowledgement of the putative role of modularity in the evolution of the Weberian apparatus, no assessment of its morphological diversity and modularity has been performed in the Siluriformes (catfishes), which display extensive morphological variation of the Weberian apparatus. For example, there are numerous variations of configuration and reductions of ossicle numbers, many group-specific modifications of vertebral centra, and enormous morphological disparities overall (see Chardon et al. 2003).

It has been suggested (Chardon et al. 2003) that the great inter-familial diversity in Weberian apparatuses of otophysans is correlated with the habitats in which particular groups evolved. Ladich (1999, 2000) provided experimental data demonstrating that not all otophysans hear equally well. Cyprinids have moderately sensitive hearing around 800 Hz, with modest Weberian apparatus diversity within the group, while catfishes are more sensitive at most frequencies while displaying the most variation among and reductions of the Weberian ossicles. As per Ladich (2023), an upper frequency of at least 5,000 Hz has been reported in about 28 species of catfishes. Chardon and Vandewalle (1997) suggested that a reduction in the size and number of Weberian ossicles among catfishes might result in an increase in the efficiency of high-frequency sound transmission from the gas bladder into the inner ear by decreasing the inertia of the ossicles. Such morphological diversity arguably has contributed to the group's ecological success in a multitude of diverse habitats.

This chapter has three objectives. Together, they provide a broad overview of the development and morphological variation of the Weberian apparatus of catfishes and serve as the foundation for future studies seeking to elucidate the evolutionary history of the apparatus within Siluriformes.

Objective 1: Redescribe the morphology/anatomy and development of the Weberian apparatus in *Ictalurus punctatus*. Here, we review and reassess the morphology and ontogeny of the Weberian apparatus in *Ictalurus punctatus*, the Channel Catfish. *Ictalurus punctatus* exhibits a full complement of Weberian ossicles and, therefore, offers a powerful model for investigating the ontogeny of the Weberian apparatus in catfishes. *Ictalurus punctatus* benefits from extensive developmental data available from prior work (Al-Rawi 1966; Grande and Shardo 2002) and abundant availability from museum collections. Understanding the ontogeny and the relative developmental timing of the Weberian apparatus in *I. punctatus* is an essential step in elucidating its evolutionary history and showing how developmental properties, such as allometry and modularity, have shaped it.

Objective 2: Assess allometry, modularity and integration within the Weberian apparatus of *Ictalurus punctatus*. Here, we use *I. punctatus* as a model species to test for modularity within the Weberian apparatus. We follow a similar approach to that used for the model cypriniform *Danio rerio* by Duclos et al. (2021) and assess patterns of modularity and integration within the Weberian apparatus of *I. punctatus* using three-dimensional (3D) geometric morphometrics.

Objective 3: Assess the modularity of vertebral elements associated with the Weberian apparatus across Siluriformes. In this final section, we use 3D geometric morphometrics to assess whether the Weberian apparatus constitutes an evolutionary module across Silurifom taxa. The modularity of elements associated with the Weberian apparatus at the scale of the order may have promoted phenotypic variation of the complex centrum and Weberian ossicles, resulting in the great diversity of "configurations" and morphologies of the Weberian apparatus seen across silurifom taxa. This study's results will serve to expand our understanding of Weberian apparatus modularity and integration within a single catfish species and across siluriform families. The results herein will contribute to our increasing knowledge of teleost axial skeleton regionalization and, more specifically, provide a deeper understanding of modularity as a mechanism for morphological diversification.

The methodology used for each objective is as follows:

Objective 1: Redescription of the morphology and development of the Weberian apparatus in *I. punctatus* is based on data taken from over two thousand specimens representing three developmental series—LU D081090, UAM F1127, and USA 040645 (Grande and Shardo 2002)—as well as a re-examination of key developmental stages and supplementary material, μ-CT scans, and morphological descriptions of Al-Rawi (1966).

Morphological nomenclature follows Al-Rawi (1966), Grande and Shardo (2002) and Chardon et al. (2003) unless specified in the text. The relative timing of the first occurrence of Weberian and dorsal-fin elements, as indicated by examining cleared and stained specimens, is reported and summarized in this chapter.

For objectives 2 and 3, specimen digitization and geometric morphometrics were used to assess variability and patterns of modularity and integration within the Weberian apparatus of *I. punctatus*. Additionally, for the modularity of the first four vertebrae with respect to other skeletal systems across Siluriformes, we assembled two datasets. We use landmark-based geometric morphometrics to measure the shape of the Weberian apparatus in *I. punctatus* and skeletal traits across Siluriformes. Landmark digitization was performed independently for both datasets. Landmarks were exported as .nts files and imported within the R statistical environment (R Core Team 2023). Procrustes superimposition of landmarks was performed separately for both datasets using the geomorph package in R (Baken et al. 2021; Adams et al. 2022). The landmarks for the two datasets were superimposed separately.

Objective 2: Assess allometry, modularity and integration within the Weberian apparatus of *Ictalurus punctatus*. For this objective, we assembled a data set comprised of 37 adult, fully ossified, alcohol-preserved specimens of *I. punctatus* (FMNH lots 16677, 16686). Specimens were scanned using a Perkin Elmer Quantum GX2 μ-CT scanner (High-resolution scans using a Copper Aluminum filter) at Loyola University Chicago. Scans were reconstructed and visualized in Stratovan Checkpoint (Version 2022.07.21.1321, Stratovan Corporation, 2022, U.S.) for geometric morphometric analyses.

A total of 33 landmarks, including five midline and 14 bilateral landmarks, were digitized on all sampled specimens of *I. punctatus* (Figure 1). In this study, landmarks 18, 19, and 20 refer to positions on the anterior part of the transverse process of vertebra 4, as per Al-Rawi (1966), Grande and Shardo (2002) and Chardon et al. (2003). Kubicek (2022) referred to this process as the outer arm of the os suspensorium. The os suspensorium of Grande and Shardo (2003), which equals Kubicek's (2022) inner arm of the os suspensorium, could not be landmarked for this study because of the difficulty in assessing its boundaries clearly in μ-CT-scans.

Individuals of *I. punctatus* are known to display differences in the relative timing of development of Weberian ossicles and associated elements due to differences in size (Grande and Shardo 2002). To test the presence of static allometry (*sensu* Klingenberg 2022) within the Weberian apparatus of *I. punctatus,* we performed a Procrustes test of the effects of size [measured as log (centroid size)] on Weberian apparatus shape, following Klingenberg (2022). This provides a model to predict shape change as a function of size changes in adult specimens *in lieu* of ontogenetic allometry. Predicted shape scores along the allometric trajectory for the Weberian apparatus of *I. punctatus* were then used to describe and illustrate shape variation of the Weberian apparatus associated with size.

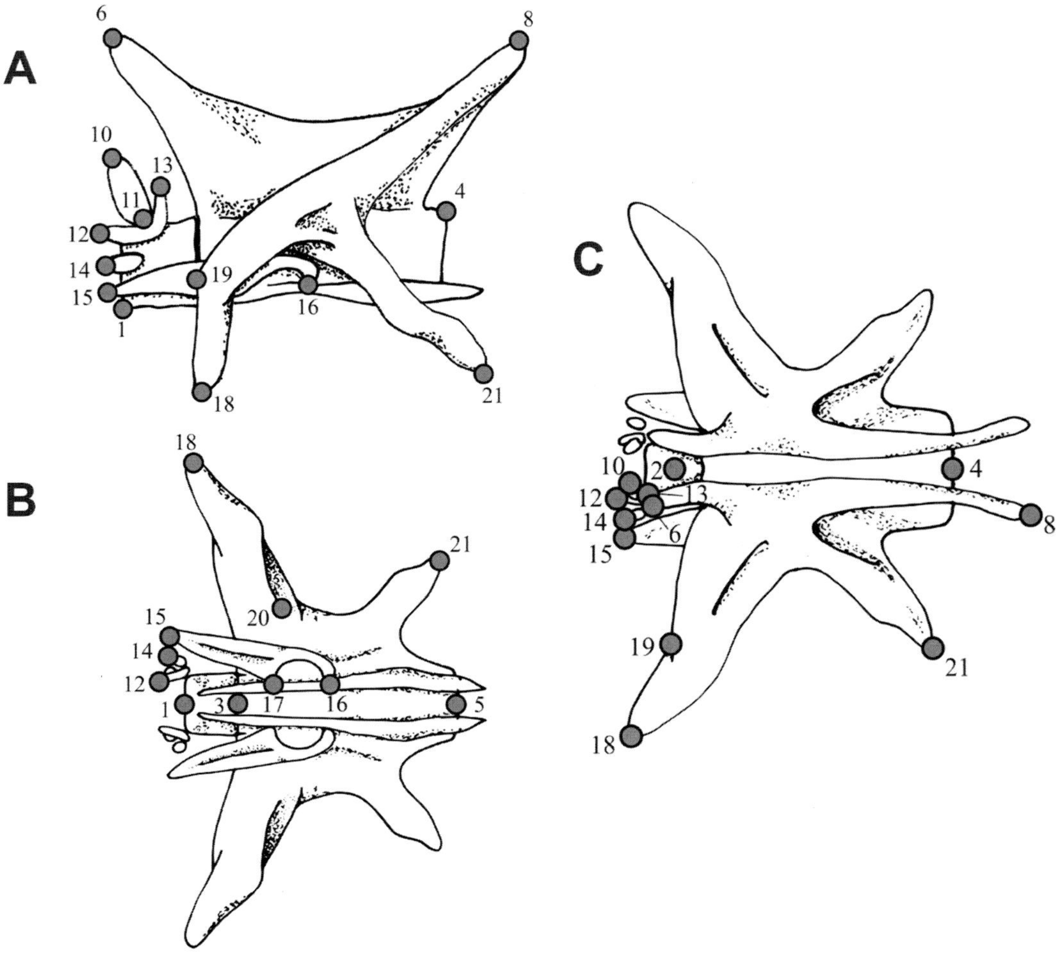

Figure 1. Landmarking scheme for the Weberian apparatus for *Ictalurus punctatus*. (A) lateral view. (B) ventral view. (C) dorsal view. Digitized landmarks: 1. Ventral midpoint of the anterior extremity of centrum 1; 2. Dorsal center of centrum 1; 3. Ventral midpoint between centra 1–2 (complex centra); 4. Dorsal center of centrum 4 (complex centrum); 5. Ventral midpoint between centra 4 and 5; 6. Dorsal extremity of the supraneural spine; 8. Dorsal extremity of supraneural; 10. Dorsal extremity of claustrum; 11. Ventral extremity of claustrum; 12. Anterior extremity of scaphium; 13. Posterior extremity of scaphium; 14. Anterior extremity of intercalarium; 15. Anterior extremity of tripus; 16. Posterior extremity of tripus; 17. Intersection of the tripus and the complex centrum; 18. Distal extremity of transverse process 4; 19. Dorsal edge of transverse process 4; 20. Ventral insertion of the distal arm of the posterior arm of the transverse process 4; 21. Distal extremity of rib 4. Landmarks 6, 8, and 10 through 21 were digitized bilaterally. Landmarks 7, 9 and 22 through 33 are corresponding right-side landmarks and are not illustrated here for ease of visualization.

We generated size-corrected shapes to assess modularity and integration within the Weberian apparatus without the confounding effects of allometric shape variation (Sidlauskas et al. 2011; Duclos et al. 2021). Allometry can have an integrating effect across multiple systems; however, integration can occur through developmental processes that are unrelated to covariation in size between traits (Hallgrímsson et al. 2019; Vasseur et al. 2022). We assessed four hypotheses of modularity within the Weberian apparatus of *I. punctatus* (Table 1). Hypotheses were derived from previous developmental and functional studies on the Weberian apparatus (e.g., Chardon et al. 2003; Duclos et al. 2021) and on the development of *I. punctatus* (Grande and Shardo 2002).

Modularity hypotheses were tested in the size-corrected dataset and the uncorrected shapes dataset (i.e., without size correction). The goodness of fit of modularity hypotheses was tested

Table 1. Hypotheses for modularity within the Weberian apparatus of *Ictalurus punctatus*.

Hypothesis	**Modules**	**Number of Modules**
Null	[C1 + C2–4 + SNS + SN4 + Cl + Sca + Int + Tri + TP4A + TP4P]	1
1. Generalized developmental pattern	[C1 + C2–4]/[SNS]/[SN4]/[Cl]/[Sca + Int]/[Tri + TP4A + TP4P]	6
2. Developmental pattern of *I. punctatus*	[C1]/[C2–4]/[SNS]/[SN4]/[Cl]/[Sca + Int]/[Tri + TP4A+ TP4P]	7
3. Functional interactions	[C1]/[C2–4 + TP4A + Tp4P]/[SNS + SN4]/[Cl]/[Sca + Int +Tri]	5
4. Alternative functional interactions	[C1]/[C2–4 + TP4A + Tp4P]/[SNS + SN4]/[Cl + Sca + Int + Tri]	4

C1, centrum 1; C2–4, centra 2–4; Cl, claustrum; Int, intercalarium; Sca, scaphium; SnS, supraneural spine; Sn4, supraneural 4; TP4A, anterior part of transverse process 4; TP4P, posterior part of transverse process 4; Tri, tripus.

using the covariance ratio (CR) (Adams 2016; Adams and Collyer 2019). The CR metric provides a measure of relative independence between subsets of landmarks. A CR below 1 usually implies that hypothesized modules are more independent from one another, while a CR above 1 implies more covariation than anticipated from separate modules (Adams 2016). Effect sizes (Z-scores) for the CR were used to compare hypothesis support (Adams and Collyer 2019). We then tested integration (the degree of covariation) between modules for the supported hypothesis using partial least squares analysis (PLS) (Rohlf and Corti 2000). Partial least squares analysis assesses correlations between traits based on the major axis of variation for each trait, thus providing a measure of covariation between modules within a set of hypothesized modules (Rohlf and Corti 2000; Adams and Collyer 2016).

Objective 3: Assess the modularity of vertebral elements associated with the Weberian apparatus across Siluriformes. To assess evolutionary modularity across siluriforms of the first four vertebrae, or the first vertebra and the complex centrum, we assembled a dataset comprised of 45 individual taxa, representing 25 families spanning the breadth of siluriform diversity (Nelson et al. 2016). CT scans were sourced from the digital repository Morphosource, which houses CT scan data of museum collection specimens (for specimen data, see supplementary material in Dryad https://doi.org/10.5061/dryad.tmpg4f548). Specimens were first visualized to ascertain whether they represented the correct species or if they presented any evidence of preservation artifacts such as bending of the vertebral column, damage to the skeleton or extreme positions of the fins with respect to the body. Damaged specimens or those exhibiting preservation artifacts were thus excluded from this study. Scans were downloaded and imported into Stratovan Checkpoint as image stacks for geometric morphometrics analyses.

A time-calibrated phylogeny (see supplemental data, Dryad https://doi.org/10.5061/dryad.tmpg4f548) of our taxon set was generated using data from the Fish Tree of Life project, a community-based project that indexes and provides taxonomic, phylogenetic and genomic information about actinopterygian fishes (Alfaro 2018; Chang et al. 2019). This time-calibrated phylogeny relies only on genetic data available for the sampled taxa and may not fully represent phylogenetic relationships across these taxa. Nonetheless, the phylogeny provides a sufficient backbone to assess whether there is a phylogenetic signal for modularity in the first four vertebrae across Siluriformes. For a different revised phylogeny of Siluriformes, see de Pinna et al. (2025). Evolutionary integration was assessed using phylogenetic PLS analysis (Rohlf and Corti 2000; Adams and Collyer 2016).

A total of 27 landmarks per specimen were digitized (Table 2). Landmarks were digitized on the left side of specimens and included the dorsal-fin complex, the pectoral fin complex, the cranium, centra 1 to 4 and associated elements, and centra 5 to 7 (post-Weberian centra). Although we attempted to limit the effect of preservation artifacts within our dataset during specimen selection, landmarks were chosen to limit the incidence of positional variability. For instance, landmarks corresponding to the fins were digitized on the insertion of the fins and not along fin rays, spines or radials. This landmarking scheme was designed to capture the shape and covariation of these elements. It is possible that some artefactual variation may have nonetheless been captured by this landmarking scheme. However, Gartner et al. (2023) assessed the impact of preservation artifacts in the relative position of jaws in detecting patterns of modularity in labrids and found that preservation artifacts had a negligible impact.

There is significant diversity in the shapes of individual Weberian ossicles as well as their presence or absence across siluriform taxa. Because of this extensive variability, many of the homologous landmarks in objective 2 could not be reliably digitized across the dataset for objective 3. However, homologous landmarks were identified on centra as well as on elements related to the Weberian apparatus that are present even when the Weberian apparatus itself is not or

Table 2. Anatomical description of digitized landmarks across siluriform taxa.

Landmark No.	Landmark Description	Anatomical Region
1	Anterior extremity of dorsal-fin complex	Dorsal-fin complex
2	Insertion of the first dorsal-fin spine	
3	Posterior extremity of dorsal-fin complex	
4	Insertion of the marginal ray of the pectoral fin	Pectoral fin complex
5	Insertion of the most medial ray of the pectoral fin*	
6	Dorsal extremity of the cleithrum	
7	Postero-lateral angle of the cleithrum	
8	Anterior extremity of the cleithrum	
9	Anterior margin of the orbit	Cranium
10	Posterior margin of the orbit	
11	Anterior extremity of the mesethmoid	
12	Posterior extremity of the pterotic**	
13	Posterior tip of the supraoccipital	
14	Anterior extremity of the quadrate	
15	Anterior extremity of the opercle	
16	Ventral midpoint of the anterior extremity of centrum 1	Weberian centra and associated elements
17	Dorsal centre of centrum 1	
18	Ventral midpoint between centra 1 and 2	
19	Dorsal extremity of neural arch 2	
20	Distal extremity of rib 4	
21	Dorsal extremity of neural arch 4	
22	Dorsal centre of centrum 4	
23	Ventral midpoint between centra 4 and 5	
24	Dorsal centre of centrum 5	Post-Weberian centra
25	Ventral midpoint between centra 5 and 6	
26	Dorsal centre of centrum 6	
27	Ventral midpoint between centra 6 and 7	

* When present, the pectoral-fin spine was landmarked at the medial insertion point of the spine, which would be landmark 4. **The post-temporal and supracleithrum are completely fused to the pterotic in some taxa; in those cases, the posterior-most tip of the bone intersecting with the cleithrum was digitized.

when it is morphologically derived. This constraint limited the number of homologous landmarks we could use because of the high disparity of cranial and post-cranial skeletal elements among siluriforms (Chardon et al. 2003).

To assess evolutionary modularity, we tested the hypothesis that the dorsal-fin complex, the pectoral-fin complex, the cranium, centra 1 to 4 and associated elements, and centra 5 to 7 (post-Weberian centra) all form independent modules using the covariance ratio (CR) and a time-calibrated phylogeny of sampled taxa (geomorph) (Adams 2016; Adams and Collyer 2019). For each dataset, landmark digitization was performed twice, and digitization error was assessed using a Procrustes ANOVA (Klingenberg and McIntyre 1998). The Procrustes ANOVA did not return significant results for either dataset (P-values of 1 and 0.593 for datasets 1 and 2, respectively), so landmark replicates were averaged between replicates following superimposition.

Supplementary material, including landmark data, R code, time-calibrated phylogeny, and data tables, can be found at Dryad https://doi.org/10.5061/dryad.tmpg4f548.

Weberian Apparatus Development and Morphology

As illustrated by Chardon and Vandewalle (1997) for an early-stage otophysan Weberian apparatus and described by Grande and Young (2004), Bird and Hernandez (2007), and Duclos et al. (2021) for the generalized cypriniform condition, the Weberian apparatus is divided into two sections. First, the pars auditum, which consists of three paired auditory ossicles (scaphium, intercalarium, and tripus), is directly involved in sound transmission. These Weberian ossicles are at least partly derived from vertebral elements (basidorsals, basiventrals, and pleural ribs), meaning that they are homologous to such elements as neural arches and parapophyzes (Rosen and Greenwood 1970; Fink and Fink 1981).

From posterior to anterior of the Weberian apparatus, the tip of the tripus (the transformator process) articulates with the anterior tunica externa, or external layer of the gas bladder, and is connected rostrally to the intercalarium by the interossicular ligament. The scaphium, which is the most anterior of the Weberian elements, ultimately transmits sound vibrations from the intercalarium to the inner ear; it consists of three processes. The concha scaphii [a horizontal process of Al-Rawi (1966)] forms the posterolateral wall of the sinus impar perilymphaticus and terminates on the posteroventral surface of the exoccipital. It serves as an attachment site for the end of the interossicular ligament. The second ascending process lies perpendicular to the concha scaphii, and the third articular process articulates with centrum 1. The ascending and articular processes allow for lateral movement of the concha scaphii. Bamford (1948) and Rosen and Greenwood (1970) reported that only the basidorsal one contributes to the formation of the scaphium. However, Matveiev (1929), Watson (1939), Vandewalle et al. (1990), Coburn and Futey (1996), Al-Rawi (1966), Bird and Mabee (2003), Grande and Shardo (2003), and Grande and Young (2004) argued that the scaphium of otophysans develops as a hybrid structure such that the concha scaphii forms from embryonic mesenchyme while the ascending process develops from basidorsal one.

The second section, the pars sustentaculum, is not directly involved in sound transmission but serves a structural role in supporting the auditory ossicles of the pars auditum. Elements of the pars sustentaculum include the centra themselves, as well as neural arches, supraneurals, parapophyzes of the first, second and fourth vertebrae, the transverse process of vertebra 4 [= parapophysis + rib 4 of Fink and Fink (1981)], and the os suspensorium. The os suspensorium, derived from the parapophysis of vertebra four, abuts the anterior chamber of the gas bladder, holding it in position (Bird and Mabee 2003). This element exhibits considerable variation among otophysans. It is most prominent in cypriniforms and most reduced among siluriforms. The claustrum, sometimes considered a fourth Weberian ossicle, is included here as part of the pars sustentaculum. When present, it is located dorsal to and separate from the scaphium and intercalarium and is not connected to the ossicles by interossicular ligaments. It thus is not part of the vibration-transmitting chain. Developmental data from cypriniform fishes, including the zebrafish, *Danio rerio* (Bird and Mabee 2003; Grande and

Young 2004; Cloutier and Grünbaum 2007), suggest that chondrification and ossification sequences of larval and juvenile individuals exhibit phylogenetic signals and that heterochronic patterns can be identified at various levels.

Ictalurus punctatus Weberian Apparatus and Development

The Weberian apparatus of *Ictalurus punctatus* is functionally complex because of its close association with the dorsal-fin skeleton (Chardon 1968). The anterior two proximal radials of the dorsal fin are expanded, elongated, and tightly articulate with the neural spine of vertebra four. In the adult, the Weberian apparatus and the dorsal-fin skeleton essentially form an interconnected functional unit (Figure 2). The development of this functional unit is discussed below.

During development in fishes as small as 10 mm TL (Stage 6; Figure 3; Table 3), as reported in Al-Rawi (1966), Grande and Shardo (2002), and Kubicek (2022), the scaphium, intercalarium and tripus are the first identifiable ossicles to form (Figure 3A). This suggests a modular linkage among these elements. These ossicles ossify quite rapidly, suggesting the importance of a functioning Weberian apparatus in larval and juvenile catfishes. All basidorsals along the vertebral column and basiventrals posterior to the second centrum have also formed. Basiventrals 1 and 2 are absent in siluroids (Fink and Fink 1981). The transformator process of the tripus appears long and threadlike. It forms a crescent-like structure on the ventral side of the complex centrum and becomes embedded in the tunica externa of the gas bladder. The transverse process of vertebra 4 has also formed.

Also, at this stage of development, the first dorsal-fin spine has formed and will eventually become the locking mechanism for the second spine in the fully formed dorsal fin. A large supraneural is also present above neural arch 3. This supraneural eventually fuses with neural arch 3.

In the dorsal fin, seven cartilaginous radials have formed (Figure 3A). The first two radials will become incorporated into the Weberian apparatus/dorsal-fin complex. The first and most proximal segment of lepidotrichia has formed in the dorsal fin. In *Ictalurus*, the anterior two lepidotrichia will develop into fin spines. Catfish spines form first as soft rays in discrete lepidotrichial segments. This is followed by adding concentric layers of dermal bone over the surface. These bony deposits cause the margins of these segments to approach each other gradually and finally fuse (Reed 1924).

In fishes of about 15–18 mm TL, the tripus appears ossified, and its anterior process has elongated, approaching the exoccipitals and extending past the anterior margin of vertebra 1. The transverse process of vertebra 4 is also ossified and continues to expand (Figure 3B). The os suspensorium has begun to form from the ventral side of the transverse process, near the connection of vertebra 4 with vertebra 5, starting with the dorsal lamina (terminology of Al-Rawi 1966). The dorsal lamina extends anteromesially between the transformator process of the tripus and the postcardinal vein and terminates in the radial nodule. The radial nodule in the adult catfish is surrounded by the crescent-shaped transformator process of the tripus (see Kubicek (2022) for an alternative nomenclature of the os suspensorium). Additionally, the intercalarium is completely ossified, and the scaphium begins to exhibit its characteristic shape as its horizontal process extends toward the exoccipitals. The large supraneural above neural arch 3 has enlarged and, although still cartilaginous, articulates with the expanded neural arches of vertebrae 3 and 4. In addition, vertebrae 2–4 have fused, thus completing the formation of the complex centrum. Vertebra 1 remains independent and never fuses with the Weberian complex centrum. In fishes of about 15 mm TL, the claustrum is observed above the scaphium. It has no physical connection to any Weberian ossicle. As discussed by Al-Rawi (1966), the claustrum rudiment is first formed by mesenchyme. It consists of two parts: a thick posterior edge formed from cartilage and a thin anterior portion formed from membrane bone. As per Al-Rawi (1966), the developing claustrum is positioned between the first spinal nerve and a small additional spinal nerve that emerges between the claustrum and the ascending process of the scaphium. Given that the number of neural arches is the same as the number of spinal ganglia (Balinsky 1961), the occurrence of the small additional spinal nerve, with its ganglion anterior to the first vertebra but posterior to the claustrum, led Al-Rawi (1966) to argue that the claustrum

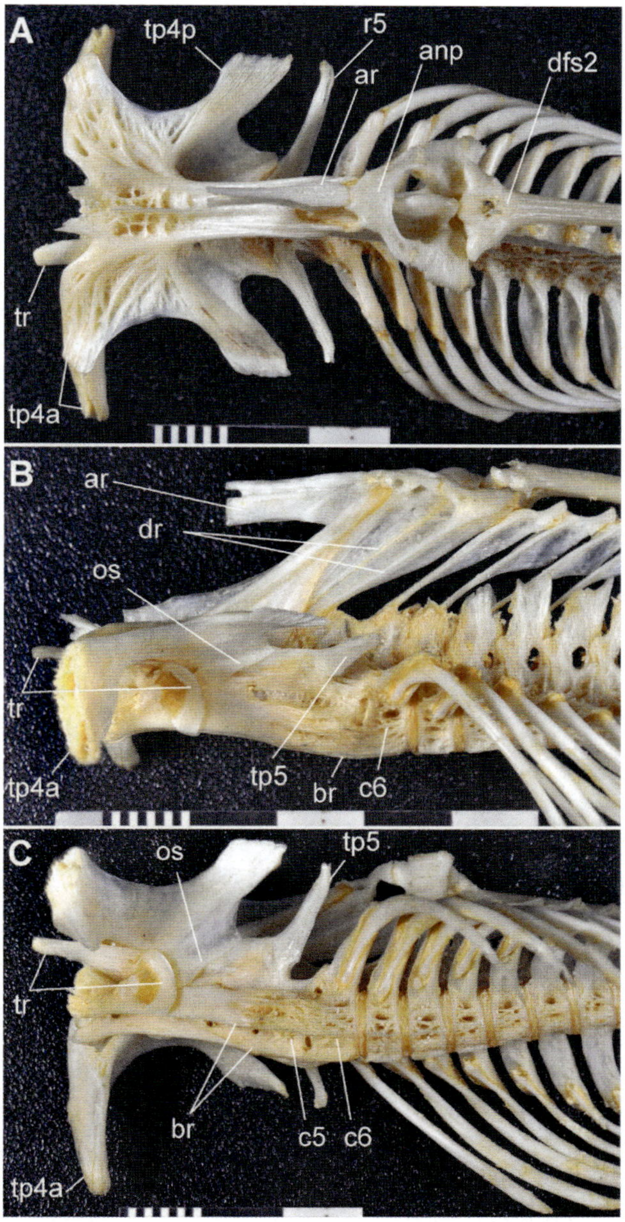

Figure 2. *Ictalurus punctatus* (FMNH 16711, partly disarticulated skeleton). (A) dorsal view. (B) lateral view. (C) ventral view. Abbreviations: anp, anterior nuchal plate; ar, anterior radial; br, bony ridge; c, centrum; dr, dorsal radial; dfs2, dorsal-fin spine 2; os, os suspensorium; tp5, transverse process of centrum 5; tp4a, anterior part of the transverse process 4; tp4p, posterior part of the transverse process 4; tr, tripus.

must be an intercalated neural arch of a lost vertebra or the neural arch of a vertebra fused with the basioccipital. He further provided supporting evidence with respect to *I. punctatus* that the number of segmented muscles in the anterior region is the same as the number of spinal nerves plus their ganglia and with the number of neural arches, but only if the claustrum is counted as a neural arch.

The early stages in the development of the Weberian apparatus/dorsal-fin complex were observed in fishes about 13–14 mm TL. The first two proximal radials of the dorsal fin lengthen ventrally; the first makes contact with neural arch 4 and is positioned between the two halves of this arch. The first dorsal-fin spine is complete, while the second spine continues to ossify along with its

Table 3. Weberian apparatus developmental staging in *Ictalurus punctatus* based on Grande and Shardo (2002) and a re-examination of developmental material. For the complete staging scheme of the post-cranial skeleton, see Grande and Shardo (2002). dpf: days post-fertilization.

Developmental Stage	Size in mm TL	dpf	First Appearances
Stage 3, embryonic	8–10 mm	4–7	Anterior basidorsals and basiventrals formed
Stage 4, embryonic	9–12 mm	5–10	Anterior protocentra mineralized; the beginning of anterior vertebral column formation
Stage 6, embryonic	10–14 mm	6–12	First appearance of tripus, scaphium and intercalarium in cartilage; transverse process of V4 formed
Stage 7, yolk-sac larval (Figure 3A)	10–14 mm	7–12	Weberian ossicles begin to ossify beginning with tripus; first dorsal-fin spine and anterior supraneural form
Stage 8, yolk-sac larval	12–14 mm	8–12	Mineralization of centra; second dorsal-fin spine forms
Stage 9, yolk-sac larval (Figure 3B)	13–15 mm	10–13	Fusion of V2 with V3
Stage 10, yolk-sac larval	14–16 mm	10–15	Claustrum forms; first two dorsal-fin radials enlarge and articulate with the spine on na4; 5 distal radials of dorsal-fin present
Stage 11, yolk-sac larval	15–17 mm	11–17	Fusion of V2+V3+V4; V5 articulates ventrally with complex centrum but autogenous dorsally
Stage 12, foraging	16–17mm	13–17	Transverse process of V4 articulates with posttemporosupracleithrum of the pectoral girdle; all neural and haemal arches ossified
Stage 13, foraging (Figure 3C)	16–27 mm	14–36	All Weberian ossicles ossified; claustrum ossified
Stage 16, foraging	31–40 mm	no data	Weberian apparatus/dorsal-fin complex complete; a horizontal process of the intercalarium is gone; a crescent process of the tripus embedded in the tunica externa of the gas bladder

corresponding dorsal-fin radial. A median cartilaginous element, called an anterior nuchal plate by some authors (e.g., Teugels 1996) or a supraneural by others (e.g., Lundberg 1982; Lundberg and McDade 1986; Grande and Lundberg 1988; Grande and de Pinna 1998), and a radial or pterygiophore by others (e.g., Brown and Ferraris 1988; Fink and Fink 1996), forms anterior to the first dorsal-fin radial. This structure, when ossified, articulates with the supraoccipital bone, forming a connection between the dorsal fin and the skull.

The remaining developmental stages are devoted to further ossifying already formed structures (i.e., making the final attachments among the Weberian ossicles, the dorsal-fin skeleton, and the skull) and growth (Figure 3C). At 30 mm TL and larger, the transverse process of vertebra 4 develops two expansions. The anterior part of the transverse process articulates with the ventral process of the mesial limb of the supracleithrum (Lundberg 1975). In individuals of more than 35 mm TL, the anterior nuchal plate has ossified, and the anterior radial has ossified fully and is now in contact with both the anterior nuchal plate and the supraoccipital. Ossification of the first proximal and distal radials is complete, and the first proximal radial forms a tight connection with the posterior margin of the neural arch complex of the Weberian apparatus. The first dorsal-fin spine helps form a locking mechanism for the second fin spine. The second proximal radial is ossified, and its proximal tip articulates with the Weberian apparatus. Eventually, the entire second radial will abut the first, which in turn is connected to the Weberian apparatus. In fishes of about 40 mm TL, proximal radials 3–6 are ossified (proximal radial 7 and distal radials 3–7 are still cartilaginous). They have elongated and extend between the two halves of neural arches 5–9, preventing them from meeting at the dorsal midline.

In a fully ossified Weberian apparatus/dorsal-fin complex (Figures 2 and 3D), a solid bony connection is made between the Weberian apparatus and the dorsal-fin skeleton via the spine of neural arch 4 and the anterior dorsal-fin radials. When fully ossified, the Weberian apparatus/dorsal-fin

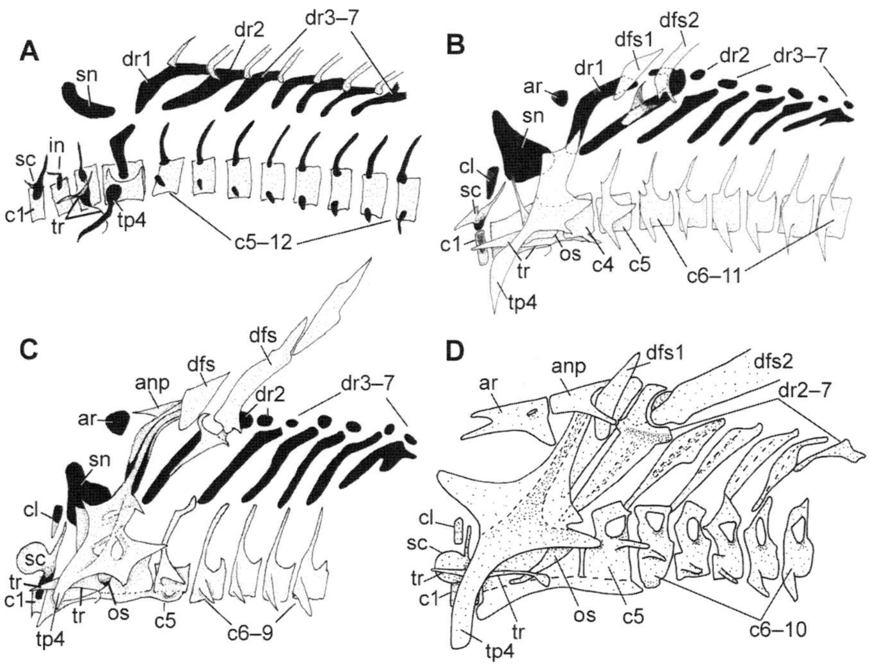

Figure 3. Development of Weberian apparatus and dorsal-fin complex of *Ictalurus punctatus*. Modified from Grande and Shardo (2002) with direct examination of specimens in developmental series LU D081090. (A) yolk-sac larva, 12.3 mm TL, stage 7, LU F.082285. (B) yolk-sac larva, 13.6 mm TL, stage 9 – early stage 10, LU F.082286. (C) foraging juvenile, 16.3 mm Tl, stage 13, LU F.082287. (D) sub-adult, FMNH 16686, 81 mm; FMNH 16711, dried skeleton, FMNH 16717, 85 mm SL. Abbreviations: anp, anterior nuchal plate; ar, anterior radial (= supraneural of Grande and Lundberg 1988); c, centrum; cl, claustrum; dfs, dorsal-fin spine; dr, dorsal-fin radial; in, intercalarium; na, neural arch; os, os suspensorium (= inner arm of os suspensorium of Kubicek 2022); sc, scaphium; sn, supraneural; tp4, transverse process of vertebra 4 (= outer arm of os suspensorium of Kubicek 2022); tr, tripus. Anterior to the left.

complex forms at least three connections with the back of the skull. The first one, as just stated, is the articulation of the anterior radial and the supraoccipital crest; the second is by means of an articulation between anterior neural arch 3 and the ventral side of the supraoccipital, and the third is via the Weberian ossicles (i.e., scaphium, and anterior process of the tripus) with the exoccipitals. The intercalarium in adult *I. punctatus* is small, imbedded in the interossicular ligament, and does not form a connection with any vertebrae. Thus, it is immovable, unlike the condition in cyprinids (Bird and Hernandez 2007; Grande and Young 2004).

Allometry Within the Weberian Apparatus of *Ictalurus punctatus*

We detected a significant effect of log (Centroid Size) on Weberian apparatus shape (p-value = 0.001, R2). This supports allometry in the shape of the Weberian apparatus in *I. punctatus*, where size explained 36.8% of the total shape variation observed. As specimens increase in size, the relative distance between the tip of the supraneural spine and supraneural 4 along the antero-posterior axis decreases as the tip of supraneural 4 extends more dorsally where it meets the dorsal fin (Figure 4A, B; Grande and Shardo 2002). The os suspensorium broadens laterally (Figure 4B) as size increases, but the distal arm of the os suspensorium appears to extend more anteriorly (Figure 4A). The claustrum, scaphium and tripus also increase in size (Figure 4A and B). As allometry was a significant contributor to shape variation in *I. punctatus*, all modularity analyses were performed on size-corrected data.

Our results indicate that morphological variation is not distributed evenly across elements of the Weberian apparatus. For example, variation in size leads to important variations in the shape of the supraneural spine and supraneural 4, which interact with the supraoccipital and dorsal fin, respectively. Size-linked variation is also evident in the shape of the os suspensorium.

A significant allometric signal in the Weberian apparatus of *I. punctatus* indicates that the morphology of particular Weberian elements may be correlated more so with the size of the fish than with age, corroborating the findings of Grande and Shardo (2002) for *I. punctatus*. A key finding of Grande and Shardo (2002) was that, while individual centra do not change much in size, Weberian ossicles become relatively larger as the fish grows larger. Interestingly, morphological and developmental variation that correlates with specimen size more than specimen age was also observed in *Danio rerio* (Grande and Young 2004), as well as in *Esox lucius* and *E. masquinongy* (Burdi and Grande 2010). This is of interest because changes to allometric trajectories are one way to increase disparity across taxa.

Allometry can act as a component of morphological integration (Vasseur et al. 2022). Indeed, correlated growth between traits and coordinated variation in shape as a function of size is likely to promote covariation in the morphology of multiple traits. As such, allometric variation can be a confounding factor when trying to detect patterns of modularity across traits. Subsequent analyses, therefore, rely on size-corrected shapes to assess patterns of modularity accounting for allometry following Duclos et al. (2021). Modularity analyses were performed simultaneously on the uncorrected shape data (i.e., without size-correction) to assess the impact of allometry on integration in *I. punctatus*.

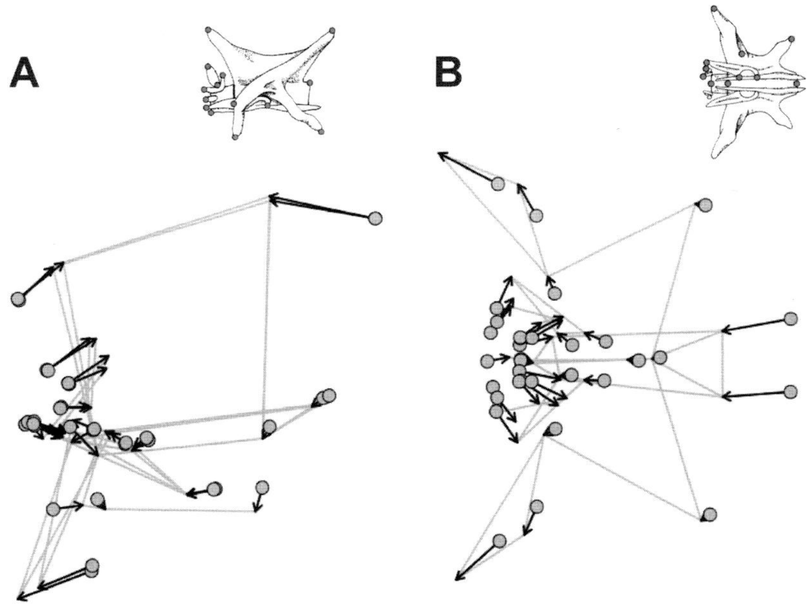

Figure 4. Morphological changes due to allometry within the Weberian apparatus of *Ictalurus punctatus*. (A) lateral view of morphological changes. (B) ventral view of morphological changes. Grey circles represent the shape of the smallest specimen, and the black arrows show the shape of the largest specimen. The magnitude of shape differences was amplified threefold to help with visualization. Illustrations with landmark placements are inset to help with visualization.

Modularity Within the Weberian Apparatus of *Ictalurus punctatus*

We found significant support for all the hypotheses for modularity within the Weberian apparatus of *I. punctatus*; however, we found a larger effect size for hypothesis 2 (Z-score: –3.65987). Pairwise comparisons between the modularity hypotheses and the null hypothesis (no modularity) significantly rejected the null; however, significant differences between hypothesis 2 and the other hypotheses were not found (Table 1). Hypothesis 1 showed the second highest level of support

(Z-score = –3.49969), while hypothesis 3 showed the lowest support (Z-score = –2.8362; see supplementary data for a detailed breakdown of results). In the uncorrected shape data for *I. punctatus*, only hypotheses 2 and 3 were supported (Z-scores of –2.07928 and –3.30248, respectively). Comparing effect sizes between hypotheses 2 and 3 of the uncorrected shape data did not, however, allow for discrimination between either hypothesis. Overall, the results herein support the modularity of the Weberian apparatus of *I. punctatus*. While it is possible that modules delineated by our hypotheses may be nested into higher-order modules or themselves decomposable into smaller subunits, our results across size-corrected and uncorrected shape data suggest that hypotheses 2 and 3 are the best descriptors of modularity within our dataset. We propose that hypothesis 2 best describes the modularity of the Weberian apparatus in *Ictalurus punctatus*, but it should be noted that allometry and correlated growth between modules, or elements therein, might change patterns of covariation, as seen in the uncorrected shape dataset (Klingenberg 2013; Vasseur et al. 2022). Hypothesis 2 describes a modular partition of the Weberian apparatus where module 1 is the first centrum; module 2 is the complex centrum; module 3 is the supraneural spine, module 4 is supraneural 4, module 5 is the claustrum, module 6 is comprised of the scaphium and the intercalarium, and module 7 is comprised of the tripus and the anterior and posterior transverse processes of centrum 4 (Figure 5A). The difference between the two most supported hypotheses, hypotheses 1 and 2, is that hypothesis 1 groups the first centrum and the complex centrum into a single module. However, the pairwise CR and PLS scores between the first centrum and the complex centrum modules supported by hypothesis 2 were low and suggest that they are indeed modular with respect to each other (CR = 0.599, PLS = 0.397).

Further decomposition of CR and PLS scores for hypothesis 2 shows a large range of values (Figure 5A). Indeed, while the pairwise CR and PLS values are, overall, low and suggest independence between hypothesized modules, the CR between the supraneural spine and supraneural 4 is 1.070.

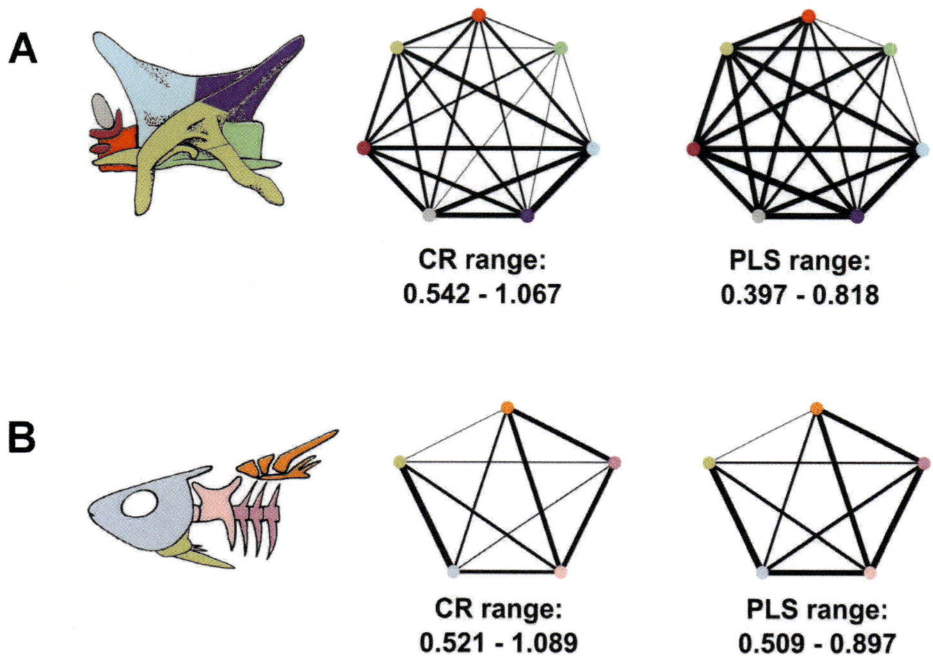

Figure 5. (A) Modularity and integration in the Weberian apparatus of *Ictalurus punctatus*. Points in the networks represent morphological modules, as illustrated on the left. Line width represents the value of the covariance ratio (CR) or partial least squares correlation (PLS) between two modules. (B) Modularity and integration across the first vertebra, the skull, the appendicular skeleton and post-Weberian apparatus across Siluriformes. Points in the networks represent morphological modules, as illustrated on the left. Line width represents the value of the covariance ratio (CR) or partial least squares correlation (PLS) between two modules.

This value indicates that the supraneural spine and supraneural 4 may be integrated within a single module. This is further supported by a PLS score of 0.809 between the two modules, which indicates significant covariation of shape between the supraneural spine and supraneural 4. Interestingly, while all the other modules are significantly independent according to the CR, they do display tendencies for covariation (Figure 5A). This is especially noteworthy between supraneural 4 and module 7 (tripus, anterior transverse processes 4, and posterior transverse process 4), where the PLS score is 0.818 (Figure 5A). As indicated above, subcomponents of different modules might covary despite relative independence between said modules. This is because different processes might promote covariation, or integration, of traits at different scales and at different times throughout ontogeny (Hallgrímsson et al. 2009; Vasseur et al. 2022; Evans et al. 2023). Modularity and integration can thus be detected simultaneously in a shape dataset, reflecting the complex network of developmental, functional and structural interactions between traits (Hallgrímsson et al. 2009; Burns et al. 2022; Evans et al. 2023).

Modularity of *Ictalurus punctatus* as a Model for the Evolution of the Weberian Apparatus in Siluriformes

It has been reported, or hypothesized, that individual ossicle formation among otophysans is variable. Although there is little obvious variation in Weberian apparatus morphology among Cypriniformes, variation within Siluriformes is extraordinary (Coburn and Grubach 1998; Chardon et al. 2003; Braun and Grande 2008; Lechner et al. 2011). For example, among siluriforms, the scaphium is present in its entirety in the more generalized catfishes such as diplomystids (Fink and Fink 1981, fig. 17) and ictalurids (Figures 2 and 3). The ascending and articular processes of the scaphium are missing in loricariids and callichyids (Chardon et al. 2003), and all Weberian ossicles associated with V1 and V2 are lost in *Corydoras* and, by extension, in callichthyids (Coburn and Grubach 1998).

Here, we advanced the first study of modularity of the Weberian apparatus in the Channel Catfish, *Ictalurus punctatus*, to serve as a model for the study of modularity across Siluriformes. While the vast diversity of Weberian apparatus morphology and configuration across catfishes is unlikely to be properly exemplified by studying phenotypic variability and modularity in *I. punctatus* alone, this study may serve as a compass in future studies of modularity in the Weberian apparatus of various siluriform taxa or across different taxonomic scales (i.e., across species, genera, or families). The results obtained here may offer new testable hypotheses regarding the role of modularity in the evolution of the Weberian apparatus within Siluriformes, and we, therefore, describe them in the context of what is known about Weberian ossicle development, function and disparity across Siluriformes.

Modules 1 and 2: Centrum 1 and the Complex Centrum

The main supporting structure for the Weberian apparatus is the anterior five centra. In catfishes, these centra exhibit a variety of fusion patterns. For example, in diplomystids, centra 2–4 are completely fused and indistinguishable. The first centrum is autogenous, and this is hypothesized as the primitive condition (Arratia 1987). In callichthyids, exemplified by *Corydoras*, V1 and V2 are lost, and the compound centrum, therefore, consists of a fusion of V3–V5 (Coburn and Grubach 1998). In the loricariid *Hypostomus*, the entire complex centrum is fused with the basioccipital (Schaefer 1987). In most other catfishes, centrum 5 is fused to the compound centrum.

In the fossil ictalurid †*Astephus*, centrum 5 is autogenous (Grande and Lundberg 1988), as it is also in diplomystids. In *Ictalurus*, centrum 5 is connected to the compound centrum along its ventral margin, but it is autogenous dorsally (Figures 2 and 3). The morphological data presented here support the first centrum and the complex centrum as forming two independent modules in *I. punctatus*. Covariation in the morphology of these two elements appears limited. Furthermore, both modules display little covariation with the other Weberian modules. This supports independence between Weberian centra and Weberian ossicles, and further corroborates the observations of

Grande and Shardo (2002), whereby Weberian centra and Weberian ossicles displayed dissimilar trajectories of shape variation throughout ontogeny.

Our results, coupled with observations on centrum morphology and development, suggest that centra may be modular with respect to Weberian ossicles across Siluriformes. Furthermore, they suggest that the first four centra (or centrum 1 and the complex centrum) may be modular with respect to non-Weberian skeletal traits (i.e., the cranial and appendicular skeleton as well as post-Weberian centra).

Modules 3 and 4: The Supraneural Spine and Supraneural 4

In *Ictalurus punctatus*, our results support the supraneural spine and supraneural 4 as being modular with respect to other Weberian elements. However, the results of both the CR and PLS analyses show integration and significant covariation between these two elements. It is thus likely that the supraneural spine and supraneural 4 form an integrated module.

Integration between the supraneural spine and supraneural might derive from shared developmental timing but could also be due to functional constraints imposed on them by other traits. Indeed, the supraneural spine contacts and articulates with the supraoccipital bone in the cranium, which, in turn, contacts the dorsal-fin complex. In parallel, supraneural 4 extends dorsally and articulates with the dorsal-fin complex early in ontogeny. Shared functional interactions with the supraoccipital and the dorsal-fin complex may drive integration of the supraneural spine and supraneural 4 in *I. punctatus*. However, it is unclear whether the same pattern may be expected across siluriform taxa.

Module 5: The Claustrum

Our results are *de facto* limited by the number of landmarks that could be reliably digitized on the claustrum in *I. punctatus*. The claustrum has a rounded shape and is not a dense ossification, severely limiting our ability to quantify shape precisely. Nonetheless, our results do support the claustrum as an independent module.

This is interesting as there is debate as to the extent to which the claustrum, as a putative Weberian ossicle, plays a role in sound transmission. Additionally, the homology of the claustrum is still being debated. According to Fink and Fink (1981), it is formed from a dissociated part of neural arch 1, or, as argued by Gayet (1985), Coburn and Futey (1996), and Chardon and Vandewalle (1997), modified from supraneural one. Al-Rawi (1966) argued that the claustrum in *I. punctatus* is a free neural arch. The claustrum is present in generalized siluroids, but it appears to be lacking in all loricarioids (Arratia 1987; de Pinna 1992).

While our results do not fully clarify the homology of the claustrum, it is worthwhile noting that the claustrum appears not to share functional and variational constraints with the tripus, the scaphium or the intercalarium. This further supports modularity as a mechanism by which the claustrum, and its presence or absence, has evolved across siluriform taxa. In other words, with the claustrum being its module, a reduction in the size of the claustrum to the point of absence in some taxa or modifications of its shape might not affect the functionality of the Weberian apparatus. Therefore, the morphological evolution of the claustrum in Siluriformes may be free of functional constraints and selective pressures surrounding sound transmission.

Module 6: The Scaphium and the Intercalarium

Integration of the scaphium and intercalarium within a single module reflects the tight functional association between these elements. These two ossicles form an integral part of the sound transmission chain of the Weberian apparatus. Although hypothesis 2 supports the tripus being in a separate module from the scaphium and intercalarium, hypothesis 3 (which was significantly supported in size-corrected data and the uncorrected shape data) supports integration of the tripus in the same module as the schaphium and intercalarium. While these three ossicles are functionally interacting within the Weberian apparatus, they are the ossicles displaying the most disparity across Siluriformes. The developmental pattern described here and by Grande and Shardo (2002) suggest that the tripus is mostly independent from the scaphium and intercalarium in its development and variation. Nonetheless, it should be noted that functional interactions between these ossicles and allometry indicate stronger tendencies of the tripus to covary with the scaphium and intercalarium. Integration of the scaphium and intercalarium also reflects the timing of development of these elements and their importance during larval and juvenile life stages in *I. punctatus*. The intercalarium, formed from basidorsal 2 (Watson 1939; Fink and Fink 1981; Vandewalle et al. 1990; Chardon and Vandewalle 1997), primitively consists of three processes: an ascending process, an anterolateral process (the manubrium), formed from mesenchyme in *I. punctaus* (Al-Rawi 1966; Grande and Shardo 2003) that serves as an attachment for the interossicular ligament, and an articular process that anchors the ossicle to the second centrum. The intercalarium is reduced in more generalized catfishes such as *Ictalurus* or lost in more advanced taxa (Fink and Fink 1981). As per Chardon et al. (2003) in many catfishes, the intercalarium is reduced to an intraligamentous ossification (the manubrium). The articular process of the intercalarium is lost in siluroids (Arratia 1987). In most siluroids, the ascending process is also lost, such that the intercalarium consists only of a nodule of bone (Fink and Fink 1981; Chardon 1968). The intercalarium is lost altogether in loricariids, trichomycterids, and nematogenyids (Arratia 1987).

Module 7: The Tripus and the Transverse Processes of Centrum 4

Our results support the tripus and the transverse processes of centrum 4 as being a module in *I. punctatus*. Integration among these three traits reflects a strong influence of developmental timing in their morphological variation. However, the three traits are involved in different functions.

The tripus is an essential component of the sound transmission chain in catfishes. In some taxa, only the tripus remains the sole functional component of the hearing chain of the Weberian apparatus. In taxa where the scaphium and the intercalarium are present, the tripus forms tight functional associations with both of these ossicles. The articular process and the proximal part of the transformator process of the tripus are formed from an enlargement of basiventral 3 in generalized catfishes, including *Diplomystes* and *Ictalurus* (Arratia 1987; Coburn and Grubach 1998; Grande and Shardo 2002). In many catfishes, the interossicular ligament, which connects the tripus and the scaphium, is shortened such that the tripus is physically connected to the choncha scaphii. The tripus and scaphium essentially become one element (Chardon et al. 2003). Baskin (1972), as per Schaefer (1987), reported the loss of the intercalarium, claustrum, and transformator process of the tripus in trichomycterids, callichthyids, astroblepids and loricariids.

Conversely, the os supsensorium and transverse processes of centrum 4 are not involved in sound transmission. No ubiquitous functional role across Siluriformes has been attributed to these elements. On the other hand, the os suspensorium plus the transverse processes of centrum 4 play a role in the anchoring of the gas bladder and articulation with the cleithrum across multiple taxa, respectively. Additionally, the important variability in the morphology of the os suspensorium among catfishes is well known. In more generalized catfishes, such as diplomystids, it is clearly prominent, but in advanced loricarioids, it is reduced to small bumps on the sides of the complex centrum. Integration between the tripus and elements of centrum 4 likely reflects developmental interactions between these traits more so than functional interactions.

Modularity of the First and Complex Centra in Siluriformes

Our results with respect to *Ictalurus punctatus* give credence to the idea that modularity might be at play for the morphological evolution of the Weberian apparatus in Siluriformes. Whether the patterns observed in *I. punctatus* are conserved across siluriform taxa will undoubtedly necessitate more in-depth studies of morphological variation within and across taxa. Nonetheless, the results presented above offer new potential avenues of investigation. The most immediately relevant result from assessing modularity in the Weberian apparatus of *I. punctatus* is the identification of centrum 1 and the complex centrum as forming modules that are relatively independent of the remainder of the Weberian ossicles and associated elements. We thus hypothesize that the Weberian apparatus in catfishes evolved from a non-modified vertebral column within a modular framework and that heterochronic events involving different modules occurred during its evolution, including both generalized and highly modified siluriform taxa.

We tested whether the associated elements of the cranial, axial and appendicular skeleton of Siluriformes displayed modularity. More specifically, we tested whether the first centrum and the complex centrum (or centra 1–4 in taxa where centra are not fused) formed an independent module from other skeletal traits and assessed evolutionary modularity across traits. Because of the diversity in morphology of the Weberian apparatus, exhaustive testing of modularity and description of variability could not be done here. Instead, we provide a first step in identifying the modularity of the Weberian centra in Siluriformes, which could have been a crucial component in the evolution of the Weberian apparatus across siluriforms, as also suggested by Duclos et al. (2021) regarding modularity in the zebrafish Weberian apparatus.

The results of the phylogenetic CR analysis support modularity across Siluriformes. We obtained a CR of 0.7787 (p-value: 0.001, Z-score: –3.9701), which indicates significant modularity in the skeleton of Siluriformes. However, pairwise CR ranged from 0.521 to 1.089 (Figure 5B). Furthermore, the results of the phylogenetic PLS analysis across Siluriformes suggest that significant covariation does occur between modules. Indeed, the overall PLS was 0.738 (p-value: 0.001, Z-scores: 5.2145), and pairwise PLS scores ranged from 0.509 to 0.897 (Figure 5).

Centra 1 through 4, which correspond to the Weberian centra in taxa with a Weberian apparatus, were significantly independent of the dorsal-fin complex (CR: 0.930), the pectoral fin complex (CR: 0.765), the cranium (CR: 0.792) and post-Weberian centra (CR: 0.936). This suggests that the first four centra in Siluriformes indeed constitute an independent module. However, significant covariation was found between centra 1 through 4 and the other modules (PLS range: 0.690–0.888).

A CR of 0.521 between the dorsal-fin complex and the pectoral-fin complex suggests that these two elements are independent of each other. The PLS value between the dorsal and pectoral fin complexes was 0.50881 and was not statistically supported (p-value: 0.072), further confirming that the dorsal and pectoral fin complexes do not covary significantly across Siluriformes.

A CR value of 1.089 was found between the cranium and the pectoral fin complex, suggesting that the cranium and pectoral girdle and fins might be more strongly integrated than anticipated. This is reflected in a PLS score of 0.897 between those two modules (p-value: 0.001). This result suggests that the pectoral girdle and the cranium might constitute a single module in which sub-modules are nested.

The Evolution of Weberian Ossicles in Siluriformes

The evolution of the Weberian apparatus has been contentiously and inconclusively debated for years. Some authors have argued that the Weberian apparatus evolved multiple times within Otophysi (Gayet 1986, 1993; Gayet and Chardon 1987), while others (Fink and Fink 1996) argued for a single origin stating, "We do indeed argue on the basis of parsimony that the most complex features known as the Weberian apparatus evolved only once." However, they state further, and most importantly, "We certainly would not argue that it evolved all at once." This conclusion, also held here, is based in part on the re-examination of key fossil otophysan taxa. For example, in a reassessment of †*Chanoides macropoma* from marine Eocene Monte Bolca deposits, Mayrinck et al. (2015) concluded that the fish had a functioning Weberian apparatus, although contrary to Fink and Fink (1996), no tripus with transformator process could be identified and was thus considered absent. They argued that the function of the tripus was assumed by a small process on the third centrum for the insertion of the interossicular ligament. This assessment, in part, supported Patterson (1984) and Chardon et al. (2003), who identified the tripus as small, probably only serving as the point of attachment for the interossicular ligament and leaving the os suspensorium to function as the tripus. The manubrium of the intercalarium is also absent in †*Chanoides*. Chardon and Vandewalle (1997) suggested that the interossicular ligament was the first element of the Weberian apparatus to evolve, and the manubrium of the intercalarium was the last. Unfortunately, all of the extinct catfishes described up to now exhibit a "modern" Weberian apparatus and cannot contribute to this evolution debate (Chardon et al. 2003; Murray and Holmes 2022).

Although there is no evidence for a functioning Weberian apparatus in Gonorynchiformes, the sister group to Otophysi, modifications in fossil and extant gonorynchiform taxa (e.g., †*Ramallichthys*, *Chanos*, †*Notogoneus*, *Gonorynchus*, and *Kneria*) seem to foreshadow the primitive condition in the Otophysi with structures that are homologous to or resemble Weberian structures. Such modifications include an expansion of anterior neural arches and supraneurals forming a roof over the neural canal, a first neural arch that abuts the back of the skull, the presence of a silvery tunic over the anterior gas bladder, the first pleural rib that is shortened and expanded laterally connecting to the tunic of the gas bladder, the gas bladder divided into two parts in *Chanos*, and saccular and lagenar otoliths positioned medially and posteriorly (Rosen and Greenwood 1970; Grande 1992, 1996; Fink and Fink 1996; Grande and Grande 2008; Grande and Arratia 2010). Rosen and Greenwood (1970) also speculatively homologized the intervertebral ligaments of the anterior vertebrae in *Chanos* with the interossicular ligament in otophysans. All of these suggest that gonorynchiform morphology might be close to that of an ancestral proto-Otophysi linkage.

Lecointre and Nelson (1996), following Eschmeyer (1966), explored the now-accepted close relationship between clupeomorphs and ostariophysans (Nelson et al. 2016). Lecointre and Nelson (1996) concluded that the Weberian apparatus could not have evolved as a single unit, stating, "If clupeomorphs and ostariophysans are related, then so too are their otophysic connections." As Eschmeyer suggested, Clupeomorpha and Ostariophysi might have evolved from the same herring-like ancestor; since they share a common history, it should not be surprising to find elements or precursors of the Weberian apparatus in some clupeomorphs, as evidenced by the presence of a "tripus" and modified neural arch elements in the clupeiforms Pristigasteroidei and some members of the Engraulidae (de Pinna and Grande 2003; Grande and de Pinna 2004, fig. 10; Braun and Grande 2008, fig. 4.3). As the two lineages diverged from a common ancestor, the clupeomorphs developed their unique acoustico-lateralis system while the ostariophysans perfected the Weberian apparatus. It, therefore, seems certain that particular Weberian ossicles/structures must have evolved long before they culminated in the functioning sound transducer seen today in otophysans.

This chapter explored Weberian apparatus evolution and axial skeletal regionalization in a novel and objective way. We hypothesized that the Weberian apparatus evolved from a non-modified vertebral column within a modular framework and that heterochronic events involving different modules occurred during its evolution, including both generalized and highly modified siluriform taxa. Results from this study indeed show that the axial skeleton of Siluriformes is modular. As demonstrated in *Ictalurus punctatus*, for example, the first four centra form an independent module from the remainder of the Weberian ossicles and associated elements. Each module theoretically evolves under different selection pressures, resulting in the enormous morphological disparity among Siluriformes and even all Ostariophysi seen today. Quasi-independence of the Weberian apparatus as a functional unit and the demonstrated sub-modules within it possibly allowed for adaptation within diverse environments. As per Chardon et al. (2003), adaptations for diversity and, in some cases, for better audition include: "shortening of the anterior centra in all catfishes and probably the loss of the first two centra in callichthyids, reduction in the size of the ossicles in loricariids and callichthyids, general loss of the articular process of the intercalarium and the scaphium in Callichthyidae, and ossicles incorporated in the neurocranium of advanced Loricarioidei." We suggest that such disparity could only have been accomplished within the framework of modularity.

Conclusion

We conclude that the Weberian apparatus of the relatively generalized siluriform *Ictalurus punctatus* is composed of seven sub-modules: first centrum, complex centrum, supraneural spine, supraneural 4, claustrum, scaphium+intercalarium, and tripus+transverse processes of centrum 4. The first centrum and the complex centrum are especially distinct. Examining modularity across representatives of 25 families of siluriforms, we find that centra 1–4 (first centrum and complex centrum) are modular relative to non-Weberian traits such as associated cranial, appendicular, and post-Weberian elements. Also, both in *I. punctatus* and across families of siluriforms, the supraneural spine and supraneural 4 are modular. These and other examples suggest that patterns of modularity among adults of the single species *I. punctatus* are broadly predictive of patterns of modularity and disparity among siluriform families.

The ostariophysan Weberian apparatus thus stands out as an excellent model for testing axial skeleton regionalization by means of modularity. Results from this study, also supported by the findings of Duclos et al. (2021) for the cypriniform *Danio rerio*, demonstrate that the Weberian apparatus is indeed a module independent from more posterior vertebrae and non-Weberian systems. Within the Weberian apparatus, quasi-independent sub-modules, evolving with respect to different selection pressures, have resulted in the enormous morphological diversity and disparity observed within Otophysi. Looking at Weberian apparatus evolution through the lens of modularity can provide insights into the mechanisms and regulation of morphological diversity. This novel approach provides an objective, repeatable, and testable framework for investigating morphological diversification at different levels of the Tree of Life.

With respect to a better understanding of Weberian apparatus modularity and evolution, the next logical step is to test axial skeleton modularity, particularly the anterior vertebral region, in non-Weberian ostariophysans (i.e., Anotophysi) such as the gonorynchiforms *Chanos* and *Gonorynchus*. Such a study, which is currently underway by the present authors and collaborators, could reveal whether regionalization of the anterior vertebral region began much earlier, somewhere in the ancestry of the Otophysi or the Ostariophysi.

Acknowledgements

We thank Hanna Karic and Claudia Kern (Loyola University) for preparing the CT scans used in the study. Special thanks to Mark Wilson (University of Alberta) for reading various drafts of the manuscript and helping prepare several of the illustrations presented within. Many thanks to Susan Mochel, Caleb McMahan, Kevin Swagel (Field Museum of Natural History), and Mark Sabaj Perez (Academy of Natural Sciences of Philadelphia) for loaning specimens examined for this study. Funding for this study came from the Colburn Foundation LUC 211155 to TG.

References

Adams, D.C. 2016. Evaluating modularity in morphometric data: challenges with the RV coefficient and a new test measure. Methods Ecol. Evol. 7(5): 565–572.

Adams, D.C. and M.L. Collyer. 2016. On the comparison of the strength of morphological integration across morphometric datasets. Evol. 70(11): 2623–2631.

Adams, D.C. and M.L. Collyer. 2019. Comparing the strength of modular signal, and evaluating alternative modular hypotheses, using covariance ratio effect sizes with morphometric data. Evol. 73(12): 2352–2367.

Adams, D.C., M.L. Collyer, A. Kaliontzopoulou and E.K. Baken. 2022. Geomorph: Software for geometric morphometric analyses. R package version 4.0.4. https://cran.r-project.org/package=geomorph.

Alexander, R.M. 1964. The structure of the Weberian apparatus in the Siluri. Proc. Zool. Soc. London. 142: 419–440.

Alexander, R.M. 1965. Structure and function in the catfish. J. Zool. (Lond.) 148: 88–152.

Alfaro, M.E. 2018. Resolving the ray-finned fish tree of life. Proc. Nat. Acad. Sci. 115(24): 6107–6109.

Al-Rawi, A.H.A. 1966. The Development of the Weberian Apparatus and Swim Bladder in the Channel Catfish, *Ictalurus punctatus* (Rafinesque). Ph.D. Thesis, University of Oklahoma, Norman, Oklahoma, USA.

Arratia, G. 1987. Description of the primitive family Diplomystidae (Siluriformes, Teleostei, Pisces): morphology, taxonomy and phylogenetic implications. Bonner Zool. Monogr. 24: 1–120.

Balinsky, B.I. 1961. An Introduction to Embryology. W.B. Saunders Co., Philadelphia.

Bamford, T.W. 1948. Cranial development of *Galeichthys felis*. Proc. Zool. Soc. Lond. 118: 364–391.

Baken, E.K., M.L. Collyer, A. Kaliontzopoulou and D.C. Adams. 2021. geomorph v4.0 and gmShiny: enhanced analytics and a new graphical interface for a comprehensive morphometric experience. Methods Ecol. Evol. 12: 2355–2363.

Baskin, J.N. 1972. Structure and Relationships of the Trichomycteridae. Ph.D. Thesis, City University of New York, USA.

Bird, N.C. and P.M. Mabee. 2003. Developmental morphology of the axial skeleton of the zebrafish *Danio rerio* (Ostariophysi: Cyprinidae). Devel. Dyn. 228: 337–357.

Bird, N.C. and L.P. Hernandez. 2007. Morphological variation in the Weberian apparatus of Cypriniformes. J. Morph. 268(9): 739–757.

Bird, N.C. and L.P. Hernandez. 2009. Building an evolutionary innovation: Differential growth in the modified vertebral elements of the zebrafish Weberian apparatus. Zool. 112(2): 97–112.

Bird, N.C., S.S. Richardson and J.R. Abels. 2020. Histological development and integration of the zebrafish Weberian apparatus. Devel. Dyn. 249(8): 998–1017.

Brown, B.A. and C.J. Ferraris. 1988. Comparative osteology of the Asian catfish family Chacidae with description of a new species from Burma. Amer. Mus. Novit. 2907: 1–16.

Braun, C. and T. Grande. 2008. Evolution of peripheral mechanisms for the enhancement of sound reception. pp. 99–144. *In*: Webb, J., R. Fay and A. Popper (eds.). Fish Bioacoustics, Springer, New York.

Burdi, A. and T. Grande. 2010. Morphological development of the axial skeleton of *Esox lucius* and *Esox masquinongy* (Esociformes: Euteleostei), with comparisons between developmental rates. pp. 411–430. *In*: Nelson, J.S., H-P. Schultze and M.V.H. Wilson (eds.). Origin and Phylogenetic Interrelationships of Teleosts. Verlag Dr. Fredrich Pfeil. München.

Burns, M.D., M.L. Collyer and B.L. Sidlauskas. 2023. Simultaneous integration and modularity underlie the exceptional body shape diversification of characiform fishes. Evol. 77(3): 746–762.

Chang, J., D.L. Rabosky, S.A. Smith and M.E. Alfaro. 2019. An R package and online resource for macroevolutionary studies using the ray-finned fish tree of life. Methods Ecol. Evol. 10(7): 1118–1124.

Chardon, M. 1968. Anatomie comparé de l'appareil de Weber et des structures connexes chez les Siluriformes. Ann. Mus. Roy. Afr. Cent., serie 8, 169: 1–277.

Chardon, M. and P. Vandewalle. 1997. Evolutionary trends and possible origin of the Weberian apparatus. Netherl. J. Zool. 47: 371–385.

Chardon, M., E. Parmentier and P. Vandewalle. 2003. Morphology, development and evolution of the Weberian apparatus in catfish. pp. 71–120. *In*: Arratia, G., B.G. Kapoor, M. Chardon and R. Diogo (eds.). Catfishes. Scientific Publishers Inc. Enfield, New Hampshire, USA.

Cloutier, R. and T. Grünbaum. 2007. Cypriniform caudal fin and Weberian apparatus: ontogeny and phylogeny. Copeia 2007: 91–92.

Coburn, M.M. and L.M. Futey. 1996. The ontogeny of supraneurals and neural arches in the cypriniform Weberian apparatus (Teleostei: Ostariophysi). Zool. J. Linn. Soc. 116: 333–346.

Coburn, M.M. and P.G. Grubach. 1998. Ontogeny of the armored catfish *Corydoras paleatus* (Siluriformes: Callichthyidae). Copeia 1998: 301–311.

Duclos, K.K., T.C. Grande and R. Cloutier. 2021. Modularity of the Weberian apparatus in the zebrafish using micro-CT technology and 3-D geometric morphometrics. pp. 59–70. *In*: Denton, J.S.S. and P. Janvier (eds.). Ancient Fishes and their Living Relatives. Verlag Dr. F. Pfeil, München, Germany.

Eschmeyer, W.N. 1966. Re-evaluation of the position of the family Dussumieriidae in clupeoid phylogeny, with comments on lower teleostean phylogeny. Forty-sixth Annual Meeting of the Society of Ichthyology and Herpetology, Miami, FL.

Evans, K.M., M. Vidal-García, V.A. Tagliacollo, S.J. Taylor and D.B. Fenolio. 2019. Bony patchwork: mosaic patterns of evolution in the skull of electric fishes (Apteronotidae: Gymnotiformes). Integr. Comp. Biol. 59(2): 420–431.

Evans, K.M., T.J. Buser, O. Larouche and M.A. Kolmann. 2023. Untangling the relationship between developmental and evolutionary integration. Sem. Cell Dev. Biol. 145: 22–27.

Fink, S.V. and W.L. Fink. 1981. Interrelationships of ostariophysan fishes (Ostariophysi). Zool. J. Linn. Soc. 72: 297–353.

Fink, S.V. and W.L. Fink. 1996. Interrelationships of ostariophysan fishes. pp. 209–249. *In*: Stiassny, M., L.R. Parenti and G.D. Johnson (eds.). Interrelationships of Fishes. Academic Press, San Diego.

Gartner, S.M., O. Larouche, K.M. Evans and M.W. Westneat. 2023. Evolutionary patterns of modularity in the linkage systems of the skull in wrasses and parrotfish. Int. Org. Biol. 5(1): obad035.

Gayet, M. 1985. Rôle de l'évolution de l'appareil de Weber dans la phylogénie des Ostariophysi, suggéré par un nouveau Characiforme du Cénomanien supérieur marin de Portugal. C. R. séances Acad. Sci. Paris, Sér. 2 300: 895–898.

Gayet, M. 1986. Problème de l'origine des osselets de Weber. Océanis 12: 357–366.

Gayet, M. and M. Chardon. 1987. Possible otophysic connections in some fossil and living ostariophysan fishes. Proc. Congr. Eur. Ichthyol. 5th, Stockholm, 1985, pp. 31–42.

Gayet, M. 1993. Rélations phylogénétiques de Gonorhynchiformes (Ostariophysi). Belg. J. Zool. 123: 165–193.

Grande, L. and J.G. Lundberg. 1988. Revision and redescription of the genus *Astephus* (Siluriformes: Ictaluridae) with a discussion of its phylogenetic relationships. J. Vert. Paleo. 8(2): 139–171.

Grande, L. and M.C C. de Pinna. 1998. Description of a second species of catfish †*Hypsidoris* and a reevaluation of the genus and the family Hypsidoridae. J. Vert. Paleo. 18: 451–474.

Grande, T. 1992. Higher Interrelationships of Fossil and Recent Gonorynchiform Fishes. Ph.D. Thesis, University of Illinois at Chicago, USA.

Grande, T. 1996. The interrelationships of fossil and Recent gonorynchid fishes and the inclusion of two Cretaceous taxa from Israel. pp. 296–315. *In*: Arratia, G. and G. Viohl (eds.). Mesozoic Fishes—Systematics and Paleoecology. Verlag Dr. Friedrich Pfeil, München.

Grande, T. and J.D. Shardo. 2002. Morphology and development of the postcranial skeleton in the channel catfish *Ictalurus punctatus* (Ostariophysi: Siluriformes). Fieldiana Zool. 1518(99): 1–30.

Grande, T. and B. Young. 2004. The ontogeny and homology of the Weberian apparatus in the zebrafish *Danio rerio* (Ostariophysi: Cypriniformes). Zool. J. Linn. Soc. 140: 241–254.

Grande, T. and M. de Pinna. 2004. The evolution of the Weberian apparatus: A phylogenetic perspective. pp. 429–448. *In*: Arratia, G. and A. Tintori (eds.). Mesozoic Fishes 3—Systematics, Paleoenvironments and Biodiversity. Verlag Dr. Friedrich Pfeil, Munich, Germany.

Grande, T. and L. Grande. 2008. Reevaluation of the gonorynchiform genera †*Ramallichthys*, †*Judeichthys* and †*Notogoneus*, with comments on the families †Charitosomidae and Gonorynchidae. pp. 295–310. *In*: Arratia, G., H.-P. Schultze and M.V.H. Wilson (eds.). Mesozoic Fishes 4—Homology and Phylogeny. Verlag Dr. Friedrich Pfeil, München.

Grande, T. and G. Arratia. 2010. Morphological analysis of the gonorynchiform postcranial skeleton. pp. 39–71. *In*: Grande, T., F.J. Poyato-Ariza and R. Diogo (eds.). Gonorynchiformes and Ostariophysan Relationships: A Comprehensive Review. Scientific Publishers, Inc., New Hampshire.

Grande, T. and C. Braun. 2012. Evolution of the Weberian apparatus. Bioacoustics 12: 120–122.

Hallgrímsson, B., H. Jamniczky, N.M. Young, C. Rolian, T.E. Parsons, J.C. Boughner et al. 2009. Deciphering the palimpsest: studying the relationship between morphological integration and phenotypic covariation. Evol. Biol. 36: 355–376.

Hallgrímsson, B., D.C. Katz, J.D. Aponte, J.R. Larson, J. Devine, P.N. Gonzalez et al. 2019. Integration and the developmental genetics of allometry. Int. Comp. Bio. 59(5): 1369–1381.

Hansen, T.F. 2003. Is modularity necessary for evolvability? Remarks on the relationship between pleiotropy and evolvability. Biosystems 69(2–3): 83–94.

Klingenberg, C.P. and G.S. McIntyre. 1998. Geometric morphometrics of developmental instability: analyzing patterns of fluctuating asymmetry with Procrustes methods. Evol. 52: 1363–1375.

Klingenberg, C.P. 2008. Morphological integration and developmental modularity. Ann. Rev. Ecol. Evol. Syst. 39: 115–132.

Klingenberg, C.P. 2013. Cranial integration and modularity: insights into evolution and development from morphometric data. Hystrix 24(1): 43.

Klingenberg, C.P. 2014. Studying morphological integration and modularity at multiple levels: concepts and analysis. Phil. Trans. R. Soc. B: Biol. Sci. 369(1649): 20130249.

Klingenberg, C.P. 2022. Methods for studying allometry in geometric morphometrics: a comparison of performance. Evol. Ecol. 36(4): 439–470.

Kubicek, K.M. 2022. Developmental osteology of *Ictalurus punctatus* and *Noturus gyrinus* (Siluriformes: Ictaluridae) with a discussion of siluriform bone homologies. Vert. Zool. 72: 661–727.

Ladich, F. 1999. Did auditory sensitivity and vocalization evolve independently in otophysan fishes? Brain. Behav. Evol. 53: 288–304.

Ladich, F. 2000. Acoustic communication and the evolution of hearing in fishes. Phil. Trans. Roy. Soc. Lond. B 355: 1285–1288.

Ladich, F. 2023. Hearing in catfishes: 200 years of research. Fish and Fisheries 24(4): 618–634. https://doi.org/10.1111/faf.12751.

Lechner, W., E. Heiss, T. Schwaha, M. Glösmann and F. Ladich. 2011. Ontogenetic development of Weberian ossicles and hearing abilities in the African bullhead catfish. PLoS ONE 6(4): e18511. doi:10.1371/journal.pone.0018511.

Lecointre, G. and G. Nelson. 1996. Clupeomorpha, sister-group of Ostariophysi. pp. 193–207. *In*: Stiassny, M.L.J., L.R. Parenti and G.D. Johnson (eds.). Interrelationships of Fishes. Academic Press, San Diego.

Lundberg, J.G. 1975. Homologies of the upper shoulder girdle and temporal region bones in catfishes (Order Siluriformes), with comments on the skull of Helogeneidae. Copeia 1975: 66–74.

Lundberg, J.G. 1982. The comparative anatomy of the toothless blindcat, *Trogloglanis pattersoni* Eigenmann, with a phylogenetic analysis of the ictalurid fishes. Misc. Pub. Mus. Zool., Univ. Mich. 163: 1–85.

Lundberg, J.G. and L.A. McDade. 1986. A redescription of the rare Venezuelan catfish *Brachyrhamdia imitator* Myers (Siluriformes, Pimelodidae) with phylogenetic evidence for a large intrafamilial lineage. Acad. Nat. Sci. Phil. 463: 1–24.

Matveive, B. 1929. Die Entwicklung der vorderen Wirbel und des Weber'schen Apparates bei Cyprinidae. Zool. Jahrb. Anat. 51: 463–534.

Mayrinck, D., P.M. Brito and O. Otero. 2015. Review of the osteology of the fossil fish formerly attributed to the genus †*Chanoides* and implications for the definition of otophysan bony characters. J. Syst. Paleo. 13(5): 397–420.

Murray, A.M. and R.B. Holmes. 2022. Osteology of the cranium and Weberian apparatus of African catfish families (Teleostei: Ostariophysi: Siluriformes) with an assessment of genera from the Palaeogene of Africa. Vert. Anat. Morph. Paleo. 9: 156–191.

Nelson, J.S., T.C. Grande and M.V.H. Wilson. 2016. Fishes of the World. John Wiley & Sons, Hoboken, New Jersey.

Patterson, C. 1984. *Chanoides*, a marine Eocene otophysan fish (Teleostei: Ostariophysi). J. Vert. Paleo 4: 430–456.

de Pinna, M.C.C. 1992. A new subfamily of Trichomycteridae (Teleostei, Siluriformes), lower loricarioid relationships and a discussion on the impact of additional taxa for phylogenetic analysis. Zool. J. Linn. Soc. 106(3): 175–229.

de Pinna, M.C.C. and T. Grande. 2003. Ontogeny of the accessory neural arch in pristigasteroid clupeomorphs and its bearing on the homology of the otophysan claustrum (Teleostei). Copeia 2003: 838–845.

de Pinna, M.C.C., L. Peixoto, V. Tagliacollo and M. Britto. 2025. Phylogenetic relationships and evolution of the major groups of Siluriformes. *In*: Arratia, G. and R.E. Reis (eds.). Catfishes, a Highly Diversified Group. Volume 2: Evolution and Phylogeny. CRC Press, Boca Raton, FL.

Raff, E.C. and R.A. Raff. 2000. Dissociability, modularity, evolvability. Evol. Devel. 2(5): 235–237.

Rasskin-Gutman, D. 2005. Modularity: jumping forms within morphospace. pp. 207–219. *In*: Callebaut, W. and D. Rasskin-Gutman (eds.). Modularity: Understanding the Development and Evolution of Natural Complex Systems. MIT Press, Boston.

R Core Team. 2023. R: A Language and Environment for Statistical Computing. R Foundation for Statistical Computing, Vienna, Austria. <https://www.R-project.org/>.

Reed, H.D. 1924. The morphology and growth of the spines of siluroid fishes. J. Morph. 38: 431–451.

Rohlf, F.J. and M. Corti. 2000. Use of two-block partial least-squares to study covariation in shape. Sys. Biol. 49(4): 740–753.

Rosen, D.E. and P.H. Greenwood. 1970. Origin of the Weberian apparatus and the relationships of the ostariophysan and gonorynchiform fishes. Amer. Mus. Novit. 2428: 1–25.

Schaefer, S.A. 1987. Osteology of *Hypostomus plecostomus* (Linnaeus) with a phylogenetic analysis of the loricariid subfamilies (Pisces: Siluroidea). Contr. Sci., Nat. Hist. Mus. Los Angeles County 394: 1–31.

Sidlauskas, B.L., J.H. Mol and R.P. Vari. 2011. Dealing with allometry in linear and geometric morphometrics: a taxonomic case study in the *Leporinus cylindriformis* group (Characiformes: Anostomidae) with description of a new species from Suriname. Zool. J. Linn. Soc. 162: 103–130.

Vandewalle, P., F. Radermaker, C. Surlemont and M. Chardon. 1990. Apparition of the Weberian characters in *Barbus barbus* (Pisces, Cyprinidae). Zool. Anz. 225: 262–376.

Vasseur, F., A.J. Westgeest, D. Vile and C. Violle. 2022. Solving the grand challenge of phenotypic integration: allometry across scales. Genetica 150(3–4): 161–169.

Von Dassow, G. and E. Munro. 1999. Modularity in animal development and evolution: elements of a conceptual framework for EvoDevo. J. Exper. Zool. 285(4): 307–325.

Watson, J.M. 1939. The development of the Weberian ossicles and anterior vertebrae in the goldfish. Proc. Roy. Soc. Lond. Ser. B. 127: 452–472.

Weber, E.H. 1820. De aure et auditu hominis et animalium, Pars I. De Aure Animalium. Aquatilium, Leipzig.

Zelditch, M.L. and A. Goswami. 2021. What does modularity mean? Evol. Devel. 23(5): 377–403.

Chapter 5

Diversity and Evolution of the Pelvic Girdle and Fin in Catfishes

Timo Moritz,[1,*] *Ann-Katrin Koch*[2] and *Philipp Thieme*[3]

With more than 4,200 species (Fricke et al. 2024), catfishes represent a species-rich taxon. They exhibit an enormous diversity of body shapes and life history strategies (Teugels 1996; Armbruster 2011; Rizzato and Almeida this volume). Accordingly, there is much diversity in their skeletal anatomy. This also holds true for the pelvic girdle, a structure often given little attention in osteological teleostean studies. Compared to other major taxa, siluriforms show an extremely high diversity in their pelvic-girdle morphology, ranging from simple rounded basipterygia to forms with several processes, from large structures that cover a significant part of the belly to completely reduced girdles, or those with additional elements of questionable homology.

A usual set of fins in primary aquatic vertebrates comprises one or two dorsal fins, a caudal fin, an anal fin, and paired pectoral and pelvic fins. The latter were the last appearing in the evolution of vertebrates in some derived jaw-less fish taxa (Larouche et al. 2017). In contrast, the ontogenetic and evolutionary development of unpaired fins can be explained as derivatives of the median fin fold, such as a precursor missing for the paired fins. The theory of lateral fin folds was and still is discussed (e.g., Thacher 1877; Balfour 1878; Shubin et al. 1997; Don et al. 2013; Abe and Ota 2017; Diogo 2020), but indeed only limited evidence can support it. Thus, the evolutionary origin of the paired fins remains nebulous.

The complexity and diversity of pelvic girdles in extant non-neoteleost actinopterygians is, in general, rather low. In *Amia* and *Lepisosteus*, the basipterygium is little more than an osseous bar with about three radials supporting the fin rays (Grande and Bemis 1998; Grande 2010). Most non-neoteleost teleosts have the posteromedial part enlarged by which the basipterygia of both sides approach or come into contact (e.g., Grande 1985; Sanford 2000; Hilton 2002). Within the Otophysi, additional structures appear: cypriniforms and some characiforms show an additional anterior process, and a posterior process can be present. In siluriforms, finally, the diversity of pelvic-girdle morphologies is extremely high.

[1] Ocean Museum Germany, Katharinenberg 14-20, 18439 Stralsund, Germany, and Institut für Biowissenschaften, Universität Rostock, Albert-Einstein-Straße 3, 18059 Rostock, Germany.

[2] Ocean Museum Germany, Katharinenberg 14-20, 18439 Stralsund, Germany, and Institut für Biowissenschaften, Universität Rostock, Albert-Einstein-Straße 3, 18059 Rostock, Germany.

[3] Ocean Museum Germany, Katharinenberg 14-20, 18439 Stralsund, Germany, and MARBEC, Université de Montpellier, IRD, CNRS, cc093, Place E. Bataillon, 34095 Montpellier Cedex 05, France.

Emails: ann-katrin.koch@meeresmuseum.de; philipp.thieme@meeresmuseum.de

* Corresponding author: timo.moritz@meeresmuseum.de

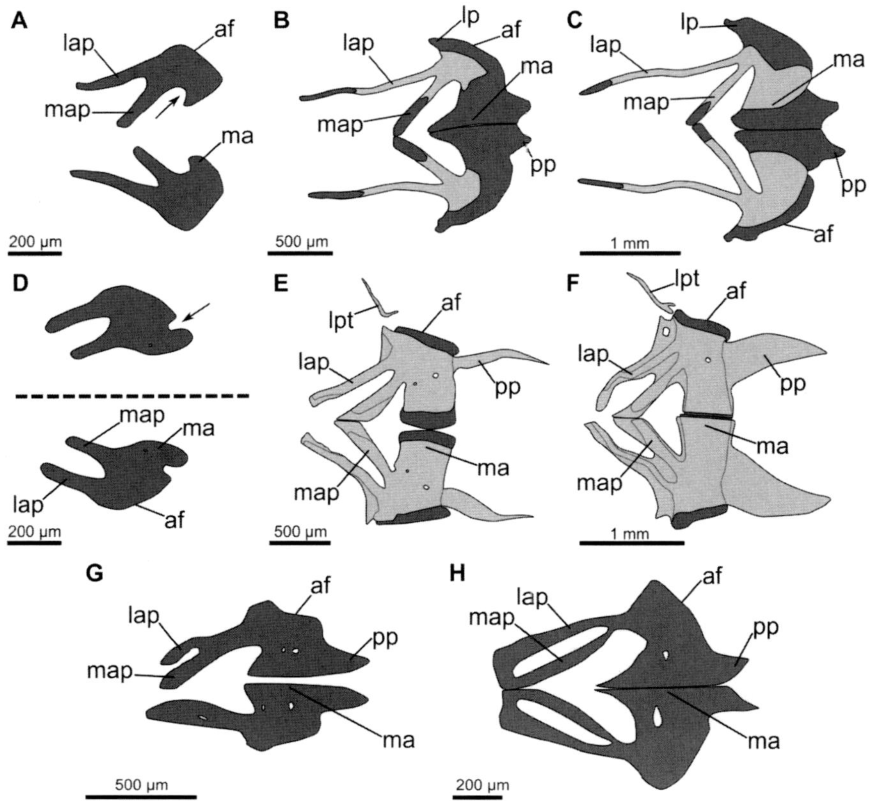

Figure 1. Ontogenetic stages of the pelvic girdle in catfishes in ventral view; fin rays omitted; light grey: bone; dark grey: cartilage. (A–C) *Clarias gariepinus* (Clariidae), 14.5 mm (A) arrow pointing to anterior notch; 30.1 mm (B); 32.3 mm (C) standard length (SL). (D–F) *Ancistrus* cf. *dolichopterus* (Loricariidae), 7.5 mm (D) arrow pointing to posterior notch, 11.2 mm (E), 14.2 mm (F) SL, dashed line in (D) indicates wide separation of both halves by yolk sack, lateropterygium displayed only on one side (E–F). (G) *Phractura clauseni* (Amphiliidae), DMM IE/11124, 11.9 mm SL. (H) *Amphilius atesuensis* (Amphiliidae), DMM IE/11706, 11.7 mm SL. Abbreviations: af: articulatory facet of fin rays; lap: lateral anterior process; lp: lateral process; lpt: lateropterygium; ma: medial arm; map medial anterior process; pp: posterior process.

For this overview, we studied 71 genera out of 33 siluriform families (Appendix 1). Most specimens belonging to different species were cleared and double stained using Alizarin red for bone and Alcian blue for cartilage (Dingerkus and Uhler 1977; Taylor and van Dyke 1985; Thieme et al. 2021). Several pelvic girdles were carefully removed from the specimens. To get information on the development, we studied the ontogenetic series of the pelvic girdle in *Corydoras*, *Ancistrus* (Loricarioidei), and *Clarias gariepinus* (Siluroidei), as well as single ontogenetic stages of additional species (Figure 1). Herein, we give an overview of the catfishes' pelvic-girdle diversity and try to trace back the possible evolution of their morphology based on recent phylogenetic hypotheses (Sullivan et al. 2006; Schedel et al. 2022; de Pinna et al. 2025).

Ontogeny

The pelvic fin, with its internal support structures, is the last fin to start developing in catfishes. Eggs of siluriforms are often large, and embryos and larvae have a big yolk sac that can persist quite long. In some species, when the basipterygial cartilage first emerges, the left and right halves can be widely separated by the yolk sac, e.g., *Ancistrus* and *Ictalurus* (Arratia 2003). With the depletion and disappearance of the yolk sac, the two halves approach each other. In other species, the yolk sac may already be depleted when the basipterygial cartilages develop, e.g., *Corydoras* and *Clarias*, and both halves are already in proximity. The first stages of pelvic-girdle development are difficult to observe, and no data on the early shape of the basipterygial cartilages is available at this point. Once the basipterygial cartilages are more developed and the first fin rays emerge, different shapes are observed in the studied species: In *Ancistrus,* a broad rectangular/trapezoid main body serves as an articulation site for the fin rays (Figure 1D). Anteriorly, two rod-shaped outgrowths indicate the Anlagen of the lateral and medial anterior processes. Medially, an oval portion is connected to the main body, representing the future medial arm. A posterior notch is present between the main part and the medial arm. The early shape of the basipterygial cartilages in *Clarias* (arrow in Figure 1A) and *Trichomycterus* (Arratia 2003, fig. 4.7C) is similar to that of *Ancistrus*. However, the medial arm is separated from the main body by an anterior notch (arrow in Figure 1D). In *Phractura* (Figure 1G) and *Amphilius* (Figure 1H), the main body of the basipterygial cartilage becomes narrower anteriorly, forming a transition zone towards the two rod-shaped outgrowths. In *Corydoras* (Huysentruyt 2008) and *Phractura* soon, these two outgrowths fuse, forming only one anterior process. In *Ictalurus*, a posterior outgrowth indicating the root of the future posterior process becomes visible around the same time the anterior processes start to emerge (Arratia 2003, fig. 4.7E [remark: anterior and posterior are switched in the drawing]; Kubicek 2022, fig. 30). As development progresses, the cartilaginous anterior processes lengthen slowly (Figure 1). Also, the medial arms grow medially towards and eventually touch each other. Further, the medial arms can widen anteriorly and posteriorly, e.g., in *Amphilius*, *Clarias*, *Phractura*, *Trichomycterus* (Arratia 2003, fig. 4.7D labelled internal anterior process). In *Ictalurus*, a cartilaginous thread extends from the posterior outgrowth of the medial arm, representing the posterior process (Kubicek 2022, fig. 30). In *Diplomystes*, a third anterior process develops in between the medial anterior process and the medial arm (Arratia 2003, fig. 4.7A, labelled medial anterior process). Further, a posterior process emerges in a rod-like shape from the posteromedial corner of the medial arm. In *Amphilius*, when the anteromedial and anterolateral processes reach their relative length, they turn towards each other and fuse later at their tips. As a result, a semicircle completes each half of the basipterygium anteriorly (Figure 1H). In *Clarias* and *Ictalurus*, a lateral process develops as an outgrowth of the main body anterior to the fin rays (Figure 1B and C). Late in development, when ossification has already started, a cartilaginous and separate radial emerges in *Phractura* between the proximal tips of the first ray.

Perichondral ossification begins in the centre of the main body and proceeds anteriorly at first and medially later. The symphyseal area between the medial arms of both halves remains cartilaginous (see below for adult morphologies). The posterior portion of the medial arm does not ossify similarly in all the species. While in *Ictalurus*, the posterior process does not ossify as part of the medial arm and, therefore, remains cartilaginous (Kubicek 2022, fig. 30), in others, such as *Trichomycterus* or *Clarias*, a larger portion of the posterior process ossifies as part of the medial arm and as a result, a notch between the ossified portion of the posterior process and articulation site of the rays occurs on the main body. In *Diplomystes nahuelbutaensis* a separate ossification of the posterior process from the posteromedial corner of the medial arm was reported (Arratia 2003, fig. 4.7C). However, such a separation is not visible in adult specimens of *Diplomystes* or ontogenetic stages of other catfishes studied so far. In loricariids like *Ancistrus* (Figure 1E) and *Hypostomus* the posterior process has no cartilaginous precursor and develops through membrane bone that first extends posteriorly and later medially.

Additionally, membrane bone can develop lateral and medial to the anterior processes and/or lateral to the main body (for more examples, see below). Often, the posterior area between both anterior processes is filled with membrane bone. The tips of the anterior processes remain cartilaginous in many species (see description of adult morphologies below); they are fully ossified, and membrane bone grows around them in *Ancistrus* and *Hypostomus*. Also, in *Ancistrus* and *Hypostomus* another element develops, the lateropterygium (Figure 1E and F; Figure 2A–D). It is positioned anterior to the first ray and lateral to the main body. On its first appearance, it is L-shaped,

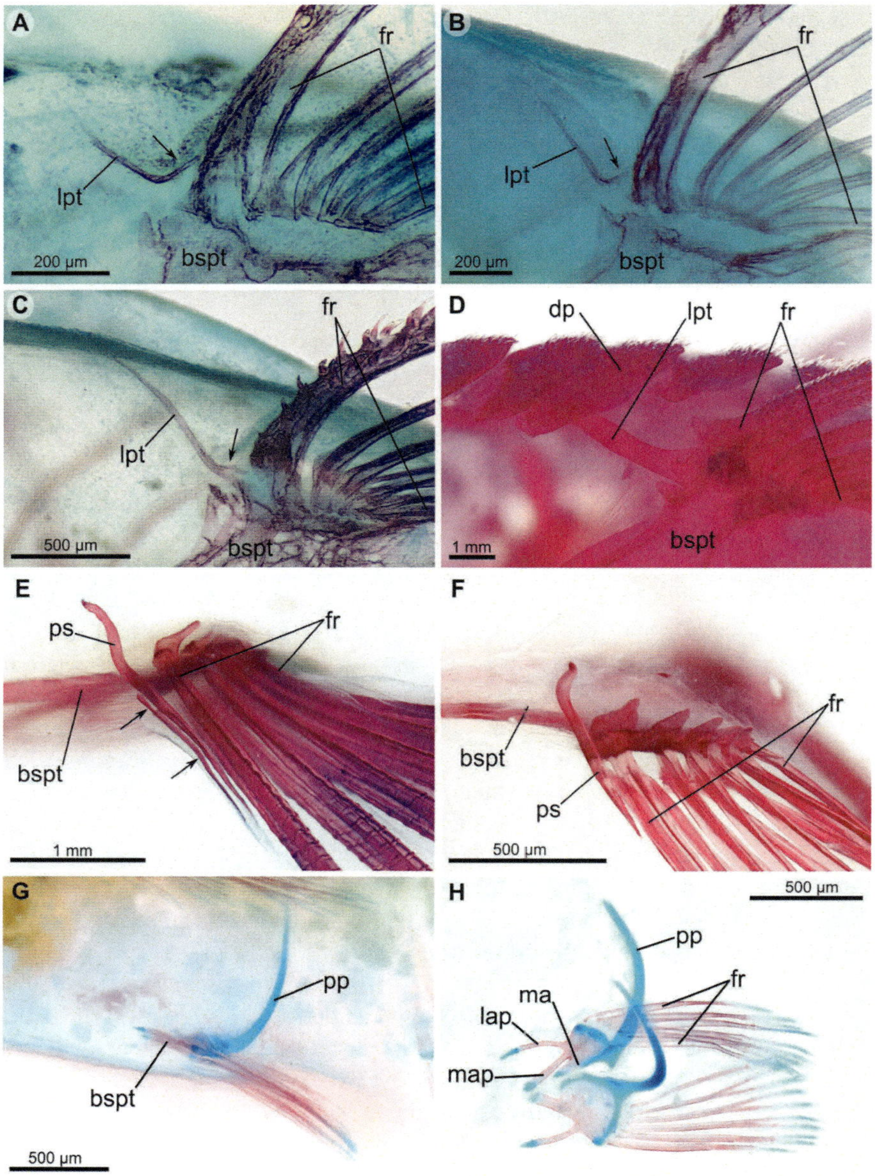

Figure 2. Details of pelvic girdles of catfishes; cleared and double-stained specimens. (A–D) Ontogeny of the lateropterygium in *Ancistrus* cf. *dolichopterus*, ventrolateral view, the arrow pointing to the distal part of the lateropterygium, 10.3 mm (A); 11.2 mm (B); 14.2 mm (C); DMM IE/11121, 53.2 mm SL. (E) *Diplomystes mesembrinus* (Diplomystidae), pelvic girdle in lateral view, DMM IE/17476, arrows pointing to supplementary splint elements. (F) *Schilbe intermedius* (Schilbeidae), pelvic girdle in lateral view, DMM IE/12209. (G–H) *Parailia pellucida* (Schilbeidae), DMM IE/16547, pelvic girdle in lateral view *in situ* (G), isolated pelvic girdle in oblique dorso-lateral view (H). Abbreviations: bspt: basipterygium; dp: dermal plate; fr: fin rays; lap: lateral anterior process; lpt: lateropterygium; ma: medial arm; map: medial anterior process; pp: posterior process; ps: pelvic splint.

with a distal part parallel to the first ray and a proximal part reaching laterodorsally (Figure 2A). During growth, the distal part becomes relatively smaller (Figure 2B, C), while the proximal part grows and later becomes a rod-like, large structure of the pelvic girdle (Figure 2D). This structure has no cartilaginous precursor and emerges clearly later than all fin rays. Its early stages are equal in shape and position to the pelvic splint of other catfish taxa (see adults below).

Pelvic Girdle Diversity in Catfishes

Diplomystidae

Diplomystids, with only nine species, have been regarded as the most primitive catfish taxon in terms of morphology and phylogenetic position (Arratia 1987; Teugels 1996; de Pinna 1998; de Pinna et al. 2025). Molecular studies did not clearly reveal diplomystids as a sister taxon to all other siluriforms but rather as sisters to the Siluroidei (Sullivan et al. 2006; Kappas et al. 2016; Schedel et al. 2022). The basal splits in catfish phylogeny, however, have only little support, and in many regards, *Diplomystes* seem to show morphological characters close to the supposed basal condition (Arratia 1987). This also seems to hold true for the pelvic-girdle morphology.

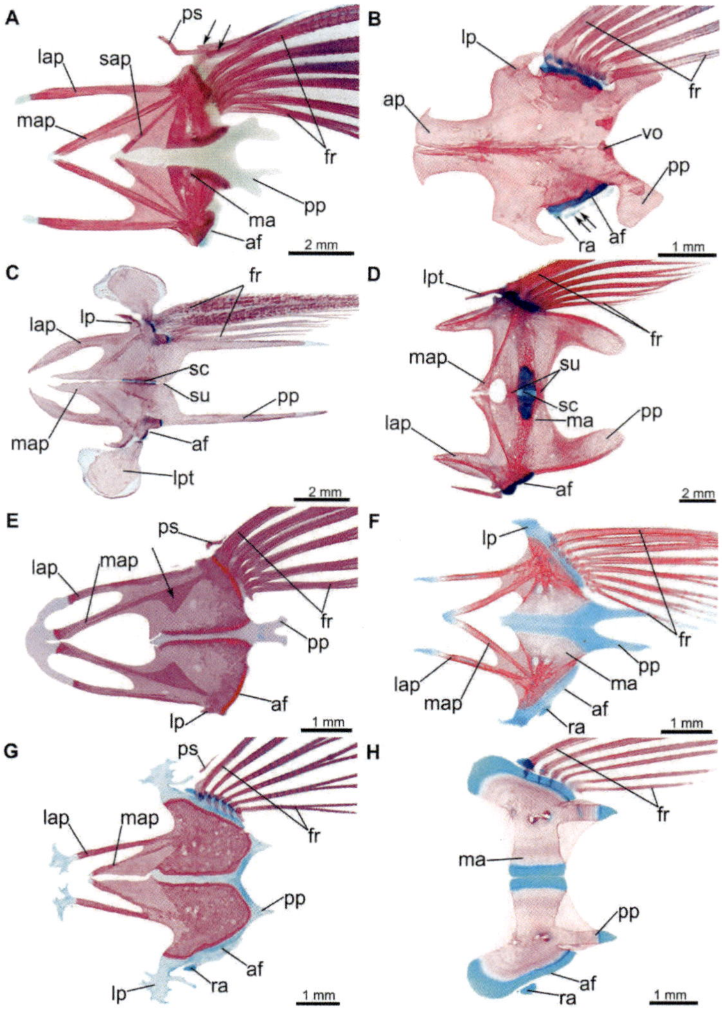

Figure 3. Pelvic girdles of catfishes; cleared and double-stained specimens in ventral view; fin rays on one side removed. (A) *Diplomystes mesembrinus* (Diplomystidae), DMM IE/17476, arrows pointing to supplementary splint elements. (B) *Corydoras paleatus* (Callichthyiidae), unreg.; arrows pointing to supplementary small cartilages. (C) *Astroblepus* sp. 'Colombia' (Astroblepidae), unreg.; (D) *Ancistrus* cf. *dolichopterus* (Hypostominae/Loricariidae), DMM IE/11119. (E) *Glyptothorax lampris* (Sisoridae), DMM IE/16488, the arrow pointing to ventral flange. (F) *Pseudauchenipterus nodosus* (Auchenipteridae), DMM IE/16996. (G) *Microglanis poecilius* (Pseudopimelodidae), DMM IE/17249. (H) *Aspredo aspredo* (Aspredinidae), DMM IE/16949. Abbreviations: af: articulatory facet of fin rays; ap: anterior process; fr: fin rays; lap: lateral anterior process; lp: lateral process; lpt: lateropterygium; ma: medial arm; map: medial anterior process; pp: posterior process; ps: pelvic splint; ra: radial; sap: supplementary anterior process; sc: synchondral cartilage; su: suture; vo: ventral outgrowth.

Diplomystids usually have three anteriorly directed rod-like processes, longer and more robust from lateral to medial (Figure 3A; Arratia 1987), although, in Fink and Fink (1981, fig. 21E), a specimen with only two anterior processes is depicted. All these processes are pre-formed in cartilage during ontogeny and keep cartilaginous tips in larger specimens (Arratia 1987). The supplementary and medial processes touch the tip of the respective process from the other side. Thin membrane bone is present at least on its bases between all processes, as well as between the supplementary one and the medial arm of the basipterygium (Figure 3A). The symphyseal cartilage is present from the supplementary process to the posterior corner or the medial arm and is in contact with the basipterygium of the other side. Also, it is continuous with the posterior process of the basipterygium, which remains cartilaginous. In our specimen, the posterior process is bifurcated (Figure 3A). In the posterolateral area, the lateropterygium forms an articulatory facet for the fin rays. A pelvic radial is reported from *Diplomystes chilensis* (de Pinna 1993, fig. 64) and from a young *Diplomystes viedmensis* (Arratia 1987, fig. 31). Otherwise, radials are not present in adult diplomystids. Arratia (1987) assumed that the radial in *Diplomystes* gets lost during growth rather than being fused to the basipteryium. *Diplomystes* has six rays, all segmented, with the most lateral one unbranched. Anterior of these regular rays there is a prominent pelvic splint (Figures 2E and 3A). The pelvic splint in *Diplomystes* features a smaller ventral counterpart, and sometimes even a very tiny pre-splint (Figure 2E) is present. Remarkable is that the pelvic splint, at least in larger diplomystids, shows some segmentation towards its tip (Figure 2E; Arratia 1987).

Loricarioidei

Within Loricarioidei, the pelvic-girdle morphology of Trichomycteridae and Nematogenyidae closely resembles that of diplomystids. Trichomycterids, by their number of species and subfamilies, provide quite some diversity, but in most cases, the basipterygium possesses two anteriorly directed processes (Figure 4A and B; Arratia 1987, 2003; Silva Pedroza 2009; Costa 2021). In some species, even a third, medial positioned process is present, which is prominent in *Trichomycterus giganteus* (Costa 2021, fig. 6A) and *T. punctulatus* (Silva Pedrosa 2009, fig. 12) or rather short in *Silvinichthys mendozensis* (Arratia 1998, fig. 11b). In most cases, both basipterygia are in tight contact medially via the symphyseal cartilage. Sometimes both halves are or appear to be firmly connected, e.g., in *Ituglanis agreste* (Lima et al. 2013) or in *Vandellia* sp. where it is even elongated anteriad (Silva Pedroza 2009, fig. 13). Posterior processes in trichomycterids are small or absent with few exceptions, e.g., in *Ochmacanthus* the posterior process is long and ossified (Silva Pedroza 2009, fig. 14). At least a part of the articulatory facet is cartilaginous, and radials are absent. A pelvic splint is usually present: it is variable in size between taxa and can be completely reduced in small species such as *Tridentopsis cahuali* (Figure 4B). Generally, in smaller trichomycterids, the pelvic girdle may be delicate. A complete loss of the pelvic girdle happened independently in several lineages in trichomycterids, e.g., *Eremophilus*, *Listura*, *Miuroglanis*, and certain species of *Trichomycterus* and *Silvinichthys* (Datovo 2014). According to the analysis of Costa et al. (2020), the pelvic-fin loss in *Cambeva* happened three times independently. Datovo (2014) reported two trichomycterids, *Ituglanis apteryx* and *Glanapteryx anguilla*, where the pelvic fin is absent in some specimens.

The pelvic morphology in *Nematogenys* fits within the spectrum of trichomycterids. The basipterygium has two anterior processes, with a more medial one meeting the tip of the respective process from the other side (Figure 4C). The posterior process is present but not prominent or absent (Figure 4C; Arratia et al. 1978; Silva Pedroza 2009). Both basipterygia meet in a straight midline with cartilaginous margins along the medial arm. There are five to six fin rays, and a pelvic splint is variably absent or present (G. Arratia pers. comm.).

In the remaining loricarioids, Astroblepidae, Callichthyidae, Loricariidae, and Scoloplacidae, the basipterygium is more massive and plate-like, and the ossified halves of both basipterygia are at least over some distance in direct contact and connected with sutures of various dimensions

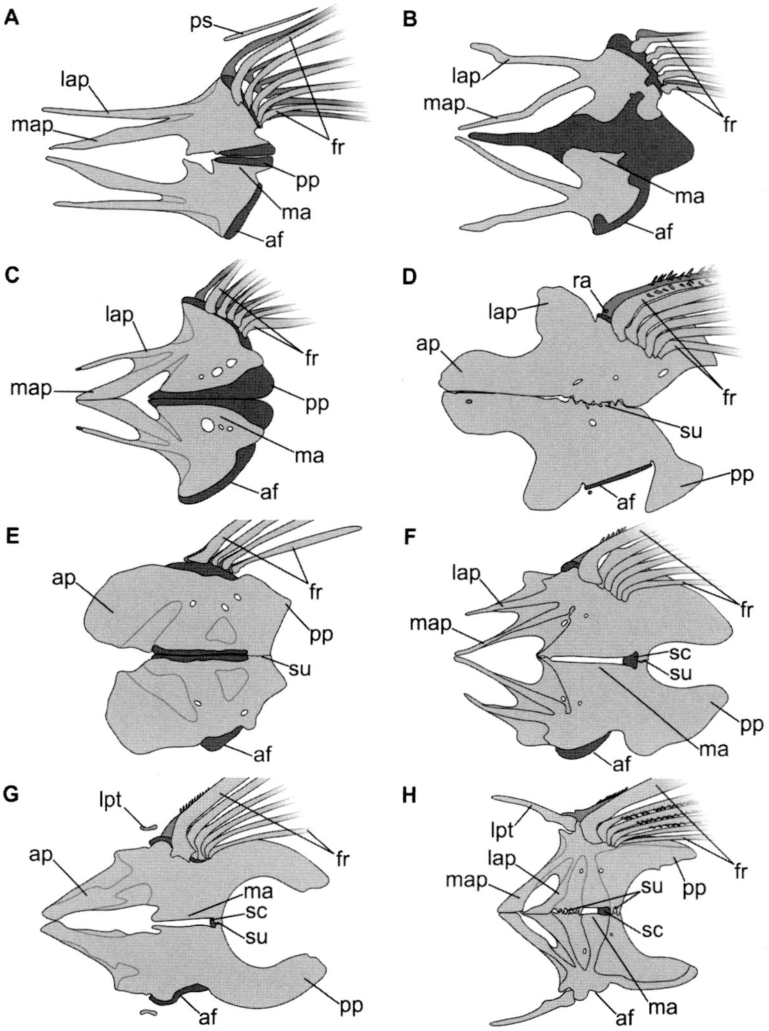

Figure 4. Pelvic girdles of loricarioid catfishes; cleared and double-stained specimens in ventral view; fin rays on one side removed; light grey: bone; dark grey: cartilage; ordinary grey: underlying halves of fin rays. (A) *Hatcheria macraei* (Trichomycteridae), DMM IE/13481. (B) *Tridentopsis cahuali* (Trichomycteridae), DMM IE/16950. (C) *Nematogenys inermis* (Nematogenyidae), USNM 440397. (D) *Aspidoras pauciradiatus* (Callichthyidae), DMM IE/17255. (E) *Scoloplax empousa* (Scoloplacidae), USNM 300327. (F) *Loricariichthys maculatus* (Loricariidae), DMM IE/17468. (G) *Loricaria cataphracta* (Loricariidae), DMM IE/16948. (H) *Paralithoxus surinamensis* (Loricariidae), DMM IE/17475. Abbreviations: af: articulatory facet of fin rays; ap: anterior process; fr: fin rays; lap: lateral anterior process; lp: lateral process; lpt: lateropterygium; ma: medial arm; map: medial anterior process; pp: posterior process; ps: pelvic splint; ra: radial; sc: synchondral cartilage; su: suture; vo: ventral outgrowth.

(Figure 4D–H). Expanded, plate-like posterior processes force the articulatory facet for the fin rays in a rather lateral position in comparison to many other catfishes.

While the pelvic girdle in callichthyids has some distinctive characteristics compared to other catfishes, there is not much diversity within this taxon (Reis 1998). The basipterygium is a massive plate with only a single anterior process, which is broad and flattened. Both basipterygia are sutured all along their medial margin and anterior process (Figures 3B and 4D). The posterior process is large, bends laterodorsally and bears a small ventrally-directed outgrowth. In callichthyids, the dorsolateral prolongation of the posterior process is at least tightly connected to the body armour via strong skin and, except for *Corydoras* and *Brochis,* even connected to a rib (Reis 1998; Silva Pedroza 2009, fig. 19). Anterior to the articulatory ridge, there is a broad lateral process which bends laterodorsally and approaches the body armour to which it connects via thick skin. This lateral process is less developed in *Hoplosternum* (Silva Pedroza 2009, fig. 20) but relatively large in *Aspidoras* (Figure 4D). Reis (1998) interpreted this process as a lateral anterior process, but ontogeny shows that both anteriorly directed processes are fused into the single anterior process (Huysentruyt 2008). Radials have not been reported from callichthyids so far (Reis 1998), but in a nicely stained specimen of *Corydoras paleateus* (Figure 3B), it becomes visible that distal to the cartilaginous margin of the articulatory facet of the basipterygium lies an additional band of connective tissue (Figure 3B) between the bases of the lepidotrichia. Within this band, there are also cartilaginous cells, sometimes aggregated into ball-like structures. The most lateral cell aggregation seems clearly differentiated, which is also visible in our *Aspidoras* specimen (Figure 4D), and we tentatively label it as radial (Figures 3B and 4D). There are six rays which can all bear odontodes (Reis 1998). The first ray is more massive than the others, is not furcated and bears more odontodes than the others. In some taxa, there are more unbranched fin rays; for example, in *C. paleatus*, the first three rays are unbranched. A pelvic splint is absent in callichthyids.

In scoloplacids with the single genus *Scoloplax*, the shape of the basipterygia is even more simplified, like a rectangular plate with rounded edges (Figure 4E). In *Scoloplax baileyi*, the anterior part of the basipterygium can be divided into lateral and medial processes (Rocha et al. 2012; Figure 4A), but in most species of the genus, it is difficult to distinguish any process (Schaefer 1990; Silva Pedroza 2009; Rocha et al. 2012). The symphysis of the basipterygia is made at least in the posterior portion of an osseous bridge connecting both halves (Figure 4E); the remaining symphysis is made by cartilage, or in some species there are additional direct bony contact zones of the basipterygia (Rocha et al. 2012; Figure 4). The articulatory facet is covered with cartilage and bears four or five rays (Schaefer et al. 1989), the first one being much stronger and bearing some odontodes.

Astroblepidae, with the single genus *Astroblepus,* is the sister group of the Loricariidae. In contrast to loricariids, *Astroblepus* is not covered by body armour. They live bottom-oriented in mountain streams and can climb rocks (Alexander 1965). For that, *Astroblepus* uses its sucker mouth and highly mobile pelvic girdle (Johnson 1912; Alexander 1965; de Crop et al. 2013). Like callichthyids and scoloplacids, the basipterygia are closely together in the symphyseal contact zone (Figure 3C). The central part of this gap is filled with continuous cartilage (not divided on the left and right sides). The direct contact between the bones of both sides is at least present in the posterior area (Figure 3C), usually also in the anterior area (Pereira and Reis 2017). Astroblepids have two anterior processes and a straight, long posterior process. Of their five rays, the first is much stronger, unbranched and may carry odontodes. A pelvic splint is missing. Anterior of the articulation of the first ray, a separate disc-like element, a lateropterygium, is present (Figure 3C; Shelden 1937, figs. 25, 50, 60; Arratia 2003, fig. 4.6; Silva Pedroza 2009, fig. 24; Pereira and Reis 2017, fig. 35). A lateropterygium is reported only for Astroblepidae and Loricariidae (Shelden 1937; Howes 1983); however, the shape of this structure in *Astroblepus* is unique. With its proximal tip, it articulates with the anterolateral corner of the articulatory facet and laterally expands into a flat, rounded structure with extensive cartilaginous margins (Figure 3C). The lateropterygium in *Astroblepus* lies flat on the ventral body side.

Loricariids comprising more than 1,000 species presently assigned to six subfamilies cover quite some diversity, but their pelvic girdles principally are similar and consist of the same elements. The basipterygia are flat and wide and connected in the body midline (Figures 2F–H and 3D). The connection is formed by bony sutures anterior and posterior, while the middle part of the basipterygial connection is prominently made of cartilage (Figure 3D; Py-Daniels 1997; Silva Pedroza 2009; Londaño-Burbano and Reis 2021). In several taxa, apparently several times convergently, the medial anterior processes and/or even the lateral anterior processes meet in the midline. If so, a second, more anterior-positioned foramen or window between both basipterygia forms.

The posterior process is flat, broad, and completely ossified. Often, its width at the base is more than half of the width of the basipterygium (Figure 2F–H). The basipterygium, especially its anterior part, may cover a large area of the belly and, in this case, can also protect the intestines, mainly of taxa with little body armour on the belly. On the other hand, in taxa with complete body armour, like in *Farlowella* or *Sturisoma*, there are only openings between bony plates for the pelvic rays.

A simplification, i.e., reduction of processes and gain of a shape with hardly delimitable regions, apparently happened several times during loricariid evolution. In some Hypoptopomatinae, the basipterygia are simplified, resembling much the ones of *Scoloplax*, e.g., *Otocinclus*, *Eurycheilichthys* (Pereira and Reis 2017) or *Plesioptopoma* (Reis et al. 2012). Similar simplifications can be found in some Loricariinae like *Loricaria* (Londoño-Burbano and Reis 2016) or *Rineloricaria* (Pereira and Reis 2017).

Loricariids have six rays, the first of which is much enlarged, spine-like and bears odontodes. In many species, this spine-like first pelvic ray looks very much like the first ray of the pectoral fin. Depending on the taxon, odontodes are often present on additional fin rays but less densely set than on the first ray. The lateropterygium in loricariids is principally rod-like (Figures 3D and 4H), originates anterior to the first ray, and approaches the body armour dorsolaterally (Figure 2D; Shelden 1937; Alexander 1965) to which it is indirectly but firmly connected via the skin. In *Ancistrus* (Figure 3D), *Neoplecostomus*, and *Hypostomus*, it is a prominent structure; in *Lamontichthys* and *Hartia*, it is much less pronounced; and in *Farlowella*, *Loricaria* (Figure 4G), *Pterosturisoma*, *Rineloricaria*, *Sturisoma* and *Sturisomatichthys* it is small (Paixão and Toledo-Piza 2009). In some Loricariinae, i.e., *Hemiodontichthys* (Londoño-Burbano and Reis 2016), *Loricariichthys* (Figure 4F), *Sturisoma*, *Pseudohemiodon* (Howes 1983), and *Pseudoloricaria* it is even absent (Paixão and Toledo-Piza 2009) or within these genera reduced to different degrees (Shelden 1937; Armbruster 2004).

The Big Asia Clade

Within the Big Asia clade (*sensu* Schedel et al. 2022), the families Bagridae and Horabagridae seem to represent the most generalised types of pelvic girdles. Their basipterygia are made of a flat main part with two anterior directed processes with cartilaginous tips (Figure 5A, B; Mo 1991; Pise et al. 2022). The tips of the medial anterior processes touch each other. The anterior processes form a straight shaft, which can be accompanied by wing-like thin membrane bone. Such membrane bone is most expressed medially from the medial anterior process. The medial arms of the basipterygia meet in the midline via symphyseal cartilage with each other. A moderate posterior process is present but not ossified or only to a small part. In the horabagrid *Paychypterus*, the posterior process is slightly irregular, with a bifurcated tip (Pise et al. 2022). In the bagrid *Sperata* (Fig. 5A) and the horabagrid *Pachypterus* (Pise et al. 2022, fig. 5b), there are six rays, of which the first is simple, although Mo (1991) reported seven rays in general for Bagridae. A rather small pelvic splint is usually present. Radials are missing.

In Sisoridae, Amblycipitidae, Akysidae and Ailiidae, the cartilaginous area between both basipterygia is continuous (Tilak 1967; de Pinna 1993; Chen and Lundberg 1994). Often this symphyseal cartilage is prolonged anteriad, forming an unpaired median tip. Compared to bagrids and horabagrids, members of Ailiidae have a slenderer pelvic girdle with long cartilaginous posterior processes. In *Clupisoma* and *Eutropiichthys*, the tips of the lateral anterior processes also meet, while in *Silonia,* the medial anterior processes do not (Tilak 1967). The medial membrane bone of the medial anterior process can be large, e.g., in *Silonia*, and even meet with that of the other side, e.g., in *Ailia* (Tilak 1967).

Sisorids are specialised hill-stream dwellers (Saxena and Chandy 1966a). Pectoral and pelvic fins play an important role in attaching to the ground in turbulent environments. Several taxa possess a specialised adhesive apparatus on their belly and the ventral surfaces of the paired fins (Saxena and Chandy 1966a). For the genus *Pseudocheneis*, even crawling habits for locomotion were described (Saxena and Chandy 1966b). With this background, highly specialised pelvic girdles could be expected in this family. Yet, the general appearance of the pelvic girdles is similar to that of bagrids and horabagrids, and only some derived characters can be found. The most prominent difference is the continuous cartilaginous connection between the tips of all anterior processes, resulting in an anterior arch (Figure 3E). This configuration results in foramina between the lateral and medial anterior processes, as well as a foramen between the medial anterior processes of both sides (Figure 3E; Saxena and Chandy 1966b, fig. 4). The anterior processes are relatively slender and long-reaching in *Pseudoecheneis*, even anterior to the pectoral girdle (Saxena and Chandy 1966b). This unusual configuration of the anterior processes seems to be an adaptation to their lifestyle. A similar condition of the pelvic girdle can be found in the not related cypriniforms *Garra* and *Psilorhynchus* (Saxena and Chandy 1966b), which live in the same habitats. In contrast to other catfishes in glyptosternin sisorids, the bony part of the anterior tip of the medial anterior process broadens (de Pinna 1996, fig. 43). Sisorids are furthermore unique in possessing a ventral flange on the base of the medial anterior process, running partially on the main body of the basipterygium (Figure 3E; Chen 1994). This flange is most expressed in *Glyptothorax* (Figure 3E) and *Pseudecheneis* (Saxena and Chandy 1966b), while in some taxa, this character is only little expressed or even lost (de Pinna 1996). The posterior process has only smaller portions ossified and, in many taxa, is even completely cartilaginous. Its shape is usually rather simple and small, but in *Bagarius*, the posterior process is elongated with posterior ramifications (de Pinna 1996). Generally, a lateral process is missing in sisorids, but Saxena and Chandy (1966b) reported one in *Pseudoecheneis*. Sisorids possess six rays, the first of which is unbranched. In *Pseudoecheneis*, the anterior margin of this ray is covered with thick pinnate skin (Saxena and Chandy 1966b). Radials are absent in sisorids. A detailed description of the musculature of the pelvic girdle in Indian sisorids was given in Saxena and Chandy (1966b).

Amblyceps mangois (Amblycipitidae) was depicted by de Pinna (1993) and Chen and Lundberg (1994). The species has two slender anterior processes and no ossified posterior process. The cartilage of the symphyseal area, posterior margin, and articulatory facet is continuous and fused between both sides. This cartilage forms a paired posterior process and an unpaired anterior median process. In *Liobagrus*, there is a gap between the cartilage of the articulatory facet and the median cartilage portion (Figure 5C; Chen and Lundberg 1994). *Liobagrus reini* has an unpaired anterior median tip formed by the symphyseal cartilage (Figure 5C), while *Liobagrus obesus* seems to have small paired such processes (Chen and Lundberg 1994). The morphology of the basipterygium in Amblycipitidae might be a topic for future studies, as the few available data are partly inconsistent and/or debatable. In *A. mangois*, the medial anterior processes meet with their tips (Chen and Lundberg 1994) or are fused (de Pinna 1993). The tips of the lateral anterior processes are simple (Chen and Lundberg 1994) or irregularly patch-like broadened (de Pinna 1993).

Chen and Lundberg (1994) identified the resulting gap at the posterior margin of the basipterygium in *Liobagrus* as the 'ischiac notch' and stated it as the plesiomorphic condition in amblycipitids, based on its presence in the fossil †*Hypsidoris* and several other catfishes (Grande 1987). Yet, †*Hypsidoris* has no close relationship to Amblycipitidae but rather seems to have a basal position within siluriforms (Grande 1987). Within the Big Asia clade, a shallow, poorly developed ischiac notch can be identified in Bagridae and Horabagridae but not in Sisoridae and Akysidae.

For the pelvic girdle in Akysidae, only limited data have been published, but these already show a high diversity (de Pinna 1993). *Akysis* (Figure 5D; de Pinna 1993, fig. 72) resembles amblycipitids with the tips of the anterior processes being connected via cartilage, *Breitensteinia* shows a remarkably broadened tip of the lateral anterior process (de Pinna 1993, fig. 68), and *Parakysis* has largely reduced the medial anterior process (de Pinna 1993, fig. 69). A posterior process is rather small and cartilaginous or completely absent. A lateral process is missing. Akysids have six rays, of which the first and, in *Breitensteinia*, also the last, are not branched. A pelvic splint is absent.

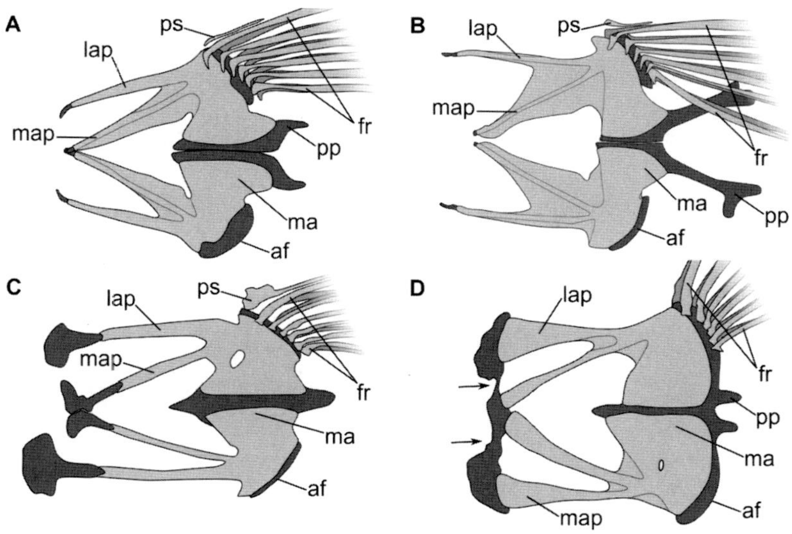

Figure 5. Pelvic girdles of catfishes of the Big Asia clade based on cleared and double-stained specimens in ventral view; fin rays on one side removed; light grey: bone; dark grey: cartilage; ordinary grey: underlying halves of fin rays. (A) *Sperata aor* (Bagridae), DMM IE/11130. (B) *Horabagrus brachysoma* (Horabagridae), unreg. (C) *Liobagrus reinii* (Amblycipitidae), USNM 71329. (D) *Akysis prashadi* (Akysidae), DMM IE/17002; arrows indicating fusion of cartilage tips of anterior processes. Abbreviations: af: articulatory facet of fin rays; fr: fin rays; lap: lateral anterior process; ma: medial arm; map: medial anterior process; pp: posterior process; ps: pelvic splint.

The Big Africa Clade

Within the Big Africa clade, the diversity is moderate: most members of the Auchenoglanididae, Claroteidae, and Mochokidae show a pelvic girdle that resembles much of the morphology as described above for Bagridae (Figure 6A–C; Tilak 1967). The basipterygium is generally quite flat to concave without dorsal or ventral outgrowths. It has two anterior processes with moderately developed membranous wings. The lateral anterior process lies almost parallel to the body axis, while the medial anterior process points medially, contacting the tip of its counterpart (Figure 6; Tilak 1967). The posterior process is moderately expressed, with a simple tip and consists mainly of cartilage. The symphyseal cartilage along the medial margin of the medial arm meets the cartilage of the other side in the midline. The cartilage of this symphysis is continuous with the cartilage of the posterior process but not with the cartilage covering the articulatory facet. A lateral process is absent or poorly developed. The first ray is simple, not spine-like.

Auchenoglanididae was formerly regarded as part of the Claroteidae, and there was much conformity between both groups. *Auchenoglanis* has six rays, and in front of the first ray is a small pelvic splint (Figure 6A). Close to the base of the first ray, there is at least in *Notoglanidium* a small radial (de Pinna 1993). The cartilage of the articulatory facet does not extend considerably anterior from the base of the first ray. In contrast, this cartilage is prolonged anteriad in the claroteids *Chrysichthys* and *Parauchenoglanis* (Figure 6B). Both genera have six rays and no pelvic splint. In one specimen of *Parauchenoglanis*, a small radial was visible close to the base of the first fin ray on only one side.

The studied mochokid specimens all have seven rays, and a small pelvic splint was found only in larger specimens of *Synodontis*. Vigliotta (2008) confirmed that principally, mochokids have seven rays, except a single *Chiloglanis* species, and that most taxa, except some *Synodontis* species and some Chiloglanidinae, have no splint. *Mochokus* and *Synodontis* bear a simple posterior process that anteriorly transitions into the symphyseal cartilage (Figure 6C). Laterally, behind the origin of the lateral anterior process, the basipterygium is slightly extended, thus forming a small lateral process that is covered by the cartilage of the articulatory ridge. A small but distinct radial is present

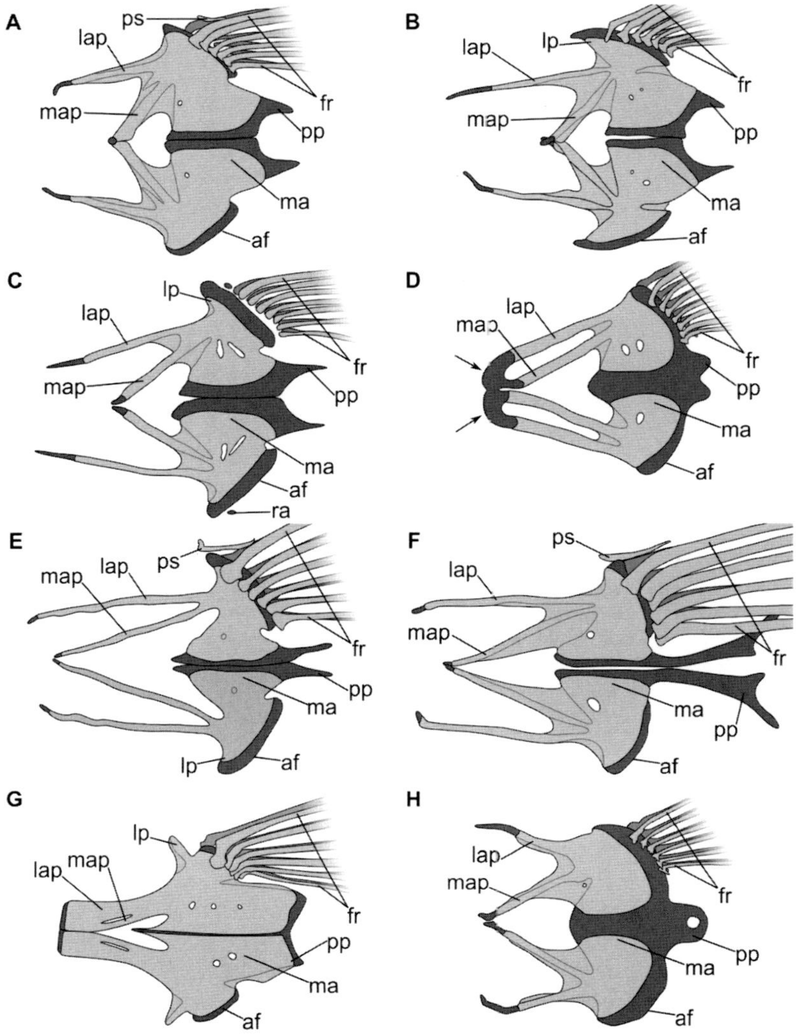

Figure 6. Pelvic girdles of catfishes of the Big Africa clade based on cleared and double-stained specimens in ventral view; fin rays on one side removed; light grey: bone; dark grey: cartilage; ordinary grey: underlying halves of fin rays. (A) *Auchenoglanis occidentalis* (Auchenoglanididae), DMM IE/15519. (B) *Parauchenoglanis monkey* (Claroteidae), DMM IE/12219. (C) *Synodontis nigrita* (Mochokidae), DMM IE/12204. (D) *Chiloglanis voltae* (Mochokidae), DMM IE/13428; arrows indicate fusion of cartilaginous tips of anterior processes. (E) *Schilbe intermedius* (Schilbeidae), DMM IE/12209. (F) *Pareutropius buffei* (Schilbeidae), DMM IE/17195. (G) *Amphilius atesuensis*, DMM IE/16552. (H) *Malapterurus electricus*, DMM IE/9277. Abbreviations: af: articulatory facet of fin rays; fr: fin rays; lap: lateral anterior process; lp: lateral process; ma: medial arm; map: medial anterior process; pp: posterior process; ps: pelvic splint; ra: radial.

in front of the first fin ray in *Synodontis* (Figure 6C). *Chiloglanis voltae* has no lateral process, no radial and only a small posterior process (Figure 6D). Hardly any part of the basipterygium is made of membrane bone. In Chiloglanidinae, the symphyseal cartilage is continuous with that of its counterpart. Also, similar to the mountain stream-dwelling sisorids, the cartilaginous tips of the medial and lateral anterior processes from one side form a continuous bow (Figure 6D; Vigliotta 2008).

Electric catfish, Malapteruridae, differ in some details from the above-described morphology of the Big Africa clade. Their pelvic girdle is relatively small in comparison to the whole-body size. Uncommon among catfishes is that their medial anterior process, at least its ossified part, extends more anteriorly than the lateral anterior process (Figure 6H; Vigliotta 2008). *Malapterurus* has a continuous cartilage that forms the articulatory ridge, the posterior process, and the symphyseal cartilage, which is also continuous with its counterpart. Further, the tips of the posterior processes are connected, forming an additional foramen (Figure 6H). While in our specimen, no radials were visible, de Pinna (1993, fig. 67) depicted on one side two, and on the other even three radials. *Malapterurus* has six rays and no splint.

With more than 100 species in three subfamilies, Amphiliidae also show some diversity in the pelvic girdle. They all have continuous symphyseal cartilage, a small ossified lateral process, and the absence of long tips on the posterior process in common (Figure 6G). Furthermore, they have all six rays (Roberts 2003), no pelvic splint (Vigliotta 2008) and, so far, no radials have been reported. Within the Big Africa clade, amphiliids are the only taxon in which the first ray has become much stronger, almost spine-like. Remarkable is also a dorsal flange (Skelton 1986). Similar to other Big Africa catfishes, the basipterygium deviates little from the horizontal plane except from this flange. In *Phractura*, it is large and contacts the body wall. Furthermore, we found a tiny bony ventral flange in the midline of the body in *Phractura*.

In amphiliids, the bony part of the posterior process is broad with a blunt posterior end; these processes of both sides are covered by a common cartilage which has no or only tiny tips (Figure 6G; de Pinna 1993). *Paramphilius* seems to represent a primitive condition within amphiliids, with the cartilages of the lateral and medial anterior process not connected (de Pinna 1993). Usually, members of Amphiliinae, as well as Leptoglaninae, have their cartilaginous tips close together or even united, resulting in a bow-like connection, as seen in chiloglanins or sisorids. In *Amphilius atesuensis,* even the bony tips of the lateral and medial anterior processes fuse (Figure 6G). *Phractura* and *Doumea* (Doumeinae; Vigliotta 2008), but also *Dolichamphilius* (Leptoglaninae; Roberts 2003) seem even more derived as their basipterygia have their lateral and medial anterior processes fused to one broad anterior portion.

Schilbeidae, with about 35 species in five genera (Fricke et al. 2024), show some differences in the pelvic girdle, which makes it worth having a closer look at them. The basic shape of their basipterygium (Figure 6E and F) is similar to that seen in claroteids, auchenoglanids or *Synodontis*, but slenderer. At least its major part lies flat and is slightly anteriorly inclined to the ventral outline of the body. Except for *Irvineia*, which has eight to ten rays, all schilbeids have six rays in the pelvic fin (de Vos 1995). There is some variability in the pelvic splint and the posterior process. Species of *Schilbe* have a pelvic splint (Figure 6E; de Vos 1995). In *S. mandibularis* and *S. intermedius*, for example, it is not small, and its proximal part is prolonged and extends dorsally in the body wall (Figure 2F). In *Siluranodon*, a pelvic splint is missing, as principally in *Parailia*. Sometimes, a very small pelvic splint can be found in *Parailia pellucida* but in some specimens, it is only present on one side. In both genera, the cartilaginous posterior process is much prolonged (Figures 2G–H and 6F), and it is not straight posteriorly but laterally and then laterodorsally along the body wall, almost in a right angle to the slightly inclined basipterygium (Figure 2G–H). In comparison to body size, the posterior process is the largest in *Parailia*. In *Schilbe*, it is relatively short but variable in length depending on the species (de Vos 1995). *Pareutropius* shows an intermediate state with a pelvic splint and prolonged posterior process present. Its posterior process is bifurcated with a small medial and longer lateral tip (Figure 6F).

De Vos (1995, fig. 94) depicted the bony parts of the pelvic girdle of *Irvineia*: this is consistent with the other schilbeids, except for the increased number of rays. *Irvineia* has a prominent pelvic splint, but there is no information on the shape and size of the cartilaginous posterior process. For *Schilbe multitaeniatus*, de Vos (1995, fig. 69) depicted a separate, cartilaginous, rod-like posterior process not connected to the basipterygium. This, however, should be critically reviewed by examining additional specimens, as it may be an artefact due to preparation or fixation; otherwise, it would be a unique state for a catfish.

Lacantunia enigmatica might belong to the Big Africa clade despite its Mesoamerican distribution. Scarce information is available on its pelvic girdle: a posterior bony process is present, and the cartilage on the posterior margin of the basipterygium is not continuous (Rodilez-Hernandez et al. 2005). It has six rays and no pelvic splint.

Remaining Siluriform Taxa

Major phylogenetic groupings among the remaining catfish families have not been established so far (e.g., Sullivan et al. 2006; Schedel et al. 2022). Therefore, lumping them should be regarded as a purely practical approach without any systematic inference. Principally, the pelvic girdles in these taxa resemble the morphology found in Bagridae and many others; few taxa are significantly derived with losses of elements and simplifications.

The basipterygium of *Austroglanis* (Austroglanididae) has two anterior processes with moderate membrane bone wings, a short, simple, largely cartilaginous posterior process, no lateral process, six rays with the first simple, not spine-like, and a moderately developed pelvic splint (Skelton 1981, fig. 21). *Heteropneustes* (Heteropneustidae) has deep incisions between the anterior processes, the medial anterior process and the medial arm (Figure 7A). Nevertheless, there is a blade-like wing of membrane bone along the medial side of the medial anterior process; the tips of these processes meet each other. There are six rays; a splint and radials are missing, but a distinct bony lateral process is present (Figure 7A).

The closely related Clariidae present a similar morphology (Figure 7B), but with incisions between anterior processes and medial arm less expressed than in *Heteropneustes*. There are also six rays, but neither splint nor radials. A lateral process is present, which is laterally covered by the cartilage of the articulatory ridge (Figure 7B). Posteriorly, there is a gap between this cartilage and that forming the symphysis. The symphyseal cartilages of both sides are largely fused and form an anterior median tip. Within clariids, there is an evolutionary trend towards body elongation which is commonly accompanied by the reduction of paired fins. In *Gymnallabes*, *Channallabes* and *Clariallabes*, pelvic fins may be reduced, depending on species or population; sometimes, the presence or absence is even variable between individuals or even within both sides of a specimen (Poll 1941; Teugels and Adriaens 2003). Respectively, there is some variation in the details of the pelvic girdle in those taxa. In *Gymnallabes*, there are still two anterior processes present (if the basipterygium is present at all), but with a maximum of four rays (Devaere et al. 2005). Some taxa like *Dolichallabes* lack pelvic fins or remnants of the pelvic girdle (Devaere et al. 2004). The anguilliform clariid *Platyallabes thioni* is an exception, with well-developed paired fins (Teugels and Adriaens 2003).

The shape of the basipterygium and its processes in the cetopsid *Helogenes* (Figure 7D) is like other catfishes, resembling *Heteropneustes*. It has two slender anterior processes with little membranous bone between their bases, and the posterior process is simple and cartilaginous but longer than in *Clarias*. *Helogenes* have no lateral process, and the symphyseal cartilages of both sides are fused anteriorly, forming a median tip (Figure 7D). In other cetopsids, the pelvic girdle is longer and slenderer (Figure 7C). Membrane bone is present between the posterior half of the anterior processes. The area between the medial anterior process and the medial arm is filled with membrane bone, resulting in a large contact zone of both basipterygia (Figure 7C). The symphyseal cartilages of both sides are fused. The posterior processes of cetopsids are special: the ossified portion is moderately developed, but there is a long cartilaginous part with multiple complex ramifications (see arrow in Figure 7C), which apparently are species-specific (de Pinna 1993, 2007). In *Bathycetopsis*, there is only a single unpaired medial posterior process (de Pinna and Vari 1995). Cetopsids, including *Helogenes*, have six rays, of which the most lateral and most medial ones are unbranched, and a well-developed splint with a long proximal part. A lateral process and radials are absent.

The pelvic girdle of heptapterids is broader and stouter than in cetopsids or clariids (Figure 7E and F). The medial anterior processes usually touch each other with their cartilaginous tips. Like cetopsids, the area between the medial anterior process and the medial arm is filled with membrane

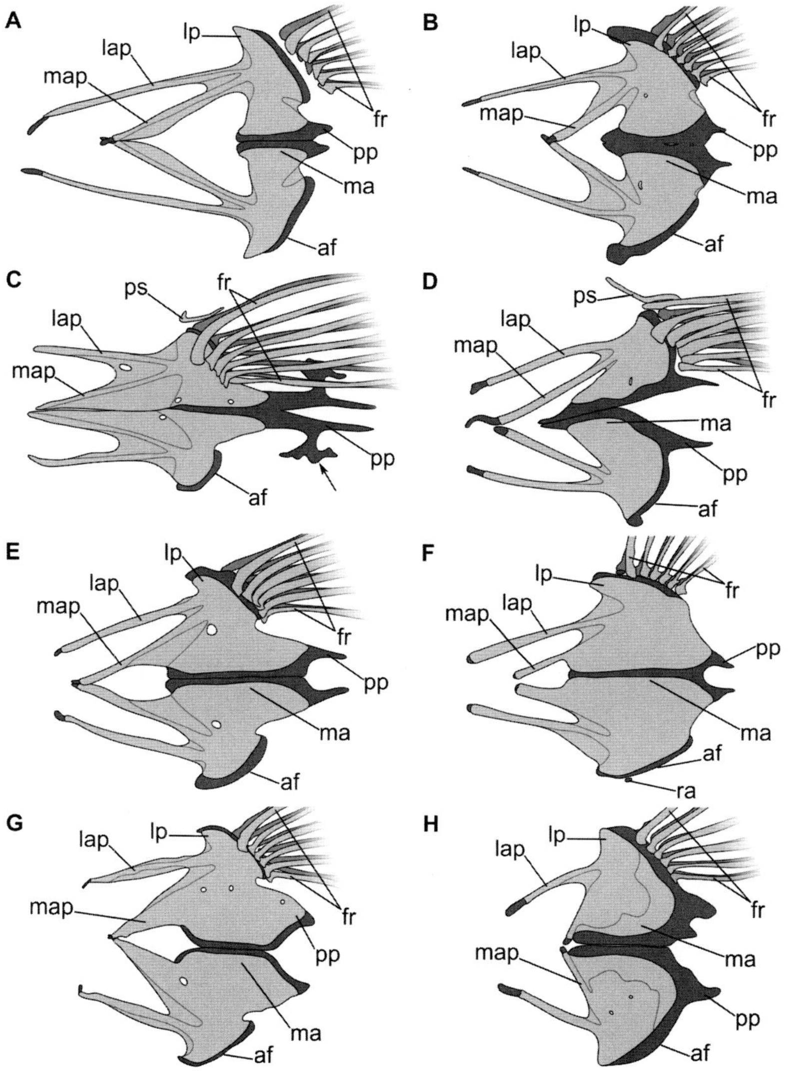

Figure 7. Pelvic girdles of catfishes based on cleared and double-stained specimens in ventral view; fin rays on one side removed; light grey: bone; dark grey: cartilage; ordinary grey: underlying halves of fin rays. (A) *Heteropneustes fossilis* (Heteropneustidae), DMM IE/13478. (B) *Clarias agboyiensis* (Clariidae), DMM IE/17474. (C) *Cetopsis orinoco* (Cetopsidae), DMM IE/17266; arrow indicating complex structure of cartilaginous posterior process. (D) *Helogenes marmoratus* (Cetopsidae), DMM IE/16946. (E) *Pimelodella megalops* (Heptapteridae), DMM IE/16945. (F) *Chasmocranus longior* (Heptapteridae), DMM IE/17467. (G) *Pimelodus pictus* (Pimelodidae), DMM IE/12016. (H) *Agamyxis pectinifrons* (Doradidae), DMM IE/12205. Abbreviations: af: articulatory facet of fin rays; fr: fin rays; lap: lateral anterior process; lp: lateral process; ma: medial arm; map: medial anterior process; pp: posterior process; ps: pelvic splint; ra: radial.

bone. While in some taxa, these membranous areas of both sides can be in contact, e.g., *Imparfinis* or *Mastiglanis* (Bockmann and Castro 2010), in most taxa, a gap remains between the membranous ossifications (Figure 7E) and a few of them even retain an indentation, e.g., *Rhamdiopsis* (de Pinna 1993; Bockmann and Castro 2010). The ossified broad portion of the posterior process is extended posteriorly. It is covered by a rather thin cartilage in continuation with the symphyseal cartilage. Each symphyseal cartilage can be united with its counterside cartilage or not. In *Rhamidopsis*, there can be a prolonged anterior tip of the symphyseal cartilage (de Pinna 1993). Such a tip, however, was not reported for *R. krugi* (Bockmann and Castro 2010). Further, the tips of the lateral anterior processes in *Rhamidopsis* were depicted as simple by Bockmann and Castro (2010) and enlarged, forming an irregular patch by de Pinna (1993). In *Chasmocranus* and *Pimelodella*, these tips are simple (Figure 7E and F), but in other members of the family, these cartilaginous ends can be significantly larger, with extreme examples in *Brachyglanis* and *Horiomyzon* (Bockmann 1998). Furthermore, heptapterids have a lateral process that is covered by the articulatory-facet cartilage. They have six rays and no pelvic splint. In *Chasmocranus*, a small independent radial is present at the base of the first ray (Figure 7F). In pimelodids, the pelvic-girdle morphology very much corresponds to that of the heptapterids *Rhamdella* and *Pimelodella*. Similarly, the membrane bone medial to the medial anterior process can either completely fill the gap between both basipterygia, as in *Propimelodus*, *Exallodontus*, *Cheirocerus* or *Pinirampus* (Parisi et al. 2006), or leave a gap anterior to the medial arms, as in *Pseudoplatystoma* (Buitrago-Suarez 2005). Within the genera *Pimelodus* and *Megalonema* both variants are found: completely filled, e.g., in *P. latissimus* (Parisi et al. 2006) and *M. orixanthum* (Lundberg and Dahdul 2008), or with gap, e.g., in *P. pictus* (Figure 7G), *P. albofasciatus* (Parisi et al. 2006), and *M. platycephalum* (Lundberg and Dahdul 2008).

In contrast, *Microglanis* (Pseudopimelodidae) has a remarkable difference from the previous two families: the cartilaginous end of the lateral anterior process is enlarged to an irregular patch (Figure 3G). More unusually, the lateral process also bears a large, irregular, and flat cartilaginous process (Figure 3G). Whether the pelvic-girdle morphology of *Microglanis* resembles that of other pseudopimelodids cannot be stated, as further information on other pseudopimelodids is presently rare. *Microglanis* has six rays, with the first unbranched but not spine-like. A small splint and a radial in the lateral position are present (Figure 3G). Grande (1987) depicted a small shallow indentation on the posterior margin of the basipterygium, the 'ischiac notch', for *M. poecilius*. We cannot confirm this notch based on the same species.

Most members of the Doradidae also have a pelvic girdle like the above-mentioned heptapterids or pimelodids: two anterior processes of which the medial ones touch each other with their tips, a broad posterior process, and a well-developed lateral process (Figure 7H). There are, however, some differences: First, the length proportion of the anterior processes compared to the main body of the basipterygium is almost equal in many genera, e.g., *Lithodoras* or *Wertheimeria*. In *Ossancora* and *Hemidoras*, the anterior processes are short, giving the pelvic girdle an even more compact appearance (Birindelli and Pérez 2011; Birindelli 2014). Further, in doradids, the bony part of the lateral process is commonly more pronounced and hook-shaped. The lateral anterior process is very slender and bears no wing-like outgrowths of membrane bone. The medial anterior process has a well-expressed lamina of membrane bone. Furthermore, doradids with about 100 species in six subfamilies show some expectable variability. In a few taxa, *Hemidoras morei*, *H. morrisi*, *Ossancora* and *Trachydoras*, this membranous lamina is enlarged, and the medial anterior process becomes indistinct from the medial arm (Higuchi 1992; Birindelli and Pérez 2011; Birindelli 2014). Five genera have six pelvic rays, with the first unbranched, but the remaining doradids have seven rays (Birindelli 2014). A pelvic splint or radials have not been reported in doradids.

Like in doradids, the pelvic girdle in Chacidae has a compact appearance (Figure 8A). The lateral anterior process is slender and short, such that the basipterygium appears broader than long. *Chaca* has a prominent lateral process, but a posterior process is hardly expressed (Figure 8A; Tilak 1971). Lateral and medial anterior processes are slender, with the medial ones touching each other. The specimen depicted by Tilak (1971, fig. 15) shows a third process more medial directly bordering the anterior margin of the medial arm. The tips of the most medial processes of both sides touch each other. In *Chaca*, continuous cartilage covers the lateral process, the articulatory ridge, the posterior margin of the basipterygium, the symphysis, and the tip of the most medial anterior process of both sides (Figure 8A; Tilak 1971). *Chaca* has six rays, which are all branched (Brown and Ferraris 1988). Radials are absent.

The enigmatic *Kryptoglanis*, placed in its own family, shares the continuous cartilaginous margin forming the articulatory facet, posterior margin, and symphyseal cartilage with the previously described Chacidae. In contrast to that family, the cartilaginous margins of both basipterygia are not fused in *Kryptoglanis* (Figure 8B; Britz et al. 2014). The basipterygium, in general, is rather

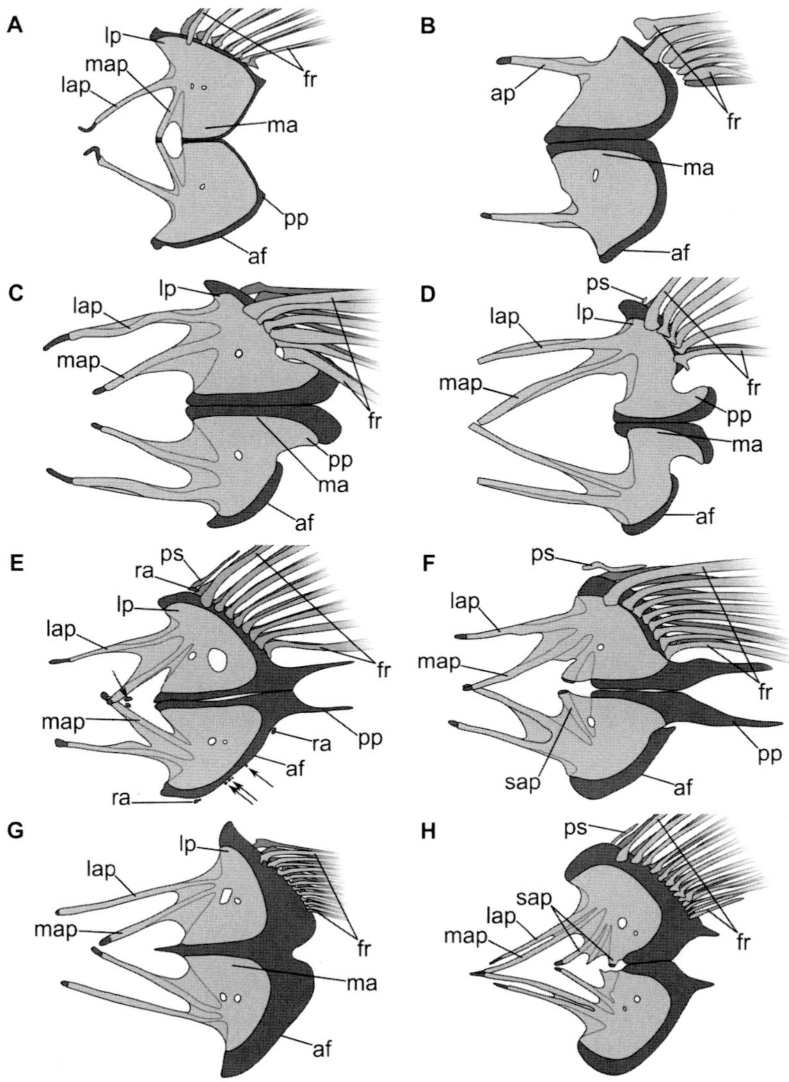

Figure 8. Pelvic girdles of catfishes based on cleared and double-stained specimens in ventral view; fin rays on one side removed; light grey: bone; dark grey: cartilage; ordinary grey: underlying halves of fin rays. (A) *Chaca chaca* (Chacidae), unreg. (B) *Kryptoglanis shajii* (Kryptoglanididae), after Britz et al. (2014, fig. 3d). (C) *Bagre bagre* (Ariidae), DMM IE/17477. (D) *Cephalocassis borneensis* (Ariidae), DMM IE/17477. (E) *Ameiurus melas* (Ictaluridae), DMM IE/16487; arrows indicating additional cartilages. (F) *Pangasianodon hypophthalmus* (Pangasidae), DMM IE/11819. (G) *Plotosus lineatus* (Plotosidae), DMM IE/14953. (H) *Silurus glanis* (Siluridae), DMM IE/15801. Abbreviations: af: articulatory facet of fin rays; ap: anterior process; fr: fin rays; lap: lateral anterior process; lp: lateral process; ma: medial arm; map: medial anterior process; pp: posterior process; ps: pelvic splint; ra: radial; sap: supplementary anterior process.

shortened and simplified. There is only a single rod-like anterior process with a little membrane wing on its base. A distinct lateral process and a posterior process are missing. *Kryptoglanis* has neither radials nor a splint. Of its six rays, the first two and the most medial one are not branched (Britz et al. 2014).

Another type of shortened basipterygium can be seen in aspredinids. In most taxa, e.g., *Amaralia*, *Aspredo*, *Bunocephalus*, *Platystachus*, *Xiliphius*, and *Hoplomyzon* (Figure 3H; Taphorn 1990; de Pinna 1993; Friel 1994), it seems that the anterior processes are usually completely reduced and that the portion anterior of the articulatory facet is mainly formed by an enlarged lateral process. In *Micromyzon*, however, the anterior margin of the basipterygium is triangular

(Carvalho et al. 2016), which could represent an anterior process. Such a structure is even more pronounced in *Ernstichthys* (Taphorn 1990). A bony posterior process with a cartilaginous tip is present in aspredinids: long in *Ernstichthys* (Taphorn 1990), moderate in *Aspredo* (Figure 3H), or not distinct in *Amaralia* (de Pinna 1993; Friel 1994). The cartilage of the posterior process is small or absent in hoplomyzontines (de Pinna 1996). In all Aspredinidae, this cartilage is not connected to any other cartilage. The symphyseal cartilage is distinct and usually not united with that of the other side, except for, e.g., *Hoplomyzon* (Friel 1994). At least in *Amaralia*, *Aspredo* (Figure 3H) and *Platystachus*, the lateral process is covered by a thick cartilaginous margin; this is not the case in *Hoplomyzon*. For most other aspredinid taxa, this data is presently missing. Where the lateral process is laterally covered with cartilage, this cartilage is continuous with that of the articulatory facet. Aspredinids have a well-developed distinct radial on the base of the first ray (Figure 3H; de Pinna 1993; Friel 1994); for *Platystachus*, even more than one radial has been depicted (Friel 1994, fig. 34). In some aspredinid taxa, the pelvic girdle is close to the pectoral girdle and partly even embraced by posteriorly directed processes of the cleithrum (Taphorn 1990)

Likely closely related to doradids and aspredinids are the Auchenipteridae, which comprise about 130 species in two subfamilies. They show a high diversity in the pelvic girdles, which cannot be comprehensively described herein. The principal composition of the basipterygium comprises a simple, largely cartilaginous, moderately elongated posterior process, a moderately developed lateral process, and two anterior processes (Figure 3F). There is a remarkable variability regarding the anterior processes. They can be slender and rod-like or broadened (Walsh 1990; Birindelli and Zuanon 2012; Birindelli 2014). The lateral anterior process can be in a quite lateral position at about the level of the lateral process, e.g., in *Auchenipterus* (Birindelli 2014), more medially, e.g., in *Asterophyus* (Silva Pedroza 2009), or clearly shifted close to the medial anterior process, e.g., in *Ageneiosus* (Walsh 1990; Birindelli 2014). In *Auchenipterus* and *Centromochlus*, the medial anterior process is continuous with the medial arm and in contact with that of the other side (Ferraris 1990; Birindelli 2014), much like in cetopsids. The shape of the lateral process shows some variability, especially when looking at its cartilaginous tip. The margin can be rather blunt, as in *Liosomadoras* (Birindelli and Zuanon 2012), or feature a small outgrowth, as in *Tetranematichthys* (Calegari et al. 2019). Additionally, a prolonged anterior-directed cartilaginous tip is often seen in auchenipterids, e.g., in *Ageneiosus* (Walsh 1990), *Asterophysus* (Silva Pedroza 2009), *Ferrarissoarensia* (Calegari et al. 2019) and *Trachycorystes* (Ferraris 1998). In *Centromochlus*, the lateral process even has an anterior and posterior directed tip (Ferraris 1998). Some auchenipterids have six pelvic rays, most have seven, and some have eight or even more. A pelvic splint is absent (Ferraris 1988). Sexually mature males of *Entomocorus* show an enormously enlarged first pelvic ray (Ferraris 1988; Reis and Borges 2006; Birindelli 2014).

The pelvic girdle of Ariidae is similar to that of heptapterids and pimelodids. The basipterygium has two prominent anterior directed processes, which are often broadened. A bony posterior process is either absent as in *Ariopsis felis* (Shelden 1937; Merriman 1940; Arratia 2003), rather small as in *Plicofolis* (Lakshmi and Rao 1989), or much expressed as in *Bagre* (Figure 8C; Merriman 1940) or *Cephalocassis* (Figure 8D). Our cleared and double-stained specimens show that there is a well-expressed cartilaginous margin on the posterior process (Figure 8C and D). Thus, maybe also species, where only the bony basipterygium was described, carry a small cartilaginous posterior process, which was so far overseen. The cartilage of the posterior process is continuous with the symphyseal cartilage but not connected to that of the articulatory facet. The symphyseal cartilages are not fused. A little-developed lateral process is present (Figure 8C and D). Ariids have six rays, with the first not spine-like and not or only once branched. A splint seems absent in most taxa, but at least a small splint is present in *Cephalocassis* (Figure 8D). Ariids present a unique sexual dimorphism: in females, a hook-like, fleshy outgrowth is expressed during the breeding season on the dorsal side of the most medial fin rays (Merriman 1940). Also, the pelvic fins are generally larger in females than in males (Merriman 1940). Most, but not all, species of ariids show this character, and its exact meaning is not yet clear. Maybe the females hold their eggs for a moment after spawning, as in ariids, the males are mouth breeders.

Like in many siluriforms, the pelvic girdle of ictalurids has two anterior processes: a distinct medial arm, a poorly developed lateral process, and a mainly cartilaginous posterior process (Figure 8E; Lundberg 1982). The posterior area between anterior arms is filled with membrane bone, and the tips of the medial anterior processes meet each other. The posterior process is rather simple but prominently elongated (Figure 8E; Shelden 1937; Vigliotta 2008). The cartilages of the symphysis, posterior process, articulatory facet, and lateral process are continuous but not fused to the counterpart side, except for *Trogloglanis* (Lundberg 1982). Ictalurids have a pelvic splint and eight or nine rays, with the first not branched (Arce-H et al. 2017). The splint is simple and parallel to the fin rays; compared with the splint of other taxa, a proximal part reaching into the body is missing. In our specimens of *Ameiurus melas*, a radial is clearly visible between the base of the first ray and the splint (Figure 8E). Additional cartilages are present along the articulatory facet in between the bases of the rays; usually, these cartilages are smaller than the first one, except for the most medial one. Furthermore, in *Ameiurus*, we found a small, paired cartilage anterior to the tip of the symphyseal cartilage and directly posterior to the medial anterior process (Figure 8E). This cartilage is not directly connected to any other structure, and its origin remains questionable.

The basipterygium of the marine plotosids resembles that of ictalurids. Plotosids have relatively longer anterior processes with membrane bone in between them; there is no bony posterior process, and the lateral process is small (Figure 8G; Tilak 1967; Grande 1987; Arratia 2003). The cartilage of the lateral process, the articulatory facet, the posterior process and the symphysis are quite broad and continuous (Figure 8G; Shelden 1937). The cartilages of the left and right basipterygia are fused, and the joint symphyseal cartilage forms an anterior-directed unpaired median tip. The posterior process is indistinct (Figure 8G). Plotosids have a high number of rays ranging from ten to 18. The first ray is not branched but also not spine-like. A pelvic splint is missing.

In members of the Pangasiidae, there can be more than two anterior-directed processes, usually three. This is similar to the pelvic girdles of diplomystids, and overall, both taxa share remarkable similarities. *Pangasianodon* has a supplementary anterior process between the anterior medial process and the medial arm, which touches the tip of the same process from the other side (Figure 8F). The supplementary process is connected with the medial arm of the basipterygium through membrane bone. In one of our specimens, the supplementary process was missing on one side. De Pinna (1993) noticed and depicted the same for a specimen of *Pangasius sanitwongsei*. In pangasiids, there is no distinct lateral process, and the long and simple posterior process is almost exclusively cartilaginous. Remarkable for the posterior process is that it is significantly broadened in the vertical plane; thus, the vertical dilatation is not visible in the ventral view illustrations (Figure 8F). All other parts of the basipterygium are in one horizontal plane. The cartilage of the posterior process is continuous with the symphyseal cartilage; both sides are not fused (Figure 8F). There is a gap between this cartilage and that of the articulatory facet. Members of the Pangasiidae usually have six rays except for the two *Pangasianodon* species, which have eight or rarely nine rays (Roberts and Vidthayanon 1992). An L-shaped pelvic splint is present, and its proximal and distal parts are almost equal in length.

The basipterygium in Siluridae shows some variability. *Silurus*, *Wallago*, *Pterocryptis*, and *Ompok* have three anterior processes: *Silurichthys* and *Hito* two, and *Belondichthys* and *Hemisilurus* have only a single anterior process (Bornbusch 1995). *Kryptopterus* species are variable with one or two anterior processes (Bornbusch 1995). In the taxa with three processes, the most medial one is a new structure, while the other two are homologous to those of other catfishes. The tips of the supplementary anterior processes meet where this structure is present (Figure 8H), while the tips of the medial anterior processes are separate in *Wallago* but meet in *Ompok* (Tilak 1967). Posterior processes are rather small and cartilaginous; lateral processes, as well as radials, are absent in silurids (Figure 8H). A pelvic splint is present. The number of rays starts from six, e.g., in *Kyptopterus geminus* (Ng 2003), but is usually higher than in other catfishes and may reach up to 14.

An interesting side note is the description of the silurid genus *Apodoglanis* by Fowler (1905), lacking pelvic fins. However, it became evident that the missing fins were just a teratological anomaly, and the genus was synonymised with *Pterocryptis* (Bornbusch 1991).

Anatomical Structures

Basipterygium

The major bone in the pelvic girdle of catfishes, the basipterygium, shows quite some morphological variability, especially if compared to the rather uniform pelvic girdles of most non-neoteleost teleosts. While usually in non-neoteleosts it is a rather simple, rod-like bone with a more or less expressed posteromedial arm (e.g., Hilton 2002, osteoglossomorphs; Grande 1985, clupeomorph; Sanford 2000, salmonids), in catfishes, the basipterygium can have several processes: a posterior, a lateral and two or more anterior processes. The basipterygium develops from a single cartilage, and the processes are extensions of this cartilage. In addition to the processes, two areas, the main body with the posterolateral articulatory facet and the medial arm with the median symphyseal cartilage, can be distinguished.

During ontogeny, the posterior process develops as an extension of the posteromedial corner of the basipterygial cartilage. In adults, it is usually a simple, tip-like appendage that is more or less developed. Often, it remains cartilaginous, e.g., Ictaluridae, Siluridae, Schilbeidae, and *Helogenes*, but ossification of the medial arm can proceed posteriorly in some taxa such that parts or almost the whole posterior process ossify. The size of a specimen apparently has an impact on the extent of this ossification. Rarely does the posterior process take other shapes than simple tips, well pronounced in cetopsids (except *Helogenes*) with highly variable ramifications (de Pinna 1993). More simple ramifications or at least blunt ends were observed, e.g., in *Pareutropius* (Figure 6F), *Diplomystes* (Figure 3A), or *Glyptothorax* (Figure 3E). The cartilage of the posterior process is sometimes in contact with the cartilage of the articulatory facet, but usually, there is a gap and often even an indentation or 'notch' in the basipterygium (Grande 1987). In most catfishes, the posterior process serves as an attachment for the *Musculus infracarnialis posterior* (Shelden 1937 [retractor ischii]; Winterbottom 1973). Where a posterior process is missing, e.g., in *Plotosidae*, this muscle attaches directly to the posterior margin of the basipterygium. Unusual in this regard is *Astroblepus*, in which the *Musculus infracarinual posterior* runs ventral of the basipterygium and attaches to the lateral anterior process (Shelden 1937). This situation is related to the high mobility of the pelvic girdle in this species. In loricariids, posterior processes can be much broadened and may also support the protection of intestines. Apparently, they largely develop as membrane bone. In callichthyids, the enlarged dorsal part of the posterior process fixes the pelvic girdle to the body armour or even a rib.

A lateral process is common in catfishes. It can be expressed in many different shapes and sizes, from a small hump to a prominent cartilaginous or bony process (Birindelli 2014). It serves as a diverter pulley for the *Musculus arrector dorsalis* running to the base of the first fin ray. This is well illustrated by Shelden (1937, fig. 47). The large lateral process in callichthyids additionally gets in contact with the body armour, thus helping in the fixation of the pelvic girdle. The possible advantages of the ramified lateral process in *Microglanis* (Figure 3G) remain unclear. The lateral process usually does not serve as an attachment point for muscle, but *Astroblepus* again presents an exception as the *Musculus infracarinual posterior* is attached there (Shelden 1937).

The majority of siluriforms have two anterior processes. *Kryptoglanis* has only a single anterior process (Figure 8B; Britz et al. 2014), and aspredinids may have lost both anterior processes. In others, the anterior processes are not easily visible as entities in adults, but close investigation or study of the ontogeny usually shows the formation of two anterior processes which are fused to each other or to the medial arm during development, e.g., in callichthyids (Huysentruyt 2008), loricariids, and amphiliids (Figure 1G and H). Additional, more medially placed processes with apparently cartilaginous precursors have been reported in diplomystids, pangasiids, silurids, some trichomycterids and *Chaca* (the latter by Tilak (1971) but not in our specimen). In most siluriforms except loricarioids, the anterior tips of the anterior processes remain cartilaginous. While these tips are generally short, they can also be elongated, e.g., *Malapterurus* (Figure 6H), or complex, e.g., *Microglanis* (Figure 3G). Mostly, the tips of the medial anterior processes touch each other. Taxa inhabiting swift mountain streams developed a connection of the cartilaginous tips of the lateral and medial anterior process several times independently (Figures 3E, 5D, and 6D). This results in a bow-like cartilage, thus in a blunt, rounded anterior end, which is apparently beneficial in strong currents, either because there are no pointed bones pointing anteriad, and/or this may result in higher stability of the basipterygium.

The lateral anterior process is usually longer than the medial one, or both processes have roughly the same length. The medial anterior process is longer in few taxa: *Malapterurus* (Figure 6H; Vigliotta 2008), some loricariids, e.g., *Loricariichthys* (Figure 4F), or the ariid *Cephalocassis* (Figure 8D). In the sisorid *Glyptothorax* (Figure 3E), the cartilaginous part of the anterior processes is fused. However, considering only the bony part, the medial anterior process is longer. Along the anterior processes, there are more or less extensive wings of membranous bone with significant differences between the taxa. These wings may lead to a complete incorporation of the anterior processes, especially the medial anterior process, into a more compact appearing basipterygium. This is the case in several doradids and auchenipterids (Birindelli 2014).

In most siluriforms, the basipterygium is a quite flat structure with processes rarely protruding from the main plane. In some taxa there is a dorsal flange, e.g., in Amphiliidae (Skelton 1986), some aspredinids (de Pinna 1996), or callichthyids (Reis 1998). In sisorids, a ventral flange is present (Figure 3E; de Pinna 1996).

In most catfishes, the left and right basipterygia are connected via cartilage, allowing a certain degree of flexibility between both halves. At least twice during evolution, i.e., in doradids and at the base of Loricariidae, Astoblepidae and Callichthyidae, both halves got in direct contact via a serrated or dentate suture connecting the left and right bony part of the basipterygium.

Radials

In catfishes, radials are largely reduced, and the fin rays articulate directly on the cartilage-capped posterior margin of the basypterygium, the articulatory facet. Alexander (1965) stated that 'pelvic radials...are absent from catfish', which was confirmed by Reis (1998). Arratia (2003) similarly wrote that pelvic radials are completely absent in Siluriformes except for diplomystids, where a medial radial can be seen in juveniles (Arratia 1987; Azpelicueta 1994). However, single cartilaginous radials have also been found in other catfish taxa like in the aspredinids (de Pinna 1993; Friel 1994), *Notoglanidium* from Auchenoglanididae (de Pinna 1993), some Pangasiidae, Siluridae and Sisoridae (Birindelli 2014), *Corydoras* and *Aspidoras* both from Callichthyidae (Figures 3B and 4D), *Pseudauchenipterus* from Auchenipteridae (Figure 3F), *Microglanis* from Pseudopimelodidae (Figure 3G), *Chasmocranus* from Heptapteridae (Figure 7F), or *Synodontis* from Mochokidae (Figure 6C). But in these taxa, the radial is positioned laterally at or close to the base of the first fin ray. In *Malapterurus*, two or even three radials are present (de Pinna 1993), and in *Ameiurus* of the Ictaluridae, we found two radials, one lateral and one medial, plus additional small radial-like cartilages irregularly distributed in between (Figure 8E). A similar band of small cartilages was also observed in *Corydoras* (Figure 3B).

Cypriniforms, characiforms, gonorynchiforms, clupeids and alepocephaliforms usually all have three clearly formed radials. Thus, in the stem group of siluriform evolution, one could expect the presence of three radials. The cartilages in siluriforms show the same position, so from an anatomical point of view, they can be called radials. However, the presence or absence of radials in siluriforms is scattered among the phylogenetic tree, and it becomes questionable if they are homologous to the radials in other otomorphs. The radial reported from diplomystids could correspond to a medial radial, and the single radials at the base of the first ray in other catfish could correspond to a lateral radial. At least the small cartilages, as found in *Ameiurus* and *Corydoras* do not seem to be homologous to a certain radial in otomophs. In our view, it seems likely that the 'radials' in siluriforms, in general, are not homologous to true radials as present in other teleosts. Due to their scattered distribution, it seems even more likely that siluriform radials appeared and disappeared several times in different taxa.

Fin Rays and Pelvic Splint

Typically, catfishes have six fin rays in the pelvic fin, with the first one being not branched and often an additional pelvic splint. Mochokids (with the exception of one species), a few doradids (e.g., *Centrodoras*, *Doras*, *Oxydoras*, *Platydoras*), some auchenipterids (e.g., *Ageneiosus*), and the fossil †*Hypsidoris* have seven rays (Grande 1987; Arratia 2003; Vigliotta 2008; Birindelli 2014). Ictalurids, *Pseudauchenipterus*, and the pangasid genus *Pangasianodon* have eight rays, and some auchenipterids (e.g., *Asterophysus*, *Auchenipterus*, and *Trachelyichthys*) have nine or even more rays (Birindelli 2014). Higher numbers are known from Siluridae with 6 to 14 rays and Plotosidae with 10 to 16 rays (Arratia 2003).

As the character state of six fin rays is widely distributed in Loricarioidei and Siluroidei and is also present in Diplomystidae, this represents the condition of the last common catfish ancestor. Further details can be added based on character distribution in present taxa: the first ray is unbranched but not spine-like, and a pelvic splint is likely present (Grande 1987). The pelvic splint shows a patchy distribution (Marceniuk et al. 2012; Birindelli 2014), so it is not unlikely that it vanished and reappeared several times in siluriform evolution. Generally, in teleost taxa, the splint has a patchy distribution (Patterson 1964; Gosline 1961). Gosline (1961) noted that the 'origin and nature of this structure seems to be unknown', and Patterson (1964) suggested that the splint 'can only be homologous' to 'a fulcral scale'. Based on our observations, there seems to be little doubt that the splint is anything other than the dorsal half of a lepidotrichia – at least in siluriforms. In *Diplomystes* (Figure 2E), the splint has as well a remnant of a ventral counterpart and at least some segmentation at its distal end, clearly identifying the serial homology of the splint with the dorsal half of a fin ray. Still, the splint differs from typical rays by later development, missing associated actinotrichia during ontogeny, and lack of associated muscles. If the pelvic splint in siluriforms is homologous to the pelvic splint in various non-neoteleost teleostean groups, further investigation is needed.

Lateropterygium

The lateropterygium is a separate ossification originating dorsally at or close to the articulation of the first pelvic fin ray with the basipterygium and (depending on size) stretching dorsolaterally towards the lateral body wall (Figure 2A–D). It is present only in Astroblepidae and Loricariidae (Shelden 1937; Howes 1983; Paixão and Toledo-Piza 2009).

In *Astroblepus* it is rounded, almost disc-shaped in its lateral portion, and its margins are covered by cartilage (Figure 3C; Shelden 1937, fig. 25; Silva Pedroza 2009, fig. 24; Pereira and Reis 2017, fig. 35; Arratia 2003, fig. 4.6). In loricariids, the lateropterygium is often rod-like, differing in relative size and slightly in shape (Armbruster 2004). The absence of the lateropterygium in a few taxa, e.g., *Hemiodontichthys* (Londoño-Burbano and Reis 2016), *Loricaria*, *Sturisoma*, *Pseudohemiodon* (Howes 1983), and *Pseudoloricaria* (Paixão and Toledo-Piza 2009) has to be regarded as a secondary loss, not only as Howes (1983) stated 'as these taxa are admitted within the Loricariidae', but also based on proposed phylogenies of Loricariidae where the most parsimonious assumption is the presence of that structure in their last common ancestor. In species with a long lateropterygium, e.g., *Hypostomus* or *Neoplecostomus*, its dorsal tip approaches the lateral body armour, to which it is connected by connective tissue, but there is no true articulation.

Shelden (1937) suggested that the lateropterygium may increase the manoeuvrability of the pelvic girdle, and it seems at first glance as a logical proposal, at least in the case of the highly mobile pelvic girdle in *Astroblepus* (Johnson 1912). Strongly connected to locomotion, but in a different way, we regard it as more likely that the lateropterygium in loricariids rather supports the fixation of the pelvic girdle relative to the body. This would allow the enlarged spine-like first fin ray of the pelvic girdle to fulfil a similar function as the alike-looking pectoral spine, i.e., protection against predators and support in locomotion (Arratia 2003) or fixation on substrate. Bressmann et al. (2021) recently showed that a fixation of the pelvic girdle to the axial skeleton via the lateropterygium facilitates terrestrial locomotion in loricariids. This observation corresponds to the observation that in species with complete armour, the lateropterygium is smaller: if the position of the pelvic girdle is fixed via the body armour, an anchorage via a separate structure is no longer needed, and the lateropterygium gets reduced in size or even completely lost.

The origin of the lateropterygium was unclear for a long time. Principally, there are four possible origins: it could be (1) a transformed pelvic splint, (2) an enlarged radial, (3) a piece of the lateral part of the basipterygium which lost connection to the main body, or (4) a completely new structure without precursor. The third hypothesis was assumed by Shelden (1937), who regarded the lateropterygium as homologous to the lateral process of other siluriforms. The first hypothesis is supported by the presence of a pelvic splint in trichomycterids, nematogenyids, diplomystids and some other taxa and its absence in the taxa, which have a lateropterygium. This view principally was already given by Regan (1904) and intensified by Howes (1983), but Arratia et al. (1978) contradicted such a possible homology. The presence-absence argument can also be made for the second hypothesis: radials occur in some siluriform taxa, but those taxa possessing a lateropterygium do not have them. Yet, radials and pelvic splints have a different ontogenetic origin. While the splint ossifies directly, radials have cartilaginous precursors and may also remain cartilaginous. When following the ontogenesis of the lateropterygium in *Ancistrus* in about the size of 10 to 14 mm (Figure 2A–C), the originating lateropterygium looks like a 'normal' pelvic splint of other siluriforms and also follows its typical developmental pattern by appearing after all other rays have formed. Thus, we regard the lateropterygium in loricariids as a derived pelvic splint and thus homologous to this structure in other catfishes. Yet, the situation seems different in *Astroblepus*. Here, a cartilaginous margin is clearly visible around the lateropterygium (Figure 3C; Shelden 1937; Silva Pedroza 2009). This cartilage identifies the lateropterygium of *Astroblepus* as a chondral bone, suggesting a different origin than that in loricariids. Consequently, even though loricariids and astroblepids are sister groups, it appears that both have independently developed a lateropterygium: on the one hand, derived from the pelvic splint, and on the other hand, from a cartilaginous origin. Ontogenetic studies of *Astroblepus* must clarify if the lateropterygium in this taxon has its origin in a pelvic radial, from the cartilage of the basipterygium itself, or if it is a completely new formation.

Evolution of the Pelvic Girdle

The pelvic fins in teleosts are primarily used for manoeuvring and stabilisation and not for propulsion. While most teleost species are free-swimming, catfish, for the most part, have a bottom-oriented lifestyle. This resulted in different or additional usages of the pelvic fins. Several siluriform taxa use their pelvic fins for other locomotion types, such as holding on to structures or 'walking' on a substrate in and sometimes even outside the water (Johnson 1912; de Crop et al. 2013; Bressmann et al. 2021). The pelvic fins can play an important role during or after mating and for parental care: In ariids, the pelvic fins somehow assist during mating or egg transfer for mouthbrooding. In some *Bunocephalus* species (Aspredinidae), females carry the eggs to the nest in a basket formed by the pelvic fins (Bruton 1996). In callichthyids, females collect freshly laid eggs with their pelvic fins for fertilisation and subsequently attach these eggs to the substrate (Zarske and Greven 2015). Ictalurids generate a water flow with their pelvic fins to ventilate the clutch (Breder 1939).

The pelvic fins in catfishes are usually rather large compared to other fish taxa and in an abdominal position, i.e., clearly positioned posterior to the pectoral fins. In most cases, the pelvic fins are positioned about halfway to the standard length. In some taxa, e.g., *Leptodoras* (Birindelli 2014), the relative position may be more anterior, but this is usually a result of a much-prolonged caudal region rather than an anterior shift. The internal supporting structures, the basipterygia, have a respective position anterior to the pelvic fins and usually lie horizontal in the ventral body wall. The pelvic girdle is embedded in musculature and not directly connected to any other skeletal element. Only in a few cases is it elongated anteriorly and approaches the pectoral girdle or is enlarged and draws near the osseous body armour, such as in loricarioids. However, most catfishes have at least one structure helping to fixate the pelvic girdle in its position. Usually, this is the proximal part of the pelvic splint or the posterior process of the basipterygium, like in callichthyids. The lateropterygium in Loricariidae, being a derived pelvic splint, seems highly developed for this purpose. While in some taxa, such as loricariids, the pelvic fins resemble much the pectoral fins, including the shape and size of the spine-like first fin ray, in free-swimming taxa, like schilbeids, silurids, or pangasiids, these fins are quite different, and the pelvic fins are always much smaller. In the latter taxa, the basipterygia can shift from a horizontal to a more inclined position.

Comparing pelvic girdles of siluriform families, some characters and character states occur more often than others, and, based on their distribution among different taxa, a generalised morphology may be deduced that includes:

1) two anterior processes, with the lateral one being larger, the medial ones touching each other, and both developing from the basipterygial cartilage;
2) a large contact zone of the symphyseal cartilages along the medial arms;
3) a (largely) cartilaginous posterior process;
4) a small lateral process;
5) a small pelvic splint; and
6) six fin rays, with the first being unbranched.

This conclusion of character states present in a generalised siluriform pelvic girdle is not based on phylogenetic data, yet it is fascinating that it is mostly represented by diplomystids (Arratia 1987), which are assumed to have diverged early from the remaining siluriforms and represent a rather conserved morphology. Despite recent molecular-based phylogenies raising questions on the position of diplomystids and placing other taxa, e.g., Loricarioidei (Schedel et al. 2022), as the earliest branching siluriform taxon, the only character that needs to be discussed in more detail is the number of anterior processes. In some diplomystids, as well as other catfish taxa such as Siluridae, Pangasiidae, and sometimes *Chaca*, three (or more) anterior processes are present. In loricarioids, at least during ontogeny, two anterior processes are present. So, it remains questionable how many anterior processes were present in the most common ancestor of all catfishes. †*Hypsidoris* could shed some light on this issue, as it is thought to represent an early branch of siluriforms. Grande (1987) placed it as the sister taxon to all siluriform families except Diplomystidae. Despite the slight shift of diplomystids in the present phylogenetic hypotheses, the idea of †*Hypsidoris* being an early branching taxon remains valid. While Grande (1987) depicted the pelvic girdle of a specimen that could have had three anterior processes (his fig. 13B) as well as a specimen which clearly had two anterior processes (his fig. 13D), Grande and de Pinna (1998) described a second species of †*Hypsidoris* which clearly has two anterior processes. This provides some evidence that the last common ancestor of catfishes may have had two anteriorly directed processes.

Altogether, the enormous diversity of catfish body shapes and lifestyles is also reflected in the diversity of their pelvic girdles. Catfishes remain a fascinating group that still raises a plethora of questions on morphology, homology and evolution. New phylogenies unravelling the intrarelationships of catfishes will help to understand further the evolution of the pelvic girdle and fin in this taxon and, in combination with ancestral character state estimations, might allow us to figure out the morphology of the pelvic girdle of the last common ancestor of siluriforms.

Acknowledgements

The authors are grateful to Philippe Laléyé (Université d'Abomey-Calavi, Benin), Jan Mol (Celos, Paramaribo, Suriname) and Hans Ho (National Museum of Marine Biology and Aquarium, Pingtung, Taiwan) for support in specimen collection and Stefan Schwabbauer (Nutrition Food, Altkalen, Germany) for providing eggs of *Clarias gariepinus*. Furthermore, we thank Annerose Goldbecher for the support in clearing and staining and Eveline Kappel for organising literature that is difficult to access.

References

Abe, G. and K.G. Ota. 2017. Evolutionary developmental transition from median to paired morphology of vertebrate fins: perspectives from twin-tail goldfish. Dev. Biol. 427: 251–257.

Alexander, R.M. 1965. Structure and function in the catfish. J. Zool. 148: 88–152.

Arce-H, M., J.G. Lundberg and M.A. O'Leary. 2017. Phylogeny of the North American catfish family Ictaluridae (Teleostei: Siluriformes) combining morphology, genes and fossils. Cladistics 33: 406–428.

Armbruster, J.W. 2004. Phylogenetic relationships of the suckermouth armoured catfishes (Loricariidae) with emphasis on the Hypostominae and the Ancistrinae. Zool. J. Linn. Soc. 141: 1–80.

Armbruster, J.W. 2011. Global catfish biodiversity. Am. Fisheries Soc. Symposium 77: 15–37.

Arratia, G., A. Chang G., S. Menu-Marque and G. Rojas M. 1978. About *Bullockia* gen. nov., *Trichomycterus mendozensis* n. sp. and revision of the family Trichomycteridae (Pisces, Siluriformes). Stud. Neotrop. Fauna Environ. 13: 157–194.

Arratia, G. 1987. Description of the primitive family Diplomystidae (Siluriformes, Teleostei, Pisces): morphology, taxonomy and phylogenetic implications. Bonner Zool. Monogr. 24: 1–120.

Arratia, G. 1998. *Silvinichthys*, a new genus of trichomycterid catfishes from the Argentinian Andes, with redescription of *Trichomycterus nigricans*. Ichthyol. Explo. Freshwaters 4: 347–370.

Arratia, G. 2003. The siluriform postcranial skeleton—an overview. pp. 121–157. *In*: Arratia, G., B.G. Kapoor, M. Chardon and R. Diogo (eds.). Catfishes. Science Publishers, Enfield, USA.

Azpelicueta, M.M. 1994. Three East-Andean species of *Diplomystes* (Siluriformes: Diplomystidae). Ichthyol. Explor. Freshwaters 5: 223–240.

Balfour, F.M. 1878. A Monograph on the Development of Elasmobranch Fishes. Macmillan and Co, London, United Kingdom.

Birindelli, J.L. and M.H.S. Pérez. 2011. *Ossancora*, new genus of thorny catfish (Teleostei: Siluriformes: Doradidae) with description of one new species. Proc. Acad. Nat. Sci. 161: 117–152.

Birindelli, J.L. and J. Zuanon. 2012. Systematics of the jaguar catfish genus *Liosomadoras* Fowler, 1940 (Auchenipteridae: Siluriformes). Neotrop. Ichthyol. 10: 1–11.

Birindelli, J.L. 2014. Phylogenetic relationships of the South American Doradoidea (Ostariophysi: Siluriformes). Neotrop. Ichthyol. 12: 451–564.

Bockmann, F.A. 1998. Análise filogenética da família Heptapteridae (Teleostei, Osteriophysi, Siluriformes) e redefinção de seu gêneros. Ph.D. Thesis, Universidade de São Paulo, São Paulo, Brazil.

Bockmann, F.A. and A.M. Miquelarena. 2008. Anatomy and phylogenetic relationships of a new catfish species from northeastern Argentina with comments on the phylogenetic relationships of the genus *Rhamdella* Eigenmann and Eigenmann 1888 (Siluriformes, Heptapteridae). Zootaxa 1780: 1–54.

Bockmann, F.A. and R. Castro. 2010. The blind catfish from the caves of Chapada Diamantina, Bahia, Brazil (Siluriformes: Heptapteridae): description, anatomy, phylogenetic relationships, natural history, and biogeography. Neotrop. Ichthyol. 8: 673–706.

Bornbusch, A.H. 1991. Redescription and reclassification of the silurid catfish *Apodoglanis furnessi* Fowler (Siluriformes: Siluridae), with diagnoses of three intrafamilial silurid subgroups. Copeia 1991: 1070–1084.

Bornbusch, A.H. 1995. Phylogenetic relationships within the Eurasian catfish family Siluridae (Pisces: Siluriformes), with comments on generic validities and biogeography. Zool. J. Linn. Soc. 115: 1–46.

Breder, C.M.Jr. 1939. Variations in the nesting habits of *Ameiurus nebulosus*. Zoologica 24: 367–378.

Bressman, N.R., C.H. Morrison and M.A. Ashley-Ross. 2021. Reffling: a novel locomotor behaviour used by neotropical armored catfishes (Loricariidae) in terrestrial environments. Ichthyol. Herpetol. 109: 608–625.

Britz, R., F. Kakkassery and R. Raghavan. 2014. Osteology of *Kryptoglanis shajii*, a stygobitic catfish (Teleostei: Siluriformes) from peninsular India, with diagnosis of a new family Kryptoglanidae. Ichthyol. Explor. Freshwaters 24: 193–207.

Brown, B.A. and C.J. Ferraris. 1988. Comparative osteology of the Asian catfish family Chacidae: with the description of a new species from Burma. Am. Mus. Novit. 2907: 1–16.

Bruton, M.N. 1996. Alternative life-history strategies of catfishes. Aquat. Living Resour. 9: 35–41.

Buitrago-Suarez, U.A. 2005. A Systematic Revision of the Catfish Genus *Pseudoplatystoma* Bleeker, 1862 (Siluriformes: Pimelodidae): Comparative Anatomy, Phylogeny, and Historical Biogeography. Ph.D. Thesis, Southern Illinois University, Carbondale, USA.

Calegari, B.B., R.P. Vari and R.E. Reis. 2019. Phylogenetic systematics of the driftwood catfishes (Siluriformes: Auchenipteridae): a combined morphological and molecular analysis. Zool. J. Linnean Soc. 187: 661–773.

Carvalho, T.P., J.G. Lundberg, J.N. Baskin, J.P. Friel and R.E. Reis. 2016. A new species of the blind and miniature genus *Micromyzon* Friel and Lundberg, 1996 (Silurifomes: Aspredinidae) from the Orinoco River: describing catfish diversity using high-resolution computed tomography. Proc. Acad. Nat. Sci. 165: 37–53.

Chen, X. 1994. Phylogenetic Studies of the Amblycipitid Catfishes (Teleostei, Siluriformes) with Species Accounts. Ph.D. Thesis, Duke University, Durham, USA.

Chen, X. and J.G. Lundberg. 1994. *Xiurenbagrus*, a new genus of amblycipitid catfishes (Teleostei: Siluriformes), and phylogenetic relationships among the genera of Amblycipitidae. Copeia 1995: 780–800.

Costa, W.J., C.R. Feltrin and A.M. Katz. 2020. A new species from subtropical Brazil and evidence of multiple pelvic fin losses in catfishes of the genus *Cambeva* (Siluriformes, Trichomycteridae). Zoosystematics Evol. 96: 715–722.

Costa, W.J. 2021. Comparative osteology, phylogeny and classification of the eastern South American catfish genus *Trichomycterus* (Siluriformes: Trichomycteridae). Taxonomy 1: 160–191.

de Crop, W., E. Pauwels, L. van Hoorebeke and T. Geerinckx. 2013. Functional morphology of the Andean climbing catfishes (Astroblepidae, Siluriformes): Alternative ways of respiration, adhesion, and locomotion using the mouth. J. Morph. 274: 1164–1179.

Datovo, A. 2014. A new species of *Ituglanis* from the Rio Xingu basin, Brazil, and the evolution of pelvic fin loss in trichomycterid catfishes (Teleostei: Siluriformes: Trichomycteridae). Zootaxa 3790: 466–476.

Devaere, S., G.G. Teugels, D. Adriaens, F. Huysentruyt and W. Verraes. 2004. Redescription of *Dolichallabes microphthalmus* (Poll, 1942) (Siluriformes, Clariidae). Copeia 2004: 108–115.

Devaere, S., D. Adriaens, G.G. Teugels, N.M. de Clerck and A.A. Postnov. 2005. Skeletal morphology of the holotype of *Gymnallabes nops* Roberts & Stewart, 1976, using micro CT-scanning. Cybium 29: 281–293.

Dingerkus, G. and L.D. Uhler. 1977. Enzyme clearing of alcian blue stained whole small vertebrates for demonstration of cartilage. Stain Technol. 52: 229–32.

Diogo, R. 2020. Cranial or postcranial—Dual origin of the pectoral appendage of vertebrates combining the fin-fold and gill-arch theories? Dev. Dyn. 249: 1182–1200.

Don, E.K., P.D. Currie and N.J. Cole. 2013. The evolutionary history of the development of the pelvic fin/hindlimb. J. Anat. 222: 114–133.

Ferraris, C.J. Jr. 1988. The Auchenipteridae: Putative Monophyly and Systematics, with a Classification of the Neotropical Doradoid Catfishes (Ostariophysi: Siluriformes). Ph.D. Thesis, The City of New York University, New York, USA.

Fink, S.V. and W.L. Fink. 1981. Interrelationships of the ostariophysan fishes (Teleostei). Zool. J. Linnean Soc. 72: 297–353.

Fowler, H.W. 1905. Some fishes from Borneo. Proc. Acad. Nat. Sci. 57: 455–523.

Fricke, R., W.N. Eschmeyer and J.D. Fong. 2024. Eschmeyer's catalog of fishes: genera/species by family/subfamily. (http://researcharchive.calacademy.org/research/ichthyology/catalog/SpeciesByFamily.asp). Electronic version accessed 30 Jan 2024.

Friel, J.P. 1995. A Phylogenetic Study of the Neotropical Banjo Catfishes (Teleostei: Siluriformes: Aspredinidae). Ph.D. Thesis, Duke University, Durham, USA.

Gosline, W.A. 1961. Some osteological features of modern lower teleostean fishes. Smithsonian Misc. Collect. 142: 1–42.

Grande, L. 1985. Recent and fossil clupeomorph fishes with materials for revision of the subgroups of clupeoids. Bull. Am. Mus. Nat. Hist. 181: 231–372.

Grande, L. 1987. Redescription of *Hypsidoris farsonensis* (Teleostei: Siluriformes), with a reassessment of its phylogenetic relationships. J. Vert. Paleontol. 7: 24–54.

Grande, L. and W.E. Bemis. 1998. A comprehensive phylogenetic study of amiid fishes (Amiidae) based on comparative skeletal anatomy. An empirical search for interconnected patterns of natural history. J. Vert. Paleontol. Mem. 4: 1–690.

Grande, L. and M. de Pinna. 1998. Description of a second species of the catfish †*Hypsidoris* and a reevaluation of the genus and the family †Hypsidoridae. J. Vert. Paleont. 18: 451–474.

Grande, L. 2010. An empirical synthetic pattern study of gars (Lepisosteiformes) and closely related species, based mostly on skeletal anatomy. The resurrection of Holostei. ASIH, Spec. Pub. 6: 1–871.

Higuchi, H. 1992. A Phylogeny of the South American Thorny Catfishes (Osteichthyes; Siluriformes, Doradidae). Ph.D. Thesis, Harvard University, Havard, USA.

Hilton, E.J. 2002. Osteology of the extant north American fishes of the genus *Hiodon* Lesueur, 1818 (Teleostei: Osteoglossomorpha: Hiodontiformes). Fieldiana Zool. 100: 1–142.

Huysentruyt, F. 2008. Early Development of *Corydoras aeneus* (Siluriformes, Callichthyidae): a Case Study for Understanding The Evolutionary Basis of Loricarioid Ontogenetic Patterning. Ph.D. Thesis, University Gent, Gent, Belgium.

Howes, G.J. 1983. The cranial muscles of loricarioid catfishes, their homologies and value as taxonomic characters (Teleostei: Siluroidei). Bull. Br. Mus. nat. Hist. (Zool.) 45: 309–345.

Johnson, R.D.O. 1912. Notes on the habits of a climbing catfish (*Arges marmoratus*) from the Republic of Colombia. Ann. N.Y. Acad. Sci. 22: 327–333.

Kappas, I., S. Vittas, C.N. Pantzartzi, E. Drosopoulou and Z.G. Scouras. 2016. A time-calibrated mitogenome phylogeny of catfish (Teleostei: Siluriformes). PLoS One 11: e0166988.

Lakshmi, K. and K.S. Rao. 1989. Postcranial skeleton of the marine catfish *Arius tenuispinis* (Osteichthyes: Siluriformes, Ariidae). J. Morph. 202: 361–377.

Larouche, O., M.L. Zelditch and R. Cloutier. 2017. Fin modules: an evolutionary perspective on appendage disparity in basal vertebrates. BMC Biol. 15: 32–58.

Lima, S.M., C.P. Neves and R.M. Campos-Paiva. 2013. *Ituglanis agreste*, a new catfish from the rio de Contas basin, northeastern Brazil (Siluriformes: Trichomycteridae). Neotrop. Ichthyol. 11: 513–524.

Londoño-Burbano, A. and R.E. Reis. 2016. Taxonomic revision and phylogenetic relationships of *Dasyloricaria* Isbrücker & Nijssen, 1979 (Siluriformes: Loricariidae), with description of a new species. Neotrop. Ichthyol. 14: e150120.

Londoño-Burbano, A. and R.E. Reis. 2021. A combined molecular and morphological phylogeny of the Loricariinae (Siluriformes: Loricariidae), with emphasis on the Harttiini and Farlowellini. PloS One 16: e0247747.

Lundberg, J.G. 1982. The comparative anatomy of the toothless blindcat, *Trogloglanis pattersoni* Eigenmann, with a phylogenetic analysis of the ictalurid catfishes. Misc. Publ. Mus. Zool. Univ. Mich. 163: 1–85.

Lundberg, J.G. and W.M. Dahdul. 2008. Two new cis-Andean species of the South American catfish genus *Megalonema* allied to trans-Andean *Megalonema xanthum*, with description of a new subgenus (Siluriformes: Pimelodidae). Neotrop. Ichthyol. 6: 439–454.

Marceniuk, A.P., N.A. Menezes and M.R. Britto. 2012. Phylogenetic analysis of the family Ariidae (Ostariophysi: Siluriformes), with a hypothesis on the monophyly and relationships of the genera. Zool. J. Linn. Soc. 165: 534–669.

Merriman, D. 1940. Morphological and embryological studies on two species of marine catfish, *Bagre marinus* and *Galeichthys felis*. Zoologica 25: 221–248.

Mo, T.-P. 1991. Anatomy, relationships and systematics of the Bagridae (Teleostei: Siluroidei) with a hypothesis of silurid phylogeny. Koeltz Scientific Books, Koenigstein, Germany.

Ng, H.H. 2003. *Kryptopterus geminus*, a new species of silurid catfish (Teleostei: Siluridae) from mainland Southeast Asia. Zootaxa 305: 1–11.

Paixão, A.C. and M. Toledo-Piza. 2009. Systematics of *Lamontichthys* Miranda-Ribeiro (Siluriformes: Loricariidae), with the description of two new species. Neotrop. Ichthyol. 7: 519–568.

Parisi, B.M., J.G. Lundberg and C. Donascimiento. 2006. *Propimelodus caesius* a new species of long-finned pimelodid catfish (Teleostei: Siluriformes) from the Amazon Basin, South America. Proc. Acad. Nat. Sci. 155: 67–78.

Patterson, C. 1964. A review of Mesozoic acanthopterygian fishes, with special reference to those of the English Chalk. Philos. Trans. R. Soc. Lond., B, Biol. Sci. 247: 213–482.

Pereira, E.H. and R.E. Reis. 2017. Morphology-based phylogeny of the suckermouth armored catfishes, with emphasis on the Neoplecostominae (Teleostei: Siluriformes: Loricariidae). Zootaxa 4264: 1–104.

de Pinna, M.C. 1993. Higher-level Phylogeny of Siluriformes, with a New Classification of the Order (Teleostei, Ostariophysi). Ph.D. Thesis, University of New York, New York, USA.

de Pinna, M.C. and R.P. Vari. 1995. Monophyly and phylogenetic diagnosis of the family Cetopsidae, with synonymization of the Helogenidae (Teleostei: Siluriformes). Smithson. Contrib. Zool. 571: 1–26.

de Pinna, M.C. 1996. Phylogenetic analysis of the Asian catfish families Sisoridae, Akysidae, and Amblycipitidae, with a hypothesis on the relationships of the neotropical Aspredinidae (Teleostei, Ostariophysi). Fieldiana, Zool. 84: 1–83.

de Pinna, M.C. 1998. Phylogenetic relationships of Neotropical Siluriformes (Teleostei: Ostariophysi): historical overview and synthesis of hypotheses. pp. 279–330. *In*: Malabarba, L.R., R.E. Reis, R.P. Vari, Z.M.S. Lucena and C.A.S. Lucena (eds.). Phylogeny and Classification of Neotropical Fishes. EDIPUCRS, Porto Alegre, Brazil.

de Pinna, M.C., C.J. Ferraris Jr and R.P. Vari. 2007. A phylogenetic study of the neotropical catfish family Cetopsidae (Osteichthyes, Ostariophysi, Siluriformes), with a new classification. Zool. J. Linn. Soc. 150: 755–813.

de Pinna, M.C., L.A.W. Peixoto, V. Tagliacolo and M.R. Britto. 2025. Phylogenetic relationships and evolution of the major groups of Siluriformes. *In*: Arratia, G. and R.E. Reis (eds.). Catfishes, a Highly Diversified Group. Volume 2: Evolution and Phylogeny. CRC Press, Boca Raton, FL.

Pise, M., S.M. Gosavi, P.A. Gorule, C.R. Verma, S.S. Kharat, L. Kalcus et al. 2022. Osteological description of Indian lepidophagous catfish *Pachypterus khavalchor* (Siluriformes: Horabagridae) from the Western Ghats of India. J. Vertebr. Biol. 71: 22021.

Poll, M. 1941. Etude systématique et morphologique d'une collection de poissons de l'Uele (Congo belge) comprenant trois espèces nouvelles. Bull. Mus. R. Hist. Nat. Belg. 17: 1–18.

Py-Daniel, L.R. 1997. Phylogeny of the Neotropical Armored Catfishes of the Subfamily Loricariinae (Siluriformes: Loricariidae). Ph.D. Thesis, University of Arizona, Tucson, USA.

Regan, C.T. 1904. III. A Monograph of the Fishes of the Family Loricariidæ. Trans. Zool. Soc. Lond. 17: 191–350.

Reis, R.E. 1998. Anatomy and phylogenetic analysis of the neotropical callichthyid catfishes (Ostariophysi, Siluriformes). Zool. J. Linn. Soc. 124: 105–168.

Reis, R.E. and T.A. Borges. 2006. The South American catfish genus *Entomocorus* (Ostariophysi: Siluriformes: Auchenipteridae), with the description of a new species from the Paraguay River basin. Copeia 2006: 412–421.

Reis, R.E., E.H. Pereira and P. Lehmann A. 2012. A new genus and species of Hypoptopomatine catfish (Siluriformes: Loricariidae) from the upper Rio São Francisco basin, Brazil. Copeia, 2012: 6–11.

Rizzato, P. and M.A. Almeida. 2025. Overview on some of the main features of the external anatomy of catfishes. pp. 19–47. *In*: Arratia, G. and R.E. Reis (eds.). Catfishes, a Highly Diversified Group. Volume 1: Their Outstanding Biology. CRC Press, Boca Raton, FL.

Roberts, T.R. and C. Vidthayanon. 1991. Systematic revision of the Asian catfish family Pangasiidae, with biological observations and descriptions of three new species. Proc. Acad. Nat. Sci. 143: 97–143.

Roberts, T.R. 2003. Systematics and osteology of Leptoglaninae, a new subfamily of the African catfish family Amphiliidae, with descriptions of three new genera and six new species. Proc. Calif. Acad. Sci. 54: 83–132.

Rocha, M., H. Lazzarotto and L.R. Py-Daniel. 2012. A new species of *Scoloplax* with a remarkable new tooth morphology within Loricarioidea (Siluriformes: Scoloplacidae) Copeia 2012: 670–677.

Rodiles-Hernández, R., D.A. Hendrickson, J.G. Lundberg and J.M. Humphries. 2005. *Lacantunia enigmatica* (Teleostei: Siluriformes) a new and phylogenetically puzzling freshwater fish from Mesoamerica. Zootaxa 1000: 1–24.

Sanford, C.P.J. 2000. Salmonid fish osteology and phylogeny (Teleostei: Salmonoidei). A.R.G. Gantner Verlag KG, Ruggell, Liechtenstein.

Saxena, S.C. and M. Chandy. 1966a. Adhesive apparatus in certain Indian hill stream fishes. J. Zool. 148: 315–340.

Saxena, S.C. and M. Chandy. 1966b. The pelvic girdle and fin in certain Indian hill stream fishes. J. Zool. 148: 167–190.

Schaefer, S.A., S.H. Weitzman and H.A. Britski. 1989. Review of the Neotropical catfish genus *Scoloplax* (Pisces: Loricarioidea: Scoloplacidae) with comments on reductive characters in phylogenetic analysis. Proc. Acad. Nat. Sci. 141: 181–211.

Schaefer, S.A. 1990. Anatomy and relationships of the scoloplacid catfishes. Proc. Acad. Nat. Sci. 142: 167–210.

Schedel, F.D.B., A. Chakona, B.L. Sidlauskas, M.O. Popoola, N.U. Wingi, D. Neumann et al. 2022. New phylogenetic insights into the African catfish families Mochokidae and Austroglanididae. J. Fish Biol. 100: 1171–1186.

Shelden, F.F. 1937. Osteology, myology, and probable evolution of the nematognath pelvic girdle. Ann. N. Y. Acad. Sci. 37: 1–95.

Shubin, N., C. Tabin and S. Carroll. 1997. Fossils, genes and the evolution of animal limbs. Nature 388: 639–648.

Silva Pedroza, W.S. 2009. Morfologia comparada da cintura pélvica de representantes da superfamília Loricarioidea (Ostariophysi; Siluriformes). Ph.D. Thesis, INPA/UFAM, Manaus, Brazil.

Skelton, P.H. 1981. The description and osteology of a new species of *Gephyroglanis* (Siluriformes, Bagridae) from the Olifants River, south west cape, South Africa. Ann. Cape Prov. Mus. Nat. Hist. 13: 217–250.

Skelton, P.H. 1986. Two new *Amphilius* (Pisces, Siluroidei, Amphiliidae) from the Zaïre River system, Africa. Rev. Zool. Afric. 99: 263–291.

Sullivan, J.P., J.G. Lundberg and M. Hardman. 2006. A phylogenetic analysis of the major groups of catfishes (Teleostei: Siluriformes) using rag1 and rag2 nuclear gene sequences. Mol. Phyl. Evol. 41: 636–662.

Taphorn, D.C. and C. Marrero. 1990. *Hoplomyzon sexpapilostoma*, a new species of Venezuelan catfish (Pisces, Aspredinidae): with comments on the Hoplomyzontini. Fieldiana, Zool. 61: 1–9.

Taylor, W.R. and G.C. van Dyke. 1985. Revised procedures for staining and clearing small fishes and other vertebrates for bone and cartilage study. Cybium 9: 107–19.

Teugels, G.G. 1996. Taxonomy, phylogeny and biogeography of catfishes (Ostariophysi, Siluroidei): an overview. Aquat. Living Resour. 9: 9–34.

Teugels, G.G. and D. Adriaens. 2003. Taxonomy and phylogeny of Clariidae—an overview. pp. 465–487. *In*: Arratia, G., B.G. Kapoor, M. Chardon and R. Diogo (eds.). Catfishes. Science Publishers, Enfield, USA.

Thacher, J.K. 1877. Median and paired fins: a contribution to the history of vertebrate limbs. Trans. Conn. Acad. 3: 281–310.

Thieme, P., P. Warth and T. Moritz. 2021. Development of the caudal-fin skeleton reveals multiple convergent fusions within Atherinomorpha. Front. Zool. 18: 20.

Tilak, R. 1967. Studies on the osteology of the nematognathine girdle in relation to taxonomy. J. zool. Soc. India 19: 101–110.

Tilak, R. 1971. A study of the osteocranium, the Weberian apparatus and the girdles of *Chaca chaca* (Hamilton): Family Chacidae, Siluroidei. Zool. Anz. 186: 417–435.

Vigliotta, T.R. 2008. A phylogenetic study of the African catfish family Mochokidae (Osteichthyes, Ostariophysi, Siluriformes), with a key to genera. Proc. Acad. Nat. Sci. 157: 73–136.

de Vos, L. 1995. A systematic revision of the African Schilbeidae (Teleostei, Siluriformes). Ann. Mus. R. Afr. Centr., Sci. Zool. 271: 1–450.

Walsh, S.J. 1990. A Systematic Revision of the Neotropical Catfish Family Ageneiosidae (Teleostei: Ostariophysi: Siluriformes). Ph.D. Thesis, University of Florida, Gainesville, USA.

Winterbottom, R. 1973. A descriptive synonymy of the striated muscles of the Teleostei. Proc. Acad. Nat. Sci. 125: 225–317.

Zarske, A. and H. Greven. 2015. Twenty years 'sperm drinking' by female catfishes – and still nothing new? An essay about the different perception of amateurs and professionals. Bull. Fish Biol. 15: 9–13.

Appendix I: Studied Material

Studied specimens for this investigation. All cleared and double-stained specimens. All sizes are given as standard length. DMM – Ocean Museum Germany; USNM – Smithsonian National Museum of Natural History, Washington, USA; for more than single vouchers, the number of specimens is given in brackets.

Akysidae: *Akysis prashadi*: DMM IE/17002[2], 25.9–28.3 mm. **Amblicipitidae:** *Liobagrus reinii*: USNM 71329[2], 36–79 mm. **Amphiliidae:** *Amphilius atesuensis*: DMM unreg.; *Phractura clauseni*: DMM IE/13426, 42.6 mm. **Ariidae:** *Aspistor quadriscutis*: USNM 207413, 119 mm; *Bagre bagre*: DMM IE/17472, 58.1 mm; *Cathorops arenatus*: USNM 286395, 95 mm; *Cephalocassis borneensis*: DMM IE/17477, 71.5 mm; *Sciades felis*: DMM IE/11739[2], 53.9–57.9 mm; *Sciades herzbergii*: DMM IE/16998, 45.1 mm. **Aspredinidae:** *Aspredo aspredo*: DMM IE/16949; *Bunocephalus coracoideus*: DMM IE/16936, 38.4 mm. **Astroblepidae:** *Astroblepus* sp.: USNM 302674, 77 mm; *Astroblepus* sp. 'Colombia': DMM IE/17478, 43.4 mm. **Auchenipteridae:** *Pseudauchenipterus nodosus*: DMM IE/16996, 38.4 mm; *Tympanopleura cryptica*: DMM IE/17473, 56.4 mm. **Auchenoglanididae:** *Auchenoglanis occidentalis*: DMM IE/15519, 68.9 mm; *Parauchenoglanis monkei*: DMM IE/12219[2], 22.2–36.7, USNM 303328, 44 mm. **Bagridae:** *Rita kuturnee*: USNM 114950, 61 mm; *Sperata aor*: DMM IE/11130[2], 48.9–57.8 mm. **Callichthyidae:** *Aspidoras pauciradiatus*: DMM IE/17255, 17.3 mm; *Corydoras brevirostris*: DMM IE/17469, 19.3 mm, DMM IE/17470, 25.5 mm, DMM IE/17471, 36.1 mm; *Corydoras palaeatus*: unreg. **Cetopsidae:** *Cetopsis orinoco*: DMM IE/17266, 64.0 mm; *Helogenes marmoratus*: DMM IE/16946, 37.6 mm. **Chacidae:** *Chaca chaca*: DMM unreg., 79.9 mm. **Clariidae:** *Clarias agboyiensis*: DMM IE/17474, 60.3 mm; *Clarias* cf. *gariepinus*: DMM IE/14371, 104.4 mm. **Claroteidae:** *Chrysichthys nigrodigitatus*: DMM IE/17479, 88.2 mm. **Diplomystidae:** *Diplomystes mesembrinus*: DMM IE/17476, 69.0 mm. **Doradidae:** *Agamyxis pectinifrons*: DMM IE/12205[2], 43.1–52.2; *Physopyxis lyra*: USNM 287438, 22–31 mm; *Rhinodoras dorbignyi*: USNM 314286, 112 mm. **Heptapteridae:** *Chasmocranus longior*: DMM IE/17467, 62.3 mm; *Imparfinis nemacheir*: USNM 434829, 52 mm; *Pimelodella megalops*: DMM IE/16945[2], 38.1–45.3 mm. **Heteropneustidae:** *Heteropneustes fossilis*: DMM IE/13478, 82.6 mm, DMM IE/13476, 102.6 mm. **Horabagridae:**

Horabagrus brachysoma: DMM unreg, 101 mm. **Ictaluridae:** *Ameiurus melas*: DMM IE/16487[2], 51.4–72.6 mm; *Noturus flavus*: USNM 347242, 70 mm; *Noturus nocturnus*: USNM 172873[3], 34–44 mm; *Prietella phreatophila*: USNM 164169; *Pylodictis olivaris*: USNM 172268, 87 mm. **Loricariidae:** *Ancistrus* cf. *dolichopterus*: DMM IE/11119, 49.2 mm; *Harttia surinamensis*: DMM IE/16943[2], 70.2–79.4 mm; *Hypostomus plecostomus*: DMM IE/16923[2], 56.1–58.2 mm; *Loricaria cataphracta*: DMM IE/16948[2], 91.5–120.9 mm; *Loricariichthys maculatus*: DMM IE/17468, 82.3 mm; *Otocinclus* sp.: DMM IE/11122, 27.1 mm; *Paralithoxus surinamensis*: DMM IE/17475, 30.2 mm SL. **Malapteruridae:** *Malapterurus electricus*: DMM unreg. **Mochokidae:** *Chiloglanis voltae*: DMM IE/13428[3], 13.7–16.0 mm; *Mochokus niloticus*: DMM IE/11012, 19.3 mm; *Synodontis nigrita*: DMM IE/12204, 40.5–46.1 mm; *Synodontis sorex*: DMM IE/11782, 41.4 mm. **Nematogenyidae:** *Nematogenys inermis*: USNM 440397, 105 mm. **Trichomycteridae:** *Hatcheria macraei*: DMM IE/13481, 49.0 mm. *Tridentopsis cahuali*: DMM IE/16950[2], 23.2–23.8 mm. **Pangasidae:** *Pangasianodon hypophthalmus*: DMM IE/11819[2], 48.4–59.9. **Pimelodidae:** *Pimelodus pictus*: DMM IE/12016, 50.8 mm; *Pseudoplatystoma* sp.: DMM IE/17466, 80.5 mm. **Plotosidae:** *Cnidoglanis macrocephalus*: USNM 219581, 83 mm; *Plotosus lineatus*: DMM IE/14953[5], 52.6–87.5 mm; *Porochilus rendahli*: USNM 173557[2], 87–93 mm; *Tandanus ater*: USNM 173551[2], 96–108 mm. **Pseudopimelodidae:** *Microglanis poecilius*: DMM IE/17249[2], 34.5–35.6 mm. **Schilbeidae:** *Parailia pellucida*: DMM IE/16547; *Pareutropius buffei*: DMM IE/17195, 32.6 mm; *Schilbe intermedius*: DMM IE/12209[2], 47.8–55.6 mm; *Siluranodon auritus*: DMM IE/11762, 66.9 mm. **Scoloplacidae:** *Scoloplax dolicholophia*: USNM 300629[2], 8–9 mm; USNM 300630, 11.2 mm; *Scoloplax distolothrix*, USNM 300623[2], 12-13 mm; *Scoloplax empousa*: USNM300626[2], 13–14 mm, USNM 300627[2], 15–16 mm. **Siluridae:** *Hemisilurus heterorhynchus*: USNM 230286, 214 mm; *Ompok bimaculatus*: DMM IE/16171, 91.5 mm; *Silurus asotus*: USNM 327171, 130 mm SL; *Silurus glanis*: DMM IE/15801, 117.5 mm. **Sisoridae:** *Erethistes hara*: DMM IE/11129, 39.6 mm; *Exostoma* sp.: USNM 130183, 138 mm; *Glyptothorax lampris*: DMM IE/16488[2]: 32.9–41.7 mm; *Glyptothorax* cf. *lampris*: USNM 288474, 69 mm.

Chapter 6

The Caudal Fin Musculature of Catfishes and the Homology of the Posterior Myomeres

Nicolás Cumplido

Since Agassiz (1833–45) and von Baer (1835), the caudal fin became a central subject in the study of fish morphology and evolution. Vogt (1842) identified a transitional phase during salmonid ontogeny, where the caudal fin resembled the heterocercal tails of chondrichthyans and early actinopterygians. This finding inspired many naturalists to characterize the series of caudal fin transformations, revealing complex and diverse morphologies (e.g., Heckel 1850; Huxley 1859; Kölliker 1860; Lotz 1864; Agassiz 1878; Balfour and Parker 1882; Mayer 1886; Ryder 1886; Harrison 1895; Hindersson 1910; Whitehouse 1910; Schmalhausen 1912, 1913). Of these works, only three focused on muscles: Hindersson (1910) compared the asymmetrical caudal fin of the pike (*Esox lucius*) to the symmetrical one of the cod (*Gadus morhua*); Schmalhausen (1912, 1913) compared teleost embryogenesis to that of basal actinopterygians and chondrichthyans; and Grenholm (1923) compared adult musculature in different teleosts. These investigations laid the foundation for our current understanding of caudal fin musculature.

Yet, most current knowledge of caudal fin anatomy derives from its skeleton (Schultze and Arratia 1989, 2013; Arratia 2015). Caudal fin osteology has served as a tool for discerning teleost interrelationships since Hollister (1936, 1937a, 1937b, 1941), Gosline (1960, 1961) and Nybelin (1963), with synapomorphies for various teleostean groups based on characters of the caudal fin skeleton (e.g., Monod 1968; Patterson 1968; Lundberg and Baskin 1969; Nybelin 1973; Eastman 1980; Fink and Fink 1981, 1996; Schultze and Arratia 1986, 1988, 1989, 2013; Fujita 1990; Arratia 1991, 2013; Arratia and Lambers 1996; Arratia and Tintori 1999; Brito 1999; Hilton and Johnson 2007; Hilton and Britz 2010; Costa 2012; Borden et al. 2013; de Santana et al. 2013; Doosey and Domke 2014; Cumplido et al. 2020; Vaz and Hilton 2020; Thieme et al. 2021). Among them, catfishes share a high intra- and interspecific variation, including reductions in fin ray number and fusions of elements in different families (e.g., Lundberg and Baskin 1969; Arratia 1976, 1982, 1983, 2003, 2018; Fink and Fink 1981; Schaefer 1987; de Pinna and Ng 2004; Cumplido et al. 2020). Despite this variation, only six works have characterized caudal fin musculature in catfishes (McMurrich 1884; Schmalhausen 1912; Grenholm 1923; Aquino and Miquelarena 2001; Huysentruyt et al. 2007; Schaefer and Provenzano 2008). Hence, examining muscle anatomy is necessary to understand how catfishes use their caudal fins to navigate their habitats and how it has evolved in their different families.

School of Veterinary Medicine, Pontificia Universidad Católica de Chile, Santiago, Chile.
Email: nicump@gmail.com

Evolution and Development of the Caudal Fin Musculature

Teleosts evolved a complex system of intrinsic muscles to control caudal fin movement (Flammang and Lauder 2008; Flammang 2014). The primitive condition in chondrichthyans and chondrosteans consisted of a narrow band of intrinsic musculature attaching to the caudal fin rays (Gemballa 2004). Hence, three sequential changes distinguished the teleost evolution of caudal fin musculature (Schmalhausen 1912, 1913; Lauder 1989; Gemballa 2004): (1) body myomeres acquired a superficial insertion onto the caudal fin rays—shared between teleosts and holosteans; (2) the intrinsic musculature broadened and separated into different muscles—also shared between teleosts and holosteans; and (3) additional intrinsic muscles evolved to control each ray separately—exclusively for teleosts. These evolutionary changes were so significant that Gosline (1971) described them as "the perfection of caudal fin locomotion," marking them as one of, if not the most, remarkable achievements of teleosts.

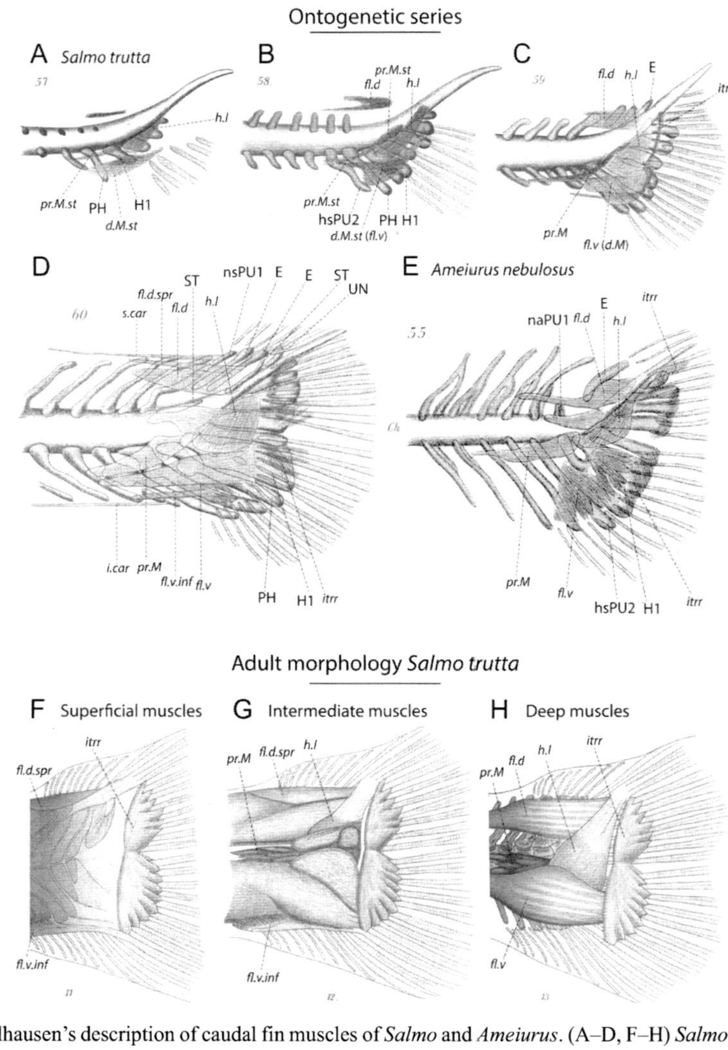

Figure 1. Schmalhausen's description of caudal fin muscles of *Salmo* and *Ameiurus*. (A–D, F–H) *Salmo trutta*; (E) *Ameiurus nebulosus*. (A) 13.5 mm length larva. (B) 16 mm. (C) 17 mm. (D) 19 mm. (A–E) scanned and modified from Schmalhausen (1912). (F–H) scanned and modified from Schmalhausen (1913). Bone terminology follows Arratia and Schultze (1992). Terminology: E, epural; H1, hypural 1; hsPU2, haemal spine of PU2 centrum; naPU1, neural arch of PU1 centrum; PH, parhypural; ST, stegural; UN, uroneural. For Muscles: *d.M.st, distaler Myogener Streifen*; *fl.d, flexor dorsalis; fl.d.spr, flexor dorsalis superior*; *fl.v, flexor ventralis*; *fl.v.inf, flexor ventralis inferior*; *h.l, hypochordal longitudinalis*; *i.car, infracarinalis posterior*; *itrr, interradiales*; *pr.M, proximaler Muskel*; *pr.M.st, proximaler Myogener Streifen*; *s.car, supracarinalis posterior.*

In his research in *Amia calva*, Schmalhausen (1912) described intrinsic musculature forming from two longitudinal stripes of mesenchyme condensing ventral to the notochord—one proximal and one distal—likely derived from body myomeres. The distal stripe formed the *flexor ventralis*, which connects the haemal spines and hypurals to the rays, and the *interradiales*, which interlink the rays. The proximal stripe, by contrast, formed a "proximal muscle." Between both stripes formed the *hypochordal longitudinalis*, running longitudinally from the hypural 1 toward the posterior-most rays (Schmalhausen 1912, fig. 56). Schmalhausen then examined the teleosts *Salmo trutta* and *Ameiurus nebulosus* (Figure 1A–E), concluding a shared ontogeny with *Amia*, with the addition of an *interradiales* on the lower caudal fin lobe, the appearance of a *flexor ventralis inferior*, and the formation of intrinsic muscles above the notochord (=epaxial)—the *flexor dorsalis* and a *flexor dorsalis superior*.

Homology of the Caudal Musculature

Classical criteria for establishing homology between parts of two or more organisms consider shared spatial relationships, ontogenetic development, innervation and/or unique structural details, among other similarities (Remane 1952; Wiley and Lieberman 2011). Regarding the caudal fin, research on osteological development has proven effective in testing homology hypotheses (e.g., Schultze and Arratia 1986, 1988, 2013; Arratia and Schultze 1992; Hilton and Johnson 2007; Wiley et al. 2015; Cumplido et al. 2020; Moritz et al. 2023). Yet, Winterbottom's (1973) influential work—standardizing muscle nomenclature for teleosts—provided broad definitions of caudal fin muscles, encompassing varying positions, orientations, points of origin and insertion, and did not provide information about their ontogeny or innervation. Hence, in contrast to its osteology, broad definitions of caudal fin muscles pose a significant challenge and hinder further testing of the homology hypothesis.

Consequently, a significant variation in caudal fin muscle terminology can be found in recent works for teleosts in general and catfishes in particular (Table 1). Hence, a revision of muscle definitions is pending and required to compare and establish muscle homology, which is fundamental to trace its sequence of evolutionary transformations. Hence, in this work, an ontogeny-based approach is used—by comparing it to Schmalhausen's work (1912)—to identify and define muscles. Moreover, motor innervation was included—a classic method for establishing muscle homology (von Ihering 1878; Goodrich 1906, 1913). Through this approach, original definitions of each muscle are discussed, and—if necessary—novel definitions are proposed.

Terminology

The nomenclature for the caudal fin skeleton follows Schultze and Arratia (2013), with additions from Lundberg and Baskin (1969) for describing catfish hypurapophyses and fusion patterns. The caudal fin endoskeleton can be subdivided into preural and ural regions (Nybelin 1963; Schultze and Arratia 1989). The preural region bears haemal arches and spines supporting procurrent rays, while the ural region—located further posteriorly—bears hypurals, unpaired haemal spines that support most principal rays. The preural-ural boundary is the most influential landmark in hypotheses of caudal fin skeletal homology (Nybelin 1977; Schultze and Arratia 1989), as it also corresponds with the exit point of the caudal artery (Nag 1967; Schultze and Arratia 2013); the last spinal nerve (Nag 1967; Schneider and Sulner 2006); and the last myomere (Gemballa 2004).

The endoskeleton supports caudal fin rays, which can be categorized in teleosts as principal and procurrent rays (Hubbs and Lagler 1947; Arratia 2008). Principal rays consist of all branched rays, along with the first unbranched ray of each lobe, which generally corresponds to one of the longest caudal fin rays. Adjacent to the principal series are the dorsal and ventral procurrent series, which bear a series of smaller rays. These rays diminish in size anteriorly and are always unbranched. The division between the uppermost principal ray and the most posterior dorsal procurrent ray corresponds to the position of the notochord or body axis. Consequently, dorsal procurrent rays

Table 1. Synonyms for caudal fin muscles in catfishes.

This Paper (*Ancistrus triradiatus*)	McMurrich 1884, fig. 9 (*Ameiurus catus*)	Schmalhausen 1912, fig. 55 (*Ameiurus nebulosus*)	Grenholm 1923, figs. 22–23 (*Auchenoglanis occidentalis*, *Ameiurus natalis* and other spp.)	Aquino and Miquelarena 2001, fig. 19 (*Hypoptopoma inexpectata*)	Huysentruyt et al. 2007, fig. 10 (*Corydoras aeneus*)	Schaefer and Provenzano 2008, fig. 24 (*Lithogenes wahari*)
Hypochordal longitudinalis	Upper division of ventral portion of deep caudal muscles (part)	*Hypochordaler Längmuskel*	*Hypochordaler Längmuskel*	*Hypochordal longitudinalis*	*Hypochordalis longitudinalis*	*Flexor dorsalis inferior* (part)
Flexor ventralis	Lower division of ventral portion of deep caudal muscles (part)	*Flexor ventralis*	*Flexor ventralis*	*Flexor ventralis*	*Flexor ventralis* (part)	*Flexor ventralis inferior*
Flexor ventralis inferior	Absent or part of *flexor ventralis*	Absent or part of *flexor ventralis*	Absent or part of *flexor ventralis*	*Flexor ventralis inferior* (part)	Absent or part of *flexor ventralis*	Flexor ventralis superior
Flexor ventralis externus	—	*Proximaler Schwanzmuskel* (part)	Absent or part of *flexor ventralis*	Absent or part of *flexor ventralis/ventralis inferioris*	Absent or part of *flexor ventralis*	Absent or part of the *flexor ventralis superior*
Flexor dorsalis	Dorsal portion of deep caudal muscles (part)	*Flexor dorsalis*	*Flexor dorsalis*	*Flexor dorsalis superior* (lower part)	*Flexor dorsalis* (part)	—
Flexor dorsalis inferior	—	Figured but not described	Figured but not described (part)	—	*Flexor dorsalis* (part)	*Flexor dorsalis superior*
Ventrilateralis	—	*proximaler Schwanzmuskel* (part)	—	*Flexor ventralis inferior* (part)	—	—
Flexor dorsalis superior	Dorsal portion of deep caudal muscles (part)	—	—	*Flexor dorsalis superior* (upper part)	*Flexor dorsalis superior*	Figured but not described (dorsalmost muscle; Figure 24b)
Flexor dorsalis accessories	—	—	—	—	—	—
Flexor ventralis accessories	—	—	—	*Infracarinalis posterior*	*Flexor ventralis inferior*	*Infracarinalis* (part)
Interradiales	Intrinsic muscles	*Interradiales* (= *plicatores*)	*Interradiales*	*Interradialis*	*Interradiales*	*Interradialis*

are always epaxial, while principal rays are hypaxial. Typically, the lowermost principal ray is associated with the haemal spine of preural centrum 2 (Schultze and Arratia 2013). The upper and lower lobes are separated by the hypural diastema, a space between hypurals 2 and 3, which establishes a mirror image symmetry between upper and lower rays among teleosts (Schultze and Arratia 2013; Desvignes et al. 2018).

Methods

For the analysis of muscles and innervation patterns, larvae and juveniles of *Ancistrus triradiatus* and *Danio rerio* were prepared following DEEP-Clear protocols (Pende et al. 2020) and CLARITY (Chung and Deisseroth 2013), respectively. These methods enable the clearing and permeabilization of whole-mount specimens. Subsequently, specific commercial antibodies—known as primary antibodies—are applied to target specific proteins within the body. In this process, a secondary antibody tagged with a known fluorophore targets the primary antibody, which in turn targets the location of the protein of interest—known as antigen. The position of this secondary antibody-primary antibody-antigen complex is then revealed through a confocal microscope or another laser-based optical imaging microscope. These laser microscopes excite the fluorophore of the secondary antibody at a specific wavelength, resulting in the emission of light at a lower energy captured by the microscope detector. Various primary and secondary antibodies can be used, allowing a wide range of combinations during this procedure.

In this study, axon tracts were visualized with an anti-neurofilament antibody (3A10, Developmental Studies Hybridoma Bank) and muscle fibers with an anti-heavy chain myosin antibody (MF-20, Developmental Studies Hybrydoma Bank). In addition, bones were stained with Alizarin Red S, which fluoresce under a 561 nm laser in red, and cell nuclei were stained with TOPRO, which fluoresce in infrared. Individual muscles and nerves were then segmented using VGSTUDIO MAX software.

Osteology

The caudal fin skeleton of *Ancistrus* exhibits a typical loricariid morphology (Figure 2A), as described by Lundberg and Baskin (1969) and Schaefer (1987). It consists of two preural vertebrae (PU2+PU3) and a terminal or compound vertebrae (CC), supporting eight upper and eight lower principal rays alongside four dorsal and five ventral procurrent rays. Preural centra bear a horizontal flange on their ventrolateral surface—above their respective haemal arches—and their dorsal postzygapophyzes, as well as the ventral postzygapophysis of PU2, fuse distally with their corresponding neural or haemal spines. The neural and haemal spines of both preural vertebrae have large anterior flanges, particularly in PU2. Additionally, all procurrent fin rays bear secondary spines on their surface, a feature also found on the lowermost principal ray.

The terminal compound centrum includes the parhypural, all hypurals, and the pleurostyle. It features an intervertebral disc only on its anterior segment, gradually narrowing and curving dorsally in its posterior portion. The pleurostyle extends posterodorsally from the centrum, partially concealed beneath the last dorsal procurrent ray and uppermost principal ray. An autogenous epural is positioned between the neural spine of the PU2 centrum and the pleurostyle, providing distal support for the two posterior procurrent rays. Two large hypural plates are fused postero-ventrally to the centrum, separated by the hypural diastema. The upper hypural plate results from the fusion of hypural 3 to 5 and supports all upper lobe principal rays. The larger lower plate—extending slightly more posteriorly than the upper—results from the fusion of the parhypural with hypurals 1 and 2 and supports all lower principal rays. It includes an anterior projection that contacts the PU2 haemal spine. A foramen located in the middle of this plate marks the preural-ural boundary, corresponding to the exit point of the caudal vasculature. The hypurapophysis, located at the base of this plate, extends as a lateral flange with a pronounced antero-lateral projection. This flange also extends

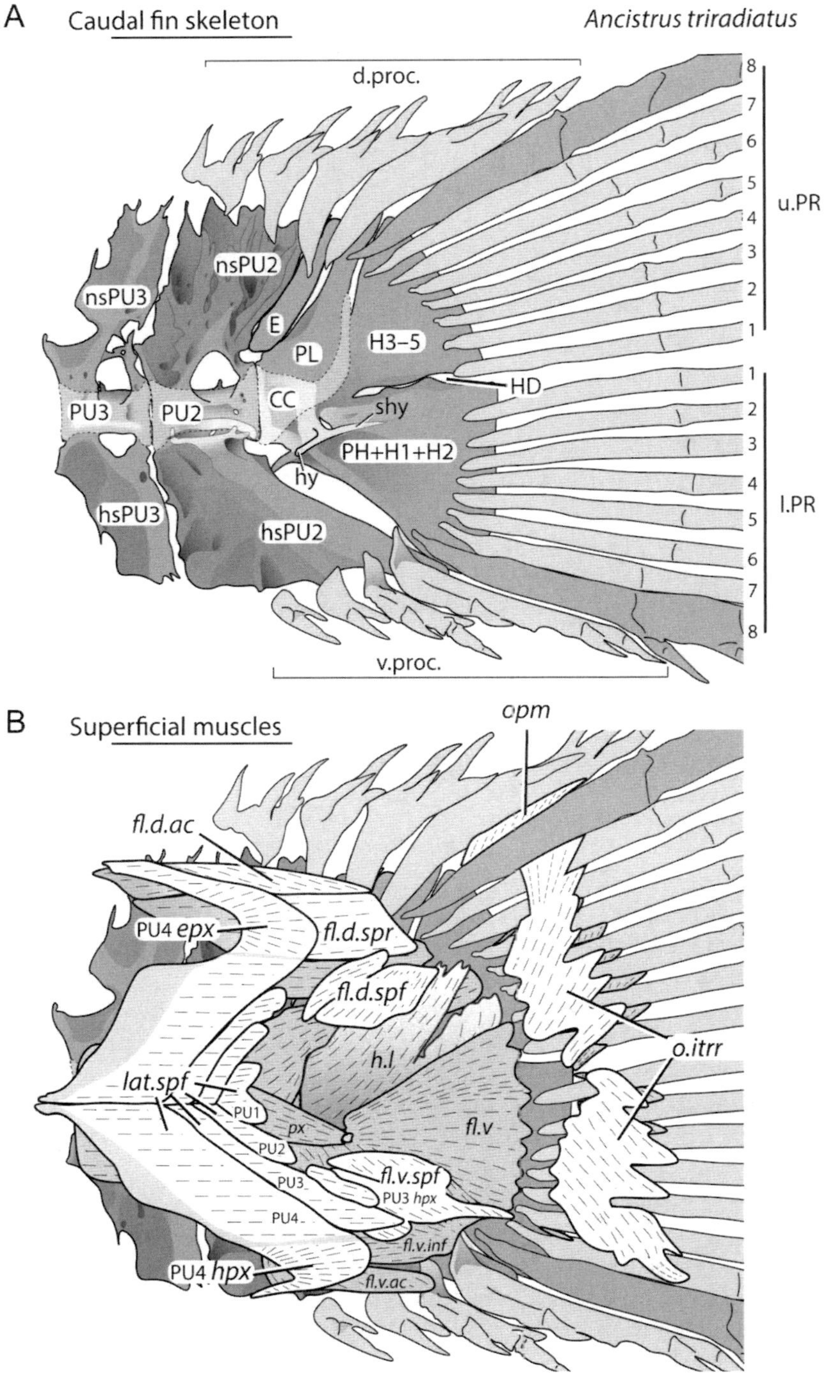

Figure 2. Skeleton and superficial muscle layers of *Ancistrus triradiatus*. (A) diagram of caudal fin skeleton. (B) diagram of superficial musculature of *Ancistrus*. Terminology for bones: CC, compound centrum; d.proc, dorsal procurrent rays; E, epural; H1–5, hypurals 1 to 5; hsPU2–3, haemal spine of PU2 or PU3 centrum; HD, hypural diastema; hy, hypurapophysis; l.PR, lower lobe principal rays; nsPU2–3, neural spine of the PU2 or PU3 centrum; PH, parhypural; PL, pleurostyle; PU2–3, preural centra 2 and 3; shy, secondary hypurapophysis; u.PR, upper lobe principal rays; v.proc, ventral procurrent rays. For Muscles: *epx, epaxialis*; *fl.d.ac, flexor dorsalis accessorius*; *fl.d.spf, flexor dorsalis superficialis*; *fl.d.spr, flexor dorsalis superior*; *fl.v, flexor ventralis*; *fl.v.ac, flexor ventralis accessorius*; *fl.v.ext, flexor ventralis externus*; *fl.v.inf, flexor ventralis inferior*; *fl.v.spf, flexor ventralis superficialis*; *h.l, hypochordal longitudinalis*; *hpx, hypaxialis*; *lat.spf, lateralis superficialis*; *o.itrr, outer layer of the interradiales*; *opm, opisthuralis*; *px, proximalis*.

posteriorly, gradually reducing its lateral extension—referred to as secondary hypurapophysis by Lundberg and Baskin (1969).

Myology

The caudal fin musculature of *Ancistrus* can be divided into different layers. The superficial muscle layer lies in the continuation of the body musculature, which progressively loses its symmetry and metameric organization posteriorly (Figure 2B). The intermediate and deep muscle layers form the intrinsic musculature of the caudal fin (Figure 3A–B), in direct contact with the caudal fin vertebrae. The caudal fin ray muscles comprise a distinct group of muscles that interconnect the caudal fin rays. A detailed description of the muscles of each layer is addressed below.

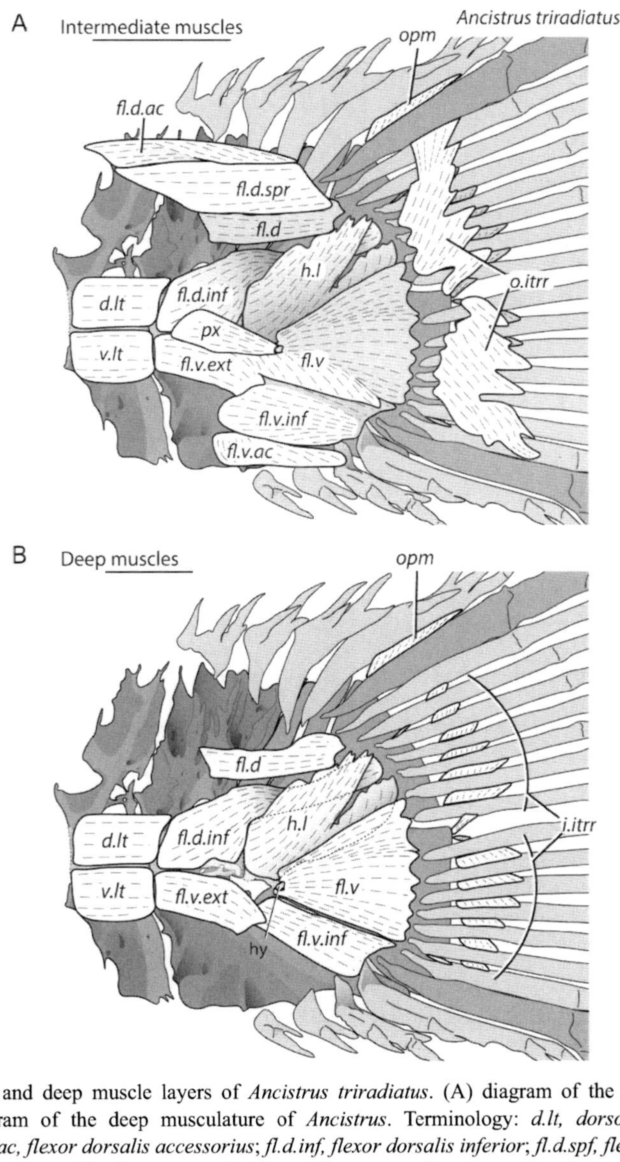

Figure 3. Intermediate and deep muscle layers of *Ancistrus triradiatus*. (A) diagram of the intermediate musculature of *Ancistrus*. (B) diagram of the deep musculature of *Ancistrus*. Terminology: *d.lt, dorsolateralis*; *epx, epaxialis*; *fl.d, flexor dorsalis*; *fl.d.ac, flexor dorsalis accessorius*; *fl.d.inf, flexor dorsalis inferior*; *fl.d.spf, flexor dorsalis superficialis*; *fl.d.spr, flexor dorsalis superior*; *fl.v, flexor ventralis*; *fl.v.ac, flexor ventralis accessorius*; *fl.v.ext, flexor ventralis externus*; *fl.v.inf, flexor ventralis inferior*; *fl.v.spf, flexor ventralis superficialis*; *h.l, hypochordal longitudinalis*; *i.itrr, inner layer of the interradiales*; *o.itrr, outer layer of the interradiales*; *opm, opisthuralis*; PU1–PU4, preural myomeres; *px, proximalis*; *v.lt, ventrilateralis*.

Superficial Muscle Layer

This layer comprises five different muscles: three that correspond to the terminal segments of the body musculature: *epaxialis*, *hypaxialis*, and *lateralis superficialis*. The other two are distinct, highly modified caudal fin muscles: *flexor dorsalis superficialis* and *flexor ventralis superficialis*, which are described below.

Caudal End of the Lateralis Superficialis (lat.spf)

The *lateralis superficialis* is a subdivision of the body musculature located laterally to the midline. It is characterized by its dark red fibers and triangular shape when viewed in a transverse section (Winterbottom 1973). Its posterior end was first described by Greene and Greene in *Oncorhynchus* (1913), covering two-thirds of the caudal peduncle surface with a broadened posterior end. It exhibits an asymmetrical posterior organization with narrowed upper myomeres.

Description: In *Ancistrus*, the caudal end of the *lateralis superficialis* consists of the last four myomeres, PU4 to PU1 (Figures 2B and 4A). It is distinguished by slim and longitudinally oriented fibers, proximal to the midline, which differentiate it from the deeper musculature, typically rich in broad and white fibers (Nag 1972).

The PU4 *lateralis superficialis* is the largest superficial muscle—indicated by dotted lines in Figure 4A—with its hypaxial portion longer than its epaxial. The PU3 *lateralis superficialis* spans nearly the same dorsoventral distance as the PU4 *lateralis superficialis*, but its fibers are notably shorter antero-posteriorly. The PU2 and PU1 *lateralis superficialis* are challenging to segment independently. It is much less elongated both dorsoventrally and anteroposteriorly than the PU4 and PU3 *lateralis superficialis*.

Innervation: The PU4 and PU3 *lateralis superficialis* are innervated by their corresponding spinal nerves (Figure 5A and C). The innervation of the PU2 and PU1 *lateralis superficialis*, however, could not be distinguished.

Ontogeny: In *Danio*, the *lateralis superficialis* is less developed than in *Ancistrus*, with its terminal PU1 and PU2 segments directed upwards, following the curvature of the upward-turned notochord (Figure 5F). This differs from the organization seen in *Ancistrus*, where the *lateralis superficialis* is more extensive and covers a significant portion of the myomere surface, separated from its dorsoposterior portion, here identified as the *flexor dorsalis superficialis* (Figures 2B and 4A).

Catfish Comparative Anatomy: This is the first description of the *lateralis superficialis* in catfishes.

Homology: The *lateralis superficialis* has a clear metameric pattern associated with the terminal segments. Further studies are required to establish a comparative analysis.

Caudal End of the Epaxialis (epx)

The *epaxialis* is the dorsal component of the body musculature above the lateral septum (Winterbottom 1973). During ontogeny, it acquires a complex tridimensional structure (Gemballa 2004), with its lower part partially covered by the *lateralis superficialis*. Greene and Greene (1913) described its terminal segments in *Oncorhynchus*, noting a substantial amount of connective tissue, with its terminal segments reduced to tendons. These tendons course beneath the *lateralis superficialis* and fasten to the aponeurosis, covering the bases of the rays in both caudal fin lobes.

Description: In *Ancistrus*, the *epaxialis* were identified up to the PU4 myomere posteriorly (Figures 2B and 4A), displaying a noticeable V-shape. It is covered ventrally by the *lateralis superficialis*, emerging dorsally onto the lateral surface, curving anteriorly. Unlike the *lateralis superficialis*, muscle fibers of the *epaxialis* are perpendicular to the curvature of the muscle arc as it turns forward, as illustrated by Gemballa (2004).

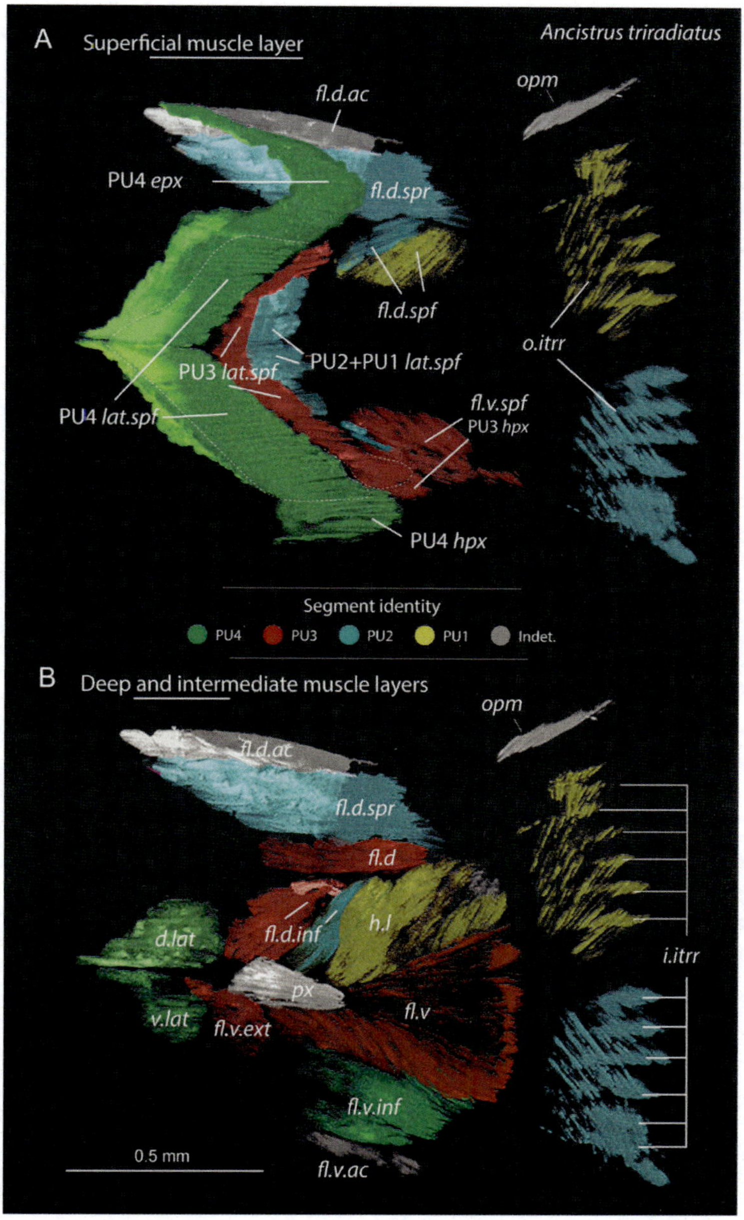

Figure 4. Muscle immunostaining and segment identity of the caudal fin muscles of *Ancistrus triradiatus*. (A–B) confocal image of an immunostaining of an 18 mm *Ancistrus triradiatus* juvenile against myosin heavy chain. (A) 3D segmentation for deep and intermediate muscle layers. (B) 3D segmentation for superficial muscle layer. Terminology: *d.lt, dorsolateralis*; *epx, epaxialis*; *fl.d, flexor dorsalis*; *fl.d.ac, flexor dorsalis accessorius*; *fl.d.inf, flexor dorsalis inferior*; *fl.d.spf, flexor dorsalis superficialis*; *fl.d.spr, flexor dorsalis superior*; *fl.v, flexor ventralis*; *fl.v.ac, flexor ventralis accessorius*; *fl.v.ext, flexor ventralis externus*; *fl.v.inf, flexor ventralis inferior*; *fl.v.spf, flexor ventralis superficialis*; *h.l, hypochordal longitudinalis*; *hpx, hypaxialis*; *i.itrr, inner layer of the interradiales*; *lat.spf, lateralis superficialis*; *o.itrr, outer layer of the interradiales*; *opm, opisthuralis*; PU1–PU4, preural myomeres; *px, proximalis*; *v.lt, ventrilateralis*.

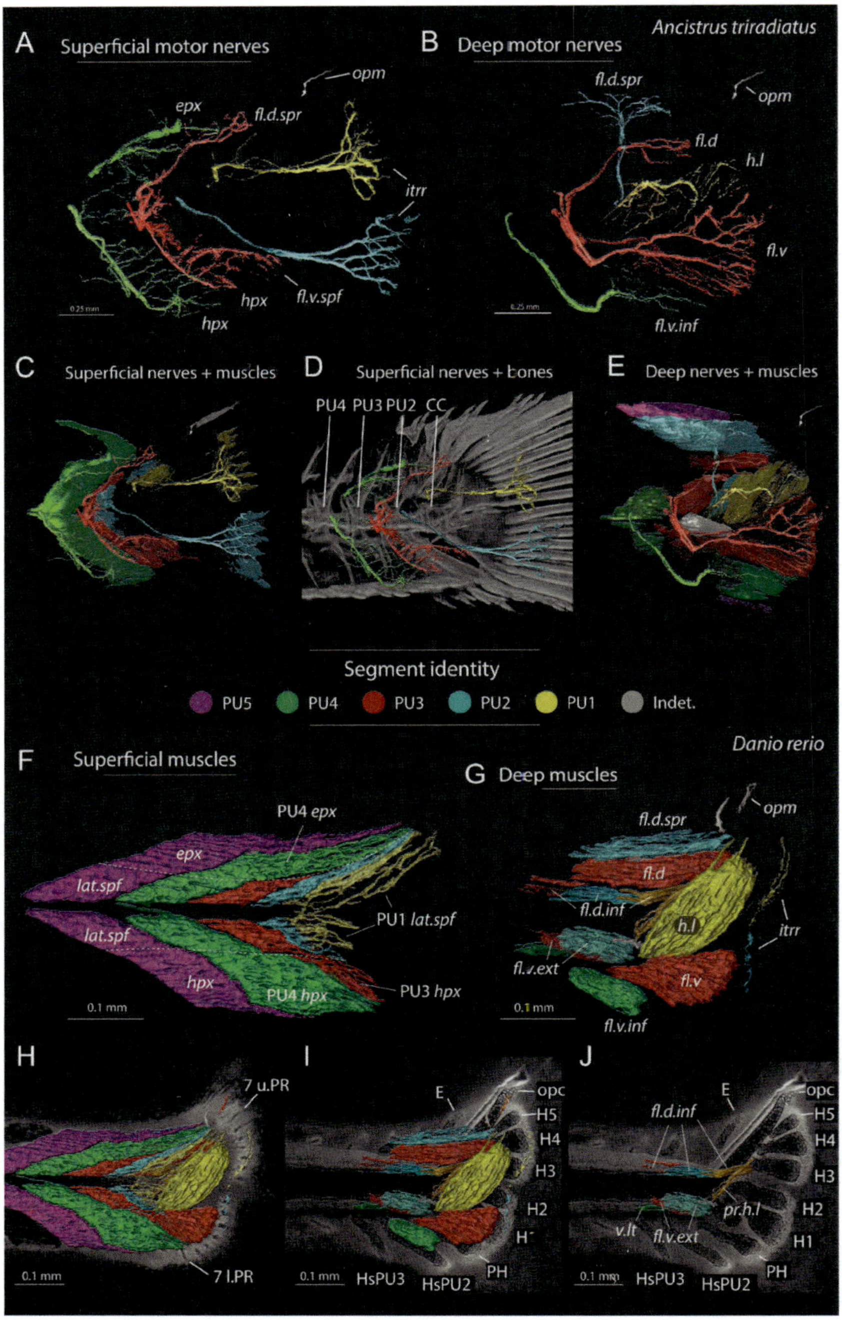

Figure 5. Ontogeny and innervation of the caudal fin muscles. (A–E) immunostaining of an 18 mm *Ancistrus triradiatus* juvenile against neurofilament to analyze muscle innervation, each motor nerve was individually segmented. (A) superficial motor nerves, labeled to the corresponding muscles innervated. (B) deep motor nerves, labeled to the corresponding muscles innervated. (C) superposition of superficial nerves and muscles; (D) superposition of the superficial muscles and bones; (E) superposition of the deep nerves and muscles. (F–J) immunostaining of a 6 mm *Danio rerio* larvae against myosin heavy chain to analyze the musculature and counterstained with TOPRO to visualize cell nuclei. (F) superficial muscles segmented. (G) deep muscles segmented. (H–J) superficial and deep muscles superposed with TOPRO nuclei staining at median (I, J) and lateral (H) confocal sagittal sections. Terminology: CC, Compound centrum; *epx, epaxialis*; *fl.d, flexor dorsalis*; *fl.d.inf, flexor dorsalis inferior*; *fl.d.spr, flexor dorsalis superior*; *fl.v*, *flexor ventralis*; *fl.v.ext, flexor ventralis externus*; *fl.v.inf, flexor ventralis inferior*; *fl.v.spf, flexor ventralis superficialis*; *h.l*, *hypochordal longitudinalis*; *hpx*, *hypaxialis*; *itrr, interradiales*; *itrr.spr, interradiales superior*; *lat.spf, lateralis superficialis*; *opm*, *opisthuralis*; PU1–PU5, preural myomeres; *px*, *proximalis*; *v.lt*, *ventrolateralis*.

Innervation: The PU4 *epaxialis* is innervated by the dorsal branch of the PU4 spinal nerve (Figure 5A).

Ontogeny: In *Danio,* the last unequivocal *epaxialis* corresponds to the PU4 myomere, as in *Ancistrus* (Figure 5F). The elongated portions of the PU2 and PU1 myomeres might be interpreted as either the *lateralis superficialis* or the *epaxialis*.

Catfish Comparative Anatomy: This is the first description of the posterior end of the *epaxialis* in catfishes.

Homology: Both *Ancistrus* and *Danio* lack *epaxialis* beyond the PU4. This observation aligns with the superficial view of *Salmo* (Figure 1F; Schmalhausen 1913), where the seemingly PU4 *epaxialis* is also the last one turning forward, originating a robust tendon that inserts onto the uppermost two branched caudal rays. A similar condition may be present in *Esox*, although it is harder to establish its segmental identity (Hindersson 1910, fig. 3). As described below, this may be associated with an internalization of the posterior *epaxialis* to form the intrinsic musculature of the caudal fin.

Flexor Dorsalis Superficialis (fl.d.spf.)

The new usage of "*flexor dorsalis superficialis*" is introduced for a distinct superficial posterodorsal muscle, illustrated by different authors. Greene and Greene (1913) described the last two dorsal myomeres as running posterodorsally, ending in a delicate flat tendon or fascia. Schmalhausen (1913) illustrated similar muscles in the lying lateral to the *hypochordal longitudinalis* in *Salmo* (Figure 1G). Nag (1967) described this muscle as the dorsal component of the "superficial dorsoventral flexor" in *Oncorhynchus*.

Description: In *Ancistrus*, the *flexor dorsalis superficialis* is a distinct superficial muscle lateral to the epural and terminal centrum (Figures 2B and 4A). It is divided into two sections with different fiber orientations. The anterior portion has fibers angled at approximately 30°, while the posterior section is about twice the size and has fibers angled at approximately 45°. Both sections converge over the pleurostyle, aligning with the uppermost branched caudal ray.

Innervation: Unidentified.

Ontogeny: In *Danio*, by its position, the *flexor dorsalis superficialis* appears to be formed from the posterior bundle of proximal fibers associated with the terminal part of the PU2 and PU1 segments (Figure 5F). These fibers are repositioned dorsal due to the upward turn of the notochord and comprise three sets: the PU2 epaxial terminal fibers and the PU1 epaxial and hypaxial fibers.

Catfish Comparative Anatomy: This muscle has not been previously identified in catfishes. It is recognized herein in *Lithogenes*, where it was misidentified as the *hypochordal longitudinalis* (Schaefer and Provenzano 2008).

Homology: The close relationship between the *flexor dorsalis superficialis* of *Ancistrus* and the terminal fibers of the *epaxialis*, as interpreted from *Danio*'s ontogeny, suggests a serial homology between both muscles, originating from PU2 and PU1 myomeres. It also coincides topologically with the two muscles described in *Salmo* (Figure 1G; Schmalhausen 1913) and in *Oncorhynchus* (Greene and Greene 1913, fig. 5; Nag 1967, fig. 8).

Caudal end of the Hypaxialis (hpx) and the Flexor Ventralis Superficialis (fl.v.spf)

The *hypaxialis* is the ventral counterpart of the *epaxialis* and a major component of the body's musculature. It is positioned beneath the lateral septum (Winterbottom 1973), and, like the *epaxialis*, it adopts a V-shape during ontogeny, with its dorsal portion partially covered by the *lateralis superficialis*.

Description: The *hypaxialis* is present in PU4 and PU3 myomeres in *Ancistrus* (Figures 2B and 4A). It is partially covered dorsally by the *lateralis superficialis*. The PU4 *hypaxialis* emerges outwards onto the lateral surface ventrally and turns forward in a much closer V-shaped configuration than its epaxial counterpart. Unlike the longitudinally arranged fibers of the *lateralis superficialis*, the *hypaxialis* fibers are oriented perpendicular to the arc as the muscle turns forward, as described for the *epaxialis*. It is positioned laterally to the anterior part of the *flexor ventralis inferior* and the *flexor ventralis accessorius*. The PU3 *hypaxialis* displays significant modifications compared to more anterior segments. Positioned just beneath the posteroventral end of the PU3 *lateralis superficialis*, it develops a slender extension toward the space between the 7th and 8th lower lobe principal ray (Figure 2B). It is located lateral to the *flexor ventralis* and just posterior to the PU4 *hypaxialis*.

Innervation: The PU4 and PU3 *hypaxialis* receive innervation by the ventral branches of the PU4 and PU3 spinal nerve, respectively (Figure 5A and C).

Ontogeny: In *Danio*, the PU4 and PU3 *hypaxialis* resemble the pattern of *Ancistrus* (Figure 5F). The PU4 *hypaxialis* is notably larger compared to the neighboring PU3 and PU5 *hypaxialis*. It is positioned lateral to the *flexor ventralis inferior* and attaches to the lowermost branched ray. In contrast, the PU3 *hypaxialis* is shorter, is lateral to the *flexor ventralis* and attaches the second lowermost branched principal ray.

Catfish Comparative Anatomy: This is the first description of the caudal fin *hypaxialis* in catfishes.

Homology: Both *Ancistrus* and *Danio* seem to lack a *hypaxialis* beyond PU3. The PU3 *hypaxialis* is highly modified in both fishes, attaching to the lowermost principal rays. Hence, due to its position opposite to the *flexor dorsalis superficialis*, it is referred to here as *flexor ventralis superficialis*. This muscle has a similar position to the ventral portion of the "superficial dorsoventral flexor" described for *Elops*, *Chanos*, and *Oncorhynchus* (Nag 1967) and was also illustrated but not described in *Salmo* (Figure 1G; Schmalhausen 1913). Grenholm (1923) referred to this muscle as "a lateral muscle that extended from a tendon sheath to the ventral rays" in *Salmo*, *Coregonus*, *Thymallus*, *Argentina*, and *Alosa*. Therefore, it is possible that the modification of the PU3 *hypaxialis* into the *flexor ventralis superficialis* could be homologous among many basal teleosts.

Caudal Fin Ray Muscles

The intrinsic caudal fin ray muscles in *Ancistrus* can be categorized into three sections: the *interradiales* of the upper and lower lobes (*itrr*), and the *opisthuralis* (*opm*) that connects the uppermost fin ray with the opisthural cartilage.

Interradiales (itrr)

The term "intrinsic muscles" was established by McMurrich (1884) for muscles supplying caudal fin rays in *Ameiurus*. Hindersson (1910) introduced the term "*interradiales*" for muscles connecting adjacent rays in *Gadus morhua*, and "*plicatores*" for the lateral, fan-shaped muscles covering the rays in *Esox lucius*. These terms later were used interchangeably and became synonymous as *interradialis* in singular or *interradiales* in plural (Grenholm 1923; Nursall 1963; Winterbottom 1973). The *interradiales* have two muscle layers: an inner layer connecting each ray to its neighbor and an outer layer spanning multiple rays. This pattern is mirrored between the upper and lower fin lobes, symmetric around the hypural diastema. Muscles in both layers originate at one ray's base and attach slightly further along the next (inner layer) or span multiple rays (outer layer). Muscle fibers angle posterodorsally above the diastema and posteroventrally below it.

Description: In *Ancistrus*, the *interradiales* comprise outer (Figures 2B, 3A, and 4A) and inner layers (Figures 3B and 4B), symmetrical above and below the hypural diastema. The inner layer includes six muscles in the upper lobe and six in the lower lobe, each connecting adjacent branched rays except between the central rays. Between the central rays, two muscle bundles link these rays to the lower hypural plate's upper tip. The outer layer covers most of the inner layer muscles, spanning over multiple fin rays (Figure 4A).

Innervation: Two spinal nerves innervate the *interradiales*. The PU1 spinal nerve innervates the upper lobe's inner and outer layers, while the PU2 spinal nerve innervates the lower lobe's inner and outer layers (Figure 5A, C, and D).

Ontogeny: In *Danio*, the *interradiales* develop as elongated muscle fibers covering the bases of the central branched principal fin rays in both the upper and lower lobes (Figure 5G and H) at the distal ends of hypurals 1 to 4 (Figure 5I). Compared to other caudal fin muscles in *Danio*, the appearance of the *interradiales* is delayed. Similarly, in *Salmo*, the *interradiales* cover the distal ends of the hypurals 1 to 4 and their appearance is delayed compared to other muscles (Figure 1A–D; Schmalhausen 1912). By contrast, in *Ameiurus* the *interradiales* display a more pronounced development (Figure 1E), covering the bases of the branched principal rays and the distal ends of the parhypural and hypurals 1 to 5.

Catfish Comparative Anatomy: The organization of the *interradiales* varies in catfishes. In *Auchenoglanis* and *Ameiurus*, the *interradiales* extend into the procurrent rays and lack an outer layer (Grenholm 1923). In *Hipoptopoma* (Aquino and Miquelarena 2001), *Lithogenes* (Schaefer and Provenzano 2008), and *Corydoras* (Huysentruyt et al. 2007), a large outer layer covers the branched principal rays; however, it is uncertain if these species have an inner layer. *Ancistrus* has a prominent outer layer and inner layers. Further descriptions of the *interradiales* will provide insights into the variation between different catfishes.

Homology: Two alternative evolutionary scenarios have been proposed for the *interradiales*. The first, proposed by Schmalhausen (1912), suggests that it originates from the distal intrinsic muscles, akin to the *flexor ventralis*. This idea was based on observations in *Amia* and *Salmo*, where a connection exists between the posterodistal part of the *flexor ventralis* and the muscles forming the *interradiales* (Figure 1B, C). Grenholm (1923) elaborated on this, noting that siluroids like *Auchenoglanis* show a *flexor ventralis* extending into ray bases, adhering to inter-ray connective tissue instead of forming tendons. The alternative hypothesis—elaborated by Grenholm (1923)—proposed that the *interradiales* originated from the superficial musculature extending into the caudal rays. He observed this pattern in *Salmo*, *Coregonus*, *Thymallus*, *Argentina*, and *Alosa,* where muscles resembling the *flexor dorsalis superficialis* and *flexor ventralis superficialis* contact the fin rays.

In *Ancistrus*, the PU1 spinal nerve innervates the *interradiales* in the upper lobe, and the PU2 nerve innervates its lower lobe counterpart, indicating a segmental antero-posterior organization. Furthermore, the hypaxial musculature of *Ancistrus* seems to lack muscles forming from PU2 and PU1 myomeres, aligning with both Schmalhausen's and Grenholm's hypotheses. In this context, the PU3 myomere would have contributed to the *flexor ventralis* (medially) and the *flexor ventralis superficialis* (laterally) (see below), while the PU2 and PU1 myomeres would have contributed instead to the *interradiales*, both to its inner and outer layers. These layers of the *interradiales* might have been developed from the internal and external myomeric musculature, respectively. As Grenholm's observed, the distinct origins of the *interradiales* might both be valid, stating that "interradiales muscles of the caudal rays can apparently be formed from several sides".

Opisthuralis (opm)

The *opisthuralis* muscle was recently described for *Salmo* (Kryvi et al. 2020). Its origin lies in the opisthural cartilage, and it inserts on the uppermost principal ray with a posterodorsal orientation.

Description: In *Ancistrus*, the *opisthuralis* is an elongated muscle that originates from the upper base margin of the uppermost principal fin ray and inserts slightly distally beneath the adjacent dorsal procurrent ray (Figures 2 and 3). It maintains a distinct separation from the *interradiales* and other caudal fin muscles.

Innervation: Unidentified

Ontogeny: In *Danio*, the *opisthuralis* develops between the uppermost principal fin ray's base and connects to the lateral surface of the opisthural cartilage (Figure 5F and G). It consists of a few elongated fibers—clearly separated from other caudal fin muscles—extending parallel to the notochordal and opisthural cartilage surfaces, in contrast to the *interradiales*. In *Salmo* and *Ameiurus* (Figure 1D and E), the *opisthuralis* exhibits a similar position, much more elongated proximo-distally than the *interradiales*.

Catfish Comparative Anatomy: Apart from *Ameiurus* (Schmalhausen 1912), this muscle has not been previously described in catfishes.

Homology: The segmental identity of the *opisthuralis* could not be determined. It likely represents the most posterior caudal fin muscle due to its association with the opisthural cartilage. A muscle resembling its position and orientation was first identified by Hindersson (1910) in *Esox* as the "*deflector pinnae caudalis profundus superior*." Grenholm (1923) recognized this muscle as part of the *interradiales* and noted its presence in various teleosts, including *Salmo*, *Argentina*, *Thymallus,* and *Abramis*, located beneath the tendons of the *hypochordal longitudinalis*. They identified it as a slender muscle originating from the base of the uppermost principal rays and inserting to the posterior-most dorsal procurrent ray and the uppermost principal rays in the same orientation as the *hypochordal longitudinalis*. Its presence in *Ancistrus* and *Danio* (this work), as well as *Abramis*, *Thymallus*, *Salmo*, *Argentina* (Grenholm 1923), *Oncorhynchus* (Winterbottom 1973) and *Esox* (Hindersson 1910), suggest a widespread presence of this muscle among basal teleosts.

Furthermore, when considering its homology among more derived groups of teleosts, Grenholm (1923) proposed that the *M. transversus* (or *adductor dorsalis sensu* Winterbottom 1973), a notably larger medial muscle found in percomorphs, evolved from proximal migration of the *opisthuralis*. This considerably larger muscle, posteroventrally oriented, originates from the upper hypurals and inserts on the lower principal rays. Further comparative analyses will unravel the function and the sequence of evolutionary transformations of the *opisthuralis*.

Intermediate Muscle Layer

The intermediate layer in *Ancistrus* consists of several muscles beneath the superficial layer (Figure 3A). This region includes the *flexor ventralis accessorius*, the *flexor dorsalis superior*, the *flexor dorsalis accessorius*, and the *proximalis*, along with the external portions of the *flexor ventralis* and *flexor ventralis inferior*.

Flexor Dorsalis Superior (fl.d.spr)

The *flexor dorsalis superior* was initially characterized by Schmalhausen (1912) in *Salmo* as a long, slender longitudinal muscle situated dorsally to the *flexor dorsalis* and aligning with the uppermost three principal rays (Figure 1D).

Description: In *Ancistrus*, the *flexor dorsalis superior* is a broad longitudinal arranged muscle that overlaps the *flexor dorsalis* laterally and dorsally (Figures 3A and 4A, B). It originates from the PU3 neural spine, with a longer dorsal portion, and extends posteriorly, crossing the upward-flexed notochord. It inserts on the uppermost three principal rays plus the posterior dorsal procurrent ray. Laterally, it aligns with the convex portion of the PU4 *epaxialis*, and dorsally, it is adjacent to the *flexor dorsalis accessorius*.

Innervation: The primary innervation comes from the motor branch of the PU2 spinal nerve. Its more superficial portion, adjacent to the PU4 *epaxialis*, also receives innervation from a portion of the PU3 spinal nerve (Figure 5A and B).

Ontogeny: In *Danio*, the *flexor dorsalis superior* is located dorsal and slightly more superficial than the *flexor dorsalis*, originating from the uppermost portion of the PU2 neural spine and attaching to the upper three principal fin rays. It is positioned beneath the anterior extensions of the PU2, PU4 and PU5 *epaxialis* (Figure 5F–I). The position, organization, and fin ray insertions of the *flexor dorsalis superior* in *Danio* match the original description of this muscle in *Salmo*.

Catfish Comparative Anatomy: A *flexor dorsalis superior* is present in *Hypoptopoma* (Aquino and Miquelarena 2001). However, in *Auchenoglanis* and *Ameiurus* probably was considered part of the superficial musculature (Schmalhausen 1912; Grenholm 1923), while in *Lithogenes*, it was mislabeled as an aponeurotic sheath, while the *flexor dorsalis superior* was not labeled (Schaefer and Provenzano 2008).

Homology: The *flexor dorsalis superior* is a teleost novelty (Schmalhausen 1912, 1913). Its position just posterior to the PU4 *epaxialis*, and its innervation mostly from the PU2 spinal nerve and partially from the PU3 spinal nerve suggest that it may have evolved from the PU2 *epaxialis*, which is further supported by the apparent absence of a PU2 *epaxialis* (Figure 2B).

Different authors have characterized it either as part of the superficial or part of the deeper musculature; more information is needed to analyze its homology and evolutionary relationship among teleosts. Yet, as the *flexor dorsalis superior* shares position, spatial relationships, and insertions in *Ancistrus*, *Danio* and *Salmo*, it seems to be homologous, at least among these groups.

Proximalis (px)

The second part described by Greene and Greene (1913) of the *flexor ventralis superficialis*, included fibers originating from the *lateralis superficialis*' tendons and fascia in the median line. These fibers extend posteriorly to form a slender tendon attaching the anterior tip of the hypurapophysis. Grenholm (1923) characterized a similar muscle, the *proximalis*, located near the midline. It originates anteriorly from lateral musculature and inserts either onto the anterior part of the hypurapophysis in certain beloniforms, such as *Belone*, *Hemirhamphus*, and *Exocoetus*, or onto the middle of a fused hypural plate in *Fundulus* (Grenholm 1923).

Description: The *proximalis* of *Ancistrus* originates anteriorly from the midline beneath the PU2 *lateralis superficialis* and lateral to the PU2 centrum (Figures 3A, 3B, and 4B). It inserts onto the anterior tip of the hypurapophysis, showing a noticeable inward and slightly ventral orientation.

Innervation: Unidentified.

Ontogeny: Unidentified.

Catfish Comparative Anatomy: This is the first description of this muscle for catfishes.

Homology: The homology of the *proximalis* remains challenging to assess without information about its innervation and ontogeny. The organization of the *proximalis* is notably similar to Grenholm's (1923) original descriptions of beloniforms and cyprinodontiforms, which may indicate a broader presence of this muscle among teleosts. Further anatomical descriptions are needed to confirm the presence and homology of the *proximalis* in basal teleosts.

Flexor Dorsalis Accessorius (fl.d.ac) (novel muscle)

Description: The *flexor dorsalis accessorius* of *Ancistrus* is positioned dorsal to the *flexor dorsalis superior* and posterior to the PU4 *epaxialis* (Figures 3A and 4A). This muscle has an elongated shape, covering the dorsolateral surface of the posterior caudal fin musculature. It originates from the tip of the PU4 neural spine and extends posteriorly, where its fibers turn laterally and ventrally, following the curvature of the preceding PU4 *epaxialis*. Its point of insertion could not be determined.

Innervation: Unidentified.

Ontogeny: Unidentified.

Catfish Comparative Anatomy: This is the first description of this muscle in catfishes.

Homology: Determining the homology of the *flexor dorsalis accessorius* is challenging due to the lack of information regarding its ontogeny and innervation. Nevertheless, its position posterior to the *epaxialis* PU4 suggests a potential correspondence with the PU3 *epaxialis*. Moreover, its shape resembles the epaxial flanking part of the last myomere described by Gemballa (2004). This epaxial flanking part results from the morphological changes when myomeres establish a W-shape and develop secondary anterior cones. Consequently, it is plausible that this same mechanism occurs in *Ancistrus*, resulting in a *flexor dorsalis accessorius* establishing from the uppermost part of a posterior W-shape myomere.

Flexor Ventralis Accessorius (fl.v.acc) (novel muscle)

Description: The *flexor ventralis accesorius* of *Ancistrus* is positioned ventral to the *flexor ventralis inferior* (Figures 3A and 4A). It has a triangular shape with an acute angle, inserting mainly on the first and largest ventral procurrent fin ray. In contrast to the lower portion of the adjacent *flexor ventralis inferior*, muscle fibers are oriented downwards. It extends slightly more posterior than the *flexor ventralis inferior*. Its origin lies in the PU2 haemal spine. Anteriorly, it appears to contact the posterior portion of the *infracarinalis posterior* medially.

Innervation: Unidentified.

Ontogeny: Unidentified.

Catfish Comparative Anatomy: This muscle is present in *Lithogenes* (Schaefer and Provenzano 2008) and *Corydoras* (Huysentruyt et al. 2007). It may also correspond to part of the posterior section of the infracarinalis shown in *Hypoptopoma* (Aquino and Miquelarena 2001). However, it appears missing in Siluroidei (McMurrich 1884; Grenholm 1923).

Homology: The *flexor ventralis accessorius,* like the *flexor dorsalis accessorius*, corresponds in shape and spatial relationships to the hypaxial flanking part described by Gemballa (2004), developing through morphological changes associated with the acquisition of secondary anterior cones during the establishment of a W-shaped myomere. Hence, the *flexor ventralis accessorius* may appear during establishing a W-shaped myomere. Its assignment to a particular myomere is pending further ontogenetic and innervations analyses.

Deep Muscle Layer

The deep muscle layer of *Ancistrus* includes a central cluster of closely arranged muscles overlying the posterior preural and terminal vertebrae (Figure 3B). In this region, eight distinct intrinsic muscles are identified: the *hypochordal longitudinalis*, the deep portions of the *flexor ventralis* and *flexor ventralis inferior*, as well as the *flexor dorsalis inferior*, *flexor ventralis externus*, *dorsolateralis* and *ventrolateralis*. Additionally, the *flexor dorsalis* is located dorsal and separated from this cluster. A detailed description of these muscles is provided in the following sections. For a list of synonymies for caudal fin muscles in catfishes, see Table 1.

Hypochordal Longitudinalis (h.l)

Schmalhausen (1912) named and described the *hypochordal longitudinalis* in the larvae of *Amia*, *Lepisosteus*, *Salmo* and *Ameiurus* (Figure 1A–E), which is positioned beneath the notochord, extending longitudinally from the last myomere. Schmalhausen illustrated the *hypochordal longitudinalis* originating from hypural 1, adjacent to a proximal stripe of mesenchymal condensations, extending toward the posterior hypurals (Figure 1A). Greene and Greene (1913) defined this same muscle in adults of *Oncorhynchus* as the *adductor caudalis ventralis* due to its function in adducting the upper lobe.

Description: In *Ancistrus*, the *hypochordal longitudinalis* can be divided into two sections: proximal and distal (Figures 3B and 4B). The distal section is smaller, originating over the hypurapophysis beneath the *flexor ventralis* (Figure 3B). It runs posterodorsally at an approximate 60° angle, with fibers extending slightly above the midline. The larger proximal portion of this muscle originates from the lateral and ventral faces of the compound centrum and connects with the posterior section ventrally. The fibers of the proximal portion are oriented posterodorsally at about 45°, aligning with the 5th to 8th principal rays of the upper lobe (Figure 3B).

Innervation: Both sections of the *hypochordal longitudinalis* received innervation from the last spinal nerve corresponding to the PU1 (Figure 5B and E), indicating a common origin from the same body segment.

Ontogeny: In *Danio*, the *hypochordal longitudinalis* originates near the base of hypural 1 (Figure 5I and J), extending posterodorsally and inserting onto the upper five principal rays (Figure 5H). It is partially divided into proximal and distal sections. The proximal section has shorter fibers that attach through a tendinous insertion to the uppermost unbranched ray. Some of these fibers deepen and run adjacent to the notochord, reaching the base of hypural 3 (*pr.h.l.*; Figure 5J). The distal section has longer fibers, extending to the distal portion of hypural 4 and attaching to the upper four branched principal rays (Figure 5I). This pattern resembles the condition described in *Salmo* by Schmalhausen (1912), where the distal fibers of the *hypochordal longitudinalis* span from hypurals 1 to 4, while proximally, a stripe of myogenic mesenchymal condensations reaches hypural 3 (Figure 1A–C). This stripe of proximal myogenic condensations closely corresponds to the position of the proximal portion of the *hypochordal longitudinalis* seen in *Danio*. In contrast, in *Ameiurus*, as well as in *Ancistrus*, the situation is reversed. The *hypochordal longitudinalis* has its longest fibers in the proximal portion (Figure 1E), extending from hypural 1 to 6. In the distal portion, the fibers originate from hypurals 1–2 and extend only as far as hypural 3–4.

Catfish Comparative Anatomy: McMurrich (1884) initially described the upper part of the ventral portion of the deep layer muscle in *Ameiurus* as a "triangular muscle, imperfectly separated into two parts." Among loricariids, Schaefer and Provenzano (2008) identified a two-section muscle over the upper hypural plate and terminal centrum of *Lithogenes*, which corresponds to the *hypochordal longitudinalis*. Aquino and Miquelarena (2001) identified a *hypochordal longitudinalis* in *Hypoptopoma*, as part of a central cluster of closely arranged muscles, as in *Ancistrus*. Grenholm (1923) mentioned a connection between the *flexor ventralis* and the *hypochordal longitudinalis* in *Auchenoglanis*, but by comparison with *Ancistrus* it appears that the small posterior portion of the *hypochordal longitudinalis* may have been misidentified as part of the *flexor ventralis*. Collectively, it seems that a *hypochordal longitudinalis* composed of two parts is shared among siluriforms and may represent a plesiomorphic condition for the group.

Homology: The evolutionary origin of the *hypochordal longitudinalis* presents an intriguing question. Its pattern of innervation and its association with hypural 1 indicates that it derives from the last myomere corresponding to the PU1. However, two main evolutionary hypotheses have been proposed regarding its origin. Schmalhausen emphasized a connection between the *hypochordal longitudinalis* and the "*proximal myogenic stripe* (*pr.M.st*)" (1912), suggesting an origin of the *hypochordal longitudinalis* from this proximal series of intrinsic muscles. In contrast, Grenholm (1923) asserted that the *hypochordal longitudinalis* differentiates from the *flexor ventralis*, which develops from the distal series of intrinsic muscles (see below). The information provided here does not favor one hypothesis. Furthermore, based on the ontogeny of *Danio*, it seems that the *hypochordal longitudinalis* has a dual origin (Figure 5G and I). Some proximal fibers of the *hypochordal longitudinalis* are slightly separated in *Danio* and end close to hypural 3

(Figure 5J), resembling the description of the proximal myogenic stripe of Schmalhausen (Figure 1A–E). This suggests that the *hypochordal longitudinalis* may originate from mesenchymal condensations derived from both the proximal and distal series. A conclusive answer to these alternative hypotheses awaits further comparative analyses.

Flexor Ventralis (fl.v)

Schmalhausen (1912) described the *flexor ventralis* of *Salmo* and *Ameiurus*, showing a pattern resembling *Amia*, by differentiating from a distal stripe of mesenchymal condensation. In *Salmo*, it extends from the base of the parhypural to the principal rays of the lower lobe (Figure 1A–C) and later extends anteriorly toward the haemal spines of PU3/PU4 (Figure 1D).

Description: In *Ancistrus,* the *flexor ventralis* is the largest caudal fin muscle, forming a distinct triangular shape that covers the lower hypural plate (Figures 3A, B and 4B). It consists of three recognizable sections. The central section, the largest, has a fan-like shape and spans from the anterior tip of the hypurapophysis to each of the branched principal rays of the lower lobe. Its insertion into these rays occurs through separate muscle bundles. The upper section, immediately adjacent and slightly beneath the central portion, originates from the upper surface of the secondary hypurapophysis (Figure 2A). It extends posterodorsally, inserting into the lowermost two rays of the upper lobe and the uppermost ray of the lower lobe.

The external portion lies lateral to the ventral margin of the central section. It consists of serially stacked fibers originating in a line that extends from the posterior end of the PU2 centrum and beneath the hypurapophysis, partially continuous with the *flexor ventralis externus* anteriorly, making both muscles hard to separate (as shown in Figure 3A). These fibers run posteroventrally and extend slightly over the deep portion of the *flexor ventralis inferior*.

Innervation: The *flexor ventralis* receives innervation from the PU3 spinal nerve (Figure 5B and E), with each of its sections innervated by a different branch.

Ontogeny: In *Danio*, the *flexor ventralis* resembles the central section of the *flexor ventralis* of *Ancistrus* (Figure 5G–I). Most fibers originate from the lateral surface of the parhypural's haemal arch, with some running beneath it and originating from its anterior surface. Posteriorly, these fibers are inserted individually onto each branched caudal fin ray of the lower lobe. In *Salmo*, the *flexor ventralis* also develops from the anterolateral surface of the parhypural (Figure 1A–B; Schmalhausen 1912). However, during ontogeny, a second external section develops, running longitudinally anteriorly and reaching the PU4 haemal spine (Figure 1C–D). In contrast, in *Ameiurus*, the *flexor ventralis* is much broader, attaching to all ventral principal and procurrent rays, and is divisible into two sections: one originating posterolaterally to hypural 1 and one anteriorly lying between the parhypural and the PU2 haemal spine (Figure 1E; Schmalhausen 1912).

Catfish Comparative Anatomy: In *Ameiurus*, McMurrich (1884) described a ventral intrinsic muscle referred to as the "lower division of the ventral portion of the deep caudal muscle"—a large muscle attaching all procurrent and lower principal rays, as well as some of the lowest upper principal rays. Grenholm (1923) identified a similar ventral intrinsic muscle (referred to as the *flexor ventralis*) among other Siluroidei, including *Auchenoglanis*, *Synodontis*, *Genidens*, *Glyptosternon*, and *Malapterurus*. This pattern also coincides with the larval condition shown by Schmalhausen in *Ameiurus* (Figure 1E), with a single large ventral *flexor ventralis*. In contrast, among the Loricaroidei, there are at least three separate deep intrinsic muscles positioned ventrally. *Ancistrus* (as described here), *Lithogenes* (Schaefer and Provenzano 2008) and *Hypoptopoma* (Aquino and Miquelarena 2001) exhibit a *flexor ventralis inferior* and a *flexor ventralis accessorius* in addition to the *flexor ventralis*. The callichthyid *Corydoras* shows at least two ventral muscles (Huysentruyt et al. 2007), a *flexor ventralis* and a *flexor ventralis accessorius*.

Considering the similarity of intrinsic muscle organization observed in *Ancistrus*, *Salmo* and *Danio*, it is likely that the presence of a single ventral intrinsic muscle in the Siluroidei represents a derived condition. The plesiomorphic state, which is maintained among the Loricariidae, involves multiple independent muscles attached to the ventral fin rays, with a *flexor ventralis* only inserted on branched principal rays. Thus, the *flexor ventralis* observed in the Siluroidei likely corresponds to a phylogenetic fusion of the ventral intrinsic muscles.

A second finding is the presence of an upper section of the *flexor ventralis* in *Ancistrus*, originating from the secondary hypurapophysis and inserted onto the principal rays of the upper lobe (Figures 2B and 4A). This upper section is absent in *Danio*, *Salmo* and most euteleosts (Winterbottom 1973). As noted by Grenholm (1923), a *flexor ventralis* inserting onto the upper lobe rays is ubiquitous across the siluroid catfishes he studied, as well as in *Ancistrus* (this work) and *Corydoras* (Huysentruyt et al. 2007), suggesting that it is a shared character of Siluriformes. Outside siluriforms, this condition has an unclear phylogenetic signal, as it is also present in the characiform *Salminus* (Grenholm 1923) and the gonorhynchiform *Chanos* (Nag 1967) but absent in cypriniforms (Grenholm 1923). Furthermore, the development of a secondary hypurapophysis may be correlated with the appearance of this upper section of the *flexor ventralis* in catfishes.

Homology: The *flexor ventralis* likely evolved from the distal intrinsic muscles (Schmalhausen 1912, 1913; Grenholm 1923; Lauder 1989; Gemballa 2004). In *Amia*, it arises from the distal stripe of mesenchymal condensations, spanning several myomeres with an evident metameric organization (Schmalhausen 1912). In contrast, in *Ancistrus*, the *flexor ventralis* is innervated by a single motor nerve, suggesting that, unlike *Amia*, it is associated with a single myomere corresponding to PU3. The similar topological connections and ontogenies of the *flexor ventralis* among *Ancistrus*, *Danio* and *Salmo* suggest its homology among basal teleosts or at least clupeocephalans. When compared to the *flexor ventralis* of *Amia*, the reduction in vertebrae number during homocercal tail evolution—concentrating fin rays in fewer vertebrae—might have derived from the division of the *flexor ventralis* into two muscles, developed from PU3 and PU4 segments: the *flexor ventralis* and *flexor ventralis inferior*. In this scenario, the *flexor ventralis* of teleost and holosteans would not be homologous.

When comparing each *flexor ventralis* section in *Ancistrus* (Figures 3B and 4B), the central section aligns with the original description of this muscle in *Salmo*'s ontogeny (Figure 1A–B). However, other sections have not been systematically compared. A segmented *flexor ventralis* has been described in *Elops*, *Oncorhynchus*, *Chanos*, *Kuhlia*, *Gnathanodon*, *Hoplopagrus* and *Lepomis*, among others (Nursall 1963; Nag 1967; Winterbottom 1973). While some of these sections may correspond to the *flexor ventralis inferior* (see below), the *flexor ventralis accessorius*, or the *flexor ventralis externus*, others could correspond to the sections observed in *Ancistrus*. Further comparisons are required to determine the homology of the *flexor ventralis* and its sections in different teleostean subgroups.

Flexor Ventralis Inferior (fl.v.inf)

Schmalhausen introduced the term *flexor ventralis inferior* (1912) to describe a muscle located beneath the *flexor ventralis* during *Salmo*'s ontogeny (Figure 1C–D). It originates from the PU2 haemal spine, associated with the "distal myogenic stripe" (Figure 1C–D), and differentiates into two sections. The deeper proximal section lies in front of the parhypural, beneath the *flexor ventralis*' external section, with fibers running in parallel to the PU2 haemal spine. The second section is more superficial and runs longitudinally (Figure 1D). Greene and Greene (1913) also identified this muscle in adult *Oncorhynchus*, originating from the haemal spines PU3 to PU1, just anterior and ventral to the *flexor ventralis*. They described its insertion on the lowermost two principal rays and some procurrent rays.

Description: The *flexor ventralis inferior* in *Ancistrus* can be divided into two sections (Figures 3A, B, and 4B). The deeper posterodorsal section is positioned adjacent to the *flexor ventralis* (Figure 3B). It has a diagonal orientation, overlying and slightly posterior to the PU2 haemal spine, originating at its base and on an anterior projection of the parhypural, inserting distally onto the last two principal rays and the first procurrent ray. Conversely, the superficial, anteroventral section runs longitudinally over the PU2 haemal spine (Figure 3B). It converges with the deeper section at insertion on the posterior-most ventral procurrent ray.

Innervation: The superficial portion of the *flexor ventralis inferior* is innervated by the PU4 spinal nerve (Figure 5B and E).

Ontogeny: In *Danio*, the flexor *ventralis inferior* (Figures 5G and I) is positioned laterally and slightly anterior to the PU2 haemal spine beneath the PU4 *hypaxialis*. It originates from the distal end of the PU2 haemal arch, where both arches converge and extend distally toward the lowermost principal ray, firmly attaching to it. This organization resembles *Salmo* (Figure 1C–D; Schmalhausen 1912), where the *flexor ventralis inferior* differentiates close to the PU2 haemal spine. In the stage analyzed in *Danio*, a superficial anterior projection had not yet been developed; nevertheless, based on the figures of *Salmo*, a thin anterior projection developed in association with its point of insertion (Figure 1D). In *Ameiurus* (Figure 1E), Schmalhausen did not distinguish a separate haemal spine; however, a portion of the *flexor ventralis* has its base located over the PU2 haemal spine, suggesting a developmental fusion between the *flexor ventralis* and *flexor ventralis inferior*.

Catfish Comparative Anatomy: The *flexor ventralis inferior* is also well developed in other loricariids such as *Hypoptopoma* (Aquino and Miquelarena 2001) and *Lithogenes* (Schaefer and Provenzano 2008). However, in *Corydoras* (Huysentruyt et al. 2007) and all Siluroidei reviewed by Grenholm (1923), it is either missing or fused to the *flexor ventralis*.

Homology: The *flexor ventralis inferior* is exclusively present in teleosts (Schmalhausen 1912, 1913). It originates from a distal mesenchymal condensation (Schmalhausen 1912) and primarily inserts on the lowermost principal ray and its neighboring rays. In *Ancistrus* it is innervated by the PU4 spinal nerve, and in *Danio* ontogeny, it lies beneath the PU4 *hypaxialis*, indicating an origin from the PU4 myomere. It probably evolved as a division of a primitive *flexor ventralis*, as in holosteans. Thus, its absence in Siluroidei could be interpreted as a reversion to a single *flexor ventralis* spanning multiple segments.

Flexor Ventralis Externus (fl.v.ext)

Greene and Greene (1913) described the *flexor ventralis superficialis* in *Oncorhynchus*, as consisting of two sections. The first section originates from the PU3 and PU4 haemal spines, extending posterodorsally toward the hypurapophysis, from where it sends tendons to the third lower principal ray. Winterbottom (1973) synonymized it as the *flexor ventralis externus* and remarked its posterior region often forms a flattened aponeurotic sheet that inserts at the fin ray bases in different teleosts.

Description: In *Ancistrus*, the *flexor ventralis externus* originates from an expanded flange at the ventrolateral portion of the PU2 centrum and its haemal arch (Figures 3A and 4B). It comprises multiple parallel fibers, oriented at approximately 45° posteroventrally. These series of fibers extend beneath the hypurapophysis, merging with the external portion of the *flexor ventralis*. Its point of insertion remains identified.

Innervation: Unidentified.

Ontogeny: In *Danio*, the *flexor ventralis externus* is part of the hypaxial proximal muscle series, situated lateral to the PU2 haemal arch and the *ventrolateralis*. It originates beneath the PU2 *lateralis superficialis* and extends anteriorly toward the haemal arch of the parhypural (Figure 5G and I). In *Salmo,* the proximal muscle (*pr.M*; Figure 1D) corresponds in orientation and position to the *flexor ventralis externus* described by Grene and Greene as *superficialis* (1913). It also aligns with the labeled proximal muscle *pr.M* in *Ameiurus* (Figure 1E) and the *flexor ventralis externus* of *Danio*.

Catfish Comparative Anatomy: This represents the first description of the *flexor ventralis externus* in catfishes. A similarly positioned muscle may be recognized in the loricariids *Lithogenes* (Schaefer and Provenzano 2008) and *Hypoptopoma* (Aquino and Miquelarena 2001). The *flexor ventralis externus* may be missing or fused to the large *flexor ventralis* in other catfishes.

Homology: The description of the *flexor ventralis externus* in *Ancistrus* matches the position but differs significantly in orientation from that of *Salmo*, *Oncorhynchus* and *Danio*. Moreover, it is challenging to discern the tendon insertion of this muscle. It is possible that the *flexor ventralis externus* has undergone substantial orientation changes in various teleosts. Nonetheless, in *Ancistrus*, *Danio*, and *Salmo*, the *flexor ventralis externus* originates from a proximal series of ventral muscles associated with the PU2 haemal arch and the PU3 and PU2 myomeres. The "external" position of the *flexor ventralis externus* may result from its location over arches rather than spines. Further research using different markers may help to resolve this issue and establish the homology of the *flexor ventralis externus* among teleosts.

Flexor Dorsalis (fl.d)

Schmalhausen (1912) described the *flexor dorsalis* as a small longitudinal stripe dorsal to the PU2 neural arches (Figures 1A–D). After notochordal flexion, it positions laterally to the epurals and the PU2–PU3 neural spines, crossing the notochord and inserting it onto the upper principal rays of the upper lobe (seemingly 3–7 u.PR).

Description: In *Ancistrus*, the *flexor dorsalis* is longitudinally oriented, originating from the base of the PU2 neural spine ventrally and from the epural dorsally, and extending posteriorly, crossing the notochord, and splitting into bundles inserting on the upper four branched principal rays of the upper lobe (4–7 u.PR) (Figures 3B and 4B). It is partially covered by the *flexor dorsalis superior* and well-separated from the slender *flexor dorsalis inferior* ventrally (Figure 3A).

Innervation: The *flexor dorsalis* in *Ancistrus* is innervated by the dorsal branch of the PU3 spinal nerve, which crosses almost perpendicularly the PU2 spinal nerve, heading toward the *flexor dorsalis superior* (Figure 5B and E).

Comparative Ontogeny: In *Danio*, the *flexor dorsalis* develops as a slender muscle that runs longitudinally between the *flexor dorsalis superior* and *inferior* (Figure 5G). It originates from the PU2 neural spine base, extends posteriorly beneath the *hypochordal longitudinalis*, crossing the notochord and attaching to the central principal rays of the upper lobe (3–5 u.PR). This pattern resembles *Salmo* (Figure 1A–D), including *flexor dorsalis*' position, shape, and insertion into the rays. In *Ameiurus* larva (Figure 1E; Schmalhausen 1912), the *flexor dorsalis* forms an L-shape, with its longer ventral portion originating from the PU3 neural arch and its dorsal portion from the epural. After crossing the notochord, it aligns with the upper principal rays of the upper lobe (4–7 u.PR).

Catfish Comparative Anatomy: The *flexor dorsalis* in *Ancistrus* resembles its original description in *Ameiurus* (Figure 1E; Schmalhausen 1912), including its L-shape with dorsal attachment to the epural, although it extended anteriorly up to PU3. In *Auchenoglanis*, the *flexor dorsalis* has a diagonal posterodorsal orientation, inserting on the upper branched principal rays (4–8 u.PR).

Homology: The *flexor dorsalis* is a teleostean novelty (Schmalhausen 1912, 1913; Gemballa 2004). Due to its pattern of innervation, the *flexor dorsalis* identity is assigned to the PU3 myomere. Notably, both in *Ancistrus* and *Danio*, there is no PU3 *epaxialis*, and the *flexor dorsalis* internally occupies this space (Figures 2B, 4A, and 5F). This suggests a possible evolutionary scenario where the *flexor dorsalis* emerged as an internalization of the posterior myomeres, specifically the PU3 *epaxialis*. As no other evolutionary hypothesis has been proposed for the evolution of the epaxial intrinsic muscles, further studies are required to test this hypothesis or provide new interpretations for the evolution of this muscle.

Flexor Dorsalis Inferior (fl.d.inf)

The *flexor dorsalis inferior* was initially described by Greene and Greene (1913) in *Oncorhynchus* as a significantly slender muscle, located ventrally and parallel to the *flexor dorsalis*. It originates from the stegural—a modified neural arch in salmonids—with certain fibers extending anteriorly toward the myotome boundary. It inserts on the lowest three principal rays of the upper lobe through a flat tendinous insertion.

Description: In *Ancistrus*, the *flexor dorsalis inferior* has a unique origin and course (Figures 3B and 4B). It originates from a broad surface on the ventral lateral face of the PU2 centrum, extending posterodorsally, then turning posteriorly and running beneath the *hypochordal longitudinalis*, before inserting onto the 3rd to 4th upper principal rays. This muscle appears to consist of fibers from at least two different axial levels, one from the anterior portion of the PU2 centrum and another from its posterior end, probably associated with the PU3 centrum.

Innervation: The posterior portion receives innervation from either the PU2 or PU1 spinal nerves (Figure 5B and E). The innervation of the anterior portion remains unidentified.

Ontogeny: In *Danio*, the *flexor dorsalis inferior* is slender, originating beneath the PU3 *lateralis superficialis*, lateral to the PU3 to PU1 neural arches and the notochord (Figure 5G, I, and J). It extends posteriorly close to the notochord, crossing it and ending at the base of the hypural 3 beneath the *hypochordal longitudinalis*, likely sending tendons toward the lowermost two principal rays of the upper lobe. In *Salmo* and *Ameiurus* (Figure 1D–E), an unlabeled muscle aligns with this description, originating from the stegural (*Salmo*) or the PU1 haemal arch (*Ameiurus*) and extending toward hypural 3's base.

Catfish Comparative Anatomy: This is the first description of this muscle among catfishes. It was illustrated in *Ameiurus* (Schmalhausen 1912), *Auchenoglanis* (Grenholm 1923) and *Lithogenes* (Schaefer and Provenzano 2008).

Homology: While some authors, such as Greene and Greene (1913) for *Oncorhynchus*, Grenholm (1923) for *Salminus*, and Le Danois (1958) for *Cyclopterus*, have distinguished a separate *flexor dorsalis inferior*, Nursall (1963) and Winterbottom (1973) synonymized it with the *flexor dorsalis*. However, ontogenetic evidence argues against this synonymy, as the *flexor dorsalis inferior* develops separately from the *flexor dorsalis* in *Salmo*, *Ameiurus*, *Danio*, and *Ancistrus*. This is a complex muscle that seems to be composed of more than one myomere, at least PU3 and PU2.

Dorsolateralis and Ventrilateralis (d.lt and v.lt)

The *dorsolateralis* and *ventrilateralis* were initially described by Grenholm (1923) as elongated muscles, either an epaxial or hypaxial, adjacent and parallel to the vertebral column, extending over several caudal vertebrae and attaching to the middle rays in *Sphyraena*, *Belone*, *Galeoides*, *Exocoetus* and *Hemirhamphus*.

Description: In *Ancistrus*, the *dorsolateralis* and *ventrilateralis* originate laterally to the PU3 centrum, positioned beneath the PU4 *lateralis superficialis* (Figures 3B and 4B). They attach to the centro-medial portion of the preceding PU5 *epaxialis* or *hypaxialis* and run longitudinally. They appear to connect with either the *flexor dorsalis inferior* or the *flexor ventralis exterior*, although their precise insertion could not be determined.

Innervation: Unidentified.

Ontogeny: In *Danio*, the *ventrilateralis* originates as small bundles of fibers that detach from the centro-medial portion of the superficial PU4 *hypaxialis* (Figure 5I and J). It corresponds to the more anterior muscle of the ventral proximal series and is positioned medial and distal to the *flexor ventralis exterior*. In *Salmo* and *Ameiurus* (Figure 1A–E), the position of the *ventrilateralis* may be established as the most anterior set of proximal fibers, which locates more ventrally, associated with the PU4 haemal spine, although later these fibers may ontogenetically fuse to the *flexor ventralis exterior*. A *dorsolateralis* was not identified in *Danio*, *Salmo* or *Ameiurus*.

Catfish Comparative Anatomy: Among other catfishes, a *ventrilateralis* is probably present in *Hypoptopoma* (Aquino and Miquelarena 2001). A *dorsolateralis* has not been previously identified in catfishes.

Homology: Besides Grenholm (1923), these muscles have not been described in the literature. Winterbottom (1973) synonymized the *ventrilateralis* as a part of the *flexor ventralis externus* or the overlying *hypaxialis*. Due to their position and ontogeny in *Danio*, *Salmo* and *Ameiurus*, these muscles likely represent the anterior portions of the inner dorsal and ventral intrinsic muscles, developing over the corresponding neural and haemal arches.

A Preural Origin for Caudal Fin Muscles

This study established the identity of the caudal fin muscles of *Ancistrus*, assigning them to specific preural body axis segments by studying their anatomy, ontogeny, and innervation patterns. While a vast number of studies has focused on the evolution of ural skeletal elements in teleosts, including hypurals, uroneurals and centra (Patterson 1968; Nybelin 1977; Schultze and Arratia 1989, 2013; Arratia 1991; Hilton and Johnson 2007), the preural region has received limited attention. Yet, during homocercal tail evolution, teleosts drastically reduced preural vertebrae number (Schultze and Arratia 1989, 2013; Cumplido et al. 2024). For comparison, more than 20 or 10 preural vertebrae support the heterocercal tails of cartilaginous fishes and chondrosteans (Bemis and Grande 1999; Little and Bemis 2004; Hilton et al. 2011), while less than eight support the fin of holosteans (Schultze and Arratia 1986; Grande and Bemis 1998; Grande 2010). This was reduced to as few as five or six preural vertebrae in teleosts in basal forms like *Elops*, *Oncorhynchus* or *Hiodon*, and even fewer in derived groups (Schultze and Arratia 1988; Arratia and Schultze 1992; Cumplido et al. 2020).

In *Amia* and *Lepisosteus* a large *flexor ventralis* extending across multiple preural and ural segments attaches to all caudal fin rays (Schmalhausen 1912, 1913; Lauder 1989; Gemballa 2004). Instead, two muscles develop in the same position in teleosts: a *flexor ventralis*, from the PU3 myomere, and a *flexor ventralis inferior*, from the PU4 myomere, both attaching to the lower lobe principal rays. Hence, these muscles likely evolved from the partition of the large *flexor ventralis* present in recent holosteans, highlighting a significant specialization during homocercal tail evolution.

Additionally, teleosts developed *interradiales* in the lower lobe, unlike *Amia*, which only has *interradiales* in the most posterior caudal rays. This change may also have been associated with modifications in the *flexor ventralis*. As the *flexor ventralis* was excluded from the posterior portion of the caudal fin, the portion associated with the PU2 myomere may have instead migrated into the lower lobe of the caudal fin. This migration could have contributed to the formation of lower lobe musculature, resembling the upper lobe musculature, enhancing control of lower lobe caudal fin movements.

Furthermore, teleosts achieved a third major evolutionary transformation by acquiring epaxial intrinsic musculature, a feature absent in holosteans (Schmalhausen 1912, 1913; Lauder 1989; Gemballa 2004). The distinct patterns of innervation and the absence of lateral musculature in posterior segments suggest that this musculature evolved through an internalization of the superficial musculature. This gave rise to the *flexor dorsalis* and the *flexor dorsalis superior*, enabling finer control of the upper lobe caudal fin movements.

In summary, the preural region played a significant role during the evolution of the homocercal caudal fin musculature. Further research focused on this region holds the potential to understand the fundamental muscular changes during the heterocercal to homocercal transition.

Conclusion

This study has provided valuable insights into understanding the morphology and identities of teleostean caudal fin muscles, providing precise definitions that can aid in assessing homologies across different teleost species. It also revisits century-old hypotheses about the evolution of these muscles, which are now open to testing with modern technologies. Future studies, particularly among basal teleosts, are crucial to gain further insights into the evolutionary changes within caudal fin musculature. Additionally, conducting ontogenetic analyses of myomere shape changes associated with the caudal fin, as shown by Gemballa (2004) for basal actinopterygians, could provide definitive answers regarding the homologies of certain caudal fin muscles. Further research is needed into the functional roles of caudal fin muscles during swimming and their contribution toward evolutionary changes in the diversity of teleost movements and behaviors. Building upon previous research, such as the works of Lauder (1982) and Flammang and Lauder (2008) on caudal fin muscle function, combined with novel phylogenetic approaches, it is now feasible to address specific evolutionary questions regarding the appearance and changes of each caudal fin muscle. By establishing testable hypotheses for muscle homologies, future studies hold the potential to deepen our understanding of the intricate relationship between form and function in the evolution of teleost anatomy and locomotion.

Acknowledgments

I would like to thank Gloria Arratia for her continuous support and enlightening discussions. I am also grateful to my funding agency, ANID, for the FONDECYT Postdoctoral grant 3210275, which made this study possible.

References

Agassiz, A. 1878. On the young stages of bony fishes. Proc. Am. Acad. Arts Sci. 14: 1–25.

Agassiz, L. 1833–45. Recherches sur les poissons fossiles. Vol. 4. Imprimerie de Petitpierre, Neuchâtel, Switzerland.

Aquino, A.E. and A.M. Miquelarena. 2001. Redescription of *Hypoptopoma inexspectata* (Holmberg, 1883), with notes on its anatomy (Siluriformes: Loricariidae). Physis 58: 134–135.

Arratia, G. 1976. Variaciones de las hipurapófisis en algunos peces Siluriformes (Familia Trichomycteridae). An. Mus. Hist. Nat. Valpso 9: 105–114.

Arratia, G. 1982. Esqueleto caudal de peces Siluriformes y sus tendencias evolutivas (Diplomystidae y Trichomycteridae). Bol. Mus. Nac. Hist. Nat. Santiago de Chile 39: 46–61.

Arratia, G. 1983. The caudal skeleton of ostariophysan fishes (Teleostei): intraspecific variation in Trichomycteridae (Siluriformes). J. Morphol. 177(2): 213–229.

Arratia, G. 1991. The caudal skeleton of Jurassic teleosts; a phylogenetic analysis. pp. 249–340. *In*: Chang, M., H. Liu and G.-R. Zhang (eds.). Early Vertebrates and Related Problems in Evolutionary Biology. Science Press, Beijing, China.

Arratia, G. and H.-P. Schultze. 1992. Reevaluation of the caudal skeleton of certain actinopterygian fishes: III. Salmonidae. Homologization of caudal skeletal structures. J. Morphol. 214 (2): 187–249. https://doi.org/10.1002/jmor.1052140209.

Arratia, G. and P. Lambers. 1996. The caudal skeleton of pachycormiforms: Parallel evolution? pp. 191–218. *In*: Arratia, G. and G. Viohl (eds.). Mesozoic Fishes – Systematics and Paleoecology. Verlag Dr. Friedrich Pfeil, München, Germany.

Arratia, G. and A. Tintori. 1999. The caudal skeleton of the Triassic actinopterygian† *Prohalecites* and its phylogenetic position. pp. 121–142. *In*: Arratia, G. and H.-P. Schultze (eds.). Mesozoic Fishes 2 – Systematics and Fossil Record. Verlag Dr. Friedrich Pfeil, München, Germany.

Arratia, G. 2003. The siluriform postcranial skeleton. An overview. pp. 121–157. *In*: Arratia, G., B.G. Kapoor, M. Chardon and R. Diogo (eds.). Catfishes. Science Publishers, Enfield, NH, USA.

Arratia, G. 2008. Actinopterygian postcranial skeleton with special reference to the diversity of fin ray elements, and the problem of identifying homologies. pp. 49–101. *In*: Arratia, G., H.-P. Schultze and M.V.H. Wilson (eds.). Mesozoic Fishes 4 – Homology and Phylogeny. Verlag Dr. Friedrich Pfeil, München, Germany.

Arratia, G. 2013. Morphology, taxonomy, and phylogeny of Triassic pholidophorid fishes (Actinopterygii, Teleostei). J. Vertebr. Paleontol. 33 (supl): 1–138. https://doi.org/10.1080/02724634.2013.835642.

Arratia, G. 2015. Complexities of early teleostei and the evolution of particular morphological structures through time. Copeia 103(4): 999–1025. https://doi.org/10.1643/cg-14-184.

Arratia, G. 2018. Otomorphs (= otocephalans or ostarioclupeomorphs) revisited. Neotrop. Ichthyol. 16(3). https://doi.org/10.1590/1982-0224-20180079.

Balfour, F.M. and W.N. Parker. 1882. VII. On the structure and development of *Lepidosteus*. Philos. Trans. R. Soc. Lond. 173: 359–442.

Bemis, W.E. and L. Grande. 1999. Development of the median fins of the North American paddlefish (*Polyodon spathula*), and a reevaluation of the lateral fin-fold hypothesis. pp. 41–68. *In*: Arratia, G. and H.-P. Schultze (eds.). Mesozoic Fishes 2 – Systematics and Fossil Record. Verlag Dr. Friedrich Pfeil, München, Germany.

Borden, W.C., T. Grande and W.L. Smith. 2013. Comparative osteology and myology of the caudal fin in the Paracanthopterygii (Teleostei: Acanthomorpha). pp. 419–455. *In*: Arratia, G., H.-P. Schultze and M.V.H. Wilson (eds.). Mesozoic Fishes 5 – Global Diversity and Evolution. Verlag Dr. Friedrich Pfeil, München, Germany.

Brito, P.M. 1999. The caudal skeleton of aspidorhynchids (Actinopterygii, Halecostomi): phylogenetic implications. pp. 249–264. *In*: Arratia, G. and H.-P. Schultze (eds.). Mesozoic Fishes 2 – Systematics and Fossil Record. Verlag Dr. Friedrich Pfeil, München, Germany.

Chung, K. and K. Deisseroth. 2013. CLARITY for mapping the nervous system. Nature Methods 10(6): 508–513. https://doi.org/10.1038/Nmeth.2481.

Costa, W.J.E.M. 2012. The caudal skeleton of extant and fossil cyprinodontiform fishes (Teleostei: Atherinomorpha): comparative morphology and delimitation of phylogenetic characters. Vertebr. Zool. 62(2): 161–180.

Cumplido, N., M.L. Allende and G. Arratia. 2020. From Devo to Evo: patterning, fusion and evolution of the zebrafish terminal vertebra. Front. Zool. 17: 18. https://doi.org/10.1186/s12983-020-00364-y.

Cumplido, N., G. Arratia, T. Desvignes, S. Muñoz-Sánchez, J.H. Postlethwait and M.L. Allende. 2024. *Hox* genes control homocercal caudal fin development and evolution. Sci. Adv. 10(3): eadj5991. https://doi.org/10.1126/sciadv.adj5991.

Desvignes, T., A. Carey and J.H. Postlethwait. 2018. Evolution of caudal fin ray development and caudal fin hypural diastema complex in spotted gar, teleosts, and other neopterygian fishes. Dev. Dyn. 247(6): 832–853. https://doi.org/10.1002/dvdy.24630.

Doosey, M.H. and N.D. Domke. 2014. Early Development of the Caudal Fin Skeleton of Capelin, *Mallotus villosus* (Osmeridae). Copeia 2014(2): 355–365. https://doi.org/10.1643/cg-13-098.

Eastman, J.T. 1980. The caudal skeletons of catostomid fishes. Am. Midl. Nat. 103(1): 133–148.

Fink, S.V. and W.L. Fink. 1981. Interrelationships of the ostariophysan fishes (Teleostei). Zool. J. Linn. Soc. 72(4): 297–353.

Fink, S.V. and W.L. Fink. 1996. Interrelationships of Ostariophysan fishes (Teleostei). pp. 209–249. *In*: Stiassny, M.L.J., L.R. Parenti and G.D. Johnson (eds.). Interrelationships of Fishes. Academic Press, San Diego.

Flammang, B.E. and G.V. Lauder. 2008. Speed-dependent intrinsic caudal fin muscle recruitment during steady swimming in bluegill sunfish, *Lepomis macrochirus*. J. Exp. Biol. 211(Pt 4): 587–98. https://doi.org/10.1242/jeb.012096.

Flammang, B.E. 2014. The fish tail as a derivation from axial musculoskeletal anatomy: an integrative analysis of functional morphology. Zoology 117(1): 86–92. https://doi.org/10.1016/j.zool.2013.10.001.

Fujita, K. 1990. The Caudal Skeleton of Teleostean Fishes. Tokai University Press, Tokyo.

Gemballa, S. 2004. The musculoskeletal system of the caudal fin in basal Actinopterygii: heterocercy, diphycercy, homocercy. Zoomorphology 123(1): 15–30.

Goodrich, E.S. 1906. Memoirs: Notes on the development, structure, and origin of the median and paired fins of fish. J. Cell Sci. 2(198): 333–376.

Goodrich, E.S. 1913. Memoirs: Metameric segmentation and homolgoy. J. Cell Sci. 2(234): 227–248.

Gosline, W.A. 1960. Contributions toward a classification of modern isospondylous fishes. Bull. Br. Mus. Nat. Hist. Zool. 6: 325–365.

Gosline, W.A. 1961. Some osteological features of modern lower teleostean fishes. Smithson. Misc. Collect. 142(3): 1–42.

Gosline, W.A. 1971. Functional Morphology and Classification of Teleostean Fishes. University Press of Hawaii, Honolulu.

Grande, L. and W.E. Bemis. 1998. A comprehensive phylogenetic study of Amiid fishes (Amiidae) based on comparative skeletal anatomy. an empirical search for interconnected patterns of natural history. J. Vertebr. Paleontol. 18(sup001): 1–696. https://doi.org/10.1080/02724634.1998.10011114.

Grande, L. 2010. An empirical synthetic pattern study of gars (Lepisosteiformes) and closely related species, based mostly on skeletal anatomy. The resurrection of Holostei. Copeia 10(2A): 1–871.

Greene, C.W. and C.H. Greene. 1913. The skeletal musculature of the king salmon. Bull. U.S. Bur. Fish. 33: 25–59.

Grenholm, A. 1923. Studien über die Flossenmuskulatur der Teleostier. Almqvist & Wiksells Boktryckeri-A.-B., Uppsala.

Harrison, R.G. 1895. Die Entwicklung der unpaaren und paarigen Flossen der Teleostier. Arch. Mikr. Anat. 46(1): 500–578. https://doi.org/10.1007/BF02906665.

Heckel, J.J. 1850. Ueber das Wirbelsäulen-End bei Ganoiden und Teleostiern. Sitzungsber. Kaiserl. Akad. Wiss. 5: 143–147.

Hilton, E.J. and G.D. Johnson. 2007. When two equals three: developmental osteology and homology of the caudal skeleton in carangid fishes (Perciformes: Carangidae). Evol. Dev. 9(2): 178–189.

Hilton, E.J. and R. Britz. 2010. The caudal skeleton of osteoglossomorph fishes, revisited: comparisons, homologies, and characters. pp. 219–237. *In*: Nelson, J.S., H.-P. Schultze and M.V.H. Wilson (eds.). Origin and Phylogenetic Interrelationships of Teleosts. Verlag Dr. Friedrich Pfeil, München, Germany.

Hilton, E.J., L. Grande and W.E. Bemis. 2011. Skeletal anatomy of the shortnose sturgeon, *Acipenser brevirostrum* Lesueur, 1818, and the systematics of sturgeons (Acipenseriformes, Acipenseridae). Fieldiana Life Earth Sci. 2011(3): 1–168. https://doi.org/10.3158/2158-5520-3.1.1.

Hindersson, H.A. 1910. Ueber die Schwanzflossenmuskulatur der Teleostier. Anat. Anz. 36(18): 465–470.

Hollister, G. 1936. Caudal skeleton of Bermuda shallow water fishes. I. Order Isospondyli: Elopidae, Megalopidae, Albulidae, Clupeidae, Dussumieriidae, Engraulidae. Zoologica 21(3): 257–290.

Hollister, G. 1937a. Caudal skeleton of Bermuda shallow water fishes. II. Order Percomorphi, Suborder Percesoces: Atherinidae, Mugilidae, Sphyraenidae. Zoologica 22: 265–279.

Hollister, G. 1937b. Caudal skeleton of Bermuda shallow water fishes. III. Order Iniomi: Synodontidae. Zoologica 22: 385–399.

Hollister, G. 1941. Caudal Skeleton of Bermuda Shallow Water Fishes, V: Order Percomorphi: Carangidae. Zoologica 26: 31–45.

Hubbs, C.L. and K.F. Lagler. 1947. Fishes of the Great Lakes. Bull. Cranbrook Inst. of Sci. 26: 1–186.

Huxley, T.H. 1859. Original communications: observations on the development of some parts of the skeleton of fishes. J. Cell Sci. 1(25): 33–46.

Huysentruyt, F., T. Geerinckx and D. Adriaens. 2007. A descriptive myology of *Corydoras aeneus* (Gill, 1858) (Siluriformes: Callichthyidae), with a brief discussion on adductor mandibulae homologies. Anim. Biol. 57(4): 433–452.

Kölliker, A. 1860. Über das Ende der Wirbelsäule der Ganoiden und einiger Teleostier. von Willhelm Engelmann, Leipzig.

Kryvi, H., K. Nordvik, P.G. Fjelldal, M. Eilertsen, J.V. Helvik, E.N. Storen et al. 2021. Heads and tails: The notochord develops differently in the cranium and caudal fin of Atlantic Salmon (*Salmo salar*, L.). Anat. Rec. 304(8): 1629–1649. https://doi.org/10.1002/ar.24562.

Lauder, G.V. 1989. Caudal fin locomotion in ray-finned fishes: historical and functional analyses. Am. Zool. 29(1): 85–102.

Le Danois, Y. 1958. Système musculaire. pp. 783–817. *In*: Grassé, P.P. (eds.). Traité de Zoologie. Tome XIII: Agnathes et Poissons. Masson, Paris.

Little, C.D. and W.E. Bemis. 2004. Observations on the skeleton of the heterocercal tail of sharks (Chondrichthyes: Elasmobranchii). pp. 563–573. *In*: Arratia, G., M.V.H. Wilson and R. Cloutier (eds.). Recent Advances in the Origin and Early Radiation of Vertebrates. Verlag Dr. Friedrich Pfeil, München, Germany.

Lotz, T. 1864. Über den Bau der Schwanzwirbelsäule der Salmoniden, Cyprinoiden, Percoiden und Cataphracten. Z. Wiss. Zool 14: 84–106.

Lundberg, J.G. and J.N. Baskin. 1969. The Caudal skeleton of the catfishes, Order Siluriformes. Am. Mus. Novit. 2398: 1–49.

Mayer, P. 1886. Die unpaaren Flossen der Selachier. Mittheilungen aus der Zoologischen Station zu Neapal 6: 217–285.

McMurrich, J.P. 1884. The myology of *Amiurus catus* (L.). Proc. Can. Inst. 2: 311–351.

Monod, T. 1968. Le complexe urophore des poissons téléostéens. Inst. Franc. Afrique Noire 81: 1–705.

Moritz, T., J. Walter, B. Grunow and P. Thieme. 2023. A true caudal fin or not? New insights in the evolution of the gadiform caudal fin. Zool. J. Linn. Soc. 199: 26–44. https://doi.org/10.1093/zoolinnean/zlad017.

Nag, A.C. 1967. Functional morphology of the caudal region of certain clupeiform and perciform fishes with reference to the taxonomy. J. Morphol. 123(4): 529–558.

Nag, A.C. 1972. Ultrastructure and adenosine triphosphatase activity of red and white muscle fibers of the caudal region of a fish, *Salmo gairdneri*. J. Cell Biol. 55(1): 42–57.

Nursall, J. 1963. The caudal musculature of *Hoplopagrus guntheri* Gill (Perciformes: Lutjanidae). Can. J. Zool. 41(5): 865–880.

Nybelin, O. 1963. Zur Morphologie und Terminologie des Schwanzskelettes der Actinopterygier. Ark. Zool. 15(35): 485–516.

Nybelin, O. 1973. Comments on the caudal skeleton of actinopterygians. Interrelationships of Fishes (eds. PH Greenwood, RS Miles & C. Patterson): 369–372.

Nybelin, O. 1977. The Polyural Skeleton of *Lepisosteus* and Certain Other Actinopterygians. Zool. Scr. 6(3): 233–244.

Patterson, C. 1968. The caudal skeleton in Lower Liassic pholidophorid fishes. Bull. Br. Mus. Nat. 16(5): 201–239.

Pende, M., K. Vadiwala, H. Schmidbaur, A.W. Stockinger, P. Murawala, S. Saghafi et al. 2020. A versatile depigmentation, clearing, and labeling method for exploring nervous system diversity. Sci. Adv. 6(22). https://doi.org/10.1126/sciadv.aba0365.

de Pinna, M.C. and H.H. Ng. 2004. The second ural centrum in Siluriformes and its implication for the monophyly of superfamily Sisoroidea (Teleostei, Ostariophysi). Am. Mus. Novit. 2004(3437): 1–23.

Remane, A. 1952. Die Grundlagen des natürlichen Systems, der vergleichenden Anatomie und der Phylogenetik. Akademische Verlagsgesellschaft Geest & Portig, Leipzig.

Ryder, J.A. 1886. On the origin of heterocercy and the evolution of the fins and fin rays of fishes. Rep. US Comm. Fish and Fisheries 1884: 981–1107.

de Santana, C.D., R.P. Vari and W.B. Wosiacki. 2013. The untold story of the caudal skeleton in the electric eel (Ostariophysi: Gymnotiformes: *Electrophorus*). PLoS One 8(7): e68719. https://doi.org/10.1371/journal.pone.0068719.

Schaefer, S.A. 1987. Osteology of *Hypostomus Plecostomus* (Linnaeus): With a Phylogenetic Analysis of the Loricariid Subfamilies (Pisces, Siluroidei). Contrib. sci, (Los Angel. Calif.) 394: 1–31.

Schaefer, S.A. and F. Provenzano. 2008. The Lithogeninae (Siluriformes, Loricariidae): anatomy, interrelationships, and description of a new species. Am. Mus. Novit. 2008(3637): 1–49.

Schmalhausen, J.J. 1912. Zur Morphologie der unpaaren Flossen. Z. Wiss. Zool 100: 509–587, 4 pls.

Schmalhausen, J.J. 1913. Zur Morphologie der unpaaren Flossen. II. Bau und Phylogenese der unpareen Flossen und inbesonders der Schwanzflosse der Fische. Z. Wiss. Zool 104(1): 1–80.

Schneider, H. and B. Sulner. 2006. Innervation of dorsal and caudal fin muscles in adult zebrafish *Danio rerio*. J. Comp. Neurol. 497(5): 702–716. https://doi.org/10.1002/cne.21038.

Schultze, H.-P. and G. Arratia. 1986. Reevaluation of the caudal skeleton of actinopterygian fishes: I. *Lepisosteus* and *Amia*. J. Morphol. 190(2): 215–241.

Schultze, H.-P. and G. Arratia. 1988. Reevaluation of the caudal skeleton of some actinopterygian fishes: II. *Hiodon*, *Elops*, and *Albula*. J. Morphol. 195(3): 257–303.

Schultze, H.-P. and G. Arratia. 1989. The composition of the caudal skeleton of teleosts (Actinopterygil: Osteichthyes). Zool. J. Linn. Soc. 97(3): 189–231. https://doi.org/10.1111/j.1096-3642.1989.tb00547.x.

Schultze, H.-P. and G. Arratia. 2013. The caudal skeleton of basal teleosts, its conventions, and some of its major evolutionary novelties in a temporal dimension. pp. 187–246. *In*: Arratia, G., H.-P. Schultze and M.V.H. Wilson (eds.). Mesozoic Fishes 5 – Global Diversity and Evolution. Verlag Dr. Friedrich Pfeil, München, Germany.

Thieme, P., P. Warth and T. Moritz. 2021. Development of the caudal-fin skeleton reveals multiple convergent fusions within Atherinomorpha. Front. Zool. 18(1): 20. https://doi.org/10.1186/s12983-021-00408-x.

Vaz, D.F. and E.J. Hilton. 2020. The caudal skeleton of Batrachoidiformes (Teleostei: Percomorphacea): a study of morphological diversity, intraspecific variation, and phylogenetic inferences. Zool. J. Linn. Soc. 189(1): 228–286.

Vogt, C. 1842. Embryologie des Salmones. Edited by L. Agassiz. *Histoire Naturelle des Poissons d'eau Douce de l'Europe Centrale*. Imprimerie d'o, Petitpierre., Neuchâtel, Switzerland.

von Baer, K.E. 1835. Untersuchungen über die Entwickelungsgeschichte der Fische: nebst einem Anhange über die Schwimmblase. bei Friedrich Christian Wilhelm Vogel, Leipzig.

von Ihering, H. 1878. Das peripherische Nervensystem der Wirbelthiere: Als Grundlage für die Kenntniss der Regionenbildung der Wirbelsäule. Verlag von F. C. W. Vogel, Leipzig.

Whitehouse, R.H. 1910. The caudal fin of the Teleostomi. Proc. Zool. Soc. Lond. 80(3): 590–627.

Wiley, E.O. and B.S. Lieberman. 2011. Phylogenetics: Theory and Practice of Phylogenetic Systematics. John Wiley & Sons, Hoboken, New Jersey.

Wiley, E.O., A.M. Fuiten, M.H. Doosey, B.K. Lohman, C. Merkes and M. Azuma. 2015. The caudal skeleton of the zebrafish, *Danio rerio*, from a phylogenetic perspective: a polyural interpretation of homologous structures. Copeia 103(4): 740–750. https://doi.org/10.1643/CG-14-105.

Winterbottom, R. 1973. A descriptive synonymy of the striated muscles of the Teleostei. Proc. Acad. Nat. Sci. Phila.: 225–317.

Chapter 7

Siluriform Venom Systems

Jacob Egge

Fish venom systems comprise glandular tissues that produce toxic compounds that are typically delivered via spines or barbs associated with the fins or operculum (Church and Hodgson 2002; Smith and Wheeler 2006; Harris and Jenner 2019). Fish venoms have evolved independently 19 times, including once in the Petromyzontiformes (buccal secretions), four times in the Chondrichthyes, once in the Anguilliformes, at least once in the Siluriformes, and 12 times in acanthomorphs (Wright 2009; Smith et al. 2016; Harris and Jenner 2019). The majority of the > 2,900 venomous fish species belong to the Chondrichthyes, Scorpaeniformes, or Siluriformes (Smith et al. 2016; Harris and Jenner 2019).

Venom glands are widespread in siluriforms and present in an estimated 1,250–1,625 species (Wright 2015). Venom glands have been reported in 23 siluriform families, but surveys of genera and species are far from comprehensive. Typically, confirmation that a species is venomous entails one or more lines of evidence, including examination of glandular structure (requiring histological preparation) or pharmacological/toxicological evidence that reports the effects of venom on target tissues or organisms in a controlled laboratory setting.

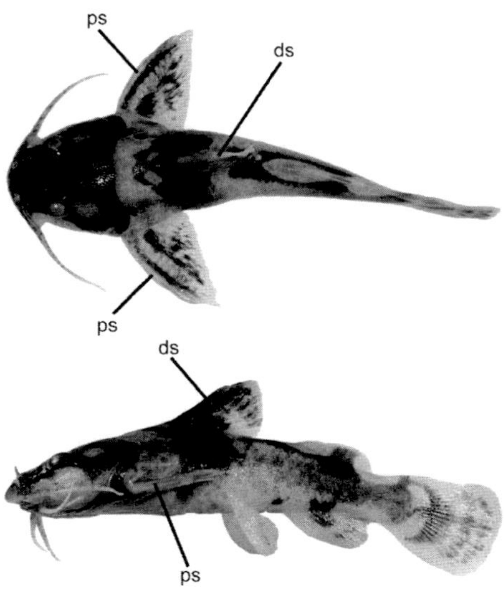

Figure 1. Image of *Noturus gladiator* (Ictaluridae) depicting pectoral and dorsal-fin spines in dorsal (top) and lateral (bottom) views. ps = pectoral spine; ds = dorsal spine. (Figure from Egge and Simons 2011. Reprinted with permission from Oxford University Press).

Department of Biology, Pacific Lutheran University, Tacoma, WA 98447, USA.
Email: eggejj@plu.edu

Siluriforms generate two types of toxic secretions: crinotoxins that disperse over the body surface and venom—a toxin that is associated with a delivery apparatus (Halstead 1970; Cameron and Endean 1973; Perrière and Goudey-Perrière 2003). Siluriform venom systems consist of pectoral and dorsal spines that serve as the delivery apparatus and their associated venom glands (Figure 1), although not all siluriforms with spines are venomous. Broadly, the effects of envenomation by the dorsal or pectoral spine venom apparatus are caused by a combination of (1) the venom itself, (2) the mechanical damage of the spine, and (3) lingering effects of the wound including secondary infection (Birkhead 1972). Understanding siluriform venom systems requires attention to pectoral and dorsal spine structure, venom-gland structure, the toxicology and composition of the venom, and the evolutionary dynamics that shape their function.

Spine Structure

Spine-locking mechanisms make siluriform spines particularly effective against gape-limited predators because they cannot be easily depressed as can spines of many other fishes (Forbes 1989). Both the dorsal and pectoral spines of siluriforms have structurally and anatomically distinct friction-locking mechanisms. The dorsal spine locking mechanism involves a modification of the first dorsal ray into a small, blunt, inverted v-shaped dorsal spine that is situated in a foramen between two nuchal plates, which are lateral extensions of the dorsal portion of the first two proximal radials (Alexander 1965; Arratia 2003). The first dorsal spine (spinelet) can be locked in place when the contraction of erector muscles brings the spine in contact with a modified dorsal radial with a roughened surface. Since the first dorsal spine has a ligamentous attachment to the second dorsal spine, this action erects the second dorsal spine, which is much longer and serves as the defensive spine.

Pectoral spines also have a friction-locking mechanism that accommodates both a binding and locking function (Fine and Ladich 2003; Ballen and de Pinna 2022 for a review of features and standardized terminology). A rough articulating surface (consisting of a dorsal, anterior, and ventral process) at the base of the pectoral spine is accommodated by an articulatory facet (a roughened groove) ventral to the posterior cleithral process (Alexander 1965; Lundberg 1970; Diogo et al. 2001). The spine is rotated into place anterodorsally and cannot be depressed with a posterior-directed force alone (Fine and Ladich 2003; Ballen and de Pinna 2022).

Pectoral spine shape and ornamentation vary among and within siluriform families. Kubicek et al. (2019) surveyed the pectoral spine morphology of 41 catfish families and only observed four that lacked pectoral spines: Kryptoglanidae, Malapteruridae, Phreatobiidae, and Trichomycteridae. The simplest pectoral spines consist of a single shaft with a sharp tip, but many taxa have more complex morphologies with a variety of associated ornamentations (Figure 2). Pectoral spines, defined as unpaired, unsegmented, spinous fin rays, begin development from the fusion of hemitrichia in the first lepidotrichium of each pectoral fin and grow through the fusion of serial hemitrichial segments that ossify and fuse distally, adding length to the spine (Arratia 2003; Vanscoy et al. 2015; Kubicek et al. 2019). Associated ornamentation can include anterior and/or posterior serrae (recurved processes that extend beyond the anterior or posterior spine margin), denticuli (processes at the anterior or posterior margin of the fin spine that form through the addition or remodeling of bone after lepidotrichial fusion), and odontodes (tooth-like structures on the spines of loricariids containing dentine and enamel; terminology follows Kubicek et al. 2019). Posterior serrations have been observed in 28 catfish families, making them the most common ornamentation (Kubicek et al. 2019). In cross-section, pectoral spines range from v-shaped to more complex anchor-shaped structures (Figure 3). Larger spines are often dorsoventrally compressed (Vansocy et al. 2015).

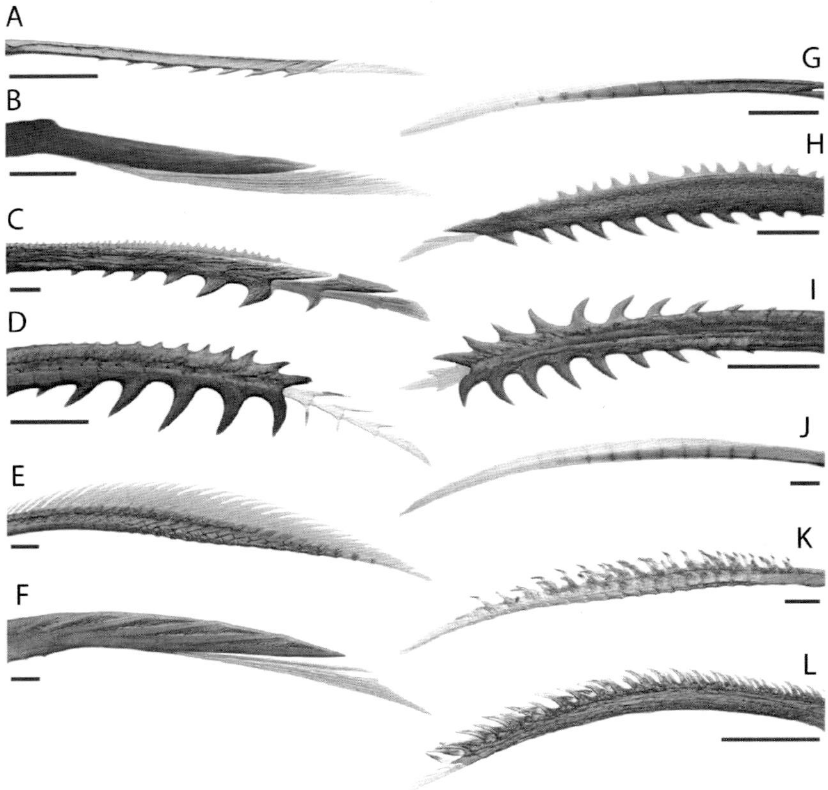

Figure 2. Cleared and stained pectoral-fin spines from multiple siluriform taxa illustrating variation in shape and ornamentation. (A) *Parailia pellucida* (Schilbeidae). (B) *Akysis vespa* (Akysidae). (C) *Ictalurus punctatus* (Ictaluridae). (D) *Pseudolaguvia kapuri* (Erethistidae). (E) *Amphilius uranoscopus* (Amphiliidae). (F) *Noturus gyrinus* (Ictaluridae). (G) *Helogenes marmoratus* (Cetopsidae). (H) *Pimelodus pictus* (Pimelodidae). (I) *Pseudobunocephalus lundbergi* (Aspredinidae). (J) *Henonemus* sp. (Trichomycteridae). (K) *Astroblepus* sp. (Astroblepidae). (L) *Ancistrus* sp. (Loricariidae). Scale bars = 1 mm. (Figure from Kubicek et al. 2019. Reprinted with permission from John Wiley and Sons).

Venom-Gland Structure

Spine-Associated Venom Glands

In most cases, the glandular tissue associated with the siluriform venom apparatus is arranged in longitudinal bundles along the pectoral and dorsal spines (Figure 3). Venom-gland cells have been explicitly observed to be associated with (some ictalurids) or between (Doradidae) posterior serrations (Wright 2009; Egge and Simons 2011). Longitudinal grooves ("furrows" *sensu* Kubicek et al. 2019) in pectoral spines that accommodate the glandular cells have been documented in Akysidae, Amblycipitidae, and some ictalurid species (Wright 2009; Egge and Simons 2011; Kubicek et al. 2019; Figure 3).

Generally, glandular tissue is most well-developed from the middle to distal tip of the spine and greatly diminishes or is absent at the spine base. The venom cells are in direct contact with bone, but are separated from epidermal cells by a layer of connective tissue (Wright 2009; Egge and Simons 2011; Figure 3). Studies that have examined the cellular structure of the venom glands and surrounding tissues have focused largely on the pectoral spines.

The venom glands have been described as holocrine, with venom components released when the venom cells rupture, as evidenced by ruptured cells within the gland (Reed 1907; Whitear et al. 1991a). There are no ducts that convey secretions or muscles associated with an injection action. The toxic components are released when the spine penetrates the victim. In addition to

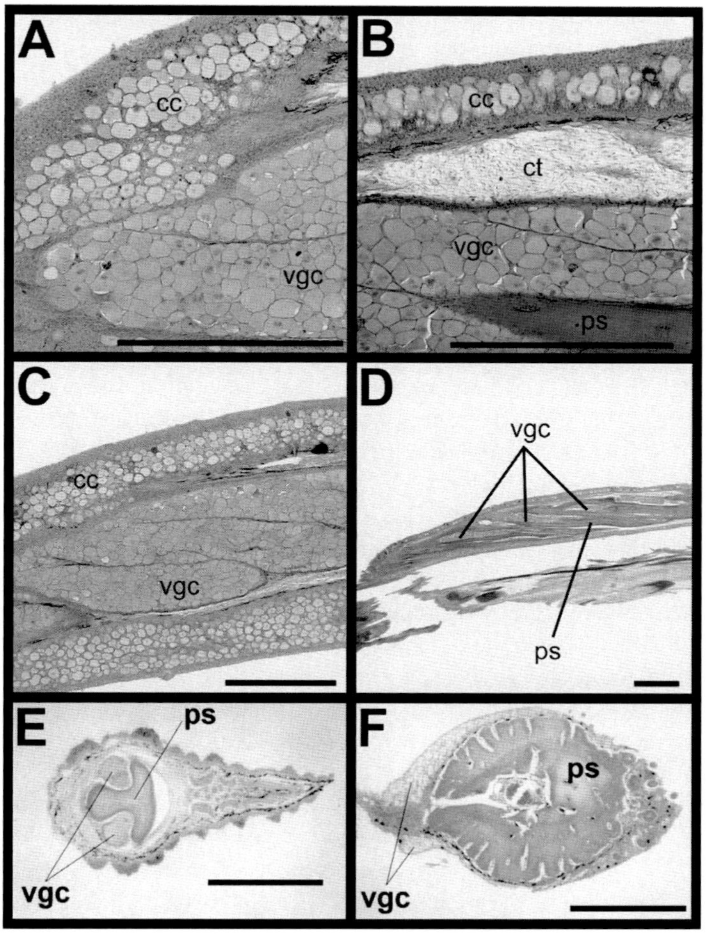

Figure 3. Histological preparations of pectoral-fin spines and associated venom glands in frontal section. (A) *Noturus insignis* (Ictaluridae), tip of venom gland tissue. (B) *Noturus insignis* (Ictaluridae), the shaft of the pectoral-fin spine visible. (C) *Noturus nocturnus* (Ictaluridae), section lateral to spine, demonstrating discrete lobes of venom gland tissue. (D) *Noturus hildebrandi* (Ictaluridae), demonstrating venom gland in grooves of the spine. (E) *Acrochordonichthys rugosus* (Akysidae), note anterior grooves that accommodate the venom gland. (F) *Dianema lonibarbis* (Callichthyidae), spine lacks grooves. cc = club cells; ct = connective tissue; psp = pectoral-fin spine; vgc = venom gland cells. Scale bars = 1 mm (A–D). Scale bars = 0.5 mm (E–F). (A–D from Egge and Simons (2011). Reprinted with permission from Oxford University Press; E–F modified from Wright 2009. © Wright 2009; Creative Commons Attribution License; licensee BioMed Central Ltd.).

epithelial cells, there are several other cell-types associated with the pectoral fin-spine, although terminology has not always been consistent among authors. Whitear et al. (1991a) offered a nomenclature based on the cellular properties of the examined venom apparatus of *Heteropneustes fossilis* (Heteropneustidae). The base of the epidermal layer surrounding the pectoral spine contains a combination of goblet cells and club cells that function as unicellular glands. Goblet cells secrete mucous, have a large central vacuole and discharge their secretion through a pore. Club cells (caudate or clavate cells of some authors) have crinotoxic secretions, perinuclear organelles, and helically coiled filaments associated with desmosomes in the outer cytoplasm (Whitear 1991a; Wright 2015). While Reed (1907) suggested that the club (clavate) cells become venom cells, Whitear et al. (1991a) demonstrated that venom cells could be definitively distinguished from club cells in the differing composition of their secretions, lack of helical filaments, lack of differentiation between perinuclear and peripheral cytoplasm, and the presence of serotonin in club cells that is absent in venom cells (Whitear et al. 1991a; Perrière and Goudey-Perrière 2003). Although venom cell size has been shown to vary between taxa (Egge and Simons 2011), venom cells are generally large polygonal cells with prominent nuclei and granular cytoplasm (Figure 3). They also tend to be more eosinophilic than club cells (Egge and Simons 2011; Carvalho et al. 2021). Whitear et al. (1991a) noted large intercellular spaces in the middle layer of the epidermis that could serve as space for the retention of secretions.

Axillary Glands

Siluriform axillary glands are subcutaneous and situated in a position perpendicular to the cleithrum on both sides of the fish (Figure 4). Unlike the venom glands associated with the pectoral and dorsal spines, axillary glands are secretory and usually connect to the skin surface through a pore located near the base of the pectoral spine, ventral to the posterior postcleithral process (Reed 1907; Halstead et al. 1953; Wright 2015). Birkhead (1972) reported that the presence of visible axillary pores diminished with size in *Ictalurus punctatus* (Ictaluridae), becoming completely absent in specimens > 24.6 mm but remaining visible across size differences in *Ameiurus melas* (Ictaluridae).

Given their proximity to the pectoral spines, early authors considered the axillary glands to be a part of the venom apparatus or "sting" (Reed 1907; Citterio 1926; Halstead et al. 1953; Birkhead 1967; Cameron and Endean 1971; Perrière and Goudey-Perrière 2003) with some suggesting that the secretions may coat the spines, thus contributing to the toxicity of the sting (Günther 1880). However, subsequent authors (Cameron and Endean 1971; Birkhead 1972; Wright 2015) found this proposition highly unlikely given the water-solubility of the secretions.

Secretions from the gland have been reported as "whitish or colourless" in callichthyids (Greven et al. 2006) and "milky" with a bitter taste in Doradidae (Sodré et al. 2021). Secretions are released during periods of stress and are accompanied by stridulatory sounds generated by the fin spines and venom-gland vibration in doradids (Sodré et al. 2021). There is a strong proteinaceous component of the secretions (Kiehl et al. 2006; Carvalho et al. 2021). Secretions of ictalurid and plotosid species have been shown to be toxic when injected into *Gambusia affinis* (Birkhead 1967) and isolated barnacle muscle, respectively (Cameron and Endean 1971); however, this method of delivery does not reflect how the secretions would be encountered by other organisms in a natural system since there is no injection apparatus directly associated with the axillary gland. While the function of axillary secretions has yet to be demonstrated conclusively, there is evidence for a bactericidal function in *Corydoras* (Callichthyidae; Kiehl et al. 2006), suggesting that the secretions may at least in part inhibit microbial growth on the skin surface. Other proposed functions include ichthyotoxic, pheromonal, or ionoregulatory effects (Kiehl et al. 2006; Wright 2015).

Axillary glands consist of several lobes encased in a sheath of connective and adipose tissue (Reed 1907; Halstead et al. 1953; Carvalho et al. 2021; Figure 4). Several lobes separated by connective tissue have been reported in most species surveyed, but tubular glands have been observed in *Corydoras aeneus* (Callichthyidae) and some doradids (Greven et al. 2006; Sodré et al. 2021). Vascularization near the walls of the gland was reported for Ictaluridae (Reed 1907; Whitear et al. 1991b), and *Acanthodoras spinosissimus* (Doradidae) glands were reported as "highly vascularized" (Sodré et al. 2021). In contrast, vascularization in the axillary gland of *Corydoras aeneus* (Callichthyidae) was reported as "poor" (Greven et al. 2006). Immature binucleate secretory cells are found on the basal epithelial layer of the gland (Reed 1907; Halstead et al. 1953; Cameron and Endean 1971; Sodré et al. 2021). As the glandular cells mature, they increase in size, and their plasma and internal membranes tend to break down, releasing their contents into a distinct glandular lumen in *Corydoras aeneus* (Callichthyidae; Greven et al. 2006) and *Noturus* spp. (Ictaluridae; Reed 1907). *Ameiurus nebulosus* (Ictaluridae) is reported to have a false lumen due to the lack of a definitive boundary between the disintegrating cells and the secretion (Whitear et al. 1991b). *Cnidoglanis macrocephalus* (Plotosidae; Cameron and Endean 1971) and *Brachyrhamdia heteropleura* (Heptapteridae; Carvalho et al. 2021) notably lack an axillary gland lumen.

While there has been no systematic survey of siluriform axillary glands, their presence has been documented in the families Akysidae, Ariidae, Diplomystidae, Callichthyidae, Heptapteridae,

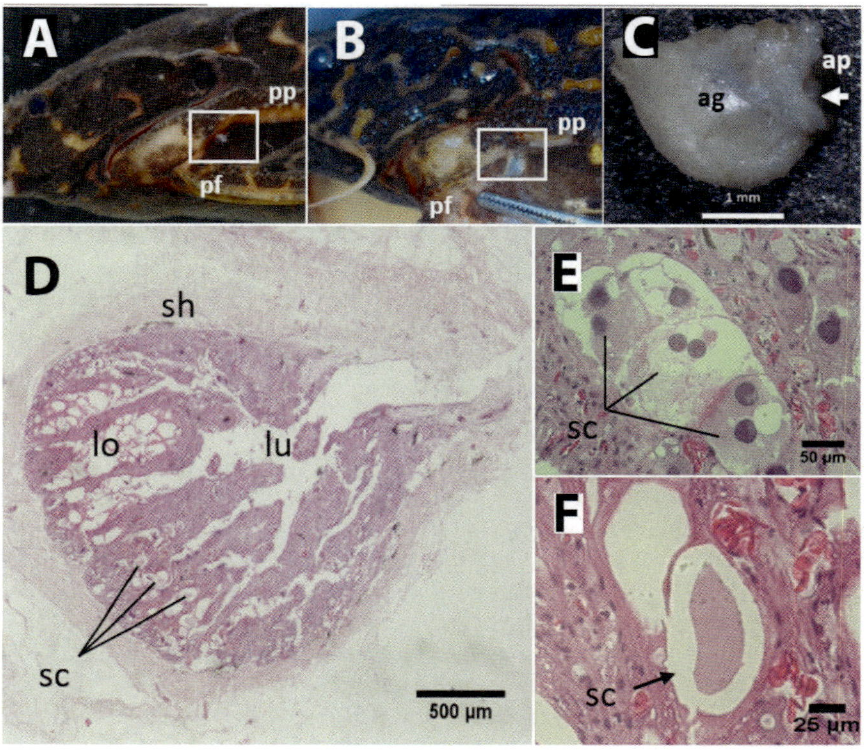

Figure 4. Axillary gland and associated pore of *Acanthodoras spinosissimus* (Doradidae). (A–B) Axillary pore is ventral to the posterior cleithral process. (C) Dissected axillary gland. (D) Longitudinal section of the axillary gland. (E) Binucleated secretory cells. (F) Filled secretory cell. pp = posterior cleithral process; pf = pectoral fin; ag = axillary gland; ap = axillary pore; sh = gland sheath; lo = lobule; lu = lumen; sc = secretory cell. (Figure modified from Sodré et al. 2021. Reprinted with permission from J.G. Sodré. © Sodré et al. 2021. Creative Commons Attribution License (CC-BY)).

Ictaluridae, Mochokidae, Plotosidae and Doradidae (Arratia 1987; Wright 2015; Carvalho et al. 2021; Sodré et al. 2021). Among doradids, other pores in the axillary region associated with glandular tissue have been reported in *Rhynchodoras*, *Doras*, *Anduzedoras*, *Hassar*, *Nemodoras*, *Tenellus*, *Trachydoras*, and some species of *Leptodoras* (Sodré et al. 2021), although the structure/function of these glands and associated pores is not well understood. Bhimachar (1944) and Whitear et al. (1991b) could not identify axillary glands in *Heteropneustes fossilis* (Heteropneustidae). Detailed anatomical work on axillary gland structure has been published for the ictalurid species *Ameiurus catus* (Citterio 1925), *Noturus gyrinus* (Reed 1924), and *Ameiurus nebulosus* (Whitear et al. 1991b); the plotosids *Cnidoglanis macrocephalis* (Cameron and Endean 1971) and *Plotosus lineatus* (Halstead and Smith 1954); the ariids *Galeichthys felis* (Halstead et al. 1953), *Arius heudeloti* (Citterio 1926), and *Arius thalassinus* (Al-Hassan et al. 1987); the callichthyid *Corydoras aeneus* (Greven et al. 2006); the heptapterid *Brachyrhamdia heteropleura* (Carvalho et al. 2021); and the doradid *Acanthodoras spinosissimus* (Sodré et al. 2021).

Functional Morphology

Functional Demands

Both the dorsal and pectoral spine apparatus are complex structures with multifunctional demands, thus imposing multiple constraints on the venom delivery apparatus. Minimally, components of the venom delivery system have functional roles in locomotion, defense (both mechanical injury and venom delivery), and, in some taxa, sound production. Since they are associated with fins, the spines necessarily play a role in locomotion, although this role differs between dorsal and pectoral fins. The spines themselves serve a defensive function by increasing the circumferential area of the fish (important for gape-limited predators), as their pointed tips and associated serrations can cause physical trauma to the attacker, either by locking in place or increasing handling time (Birkhead 1972; Emmett and Cochran 2010; Bosher et al. 2006) or pinching, a behavior that involves adduction of the pectoral spine to bring the attacker in contact with posterior serrations of the spine (Kaatz 1999; Fine and Ladich 2003). These functions are present even in species that are not venomous. Pectoral brush behavior (brushing the pectoral fin against the side of a conspecific) has also been observed in *Ictalurus furcatus* (Ictaluridae; Morgan and Fine 2020) and *Pimelodella longipinnis* (Heptapteridae; Slobodian et al. 2021) during agonistic interactions. Locomotory and defensive functions are not necessarily mutually optimal as spines in the locked position (a common defensive posture adopted while being consumed) hinder swimming, resulting in a wobbling effect (Bosher et al. 2006).

Additionally, stridulatory sound production, a form of communication, has been documented in 23 siluriform families (Rice et al. 2022). Stridulation of the pectoral spines is more common (Fine et al. 1997; Fine and Ladich 2003; Kaatz et al. 2010; Parmentier et al. 2010), but sound production is also associated with the dorsal spine in Sisoridae (*Sisor*, *Nangra*, and *Gagata*; Mahjan 1963; de Pinna 1996). Sound production is accomplished by the base of the spines and, thus, is functionally more closely linked to the locking mechanisms than to venom delivery.

In taxa with associated venom glands, the spines further function as a venom-delivery apparatus. The delivery of venom is passive, requiring puncture damage by the spine to introduce the venom into the victim's wound. Since the envenomation process requires damage to the integumentary sheath around the spine and damage to the venom gland itself, healing time is required until the venom apparatus is fully functional again (Birkhead 1972). Birkhead (1972) reported complete healing of the integument and venom gland within six days for *Ictalurus punctatus* (Ictaluridae) under experimental conditions. It is not known, however, if full healing is required to restore any form of functionality. This undoubtedly is dependent, at least in part, on how extensively the spine penetrated the wound and how much damage to the delivery structure was sustained, including varying degrees of spine breakage.

Relationship Between Fin Spines and Venom Glands

Siluriform venom systems rely on the complimentary function of the spine (delivery system) and the venom itself. Because both the dorsal and pectoral spines function as venom-delivery structures, individual fish effectively have two venom-delivery systems that operate in a similar way but have distinct functional and anatomical differences. Dorsal and pectoral spines differ in their paired versus median-fin development, anatomy, and locomotory demands.

The pectoral spine delivery system is generally more highly elaborated than the dorsal spine system, featuring ornamentation and grooves to accommodate the venom gland in some taxa. Among ictalurid species alone, four distinct morphologies of pectoral spine-associated venom delivery structure have been identified: (1) species with a smooth spine and no associated venom gland (nonvenomous); (2) species with a smooth spine with venom gland on the shaft; (3) species with a serrated spine with venom-gland tissue on the spine shaft; and (4) species with a serrated spine with venom-gland tissue on the shaft and posterior serrations (Egge and Simons 2011). Dorsal-fin spine shafts tend to have less ornamentation and exhibit less anatomical variation among taxa than pectoral spines, but that has yet to be verified in a comprehensive review of siluriform dorsal spines. These differences could have to do with the differing functional demands of the two fins, defensive or otherwise. A high degree of similarity among dorsal spine delivery mechanisms across taxa would suggest the potential for stronger functional constraints resulting in conserved form.

Attempts to describe a relationship between fin-spine morphology, venom gland shape/size, and venom potency have not yielded a clear pattern. Fin spines with serrations may increase mechanical damage versus smooth spines, but there is no direct evidence of this (Egge and Simons 2011). Serrations may also increase wound trauma and thus increase venom delivery; however, spines likely do not typically penetrate deeply into the victim. In human envenomation, where spine penetration is deep, serrated spines become embedded in the victim, preventing escape (Blomkalns and Otten 1999). However, delivery in the mouth of a gape-limited predator, like a teleost, is functionally quite different (see "Defensive Function" below). Lateral thrashing may bring serrations into contact with the victim, producing accessory wounds where additional venom could be introduced in species with venom-gland tissue associated with the posterior serrations. Conversely, smooth spines may penetrate more deeply, delivering high quantities of venom (Reed 1907; Egge and Simons 2011). However, studies on spine shape and toxicity failed to recover any relationship between venom potency and spine ornamentation (Birkhead 1972; Egge and Simons 2011).

There are no reported differences between venom composition or effects between dorsal and pectoral spine-associated glands (Birkhead 1972). However, no studies have directly addressed this because they either excluded dorsal spine venom (Wright 2009) or combined dorsal and pectoral venoms (Birkhead 1972).

Defensive Function

Most fish venoms are thought to have evolved for a defensive function. There is no evidence of siluriform venom being used for prey capture. The effectiveness of the siluriform venom apparatus at deterring predators has been tested experimentally in a few ictalurid species using the natural predator *Micropterus salmoides* (Centrarchidae). Broadly, the venom apparatus has been shown to have the following effects on the predator: increased handling time, presentation of "discomfort behaviors," and physical injury. The presence of intact dorsal and pectoral spines increases handling time relative to individuals with clipped spines and non-spiny alternative prey in *Ictalurus punctatus* and *Noturus gyrinus* (Ictaluridae; Bosher et al. 2006; Emmett and Cochran 2010; Wright 2012). "Discomfort behaviors" associated with attempts to consume catfish exhibited by piscine predators include violent head shaking, opercular flaring, and "coughing" (expansion of the buccal cavity in a rapid, repeated manner) (Bosher et al. 2006; Emmett and Cochran 2010; Wright 2011). Physical injury to the predator involves bleeding and puncture wounds to interior mouthparts (Boshar et al. 2006). Wright (2011) reported that *Noturus gyrinus* specimens with intact spines almost completely discouraged predation and increased handling time relative to individuals that had their venom glands removed, demonstrating that the presence of venom greatly enhances the defensive function versus the mechanical effects of the spines alone. An additional protein found in *N. gyrinus* venom, but lacking in sympatric *Ameiurus natalis* (Ictaluridae), may provide an enhanced protective function (Wright 2011).

Venom Composition and Action

Venom Composition

Siluriform venoms are considered to be simple in terms of chemical complexity, typically consisting of one to three active proteins or peptides [but see variation in *Ameiurus catus* (Ictaluridae) reported by Calton and Burnett (1975) and Wright (2015)]. Isolated peptide and protein fractions in the venom have been shown to produce toxic effects; however, other non-proteinaceous components may also contribute to toxicity (Wright 2015). Siluriform venoms are generally labile (heat, pH, lyophilization, and freeze/thaw effects); thus, attempts to isolate toxic components may render them functionless (Church and Hodgson 2002; Sivan 2009).

Protein components in siluriform venoms have been reported in terms of their size, with documented variation among species and families (Figure 5). A 15 kDa lethal fraction of venom (referred to as toxin-PC) has been identified in *Plotosus canius*, Plotosidae (Auddy and Gomes 1996). This and other putative toxic peptides of similar sizes (10–20 kDa) and number (1–2) have been isolated in other siluriform families including Ariidae, Callichthyidae, Horabagridae, Ictaluridae, Pseudopimelodidae, Pangasiidae, Pimelodidae, and Mochokidae via comparative electrophoretic profiles of spine venom-gland tissue extracts (Wright 2009; Figure 5). Wright (2009) also identified a slightly larger (~ 30 kDa) putative toxin component in *Pangasianodon* (*Pangasius*) *hypophthalmus* (Pangasiidae). Larger components unique to ariids have been identified in *Arius jordani* (Ariidae; ~ 39 kDa; Wright 2009) and *Cathorops spixii* (Ariidae; 65 kDa; Junqueira et al. 2007; Ramos et al. 2012), the latter of which, Wap65, was confirmed to have toxic effects and is also expressed in skin mucous.

Similar size fractions are not necessarily homologous, however, and there is variation among taxa at the species and family level with regard to the number and size of these components. *Synodontis multipuntata* and its congener *S. petricola* (Mochokidae) have highly similar venom

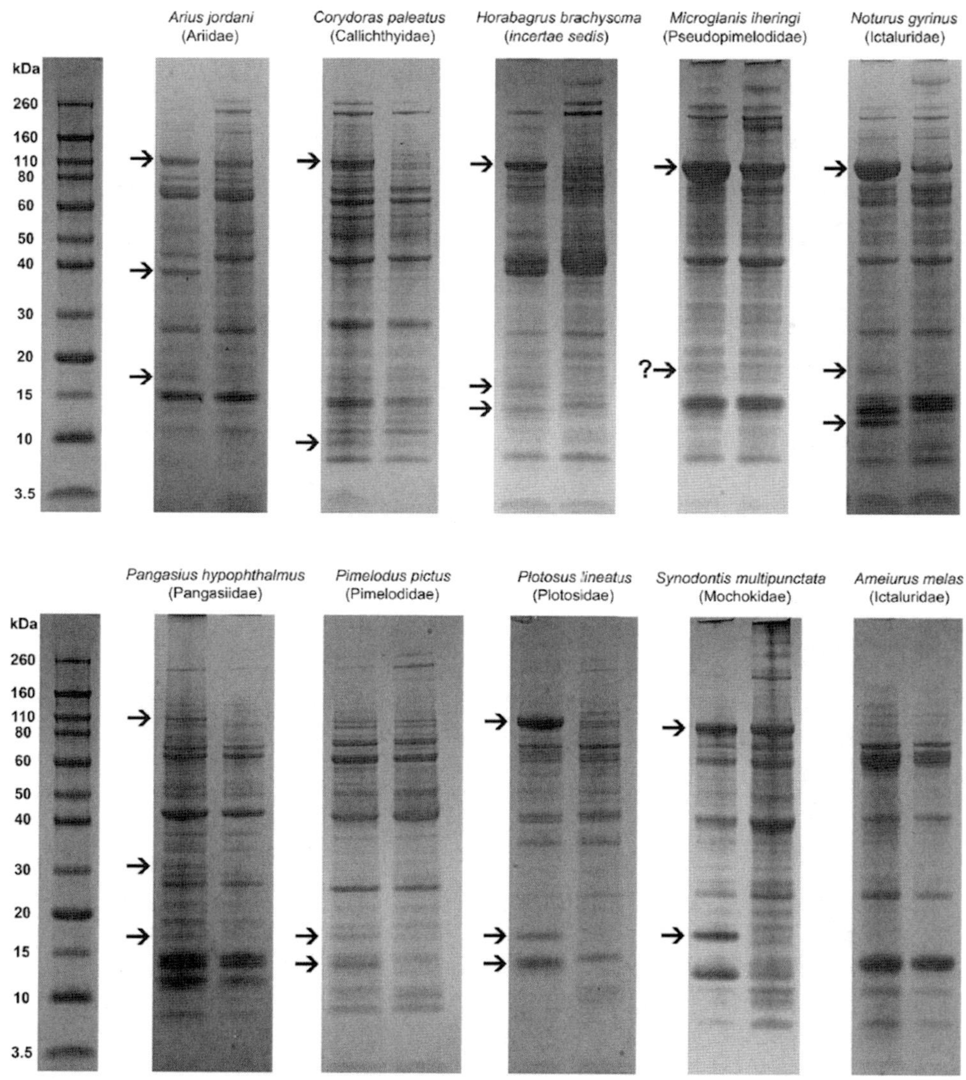

Figure 5. Results of protein size fractioning of venom extracts from multiple siluriform species using SDS-PAGE analyses. The left lanes represent venom extracts; the right lanes are from fin tissue. Arrows refer to unique venom protein bands or proteins found in greater concentration in venom extracts; (?) indicates ambiguity. *Ameiurus melas* (Ictaluridae), a putatively non-venomous species, is shown for comparison. (Figure modified from Wright 2009. © Wright 2009; Creative Commons Attribution License; licensee BioMed Central Ltd.).

composition profiles (Wright 2011). In contrast, *Noturus gyrinus* (Ictaluridae) has had three protein fractions identified, but confamilial *Ameiurus melas* (Icataluridae) has no unique putatively toxic components (Wright 2009).

Wright (2009) identified one much larger putatively toxic component with a molecular weight of 110 kDa in venomous species representing eight of nine siluriform families surveyed including Ariidae, Callichthyidae, Horabagridae, Pseudopimelodidae, Ictaluridae, Pangasiidae, Plotosidae, and Mochokidae. The component was absent in *Pimelodus pictus* (Pimelodidae), a species that contained two smaller putatively toxic components (Figure 5). Wright (2015) suggests that, while homology cannot be conclusively determined without knowing the amino acid sequence and structure, the fact that this protein is found in such a wide range of taxa may implicate it in contributing to the color loss effect noted when *Micropterus salmoides* is injected with venom extracts from a wide range of siluriform venoms (Wright 2009; Tables 1 and 2). This suggestion would require further explanation since this effect was not observed in *Pangasius hypophthalmus* (Pangasiidae) or *Plotosus lineatus* (Plotosidae), species whose venom contains the 110 kDa component. Conversely, *Pimelodus pictus* (Pimelodidae), a species whose venom does not contain the same component, did induce color loss.

In contrast with electrophoretic studies, which generally identify toxic or putatively toxic proteins based on size, transcriptomic and proteomic sequencing have identified 15 putative toxin proteins. This identification was made by comparing sequences derived from *Tachysurus fulvidraco* (Bagridae) against a database of sequences associated with known animal venoms (Xie et al. 2016). The authors recognized 15 proteins distributed in five superfamilies of toxins based on transcript similarity: Veficolin, Ink toxin, Adamalysin, Zn-α2-Glycoprotein (ZaGP), and Cysteine-Rich Secretory Proteins (CRISP) toxin. Veficolin was first identified in anguimorph lizards and snakes (Fry et al. 2010). Ink toxins have been identified in the ink secretions of sea hares and have been shown to have antimicrobial and antitumor effects (Kamiya et al. 1986; Kisugi et al. 1989). Adamalysin is a toxic protein first isolated from *Crotalus adamantues* (Kurecki et al. 1978). ZaGP inhibits tumor growth and proliferation (Hassan et al. 2008), while CRISP is expressed in many snake venoms exhibiting a range of effects, including smooth muscle contraction (an effect observed from the venoms of multiple siluriforms; Table 1), paralysis and hypothermia (Yamazaki and Morita 2004; Xie et al. 2016).

The 15 putative toxin proteins identified in this study suggest a potentially much higher complexity of protein components than previously reported, though authors stipulate that their methodology was devised primarily to recover sequences, not the number of toxins (Xie et al. 2016). It may also be the case that *T. fulvidraco*, in particular, or bagrids more generally, express an elevated number of toxic protein components, a plausible scenario given the variation observed among other siluriform species surveyed.

Venom Effects

Most lab-based empirical studies on the effects of siluriform venom have employed mammalian model targets, with mice (*Mus musculus*), rats (*Rattus norvegicus*), and guinea pigs (*Cavia porcellus*) being the most common (Table 1). The objective of many of these studies has been to understand better the composition and physiological mechanisms of action for crude venom or constituent venom components. Interpretation of these results in terms of functional performance is confounded by differing dosages (often far exceeding what would occur naturally), the delivery mechanisms used across studies and the fact that none of the mammals used represent possible predators for which the venoms have presumably evolved to protect against.

The most well-studied siluriform taxa regarding venom effects on mammalian models are *Plotosus canius* (Plotosidae) and *Heteropneustes fossilis* (Heteropneustidae). A few species of ariid (including *Arius thallansinus*, *Cathorops spixii*, and *Galeichthyes felis*) and several ictalurids (including *Ictalurus*, *Ameiurus*, and *Noturus*) have also been more broadly examined using a combination of teleost and mammalian targets. Other families completely lack data or are limited to one or a few studies (Tables 1 and 2).

Most fish venoms have cytolytic properties with cardiovascular and neuromuscular effects (Church and Hodgson 2002; Sivan 2009). Cytolytic properties can present as local necrosis at the injection site in experimental trials (Birkhead 1967, 1972); however, necrosis was not observed by Wright (2009), so secondary infection may play a role in tissue necrosis (Wright 2015). Hemorrhaging (observed in six siluriform families; Table 1) is a commonly observed effect of siluriform venoms on both teleost and mammalian targets (Table 1).

Predatory teleost fishes are likely the most common targets of envenomation in natural systems, and the effects of siluriform venom on teleosts have been examined in studies by Birkhead (1967, 1972), who injected venom from multiple ictalurid species into mosquitofish (*Gambusia affinis*), and Wright (2009), who injected venom from multiple siluriform families into largemouth bass (*Micropterus salmoides*), a known predator. Reported effects of injection of venom in teleosts include color loss, myoclonus, tetanus, hemorrhaging, edema, loss of equilibrium, necrosis, and mortality (Table 2). Of species representing the nine siluriform families surveyed by Wright (2009), only *Plotosus lineatus* (Plotosidae) caused mortality. Plotosid venom, in particular, has been shown to produce extensive cytotoxic, neuromuscular, and cardiovascular effects (Table 1). Venom from *Horobagrus brachystoma* (Horabagridae) and *Noturus gyrinus* (Ictaluridae) demonstrated the widest range of effects (four), while *Corydoras paleatus* (Callichthyidae) venom had the fewest (one) (Wright 2009; Table 2).

The most pronounced effects of the venom on mosquitofish (outside of secondary infection) generally occur within 24 hours of exposure (Birkhead 1967, 1972). For more toxic species, the level of response rose rapidly (within 24 hours) and then persisted, whereas less toxic species had symptoms decline after 24 hours (Birkhead 1967, 1972). More severe symptoms like edema, necrosis, and death became more prevalent after 24 hours, likely due to the combined effects of venom and secondary infection (Birkhead 1967, 1972).

Loss of color or chromatophore expansion in *Micropterus* and *Gambusia* were seen in seven of nine siluriform families (Tables 1 and 2) observed, suggesting there is conserved action on the nervous system as it relates to the control of chromatophore and melanophore activity (Birkhead 1967, 1972; Wright 2009). Effects on muscle contraction (myoclonus and tetanus) have also been observed in several siluriform families in a teleost target (Wright 2009; Tables 1, 2).

Neuromuscular blocking activity has been reported in from *Plotosus canius* (Plotosidae) venom (described as toxin-PC) through inhibition of muscle contraction on isolated rat and chicken muscle preparations (Auddy et al. 1994, 1995) while Fahim et al. (1996) reported severe neurotoxic symptoms in mice (via interperitoneal injection) from *Plotosus lineatus* venom. The action of the venom is thought to be presynaptic by preventing neurotransmitter release (Auddy et al. 1995; Church and Hodgson 2002). However, neuromuscular effects are not ubiquitous in siluriform venoms and have been disconfirmed in *Heteropneustes fossilis* (at least in a non-teleost model) (Datta et al. 1982; Church and Hodgson 2002; Sivan 2009). Venoms of plotosids, ictalurids, and heteropneustids produce smooth muscle contraction in guinea pig intestinal preparations, presumably because the venom either contains or releases prostaglandins (Auddy et al. 1994, 1995; Church and Hodgson 2002; Sivan 2009; Bartho et al. 2018).

Cardiovascular effects have been documented in plotosids, pimelodids, ariids, and heteropneustid venoms (Table 2). A hypertensive response was observed for *Plotosus canius* venom, while *Heteropneustes fossils* venom produced a hypotensive response in anesthetized cats (Datta et al. 1982; Auddy and Gomes 1996; Church and Hodgson 2002; Table 1). Plotosid, heteropneustid, and ariid venoms also have haemolytic effects (Church and Hodgson 2002). Inotropic effects on cardiac muscle have been reported on mammalian and amphibian muscle tissue for *Heteropneustes fossilis* venom, and *Plotosus canius* venom has been shown to cause cardiac arrest in mammalian model systems (Table 1). Pimelodid venom has a weaker effect on cardiac tissue, showing a slight decrease in contraction force on *in situ* frog hearts (Sarmiento et al. 2015; Table 1). Increased

Table 1. Reported effects of exposure to siluriform venom across multiple families on a variety of target organisms in a laboratory setting.

Family	Target	Effects	Source
Ariidae	Largemouth bass	Color loss, tetanus, and loss of equilibrium	Wright 2009
	Mosquitofish	Chromatophore expansion, hemorrhaging, edema, necrosis, mortality	Birkhead 1967
	Mice	Venular stasis, hemorrhaging, changes in arteriolar wall diameter, inflammation in post-capillary veins, increased vascular permeability	Thulesius et al. 1983; Junqueira et al. 2007; Ramos et al. 2012
Callichthyidae	Largemouth bass	Color loss	Wright 2009
Horabagridae	Largemouth bass	Color loss, tetanus, hemorrhage, loss of equilibrium	Wright 2009
Heteropneustidae	Isolated frog and guinea pig hearts	Ionotropic increase, haemolytic effects	Datta et al. 1982; Church and Hodgson 2002
	Anesthetized cats	Hypotension	Datta et al. 1982
Ictaluridae	Sailfin molly	Localized pigment darkening	Birkhead 1967
	Mosquitofish	Chromatophore expansion, inflammation, hemorrhaging, edema, necrosis, mortality	Birkhead 1967, 1972
	Largemouth bass	Color loss, myoclonus, hemorrhage, loss of equilibrium	Wright 2009
	Isolated guinea pig ileum	Smooth muscle contraction	Bartho et al. 2018
Mochokidae	Largemouth bass	Color loss, myoclonus, hemorrhage	Wright 2009
Pangasiidae	Largemouth bass	Myoclonus, loss of equilibrium	Wright 2009
Pimelodidae	Largemouth bass	Color loss, hemorrhage	Wright 2009
	in situ frog heart	Slight decrease in contraction force	Sarmiento et al. 2015
	Rats, mice	Strong nociceptive effects, edema, increased vascular permeability	Sarmiento et al. 2015; Lopes-Ferreira et al. 2014
Plotosidae	Largemouth bass	Tetanus, mortality	Wright 2009
	Isolated toad, rabbit, guinea pig hearts	Cardiac arrest, increased heart contraction	Auddy et al. 1994, 1995; Auddy and Gomes 1996
	Rat and guinea pig	Increased blood pressure, reduced respiration, death	Auddy et al. 1994, 1995
	Isolated guinea pig ileum	Smooth muscle contraction	Auddy et al. 1994, 1995
	Isolated rat phrenic nerve diaphragm, chick biventer cervicis	Irreversible blockage of twitch response	Auddy et al. 1994, 1995; Auddy and Gomes 1996
	mice, rats	Hypertension, respiratory failure, cardiac arrest, neuromuscular blockage, hyperglycemia, hemolytic and neurotoxic effects	Fahim et al. 1996; Auddy and Gomes 1996; Church and Hodgson 2002
	Ehrlisch ascites tumor (EAC) cells	Cytoxic	Fahim et al. 1996
Pseudopimelodidae	Largemouth bass	Color loss, hemorrhage	Wright 2009

Table 2. X indicates the observed effect on injection of 2 μL/g body weight of venom injected into *Micropterus salmoides*.

Species	Venom Effect					
	Color Loss	**Myoclonus**	**Tetanus**	**Hemorrhage**	**Loss of Equilibrium**	**Mortality**
Arius jordanii (Ariidae)	X		X		X	
Corydoras paleatus (Callichthyidae)	X					
Horabagrus brachysoma (Horabagridae)	X		X	X	X	
Microglanis inheringi (Pseudopimelodidae)	X			X		
Noturus gyrinus (Ictaluridae)	X	X		X	X	
Pangasius hypophthalmus (Pangasiidae)		X			X	
Pimelodus pictus (Pimelodidae)	X			X		
Plotosus lineatus (Plotosidae)			X			X
Synodontus multipunctatea (Mochokidae)	X	X		X		

Source: Table from J. Wright 2009.

vascular permeability (a common feature of many other animal venoms) has further been observed in arid and pimelodid venoms (Junqueira et al. 2007; Sarmiento et al. 2015) and likely facilitates the movement of venom proteins into surrounding tissues.

Insights From Human Envenomation

Accounts of human envenomation have been most extensively reported for ictalurids (*Ictalurus* and *Ameiurus* species in particular), *Arius thalassinus* (Ariidae), *Pseudoplatystoma* spp. (Pimelodidae), *Pimelodus maculatus* (Pimelodidae), *Plotosus lineatus* (Plotosidae), and *Heteropneustes fossilis* (Heteropneustidae). Envenomation of humans typically involves commercial or recreational fishers, home aquarists, and bathers who inadvertently step on them. In Brazil, 70% of accidents with marine fish involve catfish, and that figure rises to upwards of 90% in freshwater (Haddad and Martins 2006; Haddad 2008). Since most interactions involve handling, envenomation most commonly occurs on the hands and upper extremities, with feet and lower extremities being less common. However, it can occur when bathers step on small catfishes discharged by fishers on beaches. Envenomation of other areas of the body, including the head, is rare (Sazima et al. 2005; Haddad and Martins 2006; Aquino et al. 2016). In most cases, the spine leaves a puncture wound; however, lacerations do occur and are the most common type of wound reported in *Pseudoplatystoma* spp. envenomations (Haddad and Martins 2006; Aquino et al. 2016). In some cases, the spines remain embedded in the wound either because they break off during the encounter or the victims are unable to remove the spines and clip them off themselves (Das et al. 1995; Fredette et al. 1997; Blomkalns and Otten 1999; Negreiros et al. 2016). Presumably, spine breakage is not an adaptive mechanism of envenomation.

Common symptoms local to the wound site include pain (of varying severity), numbness, paresthesia, erythema, cyanosis, edema, numbness, lymphangitis, muscle fibrillation, cellulitis, and occasionally necrosis (Shepherd 1994; Junqueira et al. 2007; Satora et al. 2008; Aquino et al. 2016; Thalgaspitiya et al. 2017; Bentur et al. 2018). Systemic symptoms are more rarely reported but include tachycardia, weakness, sweating, arterial hypotension, dizziness, hypertension, vomiting, chills, weakness, fever, cardiac arrhythmias, and cold sweats (Thomas 1994; Satora et al. 2008; Aquino et al. 2016; Bentur et al. 2018). When systemic symptoms do occur, they typically manifest later and may result from the stress caused by the pain and secondary infection as opposed to the direct effects of the venom itself (Satora et al. 2008; Bentur et al. 2018). Bacterial infections are much more likely in untreated wounds and are most commonly polymicrobial, although the pathogenic bacteria involved differs between the freshwater and marine environments (Blomkalns and Otten 1999; Kaar et al. 2017).

Treatment typically involves immersing the wound in warm water (this is presumably effective since the venom proteins are heat labile and the vasodilation effects counteract the vasoconstricting action of the venom), pain relief (although warm water immersion itself often significantly relieves pain), thorough wound cleaning, and, in some cases, treatment with antibiotics either as prophylaxis or to combat secondary infection if established (Thomas 1994; Das et al. 1995; Blomkalns and Otten 1999; Church and Hodgson 2002; Satora et al. 2008).

Evolution

Origin Scenarios

The venom glands of siluriforms likely evolved from epidermal secretory glandular cells that are common on the skin surface. Club cells, in particular, while distinguishable from venom cells, share many morphological similarities (see "Venom Gland"). Particularly compelling evidence comes from the fact that skin secretions have been shown to exhibit toxic, antimicrobial, and/or wound-healing activity in ariids, plotosids, pimelodids, and ictalurids (Cameron and Endean 1973; Al-Hassan et al. 1983, 1985a, 1985b, 1987; Shiomi et al. 1987; Robinette et al. 1998; Perrière and Goudey-Perrière 2003; Junqueira et al. 2007). Two evolutionary scenarios, both involving the modification of epidermal secretory cells, have been proposed to explain the evolutionary origin of siluriform venom glands, one driven primarily by the toxic benefit of the secretions and the other by antimicrobial/wound healing effects.

The production of crinotoxins, non-injected toxic compounds, by epidermal secretory cells in many fishes suggests that the production of toxic compounds is a precursor to the evolution of venom delivery structures (Halstead 1970). Cameron and Endean (1973) proposed that the venom gland cells of all venomous fishes were ultimately derived from epidermal crinotoxin-secreting cells. In the case of siluriforms, individuals with a thickened epidermal layer around their pectoral and dorsal-fin spines would be at a selective advantage since such an arrangement would pair increased crinotoxin production with the ability to penetrate and deliver toxins directly into predator tissues. Subsequent selection would result in further modification of the toxin-producing cells in terms of the proteins produced and their morphology to more efficiently deliver venom to the victim (Cameron and Endean 1973; Wright 2015). Spine morphology itself would also presumably be further modified by selective pressure, resulting in features like grooves or serrations (Wright 2009, 2015). Indeed, the skin secretions of several siluriform species have been shown to share toxic components with venom. Skin secretions of *Plotosus lineatus* (Plotosidae) produce toxic effects, including hemolysis, edema, and lethality when injected into mice (Shiomi et al. 1987) and express a highly similar (13–14 kDa), likely identical, protein in the venom glands (Shiomi et al. 1988; Wright 2009; Wright 2015). Similar compelling evidence has been identified in ariids where a 65 kDa toxic protein (Warm Temperature Acclimation-Related Protein 65-kDa or Wap65) was found to be expressed in both the skin mucus and venom extract of *Cathorops spixii* (Ariidae; Ramos et al. 2012). Further, a ~ 39 kDa toxic protein component from the skin secretions of another ariid, *Arius bilineatus*, has been shown to cause agitated behavior, muscle spasms, labored breathing, and death in rabbits (Thomson et al. 1998). Wright (2009) identified a ~ 39 kDa protein fraction in the congeneric *Arius felis*, demonstrating another potential toxic protein expressed by both epidermal secretory and venom gland cells.

While crinotoxins are produced by epidermal secretory cells, their function may not, or may not always have been, predator deterrence. Experimental trials have shown that predators will repeatedly attack the same individual, apparently undeterred by any toxic secretions released (Bosher et al.

2006; Emmett and Cochran 2010; Wright 2012). Crinotoxins may have a primarily antifouling function and are well-developed in species that lack squamation (Cameron and Endean 1973; Wright 20015). In an alternative evolutionary origin scenario, Wright (2009) emphasized the wound healing and antimicrobial properties of epidermal cell secretions that have been demonstrated in ariids and ictalurids (Al-Hassan et al. 1983; Robinette et al. 1998; Wright 2015). Indeed, rubbing skin secretions on fin-spine wounds is also considered a remedy for envenomation by some fishers in North America (Das et al. 1995).

Predatory and environmental encounters typically damage the epidermal tissue surrounding the dorsal and pectoral-fin spines. Higher concentrations of epidermal secretory cells associated with the spines would lead to accelerated wound healing and reduced microbial infection - a selective advantage (Wright 2015). Under this scenario, the toxic secretions and their effects would have evolved secondarily to the primary function of wound healing. Once associated with the spines as a delivery mechanism, selection would then lead to further modification and diversification in the form and function of the spines, glandular cells, and their toxic secretions (Wright 2009, 2015). This explanation also posits that the axillary glands (also thought to be derived from epidermal secretory cells; Greven et al. 2006; Kiehl et al. 2006) evolved similarly, driven primarily by selection on the antimicrobial properties of the secretions.

Phylogenetic Patterns

Broadly, venom glands have been identified in species representing 23 different siluriform families, with three families (Lacantuniidae, Kryptoglanidae, and Phreatobiidae) remaining unsurveyed. The presence of venom glands is far more common in the Siluroidei than Loricarioidei (Wright 2009; Smith et al. 2016).

Initial reconstructions of the evolutionary history of venom gland origins on siluriform phylogenies by Wright (2009) recovered differing scenarios depending on the phylogenies used, suggesting a minimum of 2–3 unique gains. The presence of venom glands in Callichthyidae (the only instance of venom glands in the Loricarioidei) was consistently recovered as an independent origin of venom glands in both morphological and molecular phylogenies used. A separate, single origin of venom glands in the common ancestor of Siluroidei was supported by using reconstructions on morphological phylogenies recovered by Diogo (2004) and Mo (1991), although the latter was less clear, given the lack of resolution among siluroid lineages. Mapping venom gland presence/absence on a molecular phylogeny recovered by Sullivan et al. (2006) suggested, at minimum, a third independent origin in Doradidae, but once again, lack of resolution among siluroid lineages limited the interpretation (Figure 6). All of these reconstructions also suggest multiple instances of loss of venom glands in the Siluroidei. When taken together, these results suggest that multiple (minimally two, but possibly more) evolutionary origins are needed to explain the distribution of venom glands among siluriform taxa (Wright 2009).

In contrast to the multiple origins scenario, Smith et al. (2016) recovered a single origin of venom glands using a well-resolved molecular phylogeny based on a combination of mitochondrial and nuclear DNA (Figure 6). This phylogeny places the Callichthyidae as sister to Siluroidei, thus inferring homology among the venom glands of all siluriforms. Inferences based on this phylogeny also suggest seven to eight losses, although these numbers should be considered a minimum given the documented variability of venom gland presence within families and genera (Wright 2009; Egge and Simons 2011) and the fact that some families remain unsurveyed.

Regardless of the phylogeny used, losses of venom glands are far more common than independent gains. In some cases, this is accompanied by a complete loss of the entire delivery system (bony spines) in the malapterurids and some genera of Amphiliidae, Heptapteridae, Phreatobiidae, Pimelodidae, and Siluridae. Other taxa, including Aspredinidae, Auchenipteridae, Sisoridae, and some ictalurids, have lost venom glands but maintain bony spines (Wright 2009; Smith et al. 2016).

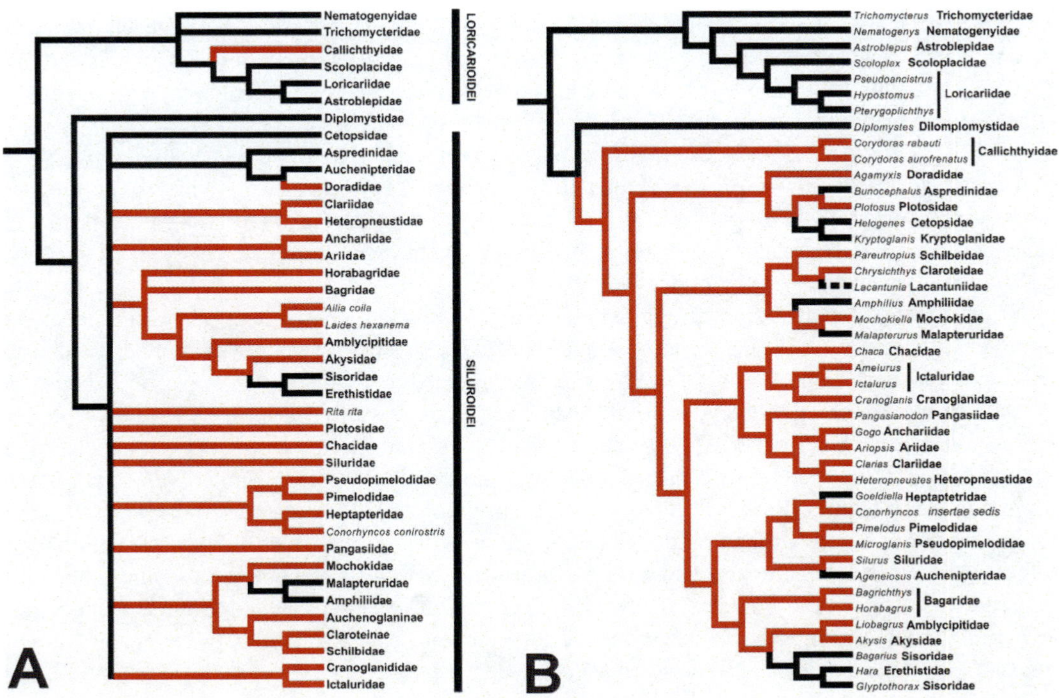

Figure 6. Alternative phylogenetic hypotheses recover multiple (A) or a single (B) origin of venom glands in siluriforms. (A) character optimization on phylogeny recovered by Sullivan et al. (2006) using two nuclear loci. (B) character optimization on phylogeny recovered by Smith et al. (2016) based on three mitochondrial and five nuclear loci. Red branches = venomous taxa; Black branches = nonvenomous taxa. Phylogenies redrawn from Wright 2009 (A) and Smith et al. 2016 (B).

The evolution of venom glands at the intrafamilial level also suggests that losses are more common than gains (Egge and Simons 2011).

Widespread losses suggest a cost to maintaining either the venom gland or the entire delivery apparatus. Absent high enough selection pressure, this system or some combination of its components is not maintained. Malapterurids lack both spines and venom glands (Wright 2009; Kubicek et al. 2019) but have evolved an alternative defensive system—an electric discharge potential that is also used in prey capture (Janetzko et al. 1987). Explaining the loss of venom glands in taxa that maintain fin spines is not as straightforward. Potential explanations might include shifts in habitat that result in the loss of predatory regimes susceptible to the venom, other functional constraints associated with the locomotory, communication, or defensive functions of the fin spines or, at the inter and intraspecific level, coevolutionary dynamics involving predators evolving resistance to venom (Wright 2009, 2012; Harris and Jenner 2019).

Selection

As noted previously, the pectoral and dorsal spines are multifunctional, so the selection on those structures is multidimensional. As such, the selection that has driven venom production, potency, and delivery is only one component of broader selective pressures on the venom delivery system in its entirety. Evidence that venom glands themselves provide a selective advantage (that is, improve survival and reproduction) comes primarily from experimental evidence cited earlier (see "Defensive Function") describing the predator deterrence effects of either the combined action of spines and venom gland (Bosher et al. 2006; Emmett and Cochran 2010; Wright 2011) or the differentiated effects of the spine alone vs the spine with a venom gland (Wright 2012). For *Noturus gyrinus* (Ictaluridae), the presence of a venom gland associated with the pectoral spine dramatically reduces predation compared with individuals who have spines but lack venom glands (Wright 2012). Even so, experimental results have been mixed since studies utilizing larger, more experienced predators demonstrated reduced predatory avoidance (Bosher et al. 2006; Emmet and Cochran 2010). Predatory learning also plays a role as predators should favor prey that presents fewer hazards for consumption, as outlined in the "dangerous prey hypothesis" (Forbes 1989).

Population-level analyses examining factors like variation in venom potency among members of the same species are lacking, limiting an understanding of the effects of selection at the population level. Furthermore, there is only a fragmented understanding of the important elements of the selective regime that might drive changes in venom potency or the delivery system in any one population or species.

Presumably, predators are a strong selective force, but which predators are the primary targets of envenomation for any one species or group that remains to be thoroughly examined. Predatory fishes would seem to be the most likely common predators for most siluriforms, but squamate, avian, and mammalian predators have all been identified as well (Birkhead 1972; Power 1984; Forbes 1989; Burr and Stoekel 1999; Hundt 2013). The venom delivery apparatus is most effective against gape-limited predators, as has been shown when comparing gape-limited versus non-gape limited avian predators (Forbes 1989). However, Birkhead (1972) speculates that mammals such as *Procyon* (a non-gape limited predator) that wade in riffles turning over rocks in search of food might impose selective pressure and explain the strong effects some siluriform venoms have on mammalian systems (see "Venom Effects"). Invertebrates are another plausible source of predation and could potentially subvert the venom delivery system entirely. Invertebrate predators are known to grasp the fin spines of *Gasterosteus aculeatus* (Scorpaeniformes), contributing to selection for their reduced size and number in some freshwater habitats (Nosil and Reimchen 2005). In any case, these predatory regimes vary across siluriform groups and impose differential selection pressures. A better understanding of the selective pressures at the population level will be important for building an improved understanding of broader patterns of evolutionary change in siluriform venom systems.

Aposematism and Mimicry

The coevolutionary interplay between venomous prey and their predators has been well documented in several vertebrate species (Heatwole and Powell 1998; Hanifin et al. 2008), but this phenomenon has received comparatively less attention among siluriforms. Siluriform venoms are thought to be non-target specific because they tend to be subject to a wide range of predation, potentially minimizing host-specific toxins seen in some other vertebrates (Harris and Jenner 2019).

While direct impacts of coevolutionary dynamics on venom toxicity or delivery are not yet documented in siluriforms, secondary evolutionary consequences are known. Aposematism and mimicry typically evolve secondarily after venom systems are in place. Both of these phenomena serve to reduce the cost of venom production by minimizing or preventing direct interaction with a predator in the first place, thus reducing the use of venom (Harris and Jenner 2019). Black and yellow banding patterns, as expressed in *Plotosus lineatus* (Plotosidae), and spotting patterns are known to be aposematic across venomous animal taxa (Harris and Jenner 2019).

Evidence for the evolution of shared patterns of aposematic coloration via Müllerian mimicry has been reported in venomous siluriform species in both Mochokidae and Callichthyidae. Mochokids endemic to Lake Tanganyika in the genus *Synodontis* (all of which are positively confirmed to be venomous) have a pattern of contrasting black dots over a solid background, and their fins exhibit a strongly contrasting black base with a white margin at some point in their development (Wright 2011). Predation experiments conducted by Wright (2011) using a naïve predator (*Micropterus salmoides*) on the Lake Tanganyikan species *Synodontis multipunctata* and *Synodontis petricola* (Mochokidae) demonstrate that the predator quickly learned to avoid both species after one or two exposures and strongly avoid attacking them subsequently. Visual cues were demonstrated to be the primary means of recognition, meaning that the shared color pattern serves as a selective advantage for predator avoidance and is maintained in Tanganyikan *Synodontis* through Müllerian mimicry, whereby all the species share the benefit of a similar phenotype. In this case, the protective aposematic coloration evolved either once in the ancestor of all Lake Tanganyikan *Synodontis*, or twice (once each in two subclades) and is maintained by selection for a selectively advantageous ancestral aposematic model (Wright 2011).

Callichthyids have more varied patterns of aposematism that can consist of contrasting black and white stripes or margins, patches of color, or distinctly colored spines (Alexandrou et al. 2011). While there is no experimental evidence of learned predator avoidance in callichthyids, the lineage has been confirmed to contain numerous venomous species (Wright 2009). Corydoradinae species tend to shoal together with similarly patterned species from different lineages (Alexandrou et al. 2011). Presumably, these shared patterns serve as common learning models for predators (both fish and bird predators). Different geographic locations tend to have different shared patterns comprising 24 mimicry rings with two to three species each (Alexandrou et al. 2011). Other sympatric species also appear to participate in the mimicry of the local Corydoradinae color pattern including species from siluriform groups known to be venomous like *Brachyrhamdia* (Heptapteridae; Slobodian and Bockmann 2013; Carvalho et al. 2021) and groups not known to be venomous, like *Otocinclus* (Loricariidae; Wright 2009; Alexandrou et al. 2011). Unlike *Synodontis*, the Müllerian mimicry among callichthyids results from shared color patterns evolving as the result of convergence among lineages in sympatry (Alexandrou et al. 2011).

References

Al-Hassan, J.M., M. Thomson and R. Criddle. 1983. Accelerated wound healing by a preparation from skin of the Arabian Gulf catfish. Lancet 321: 1043–44.

Al-Hassan, J.M., M. Ali, M. Thomson, T. Fatima and C.J. Gubler. 1985a. Toxic effects of the soluble skin secretion from the Arabian Gulf catfish (*Arius thallasinus*, Rupell) on plasma and liver enzyme levels. Toxicon 243: 532–534.

Al-Hassan, J.M., M. Thomson, K.R. Criddle, B. Summers and R.S. Criddle. 1985b. Catfish epidermal secretions in response to threat or injury. Mar. Biol. 88: 117–123.

Al-Hassan, J.M., M. Ali, M. Thomson, T. Fatima and C.J. Gubler. 1987. Prostaglandin associated mortality following intravenous injection of catfish epidermal secretions in rabbits. Prostaglandins Leukot. Med. 28: 95–102.

Alexander, R.McN. 1965. Structure and function in the catfish. J. Zool. 148: 88–152.

Alexandrou, M., C. Oliveira, M. Maillard, R.A.R. McGill, J. Newton, S. Creer et al. 2011. Competition and phylogeny determine community structure in Müllerian co-mimics. Nature 469: 84–88.

Aquino, G.N.D., C.C. Souza, V. Haddad Jr. and J. Sabino. 2016. Injuries caused by the venomous catfish *pintado* and *cachara* (*Pseudoplatystoma* genus) in fishermen of the Pantanal region in Brazil. An. Acad. Bras. Ciênc. 88: 1531–1537.

Arratia, G. 1987. Description of the primitive Family Diplomystidae (Siluriformes, Teleostei, Pisces): morphology, taxonomy, and phylogenetic implications. Bonn. Zool. Monogr. 24: 5–120.

Arratia, G. 2003. The siluriform postcranial skeleton—an overview. pp. 121–157. *In*: Arratia, G., B.G. Kapoor, M. Chardon and R. Diogo (eds.). Catfishes: Volume 1. Science Publishers, Inc., Enfield, NH, USA.

Auddy, B., M.I. Alam and A. Gomes. 1994. Pharmacological actions of the venom of the Indian catfish (*Plotosus canius* Hamilton). Indian J. Med. Res. 99: 47–51.

Auddy, B., D.C. Muhuri, M.I. Alam and A. Gomes. 1995. A lethal protein toxin (toxin-PC) from the Indian catfish (*Plotosus canius*, Hamilton) venom. Nat. Toxins 3: 363–368.

Auddy, B. and A. Gomes. 1996. Indian Catfish (*Plotosus canius*, Hamilton) Venom: Occurrence of lethal protein toxin (Toxin-PC). pp. 225–229. *In*: Singh, B.R. and A.T. Tu (eds.). Natural Toxins 2: Structure, Mechanism of Action, and Detection. Springer US, Boston, MA, USA.

Ballen, G.A. and M.C. de Pinna. 2022. A standardized terminology of spines in the order Siluriformes (Actinopterygii: Ostariophysi). Zool. J. Linn. Soc. Lond. 194: 601–625.

Barthó, L., Z. Sándor and T. Bencsik. 2018. Effects of the venom of the brown bullhead catfish (*Ameiurus nebulosus*) on isolated smooth muscle. Acta Biol. Hung. 69: 135–143.

Bentur, Y., S. Altunin, I. Levdov, D. Golani, E. Spanier, D. Edelist et al. 2018. The clinical effects of the venomous Lessepsian migrant fish *Plotosus lineatus* (Thunberg, 1787) in the southeastern Mediterranean Sea. Clin. Toxicol. 56: 327–331.

Bhimachar, B.S. 1944. Poison glands in the pectoral spines of two catfishes—*Heteropnuestes fossilis* (Bloch) and *Plotosus arab* (Forsk.), with remarks on the nature of their venom. Proc. Ind. Acad. Sci. 19B: 65–70.

Birkhead, W. S. 1967. The comparative toxicity of stings of the ictalurid catfish genera *Ictalurus* and *Schilbeodes*. Comp. Biochem. Physiol. 22: 101–111.

Birkhead, W.S. 1972. Toxicity of stings of ariid and ictalurid catfishes. Copeia 1972: 790–807.

Blomkalns, A.L. and E.J. Otten. 1999. Catfish spine envenomation: a case report and literature review. Wilderness Environ. Med. 10: 242–246.

Bosher, B.T., S.H. Newton and M.L. Fine. 2006. The spines of the channel catfish, *Ictalurus punctatus*, as an antipredator adaptation: an experimental study. Ethology 112: 188–195.

Burr, B.M. and J.M. Stoekel. 1999. The natural history of madtoms (Genus *Noturus*), North America's diminutive catfishes. pp. 51–101. *In*: Irwin, E.R., W.A. Hubert, C.F. Rabeni, H.L. Schramm, Jr. and T. Coon (eds.). Catfish 2000: Proceedings of the International Ictalurid Symposium American Fisheries Society, Symposium 24, Bethesda, Maryland.

Calton, G.J. and J.W. Burnett. 1975. Catfish (*Ictalurus catus*) fin venom. Toxicon 13: 399–403.

Cameron, A.M. and R. Endean. 1971. The axillary glands of the plotosid catfish *Cnidoglanis macrocephalus*. Toxicon 9: 345–352.

Cameron, A.M. and R. Endean. 1973. Epidermal secretions and the evolution of venom glands in fishes. Toxicon 11: 401–410.

Carvalho, T.I., J. Klaczko and V. Slobodian. 2021. Pectoral-fin glands and delivery apparatus in the catfish genus *Brachyrhamdia* Myers, 1927 (Siluriformes: Heptapteridae). Pap. Avulsos Zool. 61: e20216174.

Church, J.E. and W.C. Hodgson. 2002. The pharmacological activity of fish venoms. Toxicon 40: 1083–1093.

Citterio, V. 1925. L'apparato vulnerante di *Amiurus catus* (L.) Atti Soc. Ital. Sci. Nat. 64: 1–8.

Citterio, V. 1926. L'apparato vulnerante di *Arius heudeloti* (C. et V.) Atti Soc. Ital. Sci. Nat. 17: 24–27.

Das, S.K., M.B. Johnson and H.H. Cohly. 1995. Catfish stings in Mississippi. South. Med. J. 88: 809–812.

Datta, A., A. Gomes, B. Sarangi., P.K. Kar and S.C. Lahiri. 1982. Pharmacodynamic actions of crude venom of the Indian catfish *Heteropneustes fossilis*. Indian J. Med. Res. 76: 892–897.

Diogo, R., C. Oliveira and M. Chardon. 2001. On the osteology of catfish pectoral girdle, with a reflection on catfish (Teleostei: Siluriformes) plesiomorphies. J. Morph. 249: 100–125.

Diogo, R. 2004. Morphological Evolution, Aptations, Homoplasies, Constraints, and Evolutionary Trends: Catfishes As A Case Study On General Phylogeny and Macroevolution Enfield: Science Publishers, Inc.

Egge, J.J.D. and A.M. Simons. 2011. Evolution of venom delivery structures in madtom catfishes (Siluriformes: Ictaluridae). Biol. J. Linn. Soc. Lond. 102: 115–129.

Emmett, B. and P.A. Cochran. 2010. The response of a piscivore (*Micropterus salmoides*) to a venomous prey species (*Noturus gyrinus*). Jour. Freshw. Ecol. 25: 475–479.

Fahim, F.A., E.A. Mady, S.M. Ahmed and M.A. Zaki. 1996. Biochemical studies on the effect of *Plotosus lineatus* crude venom (*in Vivo*) and its effect on EAC-cells (*in Vitro*). pp. 343–355. *In*: Singh, B.R. and A.T. Tu (eds.). Natural Toxins 2: Structure, Mechanism of Action, and Detection. Springer US, Boston, MA, USA.

Fine, M.L., J.P. Friel, D. McElroy, C.B. King, K.E. Loesser and S. Newton. 1997. Pectoral spine locking and sound production in the channel catfish *Ictalurus punctatus*. Copeia 1997: 777–790.

Fine, M.L. and F. Ladich. 2003. Sound production, spine locking and related adaptations. pp. 249–290. *In*: Arratia, G., B.G. Kapoor, M. Chardon and R. Diogo (eds.). Catfishes: Volume 1. Science Publishers, Inc., Enfield, NH, USA.

Forbes, L.S. 1989. Prey defences and predator handling behaviour: the dangerous prey hypothesis. Oikos 55: 155–158.

Fredette, S.E., F.F. Derk and A.J. Nardozza. 1997. Catfish spine injury of the foot. J. Am. Podiatr. Med. Assoc. 87: 187–189.

Fry, B.G., K. Roelants, K. Winter, W.C. Hodgson, L. Griesman, H.F. Kwok et al. 2010. Novel venom proteins produced by differential domain-expression strategies in beaded lizards and gila monsters (genus *Heloderma*). Mol. Biol. Evol. 27: 395–407.

Glahn, J.F. and B.S. Dorr. 2002. Captive double-crested cormorant *Phalacrocorax auritus* predation on channel catfish *Ictalurus punctatus* fingerlings and its influence on single-batch cropping production. J. World Aquac. Soc. 33: 85–93.

Günther, A. 1880. An Introduction to the Study of Fishes. London: Oxford University.

Greven, H., T. Flasbeck and D. Passia. 2006. Axillary glands in the armoured catfish *Corydoras aeneus* (Callichthyidae, Siluriformes). Verh. Ges. Ichthyol. 6: 65–69.

Haddad Jr., V. and I.A. Martins. 2006. Frequency and gravity of human envenomations caused by marine catfish (suborder Siluroidei): a clinical and epidemiological study. Toxicon 47: 838–843.

Haddad Jr., V., R.A. De Souza and P.S. Auerbach. 2008. Marine catfish sting causing fatal heart perforation in a fisherman. Wilderness Environ. Med. 19: 114–118.

Halstead, B.W., L.S. Kuninobu and H.G. Hebard. 1953. Catfish stings and the venom apparatus of the Mexican catfish, *Galeichthys felis* (Linnaeus). Trans. Am. Microsc. Soc. 72: 297–314.

Halstead, B.W. and R.L. Smith. 1954. Presence of an axillary venom gland in the Oriental catfish *Plotosus lineatus*. Copeia 1954: 153–154.

Halstead, B.W. 1970. Poisonous and Venomous Marine Animals in the World, Vol. 3, Vertebrates, Osteichthyes. Washington, D.C.: US Government Printing Office.

Hanifin, C.T., E.D. Brodie Jr. and E.D. Brodie III. 2008. Phenotypic mismatches reveal escape from arms-race coevolution. PLoS Biol. 6: e60.

Harris, R.J. and R.A. Jenner. 2019. Evolutionary ecology of fish venom: adaptations and consequences of evolving a venom system. Toxins 11: 60.

Hassan, M.I., A. Waheed, S. Yadav, T.P. Singh and F. Ahmad. 2008. Zinc α2-glycoprotein: A multidisciplinary protein. Mol. Cancer Res. 6: 892–906.

Heatwole, H. and J. Powell. 1998. Resistance of eels (*Gymnothorax*) to the venom of sea kraits (*Laticauda columbrina*): a test of coevolution. Toxicon 36: 619–625.

Hundt, P.J. 2016. Double-crested cormorants (*Phalacrocorax auratus*) of Leech Lake, Minnesota: temporal variation of diets and assessment of differential prey selection in adults. The American Midland Naturalist 169: 345–353.

Janetzko, A., H. Zimmerman and W. Volknandt. 1987. The electromotor system of the electric catfish (*Malapterurus electricus*): a fine-structural analysis. Cell Tissue Res. 247: 613–624.

Junqueira, M.E P., L.Z. Grund, N.M. Orii, T.C. Saraiva, C.A. de Magalhães Lopes, C. Lima et al. 2007. Analysis of the inflammatory reaction induced by the catfish (*Cathorops spixii*) venoms. Toxicon 49: 909–919.

Kaar, C.R. and A.K. Nakanishi. 2017. Recreational and commercial catfishing injuries: a review of the literature. Wilderness Environ. Med. 28: 348–354.

Kaatz, I.M. 1999. The Behavioral and Morphological Diversity of Acoustic Communication Systems in a Clade of Tropical Catfishes (Pisces: Siluriformes). Ph.D. Thesis, State University of New York College of Environmental Science and Forestry, Syracuse, USA.

Kaatz, I.M., D.J. Stewart, A.N. Rice and P.S. Lobel. 2010. Differences in pectoral fin spine morphology between silent clades of catfishes (Order Siluriformes): ecomorphological implications. Current Zool. 56: 73–89.

Kiehl, E., C. Rieger and H. Greven. 2006. Axillary gland secretions contribute to the stress-induced discharge of a bactericidal substance in *Corydoras sterbai* (Callichthyidae, Siluriformes). Verh. Ges. Ichthyol. 5: 111–115.

Kamiya, H., K. Muramoto and M. Yamazaki. 1986. Aplysianin-A, an antibacterial and antineoplastic glycoprotein in the albumen gland of a sea hare, *Aplysia kurodai*. Experimentia 42: 1065–1067.

Kubicek, K.M., R. Britz and K.W. Conway. 2019. Ontogeny of the catfish pectoral-fin spine (Teleostei: Siluriformes). J. Morph. 280: 339–359.

Kurecki, T., M.S. Laskowski and L. Kress. 1978. Purification and some properties of two proteinases from Crotalus adamanteus venom that inactivate human α 1-proteinase inhibitor. J. Biol. Chem. 253: 8340–8345.

Lopes-Ferreira, M., E.M. Gomes, F.M. Bruni, M.J. Ferreira, P. Charvet and C. Lima. 2014. First report of interruption of mast cell degranulation and endothelial cells activation by anti-inflammatory drugs controlling the acute response provoked by *Pseudoplatystoma fasciatum* fish venom. Toxicon 90: 237–248.

Lundberg, J.G. 1970. The Evolutionary History of North American Catfishes, Family Ictaluridae. Ph.D. Thesis, University of Michigan, Ann Arbor, USA.

Mahajan, C.L. 1963. Sound producing apparatus in an Indian catfish *Sisor rhabdophorus* Hamilton. Zool. J. Linn. Soc. Lond. 44: 721–724.

Mo, T. 1991. Anatomy, relationships, and systematics of the Bagridae (Teleostei: Siluroidei) with a hypothesis of siluroid phylogeny. Theses Zoologicae 17: 1–216.

Morgan, L.D. and M.L. Fine. 2020. Agonistic behavior in juvenile blue catfish *Ictalurus furcatus*. J. Ethol. 38: 29–40.

Negreiros, M.M.D.B., S. Yamashita, T. Sardenberg, E.L. Fávero Jr., F.A.H. Ribeiro, W.T. Haddad Jr. et al. 2016. Diagnostic imaging of injuries caused be venomous and traumatogenic catfish. Rev. Soc. Bras. Med. Trop. 49: 530–533.

Nosil, P. and T.E. Reimchen. 2005. Ecological opportunity and levels of morphological variance within freshwater stickleback populations. Biol. J. Linn. Soc. 86: 297–308.

Parmentier, E., G. Fabri, I. Kaatz, N. Decloux, S. Planes and P. Vandewalle. 2010. Functional study of the pectoral spine stridulation mechanism in different mochokid catfishes. J. Exp. Biol. 213: 1107–1114.

Perrière, C. and F. Goudey-Perrière. 2003. Poisonous catfishes: venom apparatus, acanthotoxins, crinotoxins, and other skin secretions. pp. 291–314. *In*: Arratia, G., B.G. Kapoor, M. Chardon and R. Diogo (eds.). Catfishes: Volume 1. Science Publishers, Inc., Enfield, NH, USA.

de Pinna, M.C.C. 1996. Phylogenetic analysis of the Asian catfish families Sisoridae, Akysidae, and Amblycipitidae, with a hypothesis on the relationships of the neotropical Aspredinidae (Teleostei, Ostariophysi). Fieldiana Zool. 84: 1–83.

Power, M.E. 1984. Depth distributions of armored catfish: predator-induced resource avoidance? Ecology 65: 523–528.

Ramos, A.D., K. Conceição, P.I. Silva Jr., M. Richardson, C. Lima and M. Lopes-Ferreira. 2012. Specialization of the sting venom and skin mucus of *Cathorops spixii* reveals functional diversification of the toxins. Toxicon 59: 651–665.

Reed, H.D. 1907. The poison glands of *Noturus* and *Schilbeodes*. Amer. Natur. 41: 553–556.

Reed, H.D. 1924. The morphology and growth of the spines of Siluroid fishes. J. Morph. 38: 431–451.

Rice, A.N., S.C. Farina, A.J. Makowski, I.M. Kaatz, P.S. Lobel, W.E. Bemis et al. 2022. Evolutionary patterns in sound production across fishes. Ichthyol. Herpetol. 110: 1–12.

Robinette, D., S. Wada, T. Arroll, M.G. Levy, W.L. Miller and E.J. Noga. 1998. Antimicrobial activity in the skin of the channel catfish *Ictalurus punctatus*: characterization of broad-spectrum histone-like antimicrobial proteins. Cell. Mol. Life Sci. 54: 467–475.

Sarmiento, B.E., M. Rangel, J.C. Gonçalves, L. Pereira, S. Rego, L.A. Campos et al. 2015. First report of the characterization of the pathophysiological mechanisms caused by the freshwater catfish *Pimelodus maculatus* (order: Siluriformes). Toxicon 101: 55–62.

Satora, L., M. Kuciel and T. Gawlikowski. 2008. Catfish stings and the venom apparatus of the African catfish *Clarias gariepinus* (Burchell, 1822), and stinging catfish *Heteropneustes fossilis* (Bloch, 1794). Ann. Agric. Environ. Med. 15: 163–166.

Sazima, I., J. Zuanon and V. Haddad Jr. 2005. Puncture wounds by driftwood catfish during bucket baths: local habits of riverside people and fish natural history in the Amazon. Wilderness Environ. Med. 16: 204–208.

Shepherd, S., S.H. Thomas and C.K. Stone. 1994. Catfish envenomation. Wilderness Environ. Med. 5: 67–70.

Shiomi, K., M. Takamiya, H. Yamanaka and T. Kikuchi. 1987. Purification of a lethal factor in the skin secretion from the oriental catfish *Plotosus lineatus*. Nippon Suisan Gak. 53: 1275–1280.

Shiomi, K., M. Takamiya, H. Yamanaka, T. Kikuchi and Y. Suzuki. 1988. Toxins in the skin secretion of the Oriental catfish (*Plotosus lineatus*): immunological properties and immunocytochemical identification of producing cells. Toxicon 26: 253–261.

Sivan, G. 2009. Fish venom: pharmacological features and biological significance. Fish Fish. 10: 159–172.

Slobodian, V. and F.A. Bockmann. 2013. A new *Brachyrhamdia* (Siluriformes: Heptapteridae) from Rio Japurá basin, Brazil, with comments on its phylogenetic affinities, biogeography and mimicry in the genus. Zootaxa 3717: 1–22.

Slobodian, V., B. Abreu-Santos and M.N.D.L. Pastana. 2021. The rediscovery of *Pimelodella longipinnis* (Borodin, 1927), an enigmatic Atlantic Rainforest catfish species from Southeastern Brazil (Siluriformes: Heptapteridae). Pap. Avulsos Zool. 61: e20216173.

Smith, W.L. and W.C. Wheeler. 2006. Venom evolution widespread in fishes: a phylogenetic road map for the bioprospecting of piscine venoms. J. Hered. 97: 206–217.

Smith, W.L., J.H. Stern, M.G. Girard and M.P. Davis. 2016. Evolution of venomous cartilaginous and ray-finned fishes. Integr. Comp. Biol. 56: 950–961.

Sodré, J.G., T.R. Colaço-Fernandez, L.H. Rapp Py-Daniel, J.L.O. Birindelli and J. Zuanon. 2021. Putative poison gland in the thorny catfish *Acanthodoras spinosissimus* (Siluriformes: Doradidae). Acta Amazonica 51: 250–254.

Sullivan, J.P., J.G. Lundberg and M. Hardman. 2006. A phylogenetic analysis of the major groups of catfishes (Teleostei: Siluriformes) using rag1 and rag2 nuclear gene sequences. Mol. Phylogenet. Evol. 41: 636–662.

Thalgaspitiya, S.P., B.T. Wijerathne and N.P. Siriwardhana. 2017. Freshwater catfish envenoming in a tropical country. Wilderness Environ. Med. 28: 259–266.

Thomson, M., J.M. Al-Hassan, S. Fayad, J. Al-Saleh and M. Ali. 1998. Purification of a toxic factor from Arabian Gulf catfish epidermal secretions. Toxicon 36: 859–866.

Thulesius, O., J.M. Hassan, R.S. Criddle and M. Thomson. 1983. Vascular response elicited by venom of Arabian catfish (*Arius thallasinus*). Gen. Pharmacol. 14: 129–132.

Vanscoy, T., J.G. Lundberg and K.R. Luckenbill. 2015. Bony ornamentation of the catfish pectoral-fin spine: comparative and developmental anatomy, with an example of fin-spine diversity using the Tribe Brachyplatystomini (Siluriformes, Pimelodidae). Proc. Acad. Nat. Sci. 164: 177–212.

Whitear, M., G. Zaccone, M.B. Ricca and P.L. Cascio. 1991a. Fine structure and histochemistry of the venom gland in the Indian stinging catfish (*Heterpneustes fossilis*). J. Zool. 224: 479–489.

Whitear, M., G. Zaccone, S. Fasulo and A. Licata. 1991b. Fine structure of the axillary gland in the brown bullhead (*Ictalurus nebulosus*). J. Zool. 224: 669–676.

Wright, J.J. 2009. Diversity, phylogenetic distribution, and origins of venomous catfishes. BMC Evol. Biol. 9: 1–12.

Wright, J.J. 2011. Conservative coevolution of Müllerian mimicry in a group of rift lake catfish. Evolution 65: 395–407.

Wright, J.J. 2012. Adaptive significance of venom glands in the tadpole madtom *Noturus gyrinus* (Siluriformes: Ictaluridae). J. Exp. Biol. 215: 1816–1823.

Wright, J.J. 2015. Evolutionary history of venom glands in the Siluriformes. pp. 1–19. *In*: Golpalakrishnakone, P. and A. Malhotra (eds.). Evolution of Venomous Animals and their Toxins. Springer, Netherlands.

Xie, B., X. Li, Z. Lin, Z. Ruan, M. Wang, J. Liu et al. 2016. Prediction of toxin genes from Chinese yellow catfish based on transcriptomic and proteomic sequencing. Int. J. Mol. Sci. 17: 556.

Yamazaki, Y. and T. Morita. 2004. Structure and function of snake venom cysteine-rich secretory proteins. Toxicon 44: 227–231.

Chapter 8

X-Ray Computed Tomography and the Study of Catfish Morphology

Mariangeles Arce H.[1,*] and *Adam P. Summers*[2]

Computed Tomography (CT) has been around since the 1970s as a technique to produce three-dimensional (3D) images of specimens. It uses X-rays (two-dimensional (2-D) images) from multiple angles to reconstruct a 3D image that allows detailed observation of anatomical structures without compromising the integrity of the specimens. This technique complements traditional ways of studying morphology by offering a non-destructive way to investigate internal anatomy with its own set of capabilities and constraints (Ford et al. 2023).

Nuts and Bolts

Computed Tomography (CT) scanning uses X-rays to produce a three-dimensional image of a specimen. The idea of CT scanning dates to the 1970s and has at its root the premise that the 3D shape of an object can be reconstructed from many 2D shadows created by illuminating the object from different angles. These 2D shadows, called "back projections," are the raw data output of a CT scanner. A formula called the Radon transform predicts the shadow cast by an object. A computer program uses the inverse Radon transform to turn the back projections into a series of slices through the object. These slices can then be rendered, using free, open-source software, into 3D volumes or surfaces. The first CT scans of fish were made on medical scanners, where the bed of the scanner stays in place while the X-ray source and detector are whirled around the bed to create the shadow images. The resolution of these scans was excellent for diagnostic purposes in a large mammal (~ 0.3 mm), but for visualization of the internal anatomy of fishes, it was only useful for the largest specimens (> 80 cm SL). The widespread availability of micro-source CT scans, in which the X-rays come from what is effectively a point source, and the specimen spins while the source and detector remain still, has made it possible to image even the smallest vertebrates. The resolution of modern machines is about 1 um, so even a little pencil catfish like *Ammoglanis pulex* could have hundreds of slices from the tip of the snout to the operculi. The availability of CT scanners and their flexibility in imaging in three dimensions has led to an explosion of morphological data that does not leave catfishes behind.

[1] Department of Ichthyology, The Academy of Natural Sciences of Drexel University, 1900 Benjamin Franklin Parkway, Philadelphia, PA 19130, USA.

[2] Department of Biology, Friday Harbor Labs, University of Washington, Friday Harbor, WA, 98250, USA.
Email: fishguy@uw.edu

* Corresponding author: m.arce@drexel.edu

CT scanners are usually used to visualize mineralized skeletons since X-rays that generate shadow data are attenuated by elements with high atomic numbers, such as calcium and phosphate. In this context, soft tissues are relatively transparent, while bones are radio-opaque. This limitation has been overcome by adding contrast agents that either differentially bind to soft tissues or are differentially absorbed by them. We can now image muscles and organs with the same fidelity as bones. Contrast agents, like all manipulations of specimens to increase contrast, lead to artifacts, and it is important to consider this when processing tissue. The simplest and least expensive contrast agent is iodine, usually in the form of I_2KI, which is called Diffusible Iodine Contrast Enhancement CT (diceCT) scanning in the literature (Jeffery et al. 2011; Metscher 2009). A good deal of care must be taken when staining with iodine as it will minimally shrink some soft tissues a lot more than others; in particular, neural tissues can suffer if the concentration gradient is high (Vickerton et al. 2013). After the process is done, it is possible to de-stain the specimen since the binding of iodine is temporary. On occasion, the tissues might remain slightly tinted, but there is no structural change to the tissues. This is particularly important when examining materials from very rare specimens. The better images for a diceCT are obtained with long incubation times, from months to years. Alternative stains, like phosphomolybdic acid and phosphotungstic acid, provide excellent contrast, but because of their large molecular size, they will not penetrate as well as iodine. With these stains, it is possible to visualize soft tissue in embryos and other specimens in the 1–10 mm size range. They also will not leach out, leaving specimens with a permanent green cast even years after being removed from the stain, which may be an issue for rare or type material. A last technique worth mentioning is freeze drying, which adds nothing to the specimen, but removing water makes the soft tissue visible against the surrounding air. There can be shrinkage artifacts with this technique, which permanently transforms the specimen, but it has been used to estimate muscle sizes and pennation angles (Kleinteich et al. 2008).

Micro-CT was revolutionary for ichthyologists because the variable magnification of this technology allows a single machine to make images of small to medium-sized specimens. Consider the shadow of your hand at noon on a sunny day. It is the same size regardless of how far off the ground your hand is held. That is because the rays of the sun are parallel when they hit the back of your hand. In contrast, consider the shadow of your hand cast on the screen by an old-fashioned slide projector. The small light source means the rays fan out, and as your hand gets closer to the projector, the shadow of your hand will get larger. The shadow of your hand will be exactly life-size when it is directly touching the screen. Micro-source CT scanners have a small source spot size in the order of nanometers. A specimen placed close to the source will cast a large shadow on the detector, while one close to the detector will approach life-size. The distance between the source and detector is fixed, and a rotating platform with a specimen can be moved closer or further away from the source. A constraint is that the specimen cannot touch either the source or detector when it is rotated about its central axis through 180 degrees. In this way, a single CT scanner can image objects ranging in size from a papaya to the seed of a passionfruit. Or, in the vernacular of the catfish morphologist – from an adult Highwaterman Catfish (*Hypophthalmus edentatus*) to a young Sand-dwelling Catfish (*Pygidianops amphioxus*).

A micro-CT scanner typically allows the resolution to be adjusted to meet the needs of the study. However, the most extreme example of excess resolution is in synchrotron CT, where a beamline provides coherent X-rays, and the resolution can approach tens of nanometers. The volume that is imaged is small, perhaps 1 mm on the side, but the data set is inconveniently large—on the order of terabytes. This will need to be downsampled to fit into the memory of a computer, so it is sensible to skip gathering the higher-resolution data in favor of gathering data on more specimens. For most applications, including 3D morphometrics, finite element analysis, bone density, and 3D printing, a good rule of thumb is 200 slices across the skull. That means for a catfish with a width of 10 mm between the lateral edges of the opercles, the desired resolution is 50 um. This is well within the capability of low-end micro-CT scanners. This data set, assuming an aspect ratio of 10:1 (for a clariid, for example), would be ~ 100 MB, which could be loaded in a mid-level laptop computer for analysis. Deciding to go with a resolution of 25 um, which sounds twice as good, leads to a data set 8 times as large that takes at least twice as long to gather. Bumping the data set to nearly 1 GB puts it out of range of mid-level computers, and this extra resolution will not allow you to do more with the data.

Catfish Burritos Are Great

With over 11,000 specimens scanned, the #ScanAllFishes project and the NSF-funded openVertebrate (oVert) initiative have made some progress toward freely available skeletal data for all vertebrates. Much of this data is available on the Morphosource.org website. A new technique for scanning many specimens at once made the large number of datasets possible. CT scanning reconstructs a cylindrical volume that usually holds, as it does in medical CT, a single specimen. Dead catfishes do not have the tender sensibilities of live humans, so several fish can be placed in the cylindrical volume. Each scan, which is set for the optimal resolution for a fish of a particular size, may then produce data for as many as ten specimens. These specimens, wrapped together with small radio-opaque labels, are collectively a "burrito" of incipient data. Individuals must be digitally dissected from one another, but that is a simple operation that can be done off-site with free, open-source software like 3D Slicer. The advantage of the burrito is that CT is no longer an expensive endeavor per specimen, and the disadvantages of the technique are that resolution might be affected if the fish are not about the same and if there are very different levels of ossification. Between the dropping cost of the scanners, the availability of free scanners, and the capacity to scan many specimens at once, it is now possible to scan entire faunas, populations, or ontogenetic series without breaking the bank. Furthermore, there is a growing source of free, open-source data available to anyone anywhere in the world through databases like MorphoSource.org, digimorph.org and Phenome10k.org.

To this date, MorphoSource has CT scans of 356 specimens of Siluriformes representing 335 taxonomic names, and DigiMorph has 22 species representing 14 families. There are no catfishes in Phenome yet, but the database is growing. An initiative to generate a Digital Atlas of Catfish Morphology was led by John Lundberg, who created the Catfish Bones webpage (catfishbone.ansp.org), where raw CT scans for 15 silurid families can be found. Most families are represented on that page by one species, but Aspredinidae is represented by two species, and Loricariidae is represented by four species.

Why Not a Knife?

A prescient author asked in an early paper that featured CT scans of fishes why anyone would choose such an expensive, technically challenging, and abstruse method over the tried-and-true tools available to comparative anatomists since humans first flaked chert into a cutting-edge (Summers et al. 2004). The basic answer remains the same some 20 years later. First, CT scanning is a largely non-destructive way to investigate internal anatomy. As noted above, there are stains which may not be totally reversible, but a light green specimen is far more useful to posterity than one that has been completely skeletonized or from which the muscles have been removed for volumetric measurement. Non-destructive investigations are a boon when the fish is no longer extant in the wild, when studying type material, or when there are only a few specimens in collections. Perhaps less obviously, this sort of non-destructive sampling is useful when many specimens are being assayed, for example, in an ontogenetic study or population-level study. Imaging hundreds of specimens is relatively simple; dissecting them destroys specimens that might be useful in other contexts.

A second reason is that there are a host of quantitative measurements that are simple with CT data that are either impossible or time-consuming and expensive with dissection. These include measuring the volume of bones, muscles, or other soft tissues. The relative size of bones could be used for comparative taxonomic work, but these are all linear measurements. It is possible to separate taxa into completely different axes when the volume of bones is considered. Other parameters describing the skeleton are easy to reach with CT; for example, it is simple to compare the density of bones or otoliths. Within the software 3D Slicer (Fedorov et al. 2012), there is a free plug-in called Segment Geometry that will calculate the second moment of the area along any bone (Huie et al. 2022). This is a particularly relevant measure for some bones in catfishes, like opercular spines, which are likely subject to selective pressures to resist bending. Thirdly, while dissection and skeletonization make 2D morphometrics possible, 3D remains difficult. The 3D morphometrics are baked into the Slicer analysis tool in the SlicerMorph plug-in. For John Dorys (*Zeus faber*) 2D morphometrics is probably fine, but the vast preponderance of catfishes is 3D and would benefit from the inclusion of that out-of-viewing plane data that CT scanning offers (Buser et al. 2018). Yet another reason why CT beats the knife is that it is a completely digitally based analysis tool. These include 3D geometric morphometrics, the relatively simple warping analyses that offer reconstructions of ancestral states or of hypothetical morphology, and the more rarefied finite element analysis of form that makes predictions about the distribution of deformation or the likelihood of structural failure in complex structures. This engineering tool is now also available as a free, open-source piece of software in the form of FEBIO. To offer a catfish example of how this might be used, we point out the pectoral-fin spine as a multifunctional device. It might be for stabbing or for grasping. Whether a structure is suited to one or the other would fall out of a finite element analysis. The availability of open-access data increases the reproducibility of morphological data because there are often several or even many specimens of a particular species available for analysis.

What Do I Do with the Data?

The images can only be processed and visualized using specialized volumetric rendering software. There are several software options, but some of the most used in ichthyology are VG Studio Max (paid), dragonfly (paid), and 3D Slicer (open source). In general, the images are imported into the software and values for the axes x, y, and z should be entered to reflect the size of each voxel (3D pixel) and ensure that the size of the model is accurate. In the beginning, the images look quite obscure since the specimens were wrapped in some material to separate them during scanning. Before the model is created, any excess volume around the specimen is cropped out to decrease the amount of data that needs to be rendered. When the model renders, a few settings are changed. The model is set to a Phong rendering, meaning that it is set to a specific type of shading that helps smooth out the multi-surface shapes, and the camera is set to parallel rather than perspective. The histogram showing the density gradient is then adjusted to make only the skeleton visible. After this, the viewing window is set to "slices left," and each of the three cross-section windows is set to the "scene" coordinate system. The main render window is set to the same setting. Then, we proceed to unlock the scene so that the model can be rotated to guarantee that the specimen lines up as straight as possible along each axis. Once this is complete, the scene will be locked again.

The bones of the specimen are virtually dissected using regions of interest to identify each individual structure. If the structure does not intersect with other structures, the "region growing tool" can be used to make an initial selection. This is done by increasing or decreasing the tolerance until only the intended structure is selected. The "erode/dilate" tool is used to expand the selection to ensure the entire structure is selected. This selection is used to create a new region, and fine adjustments to the histogram are made to maximize the visibility of the entire structure in the rendering. The region can be assigned a color and a name. If the structure is touching another structure, the "draw" tool is used to create a buffer between them by selecting the edges of the desired structure on each slice and creating a region from that selection. The "region growing" tool can be used to grab the unselected areas of the structure. The "avoid existing regions" option needs to be selected in the "region growing" tool palette to do this. The selection is expanded with the "erode/dilate" tool, and a new region is created. The new region and region, including the edges, are combined to create a region of the entire structure. As before, this is adjusted, and a color is assigned and given a name.

Finally, it cannot be stressed enough that all data should be deposited in a publicly accessible repository as soon as possible after being gathered. Our field is immeasurably enriched by the sharing of data, and with citable document identifiers (DOIs), it is possible for depositors to get "credit" for the data they produce.

What is Next?

More recent studies have proven the use of computed tomography images to reconstruct moving morphology. The technique is known as X-Ray Reconstruction of Moving Morphology (XROMM) animation and uses a workflow of marker tracking, reconstruction, and computed tomography (CT) mesh unification. Marker X-ray trajectories are tracked using automation-assisted tracking tools (in specific software), implanting beads in the individuals. Muscle strain is measured using fluoromicrometry, which measures the distance between distant markers. Each skeletal element of interest is segmented from the CT scans and exported marker 3D coordinates in CT space. All the analysis are performed using XROMM animation workflow including marker smoothing and aligning the smoothed 3D marker coordinates with the corresponding CT marker coordinates (unification; electronic supplementary material; movie S3) to animate each skeletal CT mesh (Olsen et al. 2019; Weller et al. 2020). This combination of tools and techniques allows the researcher to create an animation to understand fish behaviors like feeding and locomotion. Some work has been done to review the feeding mechanisms and swallowing in Channel Catfish (*Ictalurus punctatus*). Olsen et al. (2019) found that motions were more coordinated (by 20–29%) during prey capture than during prey transport, and they speculate that capture is more coordinated to create a single fluid flow into the mouth, while transport is less coordinated so that the cranial elements can independently generate multiple flows to reposition prey. Weller et al. (2020) used XROMM to track prey transport in the same species (*Ictalurus punctatus*). They found that the food enters the oral cavity at high velocities as a continuation of suction and stops in the approximate location of the branchial basket before moving in a much slower, more complex path toward the esophagus. They also observe how once the prey is in the esophagus the transport is correlated with gulping motions.

CT Scanning of Fishes, Including Catfishes

Are Catfish Scans Useful?

The history of CT scans of fishes started with researchers using the technique to answer simple morphological questions relevant to one species and/or to aid with the identification of specific structures, and it evolved into answering more complex questions at a multitaxon level. Early examples include renderings of cartilaginous fishes in morphological studies with a functional cast (Summers 2000; Maisey 2001) and a detailed comparative study of vertebral microstructure (Sakashita et al. 2019). In 2000, a medical journal paper mentioned the use of CT scans to identify catfish bones obstructing the airway of a human (Lue et al. 2000). In 2010, Long and Stewart used computed tomography scans along with electron microscope microphotographs, description of techniques and a sample of otoliths to verify the identity of the otolith used for establishing the age in Channel Catfishes (*Ictalurus punctatus*). Leach et al. (2014) examined a pair of conjoined hybrid catfish (*Ictalurus punctatus* ♀ x *Ictalurus furcatus* ♂) of unequal size and development using computed radiography and computed tomography antemortem. The images revealed that the smaller fish consisted largely of bony tissues. A cavity was noted in the smaller fish that communicated with the abdomen of the larger. The specimens appear to be a typical example of parasitic twin formation in fish. Eshra and El Asely (2014) reviewed the computed tomography of four species of catfish representatives of Bagridae, Clariidae, Malapteruridae, and Mochokidae to understand and compare their feeding apparatus. The CT renderings allowed the authors to hypothesize about the feeding mechanisms based on the fish morphologies. Quan et al. (2020) reviewed the locking mechanism in one species of Thorny Catfishes (Doradidae) to characterize the internal structure of the rotation joint of the pectoral spines by using a micro-CT scan of the spine base and the pectoral girdle. In the same paper, the authors reviewed the structure of the mid-lateral scutes, and they compared the images of the micro-CT scan with those of SEM to understand their degree of mineralization and their potential cutting ability for an active defense capability.

Several researchers have used computer tomography of catfishes to illustrate the morphological features of new species, and for some of the studies, the data have been used in a phylogenetic context. In 2003, Schaefer studied the phylogenetic placement of *Lithogenes villosus*. The species was described based on only one specimen (holotype), and thus, morphological comparisons and scoring of characters were difficult to accomplish. Also, given the time of the paper, the resolution of the CT scan was not ideal for making the comparisons. Using CT scanning, Schaefer (2003) placed *L. villosus* as a sister to all other loricariids exclusive of Astroblepidae. Luckenbill and Lundberg (2009) provided some images of *Cranoglanis boudeirus* to aid in the resolution of the relationships between the Cranoglanididae and other catfishes. The authors did not provide descriptions but facilitated the images in an era in which the sharing of CT scan images was not yet broadly used.

Lundberg et al. (2014) used CT to illustrate, describe and compare the bony skeleton of *Kryptoglanis* to determine its phylogenetic position. *Kryptoglanis* was discovered and described as an unusually puzzling catfish from the ground and surface water in Kerala State, southwestern peninsular India. The catfish morphology is highly modified, making this species unique among other catfishes. The species is characterized by a combination of traits that are not unique within catfishes but that distinguish it from others. At the time, there were three specimens available in collections, thus presenting an opportunity to use CT scanning to explore its morphology and phylogenetic placement. The authors made comparisons of *Kryptoglanis* to all named living and fossil siluriform taxa, but given the resemblance of the species to members of the Siluridae, they made an intensive assessment of the synapomorphies of Siluridae relative to character states of *Kryptoglanis*. They found a few features in the species interpretable as synapomorphies and informative on its phylogenetic position, as well as many anatomically complex autapomorphies and a dense cluster of distinctive apomorphies in the snout region. Some of the examples of its complex morphology and novel conditions are in the mesethmoid and frontal, nasal, antorbital, parasphenoid, orbitosphenoid, the supraoccipital-Weberian neural arch joint, the occipital wall, occipito-vertebral joint, the long accessory processes of the compound Weberian vertebra, exoccipital-ossified transcapular ligament structure, the hyomandibular-sphenotic articulation, the quadrate, opercle, anterior ceratohyal, and the posttemporo-supracleithum joint with the fourth transverse process.

Carvalho et al. (2017) described a new species of blind banjo catfish, *Xiliphius sofiae*. Their description is based on one specimen collected in the Amazon River near Iquitos in Peru; thus, using computed tomography was paramount to include a description of the osteology for this new species without destroying the specimen for future studies. Carvalho et al. (2017) found that *X. sofiae* has the eight features proposed by Friel (1994) as synapomorphies for the genus related to the condition of the cranium, pterotic, premaxilla, supraopercle, ceratohyal, nasal, papillae on the lower jaw, and unculiferous tubercles on the skin. Observations of all these characteristics were only possible due to the use of CT. Also, the authors reported a new feature for Aspredinidae present in the species, a lateral ethmoid with an enlarged internal chamber housing an olfactory bulb.

In 2017, Lundberg et al. provided a comparative description of the osteology of *Satan eurystomus*, a North American catfish from the family Ictaluridae commonly known as Widemouth Blindcat. The species is distributed in the Edwards Aquifer in Texas but has not been collected for 30 years. There are few specimens in museum collections, making it a hard species to study and place phylogenetically within the family. The authors prepared a complete detailed anatomical revision of the osteology and compared its characteristics to those of the other blindcats in the family as well as to the flathead (*Pylodictis olivaris*). Their observations are not conclusive but provide evidence and data that could be used in a phylogenetic study.

De Pinna et al. (2018) described a new species of *Copionodon* from the Paraguaçu River basin in the Diamantina Plateau, Bahia State, northeastern Brazil. The authors included CT scan images

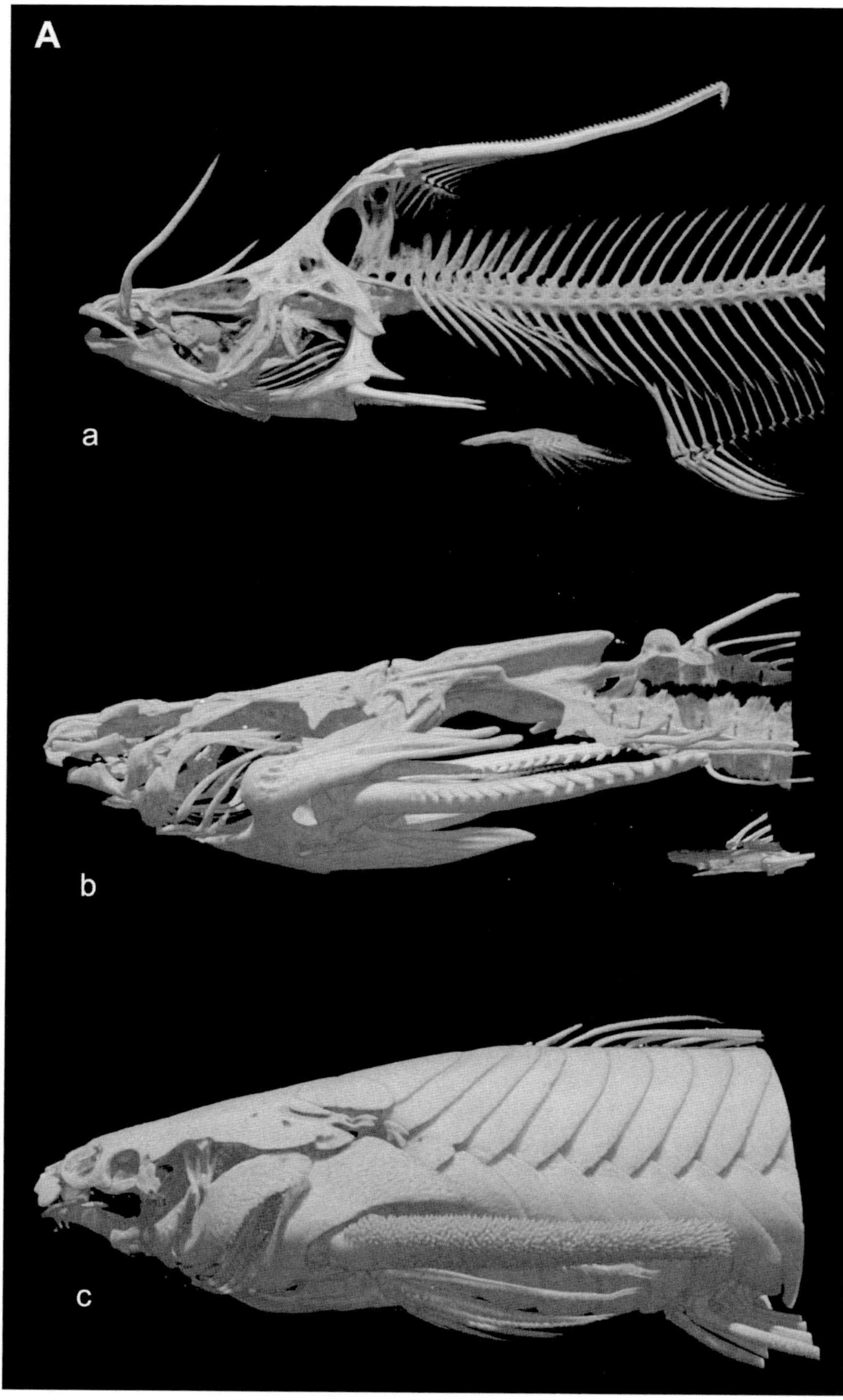

Figure 1 contd. ...

...Figure 1 contd.

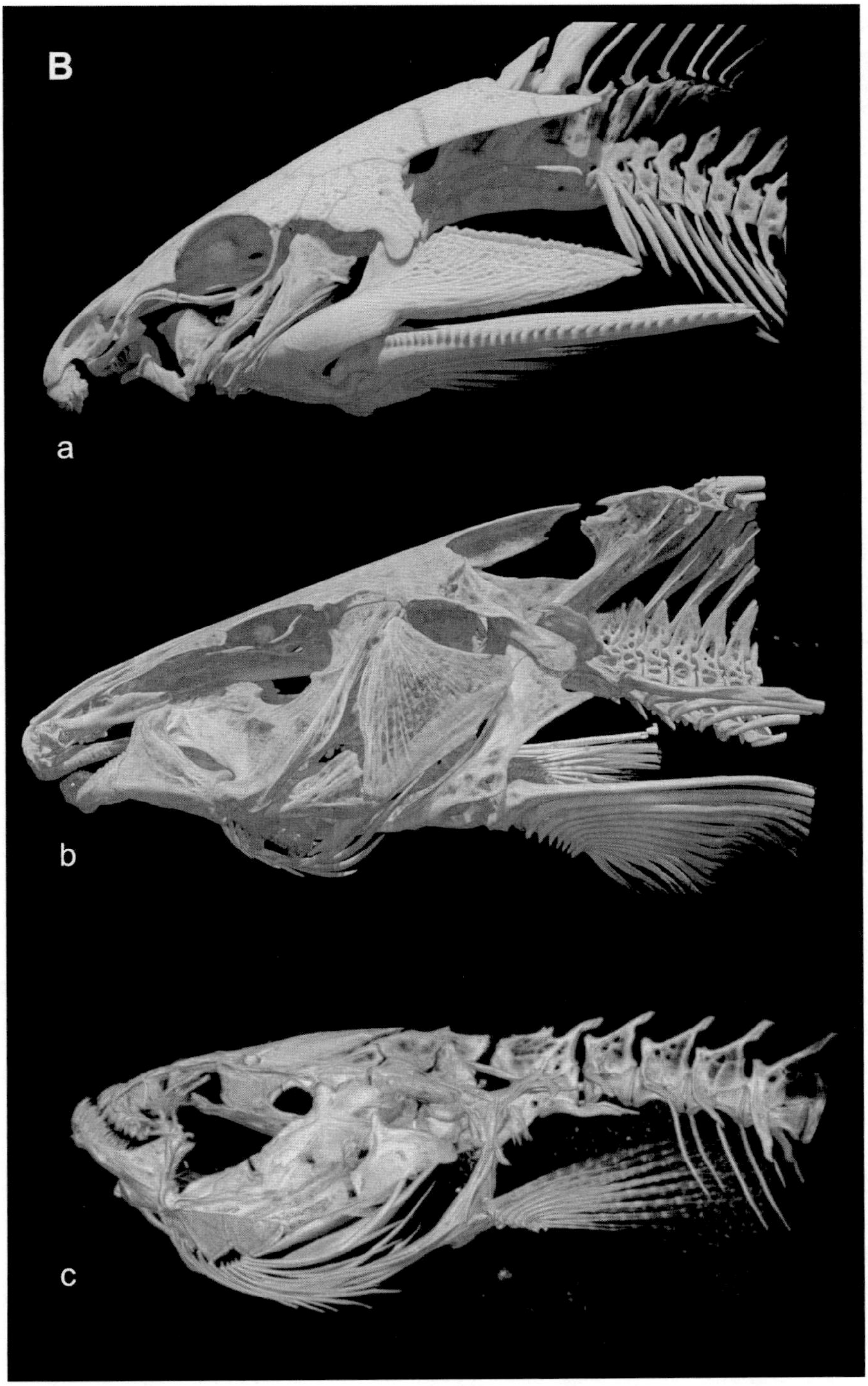

Figure 1. Lateral view of the anterior portion of the body in six families of Siluriformes. (A) a. Auchenipteridae: *Tetranematichthys wallacei* (ANSP 128688). b. Aspredinidae: *Bunocephalus* sp. (ANSP 160240). c. Callichthyidae: *Callichthys callichthys* (ANSP 53343). (B) a. Mochokidae: *Synodontis angelicus* (CU 91394); b. Pimelodidae: *Aguarunichthys torosus* (MZUSP 50398); c. Kryptoglanidae: *Kryptoglanis shajii* (ANSP 192026). Original CT scans processed by K. Luckenbill.

of the holotype, allowing an "unprecedented clarity" into details of Copionodontinae anatomy. Reis et al. (2019) described a new species of *Trichomycterus* from the Rio Doce in Southeastern Brazil. For a detailed description of the anatomy, the authors included images from a CT scan specimen. The morphology of *T. astromycterus* resembles that of the monotypic Chilean genus *Bullockia*, suggesting a possible relationship between *T. astromycterus* and *Bullockia* at a more inclusive level. However, the size of the genus *Trichomycterus* (more than 170 spp.) has made its anatomical study scattered and incomplete and additional work needs to be done to determine how these similarities reflect ancestrality. De Pinna and Keith (2019) published a species description within the genus *Mastiglanis* from French Guyana. They reviewed phylogenetically relevant characters using CT scan images. Dopazo et al. (2019) described a new species of *Pseudobunocephalus* from the lower Tocantins and Mearim River drainages. The authors presented anatomical comparisons of the new species of banjo catfish (*P. timbira*) to the other three species of *Pseudobunocephalus* and included CT images for each one of those. Carvalho and Reis (2020) described a miniature species of *Acanthobunocephalus* from the lover Purus River Basin in the Amazon basin of Brazil. The standard length for the holotype of the species measures 21.7 mm, making CT scanning a useful tool to analyze the osteological features of the species *Acanthobunocephalus scruggsi.*

Morphological Changes: How Catfishes Compare to Each Other?

The order Siluriformes includes more than 3,500 living species distributed in 40 families (Armbruster 2011). Their morphological diversity is vast, ranging from what everybody will identify as a catfish, a typical torpedo-shaped fish with no scales and long whiskers, to miniature species with depressed bodies and modified cranial structures with small whiskers, some with plates or thorny scutes and some with naked bodies. The immense morphological variation has allowed catfishes to conquer a variety of habitats and be distributed in freshwater, estuarine and marine environments. Today, they occur on every continent (excluding Antarctica, known as fossils; Grande and Eastman 1986). The body's morphology is modified in different species to their success and survival in such different environments. The use of computed tomography and micro-computed tomography has allowed researchers to explore and understand the morphological variation of many of those catfishes. In this section, we illustrate a subsample of the vast variations using CT scan images. Figure 1 shows a lateral view of the anterior portion of the body of Auchenipteridae, Aspredinidae, Callichthyidae (Figure 1A) and Mochokidae, Pimelodidae, and Kryptoglanidae (Figure 1B). In the figure, it is possible to observe variations in the body shape among catfishes, ranging from very depressed (*Bunocephalus*) to deep body (*Synodontis*) catfishes. Additionally, we can also observe the external coverage (plates) in *Callichthys* and the changes in bone density, which are sturdy in *Aguarunichthys* and poorly ossified in *Kryptoglanis* or *Tetranematichthys*.

Cranium

The cranium is composed of the cranial roof and the floor and lateral walls of the dermatocranium and endocranium. Bones in the cranial roof include the epiotics, extrascapulae, parieto-supraoccipital, pterotics, sphenotics, frontals, lateral ethmoids, mesethmoid, and nasals. Bones included in the dermatocranium and endocranium include basioccipital, exoccipitals, prootics, orbitosphenoid, pterosphenoid, mesethmoid, lateral ethmoids, vomer and parasphenoid (Arratia 2003). Major variations in the cranium, such as shape and points of suture, are used in the phylogenetic context to trace ancestrality. In catfish, the thickness of the bones can range from very thick to poorly ossified (Arratia 2003). The sutures can be smooth, retaining the sutura harmonica, or they may develop as dentate and or serrate sutures in some taxa (Arratia 2003).

The components of the cranium are similar in all catfish families; however, when it comes to shape and structure (degree of fusion and connections) of those bones, great modifications are found, and, in most cases, those can tell a story about the species including some aspects of their feeding and reproduction strategies. In Figure 2, we present the general structure of the cranium

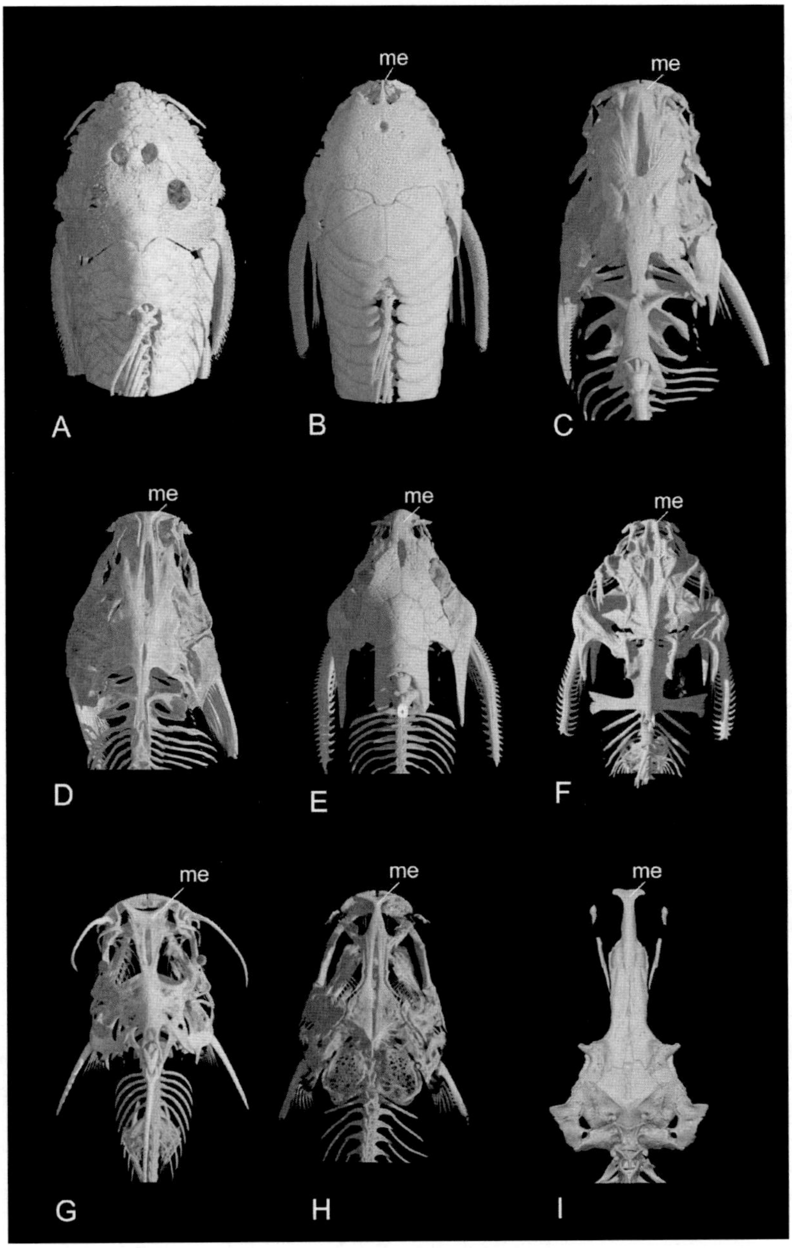

Figure 2. Dorsal view of the neurocranium of (A) Loricariidae: *Chaetostoma* sp. (AUM uncatalogued). (B) Callichthyidae: *Callichthys callichthys* (ANSP 53343). (C) Bagridae: *Horabagrus brachysoma* (SAIAB 59825). (D) Pimelodidae: *Aguarunichthys torosus* (MZUSP 50398). (E) Mochokidae: *Synodontis angelicus* (CU 91394). (F) Aspredinidae: *Bunocephalus* sp. (ANSP 160240). (G) Auchenipteridae: *Tetranematichthys wallacei* (ANSP 128688). (H) Nematogenyidae: *Nematogenys inermis* (MNRJ 22015). (I) *Trichomycterus astromycterus* (MZUSP 123341). Original CT scans processed by K. Luckenbill. Images of *T. astromycterus* published by Reis et al. (2019). Abbreviations: me – mesethmoid.

of *Chaetostoma* sp., *Callichthys callichthys*, *Horabagrus brachysoma*, *Aguarunichthys torosus*, *Synodontis angelicus*, *Bunocephalus* sp., *Tetranematichthys wallacei*, *Nematogenys inermis*, and *Trichomycterus astromycterus* in dorsal view. Differences in the structure are as simple as the covering of the bones by dermal plates (Figure 2A), the presence of multiple ridges and crests to accommodate muscles [Figure 2F and Ictaluridae (not illustrated)] and reduction of size, areas of contact and ossification degree (Figure 2G, H, and I). One could choose any bone to explore specific differences in the neurocranium and to try to understand evolution. The mesethmoid is the anterior-most bone in the neurocranium. The anterior portion of the bone bears a cornua, which supports the attachment of the premaxilla; thus, modifications on the mesethmoid shape will likely be linked to modifications on the premaxilla shape. A common shape for the mesethmoid could be observed in *Horabagrus* and *Aguarunichthys* (Figure 2C–D), where the anterior portion of the bone expands into cornua (on each side) followed by a "neck" (variable in size and shape) and with the lateral portion sutured to the lateral ethmoids and the posterior end sutured to the frontals. In families like Amphiliidae (not illustrated) or Trichomycteridae (Figure 2I), we observe an extremely modified mesethmoid with a thin posterior end, a long "neck," and reduced cornua. In Callichthyidae, we observe a fully reduced mesethmoid (Figure 2B). Another example of modification occurs in *Kryptoglanis* where the dermal component of the mesethmoid is dorsally and posteriorly reduced, as well as a truncation of the ventral floor of the mesethmoid (Lundberg et al. 2014). In *Kryptoglanis*, the mesethmoid cornua are narrow and long, bearing a subrectangular large premaxilla with narrowly conical, tall, slightly recurved, and sharp-tipped teeth.

Dentition

Catfish teeth are usually cardiform, meaning numerous, short, fine, and pointed. The bones bearing teeth are the maxilla (rarely), premaxilla, dentary and, on occasion, the pharyngeal plates. However, specialized teeth can be found in many catfish lineages. The Diplomystidae have simple, conical teeth (Arratia 1987); in some Sisoridae, the teeth are incisiviform with a concave upper margin (Adriaens 2003). In Bagridae, teeth can be molariform (Adriaens 2003), and in Scoloplacidae and Loricariidae, teeth resemble a cup shape (pers. obs). The most diverse tooth morphology can be observed in algae-scraping species, where teeth are generally elongated (Adriaens 2003).

In Figure 3, we can see dentition for *Aguarunichthys*, *Leporacanthicus*, *Chaetostoma*, *Trichomycterus*, and *Copionodon*. In the images, we can see variation from the dentition of a fish like *Aguarunichthys torosus*, where premaxillary and dentary teeth are abundant, long, and thin. *Aguarunichthys torosus* feeds on other fishes. Some modifications of teeth can be found in Loricariidae (Figure 3B and C), Trichomycteridae (Figure 3D and E), Kryptoglanidae (Figure 1B and C). Most Loricariidae will exhibit a dentition associated with a sucker mouth allowing the species to feed on algae in hard-to-reach substrates. The movement of premaxilla and dentary bones are independent, which makes them excellent scrapers. The teeth are thin, allowing the "brushing" of surfaces. However, a genus like *Leporacanthicus* has teeth specialized for scavenging the bottom of habitats in search of food (Figure 3B and C). For Trichomycteridae, we presented two different types of teeth. In *Copionodon*, the teeth are spatulated, and the dentary and premaxillary teeth are somewhat similar to each other in shape and arrangement, gradually increasing in size towards the symphysis (de Pinna et al. 2018). Other species of *Copionodon* are mainly algae eaters, but aquatic insects are also found as food items (Zanata and Primitivo 2014). In *Trichomycterus*, we observe that the dentary and premaxillary teeth are similar in shape and size, arranged in three irregular rows extending too close to the tip of the coronoid process; most jaw teeth are incisiform. We could not find any specific information on the eating habits of *T. astromycterus*. However, when looking at the diet of various *Trichomycterus* species, we found that most of them feed on microinvertebrates, insects, and larvae, while some also feed on small fish and algae (Figure 3D and E).

Hyoid Arch and Branchiostegal Rays

The hyoid arch and its associated musculature play an important role in mechanisms like respiration and suction feeding. The hyoid arch is differentiated into a dorsal hyosymplectic and a ventral hyoid bar (Adriaens and Verraes 1996). The hyoid bar and branchiostegal rays are visible in the ventral view in the CT scan images in Figure 4, and the variable morphology is evident among the species illustrated. The elements that can be observed in this view are hypohyals, ceratohyal and parurohyal. Major variations are observed in the parurohyal, ranging from a slim, long bone (Figure 4B) to a

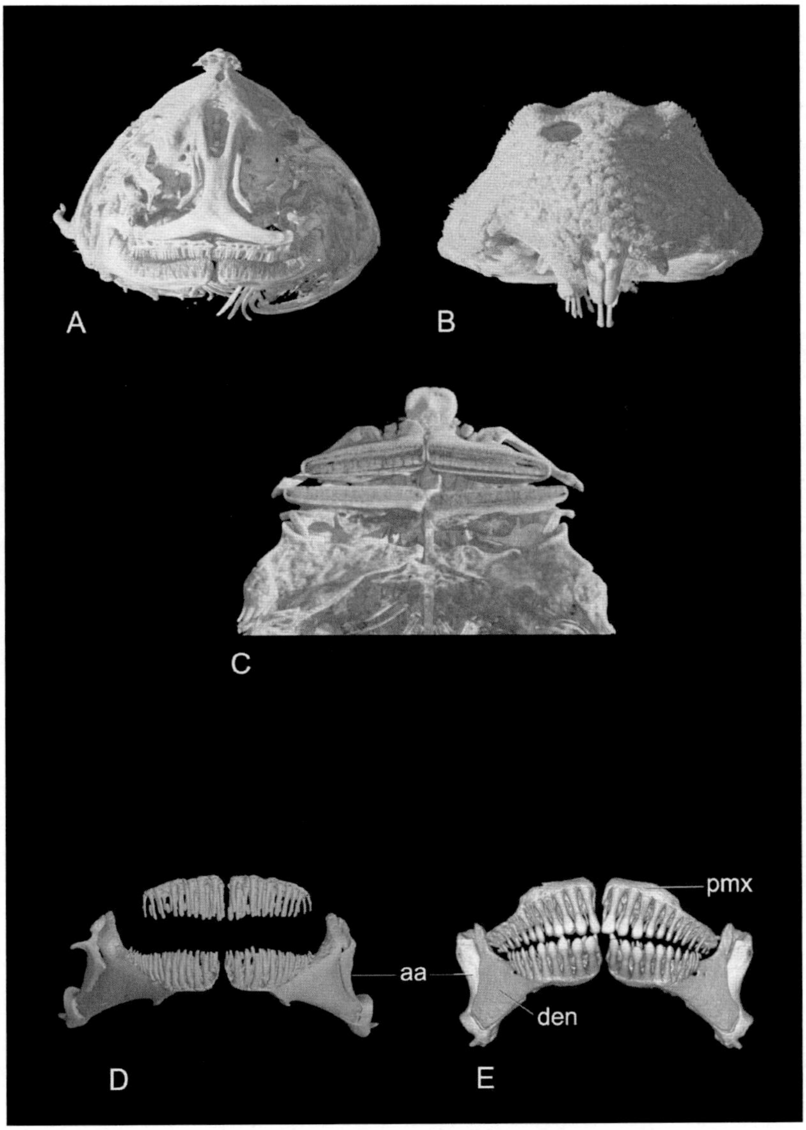

Figure 3. Detail of dentition of (A) Pimelodidae: *Aguarunichthys torosus* (MZUSP 50398). (B) Loricariidae: *Leporacanthicus joselimai* (AUM 58525). (C) Loricariidae: *Chaetostoma* cf. *milesi* (AUM uncat.). (D) Trichomycteridae: *Trichomycterus astromycterus* (MZUSP 123341). (E) Trichomycteridae: *Copionodon exotatos* (MZUSP 12063). Images of *Copionodon* published by de Pinna et al. (2018); images of *Trichomycterus* published by Reis et al. (2019). Other images were originally processed by K. Luckenbill. Abbreviations: aa – anguloarticular; den – dentary; pmx – premaxilla.

small, triangular (Figure 4A, C, D, F, G, and I), a horizontally elongated triangular (Figure 4E) and a vertically elongated form (Figure 4H).

Another conspicuous difference among the species is the position of the branchiostegal rays in relation to the ceratohyals. In *Callichthys*, the last branchiostegal ray is not in direct contact with the anterior ceratohyal and in *Synodontis* and *Chaetostoma,* the last branchiostegal ray is attached to the posterior ceratohyal instead of the anterior ceratohyal like in the other species illustrated (Figure 4). The branchiostegal rays and associated musculature contribute to the gill chamber expansion during ventilation to the suction pump process (Farina et al. 2015).

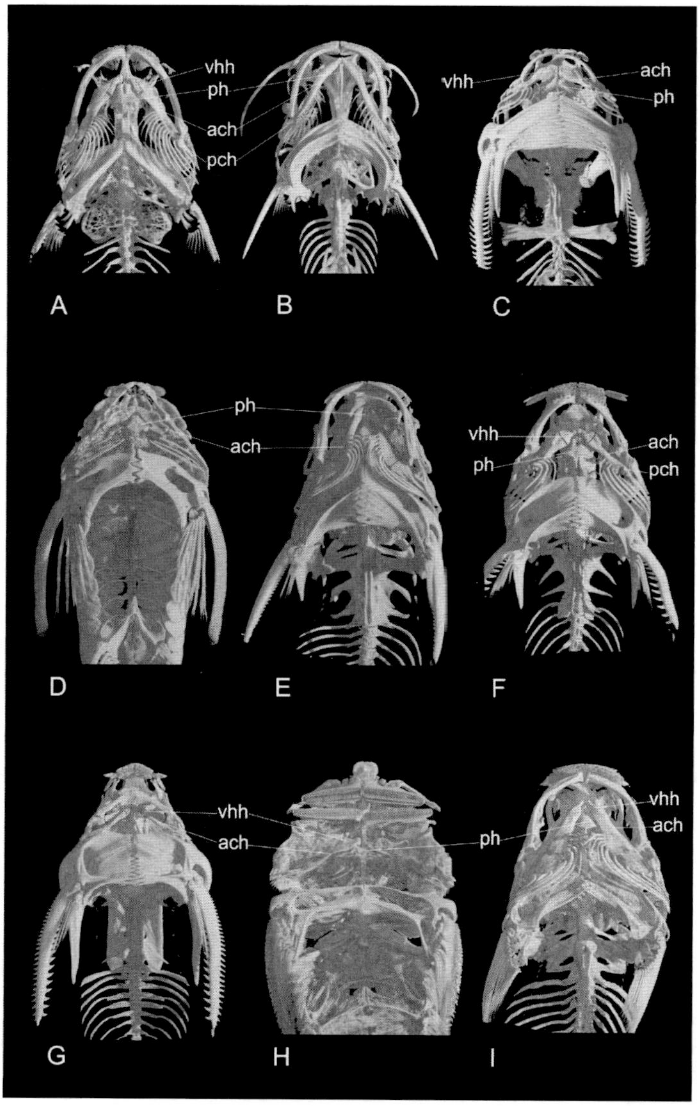

Figure 4. Ventral view of the anterior portion of the body. (A) Nematogenyidae: *Nematogenys inermis* (MNRJ 22015). (B) Auchenipteridae: *Tetranematichthys wallacei* (ANSP 128688). (C) Aspredinidae: *Bunocephalus* sp. (ANSP 160240). (D) Callichthyidae: *Callichthys callichthys* (ANSP 5334). (E) Bagridae: *Horabagrus brachysoma* (SAIAB 59825). (F) Austroglanididae: *Austroglanis gilli* (SAIAB 58357). (G) Mochokidae: *Synodontis angelicus* (CU 91394). (H) Loricariidae: *Chaetostoma* cf. *milesi* (AUM uncat.). (I) Pimelodidae: *Aguarunichthys torosus* (MZUSP 50398). Original CT scans processed by K. Luckenbill. Abbreviations: ach – anterior ceratohyal; pch – posterior ceratohyal; ph – parurohyal; vhh – ventral hypohyal. ph is wrongly labelled in *Nematogenys*.

Pectoral Girdle

The pectoral girdle consists of paired dermal posttemporo-supracleithra and cleithra and a paired chondral element, the scapula-coracoid plus mesocoracoid arch (Arratia 2003). The opposite halves of the girdle meet in a median suture ventrally, forming the pectoral symphysis. The suture can be formed by a smooth surface, or the contact surfaces may be strongly dentated or serrated (Arratia 2003). Muscles for the movement of the pectoral fins are attached to these bones, thus variation in the bones will likely be linked to the size and strength of the pectoral fins. Muscles associated with the pectoral girdle are the *abductor* and *adductor superficialis*, *abductor* and *adductor profundus*, *arrectors ventralis* and *dorsalis*, and, at least in Doradidae, the *abductor rotator* (Arce H. 2015). In some species of catfishes, the coracoid and cleithrum are ventrally covered by the *arrector ventralis*, *arrector dorsalis* and *abductor superficialis* since those bones constitute the area of origin of the muscles. In other genera (e.g., *Amblydoras*, *Bunocephalus*), the origin is in the dorsal surface of the cleithrum and coracoid, and the muscle is not exposed (Arce H. 2015).

Examples of CT scans of the pectoral girdle are illustrated in Figure 4, where variations regarding the overall shape, the shape of the posterior end of the coracoid, the size ratio between the coracoid and the cleithrum, the suture between the left and right halves of the coracoid, the crests on coracoid/cleithrum for muscles attachment and the length of the posterior cleithral process can be observed. The posterior end of the coracoid is some species forms a ~ 90-degree angle between the left and right half of coracoid (*Nematogenys* and *Aguarunichthys*, Figure 4A, I); in others, the posterior end is a semicircle (*Tetranematichthys* and *Callichthys*, Figure 4B, D) and it can also be observed as a straight edge (*Bunocephalus*, *Horabagrus*, *Austroglanis*, *Synodontis* and *Chaetostoma*; Figure 4C, E–H). The suture between the left and right halves of the coracoid (pectoral symphysis) also varies among species. It ranges from a straight line to an interdigitating suture to a suture where the halves overlap (Figure 4).

Pectoral Spine

The pectoral spine has been recognized as a useful structure in phylogenetic investigations, given its complexity and variation (Kubicek et al. 2019). Morphology of the spine varies significantly across Siluriformes, including rays with large saw-like projections, smooth, blade-like spines, cylindrical spines covered with odontodes, tooth-like structures composed of dentine with an enamel cap and flat and highly flexible structures heavily pinnate providing support to a keratinized skin pad (Kubicek et al. 2019). Some range of variation of pectoral spines is illustrated in Figure 4, where one could find smooth and sturdy spines (B and I), spines with serrae located in the posterior portion (A, E, and F), spines with serrae in the anterior and posterior portion (C and G) and spines with odontodes (D and H).

Weberian Apparatus

The Weberian apparatus is the most remarkable synapomorphy of the Otophysi (Sagemehl 1885), and it is an anatomical modification that improves the fish audition. The apparatus is composed of the compound vertebra (fused anterior vertebrae) and a paired chain of one to four ossicle and ligaments joining the swim bladder to the inner ear (Weber 1820). The Weberian apparatus makes catfishes hearing specialists since it helps them detect sound pressure and improve their hearing ability (Lechner and Ladich 2008). An enormous variability in the morphology of the Weberian ossicles and the gas bladder can be observed in Siluriformes (Lechner and Ladich 2008); some of that variability is illustrated in Figures 2 and 4. The main differences can be observed when looking at the compound centra and the anterior and posterior limb transverse processes. These two characteristics are related to the morphology of the gas bladder. In species of Callichthyidae and Loricariidae, the gas bladder is paired and tiny and, in some cases, hardly detectable (Lechner and Ladich 2008). In those families, the compound centra tends to be shorter, and the processes are thinner and shorter than in species with free gas bladders (Figure 4D and H). In the range of variation, we can observe anterior and posterior processes that are elongated and directed posteriorly (Figure 4B and G) or flat and broad, positioned at a 90-degree angle (Figure 4C), greatly expanded with well-defined canals or a porous surface (Figure 4A and I), or a combination of morphologies (Figure 4E and F).

Soft Tissues

As mentioned above, soft tissues can be observed by using different techniques during the scanning stage. This kind of data has proven successful, allowing the observation of musculature and other tissues (Figure 5). Data are being collected thanks to initiatives like the oVert project; however, the analysis and study of these structures for catfish are still awaiting publication. The resemblance

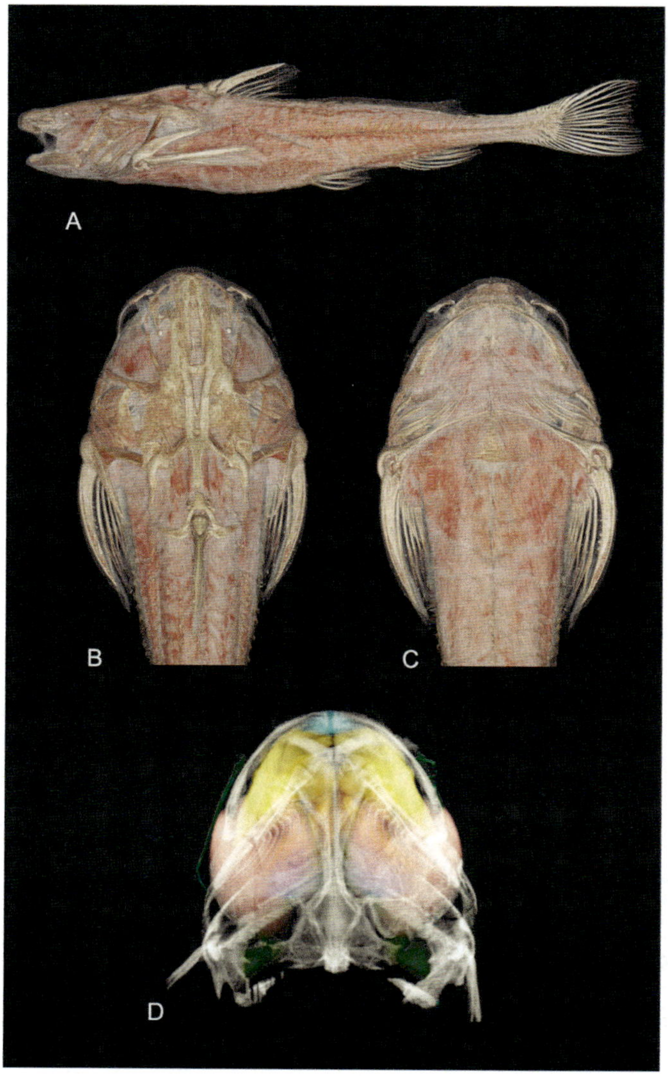

Figure 5. CT scan images of *Acrochordonichthys rugosus* (UF 167005) and *Lacantunia enigmatica* (ANSP 178695) after treatment with iodine to visualize musculature. (A) lateral view *of A. rugosus*; (B) detailed dorsal view of *A. rugosus*; (C) detailed ventral view of *A. rugosus*; (D) cranial musculature of *L. enigmatica* with and X-ray view of the bones; images prepared by K. Innamorato.

of the images to those of clear and stain specimens is undeniable, and the possibilities of studying these structures via digital dissection are exciting and will add a new layer of information for the understanding of the morphological complexity in Siluriformes.

Final Thoughts

CT scanning has been around in the medical field for about 50 years but has only seen applications in ichthyology in the last 25 years and has enormously advanced the study of morphological characteristics in different contexts. A noninvasive technique that allows both broad and deep sampling of taxa represents a sea change in our ability to make morphological inferences and to replicate internally and across prior literature. The ready availability of micro-CT scanners, which have a large upfront cost but small running costs, makes gathering these data easier and available to a more diverse cohort of scientists. The same can be said for the rise of low-cost or free software solutions for visualization and data quantitation. We foresee a near future where the skeletal anatomy of fish is an open book, and the soft anatomy is becoming more accessible. This will lead to revolutionary changes in the types of analyses that can be imagined, as the taxonomic breadth will be broad, while the detail will be very fine. Furthermore, as these data sets become more widely available, new ways of analyzing the data will arise, along with new techniques for quantification and fresh insights into interpreting the evolutionary, developmental, and morphological histories of fishes in general and catfishes in particular.

Acknowledgements

The authors thank Kyle Luckenbill for providing images and helping with some of the technical writing for the chapter. We thank Zachary Randall for preparing the images of *Acrochordonichthys rugosus*. We also thank Vitor Abrahão, Mario C.C de Pinna, and Vinicius J.C. Reis for providing images of *Trichomycterus* and *Copionodon.*

References

Adriaens, D. and W. Verraes. 1996. Ontogeny of the hyoid musculature in the African catfish, Clarias gariepinus (Burchell, 1822) (Siluroidei: Clariidae). Zool. J. Linn. Soc. 121: 105–128.

Adriaens, D. 2003. Feeding mechanism in catfishes. pp. 3–46. *In*: Arratia, G., B.G. Kapoor, M. Chardon and R. Diogo (eds.). Catfishes. Science Publishers, Inc; Enfield.

Arce H., M. 2015. Mandibular, hyoid and pectoral musculature of thorny catfishes (Siluriformes: Doradidae). Proc. Acad. Nat. Sci. Phila. 164: 229–277.

Armbruster, J.W. 2011. Global catfish diversity. Am. Fish. S. S. 77: 15–37.

Arratia, G. 1987. Description of the primitive family Diplomystidae (Siluriformes, Teleostei, Pisces): morphology, taxonomy and phylogenetic implications. Bonner Zool. Monogr. 24: 1–120.

Arratia, G. 2003. Catfish head skeleton— An overview. pp. 3–46. *In*: Arratia, G., B.G. Kapoor, M. Chardon and R. Diogo (eds.). Science Publishers, Inc, Enfield.

Buser, T.J., B.L. Sidlauskas and A.P. Summers. 2018. 2D or Not 2D? Testing the Utility of 2D Vs. 3D Landmark Data in Geometric Morphometrics of the Sculpin Subfamily Oligocottinae (Pisces; Cottoidea). Anat. Rec. 301: 806–818.

Carvalho, T.P., R.E. Reis and M.H. Sabaj. 2017. Description of a new blind and rare species of *Xyliphius* (Siluriformes: Aspredinidae) from the Amazon basin using high-resolution computed tomography. Copeia 105: 14–28.

Carvalho, T.P. and R.E. Reis. 2020. A New Miniature Species of *Acanthobunocephalus* (Silurifomes: Aspredinidae) from the lower purus river basin, Amazon Basin, Brazil. Copeia 108: 347–357.

Dopazo, V., T.P. Carvalho, R.E. Reis and W.B. Wosiacki. 2019. A new species of *Pseudobunocephalus* Friel, 2008 (Siluriformes: Aspredinidae) from the lower Tocantins and Mearim river drainages, North and Northeast of Brazil. Zootaxa 4586: 109–126. DOI: https://doi.org/10.11646/zootaxa.4586.1.5.

Eshra, E.A. and A.M. El Asely. 2014. Comparative osteology and three-dimensional computed tomography of Nile catfishes, in relation to feeding mechanisms. J. Morph. Sci. 3: 162–170.

Farina, S.C., T.J. Near and W.E. Bemis. 2015. Evolution of the branchiostegal membrane and restricted gill openings in actinopterygian fishes. J. Morph. 276: 681–694.

Fedorov, A., R. Beichel, J. Kalpathy-Cramer, J. Finet, J-C. Fillion-Robin, S. Pujol et al. 2012. 3D Slicer as an image computing platform for the quantitative imaging network. Magn. Reason. Imaging 30: 1323–1341.

Ford, K.L., J.S. Albert, A.P. Summers, B.P. Hedrick, E.R. Schachner, A.S. Jones et al. 2023. A new era of morphological investigations: reviewing methods for comparative anatomical studies. Integr. Org. Biol. 5: 1–13. https://doi.org/10.1093/iob/obad008.

Friel, J.P. 1994. A Phylogenetic Study of the Neotropical Banjo Catfishes (Teleostei: Siluriformes: Aspredinidae). Doctoral Dissertation. Durham, North Carolina, Duke University, 255 p.

Grande, L. and J.T. Eastman. 1986. A review of Antarctic ichthyofaunas in the light of new fossil discoveries. Paleontology 29: 113–137.

Huie, J.M., A.P. Summers and S.M. Kawano. 2022. SegmentGeometry: A tool for measuring second moment of Area in 3D Slicer. Integr. Org. Biol. 4: 1–10. https://doi.org/10.1093/iob/obac009.

Jeffery, N.S., R.S. Stephenson, J.A. Gallagher, J.C. Jarvis and P.G. Cox. 2011. Micro-computed tomography with iodine staining resolves the arrangement of muscle fibers. J. Biomech. 44: 189–92. doi: 10.1016/j.jbiomech.2010.08.027.

Kleinteich, T., A. Haas and A.P. Summers. 2008. Caecilian jaw closing mechanics - integrating two muscle systems. J. Roy. Soc. Interface 5: 1491–1504.

Kubicek, K.M., R. Britz and K.W. Conway. 2019. Ontogeny of the catfish pectoral-fin spine (Teleostei: Siluriformes). J. Morph. 280: 339–359. doi: 10.1002/jmor.20947.

Leach, C.A., M.W. Thomas, A.J. Cooley, G.C. Waldbieser, B.G. Bosworth, E.L. Torrans et al. 2014. Histological and computed tomographic evaluation of a parasitic conjoined twin in hybrid catfish (*Ictalurus punctatus* [Rafinesque] X *Ictalurus furcatus* [Lesueur]). J. Fish Dis. 37: 521–526. doi: 10.1111/jfd.12140. Epub 2013 Sep 2. PMID: 23991936.

Lechner, W. and F. Ladich. 2008. Size matters: diversity in swimbladders and Weberian ossicles affects hearing in catfishes. J. Exp. Biol. 211: 1681–1689. doi:10.1242/jeb.016436. PMID: 18456895.

Long, J.M and D.R. Stewart. 2010. Verification of otolith identity used by fisheries scientists for aging channel catfish. T. Am. Fish. Soc. 139: 1775–1779

Lue, A.J., W.D. Fang and S. Manolidis. 2000. Use of plain radiography and computed tomography to identify fish bone foreign bodies. Otolaryngol Head Neck Surg. 123: 435–438. doi: 10.1067/mhn.2000.99663. PMID: 11020181.

Luckenbill, K.R. and J.G. Lundberg. 2009. CAT scan-based images of the skeleton of the Asian Catfish *Cranoglanis* (Siluriformes: Cranoglanididae). Proc. Acad. Nat. Sci. Phila. 158: 297–299. https://doi.org/10.1635/053.158.0115.

Lundberg, J.G., K.R. Luckenbill, K.K.S. Babu and H.H. Ng. 2014. A tomographic osteology of the taxonomically puzzling catfish *Kryptoglanis shajii* (Siluriformes, Siluroidei, incertae sedis): Description and a first phylogenetic interpretation. Proc. Acad. Nat. Sci. Phila. 163: 1–41. DOI: 10.1635/053.163.0101.

Lundberg, J.G., D.A. Hendrickson, K.R. Luckenbill and M. Arce H. 2017. *Satan's* skeleton revealed: a tomographic and comparative osteology of *Satan eurystomus*, the subterranean Widemouth Blindcat (Siluriformes, Ictaluridae). Proc. Acad. Nat. Sci. Phila. 165: 117–173.

Maisey, J.G. 2001. CT-scan reveals new cranial features in devonian chondrichthyan "*Cladodus*" *wildungensis*. J. Vert. Paleontol. 21: 807–810. https://doi.org/10.1671/0272-4634(2001)021[0807:CSRNCF]2.0.CO;2.

Metscher, B.D. 2009. Micro-CT for comparative morphology: simple staining methods allow high-contrast 3D imaging of diverse non-mineralized animal tissues. BMC Physiol 9: 11. https://doi.org/10.1186/1472-6793-9-11.

Olsen, A.M., P. Hernández, A.L. Campand and E.L. Brainerd. 2019. Channel catfish use higher coordination to capture prey than to swallow. Proc. Roy. Soc. B.–Biol. Sci. 286: 1–10. http://doi.org/10.1098/rspb.2019.0507.

de Pinna, M.C.C., V. Abrahão, V. Reis and A. Zanata. 2018. A new species of *Copionodon* representing a relictual occurrence of the Copionodontinae (Siluriformes: Trichomycteridae), with a CT-scan imaging survey of key subfamilial features. Neotrop. Ichthyol. 16: e180049. https://doi org/10.1590/1982-0224-20180049.

de Pinna, M.C.C. and P. Keith. 2019. *Mastiglanis durantoni* from French Guyana, a second species in the genus (Siluriformes: Heptapteridae), with a CT scan survey of phylogenetically-relevant characters. Cybium 43: 125–135.

Quan, H., W. Yang, Z. Tang, R.O. Ritchie and M.A. Meyers. 2020. Active defense mechanisms of thorny catfish. Mater. Today 38: 35–48.

Reis, V.J.C., M.C.C. de Pinna and T.C. Pessali. 2019. A new species of *Trichomycterus* Valenciennes 1832 (Trichomycteridae: Siluriformes) from the Rio Doce drainage with remarkable similarities with *Bullockia* and a CT-scan survey. J. Fish Biol. 95: 918–931. https://doi-org.ezproxy2.library.drexel.edu/10.1111/jfb.14089.

Rodiles-Hernández, R., D.A. Hendrickson, J.G. Lundberg and J.M. Humphries. 2005. *Lacantunia enigmatica* (Teleostei: Siluriformes) a new and phylogenetically puzzling freshwater fish from Mesoamerica. Zootaxa 1000: 1–24.

Sagemehl, M. 1885. Beiträge zur vergleichenden Anatomie der Fische. III. Das Cranium der Characiniden nebst allgemeinen Bemerkungen über die mit einem Weber'schen Apparat versehenen Physostomenfamilien. Morph. Jahrb. 10: 1–119.

Sakashita, M., M. Sato and S. Kondo. 2019. Comparative morphological examination of vertebral bodies of teleost fish using high-resolution micro-CT scans. J. Morph. 280: 778–795. https://doi.org/10.1002/jmor.20983.

Schaefer, S.A. 2003. Relationships of *Lithogenes villosus* Eigenmann, 1909 (Siluriformes, Loricariidae): Evidence from high-resolution computed microtomography. Am. Mus. Novit. 3401: 1–55.

Summers, A.P., R. Ketcham and T. Rowe. 2004. Structure and function of the horn shark (*Heterodontus francsisi*) cranium through ontogeny – the development of a hard prey crusher. J. Morph. 260: 1–12.

Vickerton, P., J. Jarvis and N.S. Jeffery. 2013. Concentration-dependent specimen shrinkage in iodine-enhanced microCT. J. Anat. 223: 185–193.

Weber, E.H. 1820. De Aure et Auditu Hominis et Animalium. Pars I. De Aure Animalium Aquatilium. Leipzig: Gerhard Fleischer.

Weller, H.I., A.M. Olsen, A.L. Camp, A.R. Manafzadeh, L.P. Hernandez and E.L. Brainerd. 2020. An XROMM study of food transport and swallowing in Channel Catfish. Integr. Org. Biol. 19, 2: 1–13. doi:10.1093/iob/obaa018.

Zanata, A.M. and C. Primitivo. 2014. Natural history of *Copionodon pecten*, an endemic trichomycterid catfish from Chapada Diamantina in northeastern Brazil. J. Nat. Hist. 48: 3–4. 203–228, DOI:10.1080/00222933.2013.809168.

Section 3
Biology of Catfishes

Chapter 9

Miniature Catfishes

Diversity, Ecology and Evolution

Elisabeth Henschel,[1,2,*] *Axel M. Katz*[1] and *Pedro H. N. Bragança*[3,4]

"Nothing is known of their habits and they are so small (the largest known specimen is but 27 mm long) that it is a wonder that any of them have arrived in the bottles of the naturalist."

Eigenmann (1918, p. 261)

Miniaturization, i.e., the extreme reduction in adult body size within a lineage, is an evolutionary process that spans across distinct metazoan taxa, from invertebrates such as crustaceans, mollusks and annelids, to vertebrates, with teleost fishes being among the most specious lineage including miniaturized organisms within this subphylum. In his pioneering work on the family Pygidiidae (currently recognized as the pencil catfish family—Trichomycteridae), German-American ichthyologist Carl H. Eigenmann expressed surprise at having specimens of Tridentinae in hand. This subfamily of trichomycterid is known for its minute size and little-known habits. Despite museum specimens indicating that the first miniaturized specimens, belonging to different Neotropical fish lineages, were sampled during Louis Agassiz's Thayer Expedition (1865–1866), their existence went unnoticed by most naturalists until the early twentieth century, probably due to their minute dimensions. However, Eigenmann was exceptional not only in describing 195 genera and about 600 species but also in being among the first to describe a miniaturized fish species from the Neotropical region – the callichthyid *Gastrodermus hastatus* (Eigenmann and Eigenmann, 1888). This description was based on material collected during the Thayer Expedition (Eigenmann and Eigenmann 1888). Shortly after, the first descriptions of small-sized species of different groups set the stage for a later flourishment of this field of study that began in the 1960s and continues until today, fostering vivid discussions on the definitions of a miniature fish.

Among the first works dedicated to understanding miniaturization in fishes as a process beyond merely highlighting small size in species descriptions, Myers (1958) approached morphological

[1] Laboratory of Systematics and Evolution of Teleost Fishes, Institute of Biology, Federal University of Rio de Janeiro, 21641-599 Rio de Janeiro, RJ, Brazil.
[2] Laboratory of Molecular Systematics (BEAGLE), Department of Animal Biology, Federal University of Viçosa, 36570-000 Viçosa, MG, Brazil.
[3] South African Institute for Aquatic Biodiversity, Private Bag 1015, Makhanda, South Africa 6140.
[4] Department of Ichthyology, American Museum of Natural History, 10024-5102 New York, NY, United States.
Emails: axelmk@gmail.com; pbraganca@amnh.org
* Corresponding author: elisabeth.henschel@ufv.br

trends observed in small-sized fishes. He considered these to be species that range between 10 and 25 mm of Standard Length (SL) as adults. Although not formally proposing a concept for miniaturization, this was the first publication to consider a size range for small fishes and to observe the high frequency of reductions and paedomorphosis in their anatomy. Among the common changes in body morphology, Myers (1958) noticed reductions in the number of scales, fin rays, teeth, and simplification of the latero-sensory system and skull bones (e.g., less ossified when compared to larger related fishes), as well as the retention of the pronephros as a functional kidney in adults. Thirty years later, inspired by Myers's (1958) ideas, Weitzman and Vari (1988) proposed the first explicit definition of miniaturization in fishes with a deep focus on freshwater South American species. The authors considered miniature species that do not surpass 26 mm of SL when adults or that reach sexual maturity under 20 mm SL and exhibit a myriad of reductive character states like those cited by Myers (1958). The arbitrariness of choosing 26 mm SL as a cutoff for miniature fishes was even recognized by Weitzman and Vari (1988), who established this threshold based on the usual presence of paedomorphic features among the small-sized species examined. Another main point acknowledged by them, however, was the existence of species that slightly surpassed 26 mm SL but still exhibited a series of reductions and paedomorphic character states. Their study established the long-needed formal conceptualization of what constitutes a miniature fish and provided a proper context for a new wave of studies dedicated to small fish taxa. This body of research has been growing systematically from the early 1990s until today, with all studies adopting the maximum 26 mm SL as a cutoff standard for miniatures.

Weitzman and Vari (1988) also stressed the importance of systematic knowledge to the study of miniatures, but at the time when their miniaturization concept was proposed, the state of phylogenetic relationships among most fish taxa was very incipient, hampering a more accurate approach to how many times each major Neotropical fish group have undergone miniaturization. This scenario is far distant from what is seen today, with many recent studies challenging previously widely accepted relationship hypotheses between major fish taxa (Betancur-R. et al. 2013, 2017) and other diverse groups such as the Trichomycteridae subfamily Microcambevinae (Costa et al. 2020).

At first glance, miniaturized species can be considered more closely related to each other than other species (i.e., that a single miniaturization event occurred at the most recent common ancestor of that lineage) due to their apparent share of similar morphological traits. However, as is becoming more evident from molecular-based phylogenies, the assumption that the presence of reductive traits in different miniature species is due to the sharing of a miniaturized recent common ancestor is not always confirmed, with convergence also playing a significant role in the evolution of miniature species. Considering the importance of including phylogenetic relationships in the study of miniatures, Bragança and Burns (*in press*) proposed the following broader concept of miniaturization: miniature fishes are those that have undergone a reduction in adult body size when compared to their sister group and exhibit morphological simplifications (i.e., reductions and losses). This idea that avoids the use of arbitrary measurements will probably reveal groups that were never previously recognized as miniature.

According to the last published list, 273 species of miniaturized Neotropical fishes are recognized (Bragança and Burns *in press*). Estimates on the number of miniature fishes have been focused historically on the Neotropics (Weitzman and Vari 1988; Toledo-Piza et al. 2014; Bragança and Burns *in press*), with few studies listing Afrotropical miniatures (Conway and Moritz 2006; Bragança et al. 2023), and Kottelat and Vidthayanon (1993) not recording any miniaturized catfish in Asia, despite their existence. Miniatures are found in freshwaters of South America, Africa and Asia, which include heterogeneous environments, water types, and a higher degree of ecosystemic complexity, with more specific microhabitats and resources available for exploration of unusual ecological niches (Roberts 1972; Leitão et al. 2016; Bragança and Burns *in press*). Until the development of this book chapter, the order Siluriformes comprised 75 Neotropical and nine Afrotropical miniature species, which account for approximately 2% of the whole order species-level diversity, being then the second-most representative order of miniatures behind the Characiformes (Bragança and Burns *in press*; Bragança et al. 2023).

This chapter is dedicated to the study of distinct aspects of miniature siluriforms: (1) their diversity, considering the concepts of miniaturization by Weitzman and Vari's (1988) and by Bragança and Burns (2023); (2) their unique morphological and ecological aspects; (3) global inequalities in the study of miniature catfishes; and (4) knowledge gaps and future research perspectives focused on small-sized catfishes. We present two updated lists of miniature catfishes according to each miniaturization concept, considering all biogeographic domains. We discuss the main morphological and ecological specializations of miniature catfishes, also taking into account their conservation status. Finally, we approach the historical disproportions in the study of small-bodied siluriforms relative to their localities, and the impact of Weitzman and Vari's (1988) work in biogeographic realms other than the Neotropics. Finally, we highlight knowledge gaps in research on potential miniature catfishes, with the hope that future studies will be dedicated to these taxa and a better elucidation of miniaturization across Siluriformes.

Miniaturization in Siluriformes

Siluriformes are the second-most diverse order of vertebrates, comprising about 4,217 valid species distributed within 40 families, only being rivaled in terms of diversity by the Cypriniformes (Fricke et al. 2023). Catfishes vary greatly in size, ranging from the Mekong giant catfish *Pangasianodon gigas* Chevey, 1931, which reaches up to 3 m in total length, to the smallest known species, the spiny-dwarf catfish *Scoloplax dolicholophia* Schaefer, Weitzman and Britski, 1989, which grows to a maximum adult size of 12 mm SL. Considering the high species-level diversity and the existing unmatched magnitude of variation in size within the order, a great number of miniature species would be expected. However, according to the last two lists of miniature fishes, Siluriformes only comprised 84 miniature species, of which 75 are from the Neotropics (Bragança and Burns *in press*) and nine from the Afrotropics (Bragança et al. 2023), with Asian catfish species missing from the only list published on Asian small-sized fishes (Kottelat and Vidthayanon 1993). Currently, the total number of catfish miniature species is less than the total of Neotropical characiform miniatures (103; Toledo-Piza et al. 2014), probably an underestimated number considering that the Characiformes is less diverse with about 60% of the species-level diversity of Siluriformes (Fricke et al. 2023).

Aiming to access the actual number of miniaturized catfish species properly, Tables 1 to 3 represent an updated list of Neotropical, Afrotropical and Asian siluriform miniature species, respectively, separated by family and following the cutoff size of 26 mm SL as proposed by Weitzman and Vari (1988). In addition to the maximum body size attained by the species, these tables also present information on their microhabitat and dietary preferences when available and include the species' conservation status according to IUCN criteria. According to Weitzman and Vari's (1988) criterion, 118 miniature catfish species are recognized, of which 92 are from the Neotropics, nine from the Afrotropics and 17 from Asia. This represents an addition of 14 miniature Neotropical catfish species to the last list by Bragança and Burns (*in press*) and *Cetopsidium minutum* (Eigenmann, 1912), which was never cited in previous lists (Weitzman and Vari 1988; Costa and Le Bail 1999; Toledo-Piza et al. 2014; Bragança and Burns *in press*). This is also the first effort directed toward comprehending the diversity of small-sized Asian catfishes inserted within a formal miniaturization context.

Table 1. Updated list of miniature catfish species from the Neotropics according to Weitzman and Vari's (1988) criterion of miniaturized fish. Conservation status without asterisk indicates that the evaluation was made by IUCN and when followed by an asterisk indicates data retrieved from the Brazilian ICMBIO (2018) Red List.

	SL (max)	**Reference**	**Weitzman and Vari (1988)**	**Costa and Le Bail (1999)**	**Toledo-Piza et al (2014)**	**Bragança and Burns (2023)**	**Present Study**	**Microhabitat**	**Diet**	**Conservation Status**
Aspredinidae										
Acanthobunocephalus nicoi	19.7	Costa and Le Bail 1999		X				Sand/leaf packs	Benthic invertivore/ meiovores/microvore	LC
Acanthobunocephalus scruggsi	22.1	Carvalho and Reis 2020				X		Sand/leaf packs	Benthic invertivore/ meiovores/microvore	VU
Hoplomyzon atrizona	25.3	Schultz 1944				X		Sand	Benthic invertivore/ meiovores/microvore	NE*
Hoplomyzon cardosoi	18.5	Carvalho et al. 2017				X		Sand/gravel rocks	Benthic invertivore/ meiovores/microvore	NE
Hoplomyzon papillatus	16.9	Weitzman and Vari 1988	X					Mud-silt/ sand	Benthic invertivore/ meiovores/microvore	NE*
Dupouyichthys sapito	23	Schultz 1944				X		Mud-silt/ sand/gravel rocks	Benthic invertivore/ meiovores/microvore	NE*
Micromyzon akamai	15.8	Costa and Le Bail 1999		X				Mud-silt/ sand	Benthic invertivore/ meiovores/microvore	LC
Micromyzon orinoco	15.6	Carvalho et al. 2016				X		Mud-silt/ sand	Benthic invertivore/ meiovores/microvore	NE
Ernstichthys taquari	22.8	Dagosta and de Pinna 2021				X		Sand/gravel rocks	Benthic invertivore/ meiovores/microvore	NE
Astroblepidae										
Astroblepus chimborazoi	25	Toledo-Piza et al. 2014			X			Gravel rocks	Benthic invertivore/ meiovores/microvore	NE*
Auchenipteridae										
Gelanoglanis nanonocticolus	22.2	Toledo-Piza et al. 2014			X			Surface	Benthic invertivore/ meiovores/microvore - surface inverts	LC
Gelanoglanis pan	24.7	Calegari et al. 2014				X		Surface	Benthic invertivore/ meiovores/microvore - surface inverts	NE
Gelanoglanis varii	17.7	Calegari and Reis 2016				X		Surface	Benthic invertivore/ meiovores/microvore - surface inverts	NE
Tatia marthae	23.1	Toledo-Piza et al. 2014			X			Surface	Surface inverts	DD*
Callichthyidae										
Aspidoras brunneus	21.3	Weitzman and Vari 1988	X					Submerged wood, vegetation	Detritivores	LC

Table 1 contd. ...

...Table 1 contd.

	SL (max)	Reference	Weitzman and Vari (1988)	Costa and Le Bail (1999)	Toledo-Piza et al (2014)	Bragança and Burns (2023)	Present Study	Microhabitat	Diet	Conservation Status
Aspidoras carvalhoi	25.4	Weitzman and Vari 1988	X					Submerged wood, vegetation	Detritivores	DD
Hoplisoma boehlkei	25.7	Toledo-Piza et al. 2014			X			Gravel rocks	Benthic invertivore/ meiovores/microvore	VU*
Hoplisoma cochui	25	Toledo-Piza et al. 2014			X			Gravel rocks	Benthic invertivore/ meiovores/microvore	LC
Gastrodermus gracilis	23.2	Weitzman and Vari 1988	X					Gravel rocks	Benthic invertivore/ meiovores/microvore	LC
Hoplisoma habrosum	20.1	Weitzman and Vari 1988	X					Gravel rocks	Benthic invertivore/ meiovores/microvore	NE*
Gastrodermus hastatus	24	Weitzman and Vari 1988	X					Gravel rocks	Benthic invertivore/ meiovores/microvore	NE*
Gastrodermus pygmaeus	23.7	Weitzman and Vari 1988	X					Gravel rocks	Benthic invertivore/ meiovores/microvore	LC
Cetopsidae										
Denticetopsis royeroi	18	Toledo-Piza et al. 2014			X			Water column	Piscivores	DD*
Denticetopsis sauli	21	Toledo-Piza et al. 2014			X			Water column	Piscivores	DD*
Doradidae										
Physopyxis ananas	22	Toledo-Piza et al. 2014			X			Sand/ Submerged wood, vegetation/ leaf packs	Omnivorous	LC
Physopyxis cristata	22.7	Toledo-Piza et al. 2014			X			Sand	Omnivorous	LC
Heptapteridae										
Gladioglanis anacanthus	26.3	Rocha et al. 2008					X	Leaf litter	unknown	DD
Horiomyzon retropinnatus	24	Weitzman and Vari 1988	X					Sand/gravel rocks	Benthic invertivore/ meiovores/microvore	NE*
Loricariidae										
Hypoptopomatinae										
Corumbataia liliai	24.7	Silva et al. 2018				X		Submerged wood, vegetation	Algivores	NE
Hisonotus bocaiuva	25.6	Roxo et al. 2013				X		Gravel rocks/ submerged wood, vegetation	Algivores	LC

Table 1 contd. ...

...Table 1 contd.

	SL (max)	Reference	Weitzman and Vari (1988)	Costa and Le Bail (1999)	Toledo-Piza et al (2014)	Bragança and Burns (2023)	Present Study	Microhabitat	Diet	Conservation Status
Nannoplecostomus eleonorae	22.2	Weitzman and Vari 1988			X			Gravel rocks	Algivores	LC
Otocinclus tapirape	24.1	Weitzman and Vari 1988			X			Submerged wood, vegetation	Algivores	LC
Rhinotocinclus britskii	23	Garavello1988 *Parotocinclus aripuanensis*			X			Submerged wood, vegetation	Algivores	LC
Rhinotocinclus bockmanni	23.5	Carvalho and Datovo 2012				X		Gravel rocks/ submerged wood, vegetation	Algivores	DD
Rhinotocinclus dinizae	23.3	Ribeiro-Silva et al. 2020				X		Submerged wood, vegetation	Algivores	NE
Rhinotocinclus collinsae	25.5	Toledo-Piza et al. 2014			X			Submerged wood, vegetation	Algivores	NE*
Rhinotocinclus halbothi	19.9	Toledo-Piza et al. 2014			X			Gravel rocks/ submerged wood, vegetation	Algivores	LC
Rhinotocinclus hardmani	25.1	Lehmann et al. 2022				X		Gravel rocks	Algivores	LC
Rhinotocinclus pentakelis	23.5	Roxo et al. 2019				X		Submerged wood, vegetation	Algivores	NE
Pseudopimelodidae										
Microglanis oliveirai	26.3	Ruiz and Shibatta 2011					X	unknown	unknown	DD
Microglanis robustus	23.3	Ruiz and Shibatta 2010				X		Submerged wood, vegetation	Benthic invertivore/ meiovores/microvore	CR
Microglanis zonatus	20	Toledo-Piza et al. 2014			X			Submerged wood, vegetation	Benthic invertivore/ meiovores/microvore	LC*
Scoloplacidae										
Scoloplax baileyi	15.4	Toledo-Piza et al. 2014			X			Leaf packs	Benthic invertivore/ meiovores/microvore	LC
Scoloplax baskini	16.1	Toledo-Piza et al. 2014			X			Leaf packs	Benthic invertivore/ meiovores/microvore	LC

Table 1 contd. ...

...Table 1 contd.

	SL (max)	Reference	Weitzman and Vari (1988)	Costa and Le Bail (1999)	Toledo-Piza et al (2014)	Bragança and Burns (2023)	Present Study	Microhabitat	Diet	Conservation Status
Scoloplax dicra	13.8	Weitzman and Vari 1988	X					Leaf packs	Benthic invertivore/ meiovores/microvore	LC
Scoloplax distolothrix	17.9	Costa and Le Bail 1999		X				Leaf packs	Benthic invertivore/ meiovores/microvore	LC
Scoloplax dolicholophia	12	Costa and Le Bail 1999		X				Leaf packs	Benthic invertivore/ meiovores/microvore	LC
Scoloplax empousa	19.9	Costa and Le Bail 1999		X				Leaf packs	Benthic invertivore/ meiovores/microvore	LC
Trichomycteridae										
Glanapteryginae										
Pygidianops cuao	18.7	Toledo-Piza et al. 2014			X			Sand	Benthic invertivore/ meiovores/microvore	NE*
Pygidianops eigenmanni	23	Weitzman and Vari 1988	X					Sand	Benthic invertivore/ meiovores/microvore	LC
Pygidianops magoi	12.6	Toledo-Piza et al. 2014			X			Sand	Benthic invertivore/ meiovores/microvore	NE*
Typhlobelus macromycterus	21.9	Toledo-Piza et al. 2014			X			Sand	Benthic invertivore/ meiovores/microvore - detritivores	DD
Typhlobelus guacamaya	22.3	Toledo-Piza et al. 2014			X			Sand	Benthic invertivore/ meiovores/microvore - detritivores	NE*
Microcambevinae										
Microcambeva barbata	26.1	Costa and Bockman 1994					X	Sand	Benthic invertivore/ meiovores/microvore	NT
Microcambeva draco	24.6	Toledo-Piza et al. 2014			X			Sand	Benthic invertivore/ meiovores/microvore	EN
Microcambeva mucuriensis	25.2	Costa et al. 2019a				X		Sand	Benthic invertivore/ meiovores/microvore	NE
Sarcoglanidinae										
Ammoglanis amapaensis	17.9	Toledo-Piza et al. 2014			X			Sand	Benthic invertivore/ meiovores/microvore	LC
Ammoglanis diaphanus	18.7	Costa and Le Bail 1999		X				Sand	Benthic invertivore/ meiovores/microvore	DD
Ammoglanis multidentatus	19.7	Costa et al. 2019b				X		Sand	Benthic invertivore/ meiovores/microvore	NE
Ammoglanis natgeorum	12.9	Henschel et al. 2020a				X		Sand/leaf packs	Benthic invertivore/ meiovores/microvore	NE
Ammoglanis obliquus	15.5	Henschel et al. 2020b				X		Sand	Benthic invertivore/ meiovores/microvore	NE
Ammoglanis pulex	14.9	Toledo-Piza et al. 2014			X			Sand	Benthic invertivore/ meiovores/microvore	NE*

Table 1 contd. ...

...Table 1 contd.

	SL (max)	Reference	Weitzman and Vari (1988)	Costa and Le Bail (1999)	Toledo-Piza et al (2014)	Bragança and Burns (2023)	Present Study	Microhabitat	Diet	Conservation Status
Malacoglanis gelatinosus	19.9	Weitzman and Vari 1988	X					Sand	Benthic invertivore/ meiovores/microvore	LC*
Stenolicmus ix	19	Wosiacki et al. 2011				X		Sand	Benthic invertivore/ meiovores/microvore	LC
Sarcoglanis simplex	21	Weitzman and Vari 1988	X					Sand	Benthic invertivore/ meiovores/microvore	LC
Stegophilinae										
Schultzichthys gracilis	26	Toledo-Piza et al. 2014			X			Sand	Ectoparasite	NE*
Trichomycterinae										
Ituglanis compactus	25.8	Castro and Wosiacki 2017				X		Sand	Benthic invertivore/ meiovores/microvore	NE
Tridentinae										
Miuroglanis platycephalus	12.3	Weitzman and Vari 1988	X					Sand	unknown	LC
Tridens chicomendesi	21.2	Henschel et al. 2023					X	Mud/ vegetation	Benthic invertivore	DD
Tridens vitreus	19.8	Henschel et al. 2023					X	Sand	unknown	DD
Tridensimilis brevis	30	de Pinna and Wosiacki 2003	X					unknown	unknown	DD
Tridensimilis venezuelae	25	Weitzman and Vari 1988	X					Sand	unknown	DD*
Tridentopsis cahuali	22.2	Toledo-Piza et al. 2014			X			Sand	Benthic invertivore	NE*
Tridentopsis pearsoni	23	Weitzman and Vari 1988	X					Sand	Benthic invertivore	LC
Tridentopsis tocantinsi	23	Weitzman and Vari 1988	X					Sand	Benthic invertivore	DD
Vandelliinae										
Paracanthopoma saci	21.8	Dagosta and de Pinna 2021				X		Sand	Ectoparasite	NE
Paracanthopoma cangussu	15.7	Henschel et al. 2021a				X		Sand	Benthic Invertivore/ ectoparasite	NE
Paracanthopoma alleynei	26	Henschel et al. 2021b					X	Collected from host	Ectoparasite	NE
Paracanthopoma ahriman	20.1	de Pinna and Dagosta 2022					X	Sand	Ectoparasite	NE
Paracanthopoma capeta	18.7	de Pinna and Dagosta 2022					X	Sand	Ectoparasite	NE
Paracanthopoma carrapata	23.3	de Pinna and Dagosta 2022					X	unknown	Ectoparasite	NE

Table 1 contd. ...

...Table 1 contd.

	SL (max)	Reference	Weitzman and Vari (1988)	Costa and Le Bail (1999)	Toledo-Piza et al (2014)	Bragança and Burns (2023)	Present Study	Microhabitat	Diet	Conservation Status
Paracanthopoma daemon	20.4	de Pinna and Dagosta 2022					X	Sand	Ectoparasite	NE
Paracanthopoma irritans	17.2	de Pinna and Dagosta 2022					X	Sand	Ectoparasite	NE
Paracanthopoma malevola	20	de Pinna and Dagosta 2022					X	unknown	Ectoparasite	NE
Paracanthopoma satanica	25.4	de Pinna and Dagosta 2022					X	unknown	Ectoparasite	NE
Paracanthopoma vampyra	22.7	de Pinna and Dagosta 2022					X	Sand	Ectoparasite	NE
Incertae sedis										
Potamoglanis anhanga	13.1	Toledo-Piza et al. 2014			X			Sand/ vegetation	unknown	LC
Potamoglanis hasemani	18	Weitzman and Vari 1988	X					Roots of aquatic vegetation	Benthic invertivore	LC
Potamoglanis johnsoni	16	Weitzman and Vari 1988	X					Roots of aquatic vegetation	Benthic invertivore	LC
Potamoglanis wapixana	15.7	Henschel 2016				X		Leaf packs/ sand/ submerged wood	Benthic invertivore	LC

When considering the recently proposed phylogenetic concept of miniaturization in fishes by Bragança and Burns (*in press*) (i.e., species that have undergone a significant reduction in adult body size when compared to related lineages and that present reductions and structural simplifications), the number of Neotropical miniature catfishes increases from 92 to 106 (Table 4). Despite their small sizes, the species listed in Table 4 have maximum recorded standard length sizes that surpass the threshold proposed by Weitzman and Vari (1988). However, they were recognized as miniatures by the presence of paedomorphic traits and knowledge of body sizes from closely related taxa (phylogenetic information). Table 4 also exhibits relevant data on microhabitats, dietary preference and conservation status when available.

For example, the Asian amblycipitid *Amblyceps serratum* Ng and Kottelat, 2000 is the smallest species of the genus, reaching up to 45.3 mm SL, whereas its congeners can grow up to 60 mm SL, and it has fewer anal-fin rays and vertebrae when compared to other species of *Amblyceps* Blyth, 1858. The genus *Hyalobagrus* Ng and Kottelat, 1998 comprises three small-sized species that exceed 26 mm SL but are distinguished from other bagrids by a combination of character states that involve reductions and losses and typical morphological features of miniatures, such as translucent body, reduction of the extrascapular bone and absence of the entopterygoid and first pharyngobranchial bones (Ng and Kottelat 1998). The sisorids *Erethistoides vesculus* Ng, Ferraris and Neely, 2012 and *Glyptothorax panda* Ferraris and Britz, 2005 are the smallest species of their

Table 2. Updated list of miniature catfish species from the Afrotropics according to Weitzman and Vari's (1988) criterion of miniaturized fish

	SL (max)	Reference	Bragança et al. (2023)	Present Study	Microhabitat	Diet	Conservation Status
Amphiliidae							
Zaireichthys kafuensis	19.7	Eccles et al. 2011	X		Sandy/gravel substrate	unknown	NE
Zaireichthys kunenensis	22.1	Eccles et al. 2011	X		Sandy/gravel substrate	unknown	NE
Zaireichthys lacustris	25.3	Eccles et al. 2011	X		Sandy/gravel substrate	unknown	LC
Zaireichthys pallidus	18.5	Eccles et al. 2011	X		Sandy/gravel substrate	unknown	NE
Zaireichthys wamiensis	16.9	Eccles et al. 2011	X		Sandy/gravel substrate	unknown	VU
Zaireichthys zonatus	23	Roberts 2003	X		Sandy/gravel substrate	unknown	DD
Psammphiletria delicata	15.8	Roberts 2003	X		Sandy/gravel substrate	unknown	NE
Psammphiletria nasuta	15.6	Roberts 2003	X		Sandy/gravel substrate	unknown	NE
Tetracamphilius clandestinus	22.8	Roberts 2003	X		Sandy/gravel substrate	unknown	DD

Table 3. Updated list of miniature catfish species from Asia according to Weitzman and Vari's (1988) criterion of miniaturized fish.

	SL (max)	Reference	Present Study	Microhabitat	Diet	Conservation Status
Bagridae						
Nanobagrus lemniscatus	23.4	Ng 2010	X	Sand/leaf litter/vegetation	unknown	VU
Akysidae						
Akysis heterurus	26.1	Ng 1996	X	unknown	unknown	DD
Akysis ephippifer	26.6	Ng and Kottelat 1998a	X	unknown	unknown	DD
Akysis recavus	26.2	Ng and Kottelat 1998a	X	unknown	unknown	DD
Akysis fuliginatus	22	Ng and Rainboth 2005	X	Submerged wood, vegetation	unknown	DD
Akysis fontaneus	23.5	Ng 2009a	X	Submerged wood, vegetation	unknown	DD
Parakysis notialis	21.9	Ng and Kottelat 2003	X	unknown	unknown	CR
Sisoridae						
Erethistes jerdoni	21	Ng and Kottelat 2007	X	Sand	unknown	LC
Erethistes nareshi	25	Mahapatra and Kar 2015	X	unknown	unknown	NE
Erethistoides luteolus	26.1	Ng et al. 2012	X	unknown	unknown	NE
Pseudolaguvia ribeiroi	25.7	Ng 2006	X	unknown	unknown	LC
Pseudolaguvia muricata	25.7	Ng 2005	X	Submerged wood (debris)	unknown	DD
Pseudolaguvia ferula	25.4	Ng 2006	X	Sand/gravel rocks	unknown	DD
Pseudolaguvia flavida	24.6	Ng 2009c	X	Sand	unknown	DD
Pseudolaguvia assula	23.7	Ng and Conway 2013	X	unknown	unknown	NE
Pseudolaguvia fucosa	27.4	Ng et al. 2016	X	Sand/gravel rocks	unknown	NE
Pseudolaguvia permaris	26.6	Vijayakrishnan et al. 2023	X	Vegetation	unknown	NE

Table 4. List of species that can be newly considered miniatures according to Bragança and Burns (*in press*) criterion. **Gastrodermus pauciradiatus* was included in the first list by Weitzman and Vari (1988) and later removed by Toledo-Piza et al. (2014) because it was larger than 26 mm SL, but it is herein considered a miniature. ***Cetopsidium minutum* reaches up to 21 mm SL according to Vari et al. (2005), but in Vari and Ferraris Jr. (2003), there is an indication that it can reach up to 49 mm SL. However, the specimen with that size was never used in the family's recent studies.

		SL (max)	Reference	Microhabitat	Diet	Conservation status
Amblycipitidae						
	Amblyceps serratum	45.3	Ng and Kottelat 2000	Gravel substrate	unknown	DD
Bagridae						
	Hyalobagrus ornatus	30	Vidthayanon 2002	unknown	unknown	NE
	Hyalobagrus flavus	38.6	Ng and Kottelat 1998b	Leaf litter	unknown	DD
	Hyalobagrus leiacanthus	36.7	Ng and Kottelat 1998b	Muddy substrate	unknown	DD
	Nanobagrus fuscus	38.3	Ng 2018	Water column	unknown	DD
	Nanobagrus armatus	28.3	Tan and Ng 1999	unknown	unknown	LC
	Nanobagrus nebulosus	36.2	Ng 2008	unknown	unknown	DD
	Nanobagrus stellatus	34	Tan and Ng 1999	unknown	unknown	LC
	Nanobagrus immaculatus	29.3	Ng 2008	unknown	unknown	VU
Callichthyidae						
	Aspidoras velites	27.9	Britto et al. 2002	Submerged vegetation	unknown	DD
*	*Gastrodermus pauciradiatus*	29*	Reis et al. 2003	Gravel rocks	Benthic invertivore/ meiovores/microvore	LC
Cetopsidae						
**	*Cetopsidium minutum*	21 **	Vari and Ferraris et al. 2003	Submerged wood, vegetation	unknown	LC
Heptapteridae						
	Gladioglanis conquistador	38.6	Lundberg et al. 1991	Leaf litter, sand	unknown	NE
	Gladioglanis machadoi	33.5	Ferraris and Mago-Leccia 1989	Leaf litter	Benthic invertivore	NE
Trichomycteridae						
	Ituglanis macunaima	33	Datovo and Landim 2005	Submerged wood/ vegetation	unknown	NT
	Microcambeva bendego	28.1	Medeiros et al. 2020	Sand	Benthic invertivore/ meiovores/microvore	NE
	Microcambeva jucuensis	27.2	Costa et al. 2019a	Sand	Benthic invertivore/ meiovores/microvore	NE
	Microcambeva watu	27.3	Medeiros et al. 2021	Sand	Benthic invertivore/ meiovores/microvore	NE
	Tridens melanops	27	de Pinna and Wosiacki et al. 2003	unknown	unknown	NE
Sisoridae						
	Erethistes minusculus	30.2	Ng and Kottelat 2007	unknown	unknown	
	Erethistoides vesculus	29.2	Ng et al. 2012	unknown	unknown	NE
	Glyptothorax panda	32	Ferraris and Britz 2005	unknown	unknown	DD
	Microglanis xylographicus	27.8	Ruiz and Shibatta 2011	unknown	unknown	DD

respective genera, with the first exhibiting some reductions (e.g., in the number of vertebrae) when compared to congeners.

Within the Neotropics, the callichthyid *Aspidoras velites* Britto, Lima and Moreira, 2002 exceeds the limit proposed by Weitzman and Vari (1988) by only 1.9 mm SL (it reaches a maximum size of 27.9 mm SL) yet exhibits a myriad of reductive character states when compared with congeners: a naked predorsal area, lacking bony plates; first pectoral- and dorsal-fin rays not modified into spines; a reduced and narrow posterior process of supraorbital; a poorly ossified hypobranchial 2 and reduced infraorbital flanges. Among the Trichomycteridae, representatives of two distinct subfamilies are now considered miniatures under Bragança and Burns (*in press*) criterion. The elongate tridentine *Tridens melanops* Eigenmann and Eigenmann, 1889 slightly surpasses 26 mm SL, but the paedomorphic nature of its morphology is indicated by the anatomy of its recently described congeners: *T. vitreus* Henschel Ohara and Costa, 2023 and *T. chicomendesi* Henschel Ohara and Costa, 2023. Both have a wide cranial fontanel, a mostly cartilaginous mesethmoid, and several other reductions on their branchial baskets, caudal skeleton, opercular suspensorium, and neurocranium. *Tridens melanops* was not listed in Weitzman and Vari's (1988) work and was first considered an "elongate miniature" by de Pinna (1989); further studies focused on this species might confirm shared morphological reductions with its miniature congeners. The trichomycterine *Ituglanis macunaima* Datovo and Landim, 2005 is among the smallest species of the genus, reaching approximately 30 mm SL, being rivaled only by the undoubtedly miniature *Ituglanis compactus* Castro and Wosiacki, 2017. *Ituglanis macunaima* exhibits a reduced number of pectoral rays, branched caudal rays, pleural ribs, vertebrae, interopercular odontodes, and the cephalic latero-sensory system pores, being therefore considered here as a miniature.

After unifying both Weitzman and Vari's (1988) and Bragança and Burns (*in press*) concepts of miniaturization, we now recognize 95 miniature catfish species for the Neotropical region, nine for the Afrotropical region and 29 for the Asian realm, totalizing 135 species. Based on data illustrated in Tables 1–4, we developed graphs that will allow proper visualization of habitat and dietary preferences and the conservation status of miniature catfishes. These topics will be considered below.

Ecology and Conservation

Considerations of the ecology and conservation of small-sized fish are usually scarce in the literature, being only more explicitly approached in recent species descriptions and revisions in a qualitative form or quantitively for a few species in a restricted geographic context (Perkin et al. 2022). The availability of information for Neotropical species is much more evident compared to Afrotropical and Asian species (Tables 1–4): of the 95 miniature catfish species from the Neotropics, only 4% are lacking microhabitat data, and 7% are data deficient for feeding habits. In contrast, the habitat preferences of 48% of miniature Asian catfishes are unknown, and their dietary preferences are not widely available in the scientific literature. When taking a closer look at the conservation status of miniaturized siluriforms, the lack of knowledge is even clearer. Approximately half of the Neotropical, 55% of the Afrotropical, and 20.5% of the Asian species were not evaluated according to IUCN's criteria, a worrying panorama that hampers further evolutionary studies and conservation policies directed to small-sized catfishes. This section is dedicated to presenting and discussing new data on the ecological and conservational aspects of miniaturized siluriforms.

A habitat preference for complex substrates in highly elaborated ecosystems by miniaturized fishes was first hypothesized by Weitzman and Vari (1988). Greater habitat structural complexity is associated with a higher number of vacant niches due to increased microhabitat availability; therefore, the more complex a habitat is, the greater its species richness. Although some marine habitats exhibit some degree of complexity (e.g., coral reefs and mangroves), a wider mosaic of habitats is mostly found in freshwaters. Moreover, freshwaters (such as streams and rivers) are unique in being characterized by high levels of environmental heterogeneity in abiotic factors (e.g., types

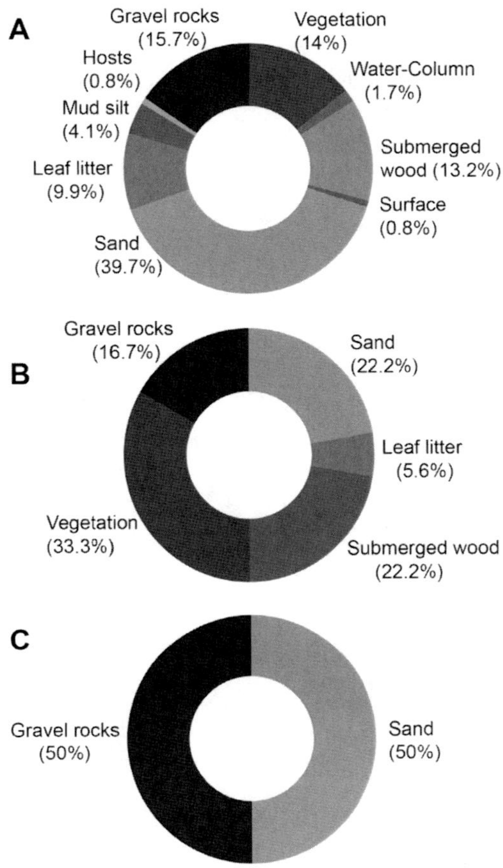

Figure 1. Habitat occupation of miniature species of Siluriformes according to the biogeographic realm. (A) Neotropical. (B) Afrotropical. (C) Asian.

of substrates, temperature gradient, distinct depths, water flow speed, etc.), leading to the presence of more complex habitats (Giller and Malmqvist 1998). Thus, beyond the fact that Siluriformes is a primarily freshwater lineage and that its marine forms are less diverse, it is not surprising that no miniature catfish species were found in marine environments. Overall, the habitat preference of miniature catfish species from the Neotropics, Afrotropics, and Asia was mostly directed toward sand (psammophily), followed by gravel, submerged wood, vegetation, leaf litter, mud slit, water column and surface and, lastly, by hosts in the case of parasitic species (Figure 1).

Interstitial habitats are characterized by the presence of tiny spaces between their components, providing shelter, protection, food, and high levels of oxygen (due to the water flow between the sand grains). Additionally, this environment provides lower dragging activity and is used for foraging and spawning for its associated fauna, allowing a higher degree of ecosystemic complexity. Thus, there is a widespread preference among most miniature catfish for different interstitial substrates, such as sand bottom, gravel substrate, and leaf litter. Interstices are abundant in alimentary resources: they allow biofilm and periphyton development, accumulation of organic matter, and shelter a vast meiofauna (invertebrates smaller than 0.5 mm), which are the main food items of most miniature catfishes (Tables 1–4). Among the seven listed microhabitats where miniaturized siluriforms are found, five are interstitial (sand, gravel, submerged wood, leaf litter, and mud slit), with psammophily (sand-dwelling habits) being the most common among miniature catfish.

Despite all the aforementioned complexity and resource abundance of interstitial environments, they are also challenging in many ways. Substrate grains are highly liable to turbulences caused by water flow that strongly influence the transport and deposition of particles of distinct sizes—impacting the dimensions of interstices—and affect the whole environment by determining factors such as oxygen availability and pH. Although the consequences of water currents on the biology of miniature catfishes are unknown, associations between their morphologies and tolerance to turbulence in interstitial habitats can be made. Interstitial invertebrates, the most studied burrowing animals, exhibit body forms highly adapted to this type of habitat, with adhesive organs, anchoring structures, body form, and many kinds of integuments that protect them against abrasion caused by the movement of sediments (Boaden 1968). The psammophilic habit also reduces the dragging force caused by fast-flowing waters. Among siluriforms, protective and anchoring structures are widespread regardless of their size: bony plates are found in loricariids, callichthyids, scoloplacids, hoplomyzontines and doradids; suctorial mouth disks in astroblepids and loricariids; different kinds of fin spines that arose independently in many lineages; and highly organized opercular and interopercular patches of odontodes, as the ones seen in trichomycterids, allows the existence of catfishes in turbulent waters of distinct dimensions – from torrential waterfalls in the Andes to interstitial habits. Moreover, the ancestral body form of Siluriformes is a depressed and flattened body, highly adapted to a substrate-oriented life (Alexander 1966), consistent with the overall morphology and habitat preference of miniature catfishes. In general, microhabitats occupied by interstitial miniature catfishes are grossly classified according to the type of substrate, i.e., sand, mud, gravel, leaf litter and others, neglecting a crucial aspect: their porosity and granulometry. The types of grains that compose these microhabitats vary in terms of size, packing (sedimentation level), which impacts the oxygen concentration in the interstice, and mineral composition, all these elements critical to the microhabitat occupation by miniature fishes. Hence, further refinement of ecological variables observed in the field might reveal even higher degrees of substrate preference and habitat specialization among miniature catfishes.

The preference for vegetated environments is also significant, accounting for 13.6% and 25% of the habitat preference among miniature Neotropical and Asian catfishes, respectively. Aquatic vegetation plays a key role in freshwater ecosystems by increasing their complexity and heterogeneity, serving as a shelter and food source and acting on nutrient stocking and cycling. Aquatic macrophytes are photosynthetic organisms (ranging from macroalgae to vascular plants) that spend at least part of their life cycle in aquatic environments. Given the extent of taxonomic groups that can be considered aquatic macrophytes, these plants are formally classified by their distribution in the water column into seven types: fixed submerged, free submerged, fixed floating, free floating, emerging, amphibious, and epiphytes. However, similarly to the gross classification seen in interstitial habitats, detailed information on the preferred vegetation type and species inhabited by miniature catfishes is usually scarce, being limited to terms such as "marginal vegetation," "submerged vegetation," and "roots of trees," even when photographs of habitats are available. Moreover, the extent and types of the ecological relationships between miniature catfishes and aquatic macrophytes remain obscure due to the aforementioned simplified descriptions and also the lack of observations *in situ*. For example, specimens of *Potamoglanis johnsoni* (Fowler, 1932) and *P. hasemani* (Eigenmann, 1914) were collected exclusively in roots of the free-floating water lettuce *Pistia stratiotes* in the Paraguay and Tapajós River basins, respectively (E. Henschel, pers. obs.); however, it remains unknown whether these roots act as temporary or permanent shelters, or if these species forage elsewhere.

Miniature catfishes inhabiting the water column or the surface of a stream are restricted to the Neotropical region and account for only 1.7% each of the total miniatures. This pattern opposes the pattern commonly seen for the other two orders with more miniature fishes, the Characiformes and Cyprinodontiformes, where the majority of miniatures have a preference to inhabit the water column or the surface of lentic and slow-flowing streams as most of their bigger relatives (Bragança and Burns *in press*; Bragança et al. 2023). The two miniaturized siluriforms that are found in water columns are the cetopsids, *Denticetopsis royeroi* Ferraris, 1996 and *D. sauli* Ferraris, 1996. They exhibit the typical body morphology of Cetopsidae: a robust and cylindrical body with a slightly curved head that allows whale catfishes, as they are commonly known, to be effective free swimmers. The only miniaturized catfishes inhabiting water surfaces are species of the Auchenipteridae. *Tatia marthae* Vari and Ferrraris, 2013 also exhibits a robust body and shares the same habits with other species of *Tatia* Miranda-Ribeiro, 1911, hiding in submerged trunks and gravel rocks interstices during the day and foraging close to the water surface at night. It is important to emphasize that miniature water columns and surface swimmers are not found in main river channels or even medium-sized rivers and streams but in still or slow-flowing waters in shallow first to third-order streams. However, miniature species of *Gelanoglanis* Böhlke, 1980 are found on the open surface of main river channels during the night, and their diurnal habitat is still unknown (Reis 2023, pers. comm.). *Denticetopsis sauli* was collected in outflow streams from a series of morichales (= buriti palm groves), and *D. royeroi* and *T. marthae* were found in two distinct small streams (caños) in Venezuela.

Finally, 0.8% of Neotropical miniature catfishes found in association with hosts are the ectoparasitic candirus (or *carneros*, in Spanish). The candirus are trichomycterid species allocated in three different subfamilies: the Tridentinae, the Stegophilinae (mucus, scales, skin and carrion feeders) and the Vandelliinae (vampire catfishes – blood and insect feeders). At this point, however, no Tridentine was ever collected in association with hosts, being mostly found in interstitial habitats (Table 1), although they share many morphological specializations to a parasitic lifestyle with stegophilines and vandelliines, indicating they might also be ectoparasites. Among species of candirus found attached to their hosts (i.e., Stegophilinae and Vandelliinae), miniaturization occurred in the stegophiline *Schultzichthys gracilis* Dahl, 1960 and in the Vandelliinae widespread genus *Paracanthopoma* Giltay, 1935 that currently comprises 13 species, of which 11 are miniaturized. Curiously, there is no apparent correlation between miniaturization and a parasitic lifestyle – in fact, miniature candirus only account for a small portion of parasitic trichomycterids known, with some species reaching up to 200 mm SL [species of the genus *Vandellia* Valenciennes, 1846]. Moreover, parasite size is also not deterministic in terms of host selection. Miniature species of *Paracanthopoma* were found attached to a wide variety of hosts: the pimelodids *Brachyplatystoma vaillantii* (Valenciennes, 1840), *Phractocephalus hemioliopterus* (Bloch and Schneider, 1801) and *Zungaro zungaro* (Humboldt, 1821), and the doradids *Doras micropoeus* (Eigenmann, 1912) and *Doras phlyzakion* Sabaj-Pérez and Birindelli, 2008 – all similar-sized large species that are sometimes also parasitized by larger vandelliines. Interestingly, feeding might not be the only reason why candirus is associated with larger fishes. The *in-loco* observation of individuals of *Paracanthopoma* parasitazing *Z. zungaro* led to the first hypothesis of hitch-hiking by attaching to other fish species. This observation was based on the fact that these specimens were fixed to various body parts (e.g., caudal and pectoral fins and dorsal-fin base), as opposed to the feeding locations typically observed with hematophagous candirus (Zuanon and Sazima 2005). It should be noted that candirus are nocturnal species that forage and seek prey at night. During the day, several species are found buried in sandbanks (psammic environments), such as the miniatures *P. saci* Dagosta and de Pinna, 2021, *P. cangussu* Henschel, Katz and Costa, 2021, and *P. daemon* de Pinna and Dagosta, 2022.

Another extremely relevant ecological aspect of miniature catfishes directly related to habitat occupation is the wide spectrum of their feeding habits. Most species descriptions do not provide information on these fishes' trophic ecology, and there are few published reports on the dietary habits of Afrotropical miniatures (Eccles et al. 2011) but none on Asian miniature siluriforms. Reliable data were only available for Neotropical species (Figure 2), and we classified their dietary preferences into nine categories: algivores (bottom-dweller algae eaters), benthic invertivores (feeding on benthic aquatic insects or bottom-dwelling macroinvertebrates), detritivores (detritus feeders), ectoparasites (feeding on body parts of other aquatic organisms), meiovores (eating invertebrates that do not surpass 0.5 mm), microvores (feeding on microscopic organisms, mostly protists, that inhabit interstitial environments), omnivorous (plant- and animal matter eaters), piscivores (feeding on other fishes) and surface invertebrates' eaters (consuming invertebrates that inhabit the water surface).

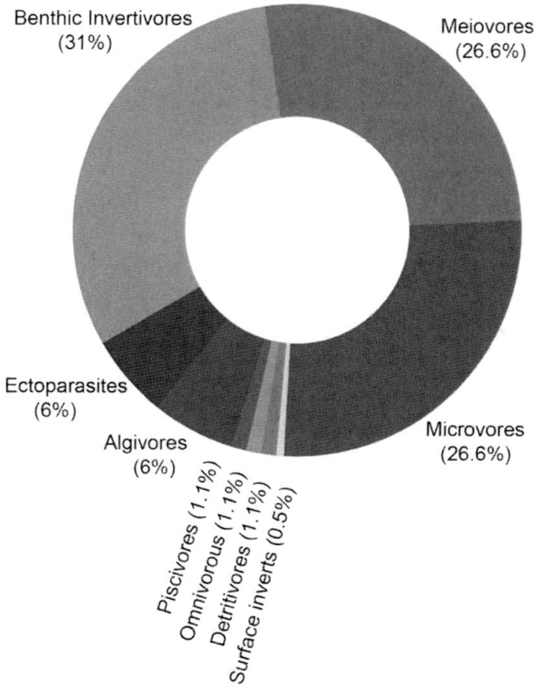

Figure 2. Dietary preferences of miniature species of Siluriformes from the Neotropical Region.

As presented in Tables 1–4, most miniature catfish species are microvores, meiovores, and benthic invertivores simultaneously, i.e., they feed on small-sized invertebrates that live closely with the substrate. Together, these three dietary preferences account for 84% of the feeding habits of all miniature siluriforms. This is not surprising since 96.6% of them are interstitial species, an amount represented by the sum of habitat preference for sand, gravel rocks, submerged wood and vegetation, leaf litter, and mud slit. As aforementioned, interstitial habitats are home to a myriad of animal species, being rich and complex environments in terms of ecological relations, especially rich in food provision for miniature fish species given that many species of small invertebrates are found within and above sediments. Also, most catfish species, not only miniatures, are predators and aquatic invertebrates (mostly insects, crustaceans, and mollusks) also correspond to a great part of their food intake, with species spending most of the time foraging and resting/hiding near the substrate. Thus, most miniature siluriforms share with their medium- or large-sized relatives a diet that is mostly based on aquatic invertebrates that are associated with a largely benthic or psammophilic lifestyle, but the extreme reduction in body size allows the capture of much smaller prey. Data on the feeding preferences of the African sand-dwelling catlets *Zaireichthys* Roberts, 1968 indicate a preference for small interstitial invertebrates such as ostracods, cladocerans, copepods, chironomid, and small trichopteran larvae (Eccles et al. 2011).

Ectoparasitism by the miniaturized trichomycterid candirus *Schultzichthys gracilis* and species of *Paracanthopoma* is another significant feeding habit, corresponding to a 6% share of the total. Albeit records of stegophilines feeding on mucus, scales, skin, flesh, and even dead matter of other fishes are relatively common in the literature, specifically, no data on the diet of *S. gracilis* is available. Its parasitic lifestyle was inferred because it shares many morphological features clearly specialized to parasitism (e.g., a wide inferior semicircular mouth) as is seen in other parasitic stegophiline species rather than on empirical observations. On the other hand, as mentioned before, hosts of some species of *Paracanthopoma* were identified in the field, all of them being other siluriform species. Recently, stomach content identification based on metabarcoding of both Stegophilinae and Vandelliinae, allowed precise species-level identification of their hosts (Bonato et al. 2018, 2022). However, these approaches have not included any miniaturized candirus, probably due to their minute size and consequent difficulties in obtaining high-quality gut content from dissections, a commonly destructive procedure.

Interestingly, miniaturized species of *Paracanthopoma* exhibit markedly distinct skull morphologies (albeit within the limits imposed by the Vandelliinae parasitic *bauplan*, such as the presence of claw-like teeth), varying in meristics and anatomy of their mandibular arches and opercular structures, suggesting that each species might feed more commonly on particular host species. Although specimens of *Paracanthopoma* are available in scientific collections throughout South America, allowing future ecomorphological studies, that is not the case for *S. gracilis*: lots of this species are scarce and, when available, restricted to no more than five specimens. Finally, it is worth mentioning that the diet of candirus might not be exclusively based on parts of their hosts. On the description of the miniature *P. cangussu*, Henschel et al. (2021a) identified Chironomidae (Insecta: Diptera) larvae in the guts of cleared and stained specimens, this being the first record of insect feeding by candirus. In this way, further studies might reveal additional food items in the guts of ectoparasitic siluriforms.

Algivory, accounting for 6% of the total, is the preferred diet by the 11 miniaturized species of Loricariidae and is a widespread feeding habit among loricariids. All these miniature armored catfishes (popularly known in Portuguese as cascudinhos) occupy habitats that contain a great amount of vegetal matter (mostly submerged wood and vegetation; Table 1) and are allocated in the subfamily Hypoptopomatinae, which most recent common ancestor was probably a small-sized fish that used to inhabit slow- and fast-flowing vegetated habitats and rapids with rocky substrates (Roxo et al. 2017). Physiological adaptations are needed for successfully feeding on algae, given the amount of complex structural molecules in vegetal matter, such as cellulose, and herbivores usually rely on symbiotic gut microbes or extracorporeal predigestion made by microorganisms in the environment. Studies based on the ecomorphology of wood-eating grazer loricariids from the subfamily Hypostominae revealed that their gut microbiomes are distinct from that of environmental wood samples and vary according to host identity (McCauley et al. 2020). Moreover, oral structures (especially teeth) also exhibit variation according to dietary preferences. Further studies might access exactly the types of algae these small hypoptomatines are feeding on and how digestion of vegetal matter occurs.

Detritivory, omnivory, and piscivory combined are the feeding strategies adopted by 1.1% of all miniaturized siluriforms (Figure 2). Detritus corresponds to decaying organic matter in different stages of decomposition, being composed of physically and chemically processed animal and plant matter, and is a widespread feeding habit among fishes in South America. Despite its common appearance in the guts of many fishes, especially grazers, it does not necessarily mean that the fish is detritivore, i.e., that the individual is obtaining nutritional value from this source of food. Detritivorous fishes can eat by four distinct strategies: biting into the bottom and posteriorly swallowing chunks of detritus, sucking aggregates of detritus from the substrate, scooping mud, or filtering water particles from the surrounding means. Considering the mouth anatomy and suction-feeding habits of callichthyids, we assume that miniature species of the genus *Aspidoras* Ihering, 1907 that feed on decaying organic matter also have suctorial habits. Very few data are available on the ecological preferences of the small-sized thorny catfishes of the genus *Physopixis* Cope, 1871, and it seems to share with other doradids an omnivorous habit, feeding on both plant and animal matter. As opposed to the insectivorous habit of most species of Cetopsidae, the miniatures *Denticetopsis royeroi* and *D. sauli* are piscivores. Interestingly, these two species were the basis for Ferraris (1996) to develop the original designation of *Denticetopsis* Ferraris, 1996 that, among other character states, was initially differentiated from other cetopsids by the presence of a cluster of caniniform teeth at the dentary symphysis that probably allow these two species to feed on other fishes. A broader concept of *Denticetopsis*, developed by Vari et al. (2005), does not consider the presence of this patch of specialized teeth as diagnostic for the genus. Indeed, *D. praecox*, described by Vari et al. (2005) in the same study, lacks this structure and was found to be insectivorous.

The four miniature species of Auchenipteridae are the only known small-sized siluriforms that feed on surface invertebrates, accounting for only 0.5% of all dietary preferences. Surface invertebrates are a dietary category that encompasses both free-swimming aquatic and terrestrial invertebrate species that accidentally fall into the water body. Although knowledge of several taxonomic, evolutionary, and even ecological aspects of auchenipterids increased in the last decades, data on their feeding habits remain scarce and biased, restricted to few species. On the other hand, miniature species of *Gelanoglanis* can be found associated with the substrate and feed on micro- and meiofauna, benthic invertebrates, and surface invertebrates. *Tatia marthae* is the only miniaturized surface swimmer that eats invertebrates found in surface waters.

The overall lack of knowledge about miniature catfishes is reflected not only in the scarcity of data relative to their trophic ecologies, as discussed above, but also in the challenges of accurately assessing their conservation status. Figure 3 is based on data from Tables 1–4, considering the total amount of miniature catfish species from the Neotropical, Afrotropical and Asian realms. It depicts both assessed species vs. non-assessed species under the conservational status criteria established by the International Union for Conservation of Nature (IUCN).

Out of the 135 herein-recognized miniature catfish species, approximately 40% were not evaluated, which means that they might be facing threats, but these are not clear since a basic conservation step is the species assessment *per se*. This is of special concern considering that the high degree of habitat specialization (i.e., use of restricted resources and/or habitats) by miniature species is usually correlated to higher extinction risks, as seen in other groups of organisms, such as beetles, dragonflies, birds, and corals (Brandl et al. 2018). Moreover, habitat specialization commonly leads to restricted geographic ranges. Geographically restricted species are especially prone to populational declines, even more liable to habitat loss by anthropogenic actions.

Another problematic scenario is the relatively high proportion of Data Deficient species: of the 85 miniature catfish species, 23 (about 25%) were allocated in this category. Data Deficient taxa lack adequate information on distribution and abundance to allow proper extinction risk assessment, emphasizing the urge for miniature-focused studies dedicated to filling this conservational gap. Several factors underlie the Wallacean shortfall observed among miniaturized species, such as (1) difficulties in collecting small-sized species that demand specialized equipment and fixation methods; (2) lack of understanding of the species' geographic distribution range (especially the ones from remote areas that might have a wider or more restricted distribution then is known); and (3) lack of quantitative studies focused on ecological aspects of miniatures.

Together, the categories "Vulnerable," "Endangered," and "Critically Endangered" are labeled as "Threatened." This wider classification comprises taxa with strong evidence for the populational decline over time, limited occurrence range or area of occupancy, a combination of small population size and inferred decline trend, very small population, or quantitative analysis that indicates a probability of extinction (IUCN 2012). Nine miniature siluriform species are considered threatened, a clearly underestimated number considering the high amount of species that have not been evaluated and the Data Deficient species. Following additional studies directed to address the present taxonomic gaps relative to miniature species, even taxa allocated in the Least Concern category might eventually be re-evaluated and transferred to one of the threatened categories. The increase in the number of species delimitation studies based on integrative data (morphology, genetics, ecology, and distribution, for example) will probably allow the identification of hidden diversity within populations that are currently considered to represent the same species, directly impacting conservation status assessments.

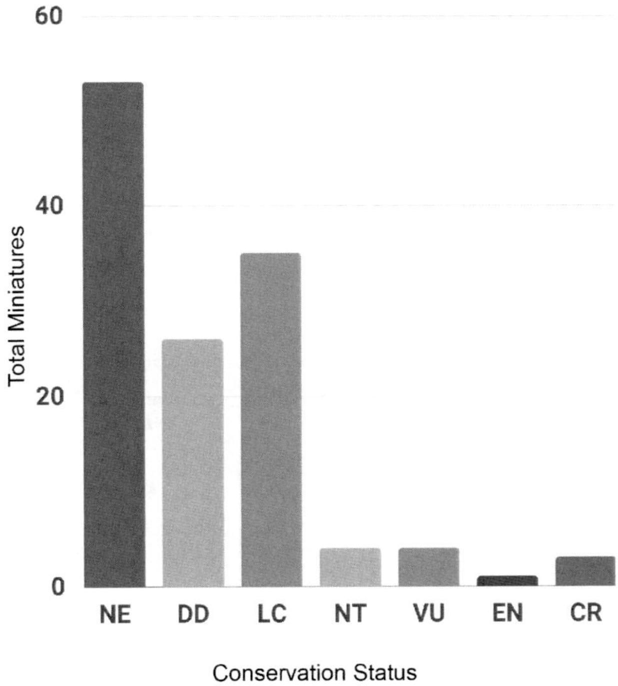

Figure 3. Conservation status of miniature species of Siluriformes from all biogeographic regions based on IUCN (2012) criteria. NE, Not Evaluated; DD, Data Deficient; LC, Least Concern; NT, Near Threatened; VU, Vulnerable; EN, Endangered; CR, Critically Endangered.

Geographic Inequalities in the Study of Miniaturized Catfishes

Nested within the generalized lack of knowledge on miniature catfish species is an even finer shortfall layer: geographic discrepancies between the current understanding of Neotropics, Afrotropics, and Asia miniatures are clearly seen by a comparison between the rate and number of miniature siluriforms described. Figure 4A compares the total number of species from these three distinct biogeographic realms was and illustrates how different the historical pace was in describing miniature catfish species'.

Although the first miniaturized Siluriformes described was *Erethistes jerdoni* (Day, 1870) from Bangladesh, the description of the miniature callichthyid *Gastrodermus hastatus* 18 years later was the ground zero of the upcoming two centuries that were marked by an ever-growing number of descriptions of small-sized Neotropical fishes – not only minute catfishes, but also members of Clupeiformes, Characiformes, Cyprinodontiformes, Perciformes, and Cichliformes. It is notable that after the publication of Weitzman and Vari's (1988) concept of miniaturization in fishes, the rate of description of miniaturized catfish species from the Neotropical region increased by about 700%. The impact of their work was immediately felt by ichthyologists studying Neotropical fauna. Published in August 1988, it first cited less than a year later, in February 1989, with the description of a new genus and species of catfish – the pimelodid *Gladioglanis machadoi* Ferraris and Mago-Leccia, 1989 that is now known to reach up to 34 mm SL, surpassing the threshold established by Weitzman and Vari (1988). Ever since its publication, this work has been widely cited and adopted in most works focused on small-sized fishes because it provided a formal context defining a miniature fish. This formal definition allowed knowledge accumulation and a good amount of scientific literature that fosters new insights and questions on the evolution of minute fishes – in the early days after this concept was created, most works were concentrated on species description based on morphological data and a few ecological observations. Today, it is much more common that

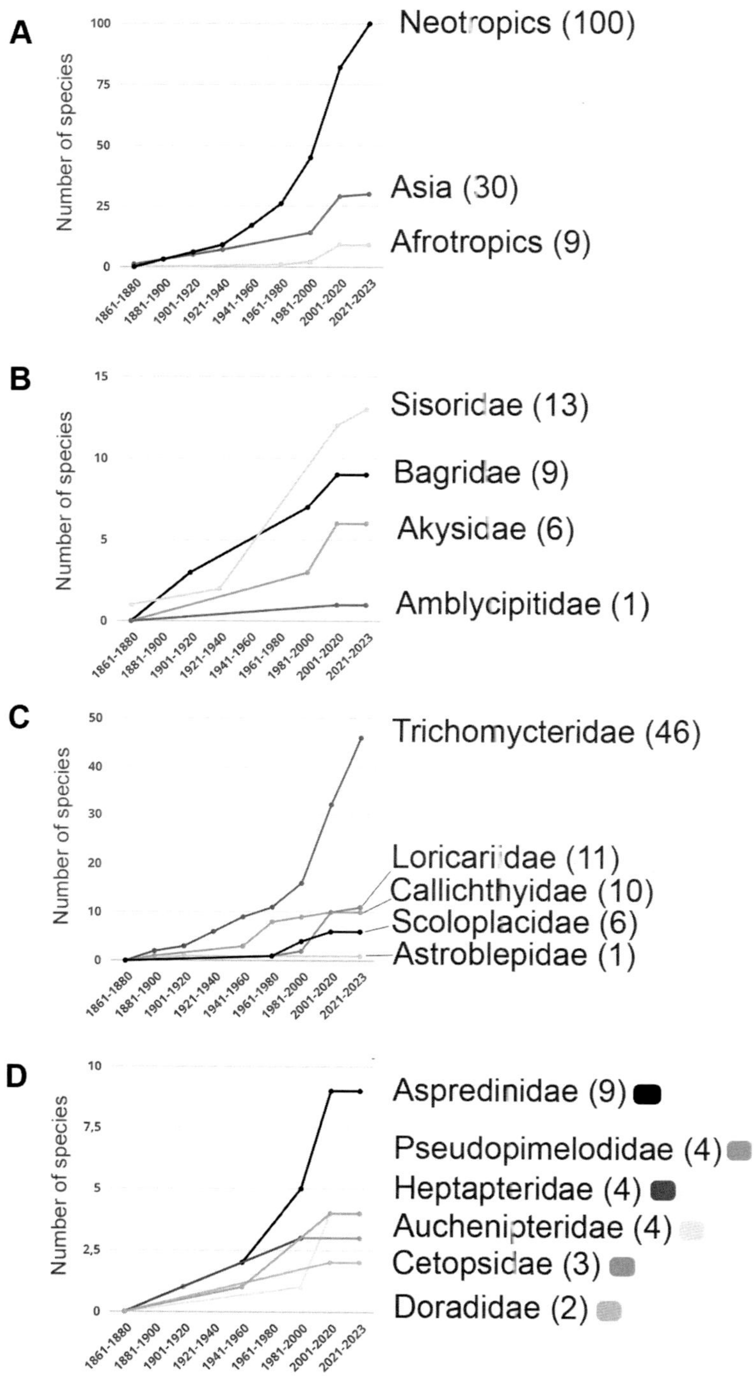

Figure 4. (A) Cumulative number of miniature catfish species described for the Neotropical, Asian and Afrotropical regions per 20-year period. (B) Species descriptions per family of Asian miniature species. (C) Species descriptions per family of Neotropical miniature Loricarioidei. (D) Species descriptions per family of miniature Siluroidei.

studies approaching miniature fishes (not only catfishes) raise questions looking retrospectively at knowledge gaps and try to answer them through multidisciplinary approaches based on morphology, ecology and genetics.

However, Weitzman and Vari's (1988) concept of miniaturization seems to have had a different impact on the knowledge and research conducted on Asian and African miniatures. At the same time, detailed studies on the development and the morphology of African and Asian miniature species have been published; apparently, no impact was seen on the rate of studies investigating the alpha taxonomy of these little fishes. A contrasting difference between these three highly diverse regions consists in the different taxonomic impediments they face, with the Neotropical region being the only one with a wide and active community of local ichthyologists and natural history collections. Most studies on Neotropical ichthyology, including those on the diversity of miniature species, are led by local taxonomists, whereas those on African and Asian miniatures are conducted by a few researchers from Europe and the US. This historical pattern is seen not only in the study of miniatures but in the taxonomic study of African fishes as a whole. For example, the formation of a future generation of taxonomists in Africa, which is the only way to speed up biodiscovery, is extremely difficult without the creation of regional and local natural history collections, which will allow the absorption of knowledge transfer from many key exchange and training programs funded by the Global North. As described in detail by Skelton and Swartz (2011), within a biodiversity crisis and a climate change scenario, it is critical that African countries develop their own capacities for fish taxonomy, balancing between increasing scientific capacity and limited economic resources. Advances in taxonomic knowledge are probably the only way to promote biodiversity and ecosystem conservation, sustainable use of natural resources, food security, and secure the well-being of millions of people. Ultimately, the low number of miniature species known from Africa and Asia have deep historical roots, and only with more local taxonomists unbiased comparative studies on the diversity, richness, and estimates of miniatures between all three regions will be possible.

As seen in Figure 4A, the rate of description of miniature catfish species from the Neotropics has been increasing, especially since the year 2000, while the rate of new miniature species described from the Afrotropics and Asia is stagnant. In South America, there are more than ten research groups from distinct institutions focusing their efforts on the systematics and taxonomy of Siluriformes, conducting fieldwork and consistently publishing papers – a much higher number than African or Asian-based institutions. Within the Neotropical region, most of this production is reflected in the high number of species of the superfamily Loricarioidei when compared to other families allocated in Siluroidei (Figure 4B, C and D). The family Trichomycteridae accounts for the vast majority of miniature catfish species (a total of 46, almost half of all Neotropical miniature catfishes), and the rate of description keeps increasing, being then the greatest responsible for the growing pace of new species discoveries seen among Neotropical catfish species. The most recent miniature catfish species from Asia was described recently after a seven-year gap, the sisorid *Pseudolaguvia permaris* Vijayakrishnan, Praveenraj and Mishra, 2023. Sisoridae comprises more miniaturized catfishes than all other Asian families, with most of its small-sized species being described in the 2000s. The knowledge gap is even bigger for Afrotropical minute catfishes: the most recent descriptions date from 2011. However, this evident Linnean shortfall in the study of miniature and the apparent stagnant rate of descriptions of species from Africa and Asia does not mean that their diversity is necessarily lower when compared to the Neotropical ichthyofauna. The use of multidisciplinary approaches based on morphology, ecology and genetics will reveal a myriad of undescribed miniature catfishes in the upcoming years in the Afro- and Neotropics (Henschel and Bragança pers. obs.) and might do the same for Asian species.

Knowledge Gaps and the Future of Research of Miniature Siluriforms

Many shortfalls hamper a proper assessment of the diversity, evolution, ecology and conservation of miniature catfishes as herein reviewed. Table 5 exemplifies minute siluriform species that can

Table 5. Small-sized catfish species that slightly surpass the cutoff size of 26 mm SL but lack comparative data (potential miniatures).

	SL (Max)	Reference
Akysidae		
Akysis microps	28.8	Ng and Tan 1999
Akysis pulvinatus	29.6	Ng 2007
Akysis bilustris	27.2	Ng 2011
Parakysis hystriculus	28.2	Ng 2009b
Pseudobagarius fuscus	28.8	Ng and Lim 2008
Bagridae		
Batasio spilurus	42	Ng and Lim 2008
Sisoridae		
Erethistes minusculus	31	Ng and Kottelat 2007
Pseudolaguvia shawi	28.3	Hora 1921
Pseudolaguvia kapuri	28.1	Tilak and Husain 1975
Pseudolaguvia tenebricosa	31.5	Britz and Ferraris 2003
Pseudolaguvia inornata	27.4	Ng 2005
Pseudolaguvia ferruginea	28.9	Ng 2009c
Pseudolaguvia viriosa	27.2	Ng and Tamang 2012
Pseudolaguvia lapillicola	28.7	Britz et al. 2013
Pseudolaguvia jiyaensis	31.2	Tamang and Sinha 2014
Pseudolaguvia nepalensis	27.4	Rayamajhi et al. 2016
Pseudolaguvia meghalayaensis	28.7	Lokeshwor and Marak 2022
Callichthyidae		
Aspidoras psammatides	28.5	Britto et al. 2005
Aspidoras velites	31	Britto et al. 2002
Pseudopimelodidae		
Microglanis leniceae	33	Shibatta 2016

be potentially considered miniatures, considering both miniaturization concepts: the size-based one by Weitzman and Vari (1988) and the phylogenetic one by Bragança and Burns (*in press*). This means that the listed species are usually small-sized, surpassing the 26 mm SL limit or are among the smallest of their respective genera. However, the general lack of information on their osteology and/or phylogenetic relationships makes it, at this point, impossible to establish whether they are miniatures or not.

The current underestimated number of described miniaturized siluriforms is the main impediment to any other related study – of the globally known nearly 4,200 species of Siluriformes, only 135 are miniatures, a number that might increase in the near future. The initial barriers to research focused on small fishes were the lack of collection efforts directed to specific microhabitats (such as sandbanks, gravel rocks, vegetation, etc.), the lack of scientists able to identify adult miniatures as actual mature specimens, with many specimens being mistakenly identified as juvenile stages of larger species, and curatorial issues with fixing and preserving very small specimens. Nowadays, as reflected in the growing number of descriptions of miniatures, especially in the Neotropics (Figure 4A), collection efforts are more directed toward specialized microhabitats and rely on proper and finer mesh-sized nets, including tissue sampling for DNA sequencing. This leads to a wider availability of small-sized specimens for destructive but informative procedures, such as anatomical preparations for osteology and dissection for musculature and examination of stomach content.

Historically, morphology was the main source of information for miniature fishes as a whole. Although most of the first descriptions of miniaturized fishes were limited to taxonomic descriptions of taxa, many of these early studies provided interesting insights into the likely morphological effects of miniaturization among fishes. Some anatomical effects of miniaturization in fishes, such as reduction in the latero-sensory canal system, number of vertebrae, fin rays and body scales (bony plates, in the case of Siluriformes) and lower sculpturing of skull bones were first reported by Myers (1958) and later reinforced by Weitzman and Vari (1988). Reduction in the pores of the latero-sensory canal system reaches an extreme condition in the trichomycterid *Potamoglanis anhanga* Dutra, Wosiacki and de Pinna, 2012 that entirely lacks the cephalic portion of the system, being restricted to two lateral line pores. However, it is important to emphasize that pores are extremely difficult to see in such small specimens, thus examining both cleared and stained and ethanol-fixed specimens under different lightning conditions might reveal additional pores that were not previously noticed. Despite a common reduction in vertebrae and/or fin ray counts when compared to larger non-miniature related species, it is not necessarily seen among all miniature catfishes; some species have more vertebrae and fin rays than other closely related and larger species. Moreover, selection pressures toward reducing appendages might be more related to a burrowing lifestyle than to miniaturization, given that larger interstitial species also exhibit fewer fin rays or even the entire loss of some fins (Adriaens et al. 2010). A lower degree of skull bone sculpturing usually reflects on non-ossified cranial bones, with large cartilaginous areas (or even entirely cartilaginous elements), in the presence of wide cranial fontanels (reduction in the skull roof area) and the loss of whole or part of structures, such as metapterygoid, pterosphenoid, fin radials, and processes in larger bones. Other osteological characters are also subject to structural simplification due to miniaturization, such as the pelvic- and caudal-fin skeletons, which, in the first case, is commonly lost among miniature catfishes, while many hypural plate elements might be cartilaginous. It is important to note that, in spite of having many reductive/paedomorphic morphological character states, miniatures may also exhibit newly acquired morphological innovations (i.e., derived traits) that differentiate them from other lineages. This is the case of the spiny-dwarf catfish species of the family Scoloplacidae, which share a unique dermal plate pattern, a large shield-shaped rostral plate that bears odontodes, and distinctive mesethmoid and lateral ethmoids. Another marked characteristic of miniaturized fishes is their coloration pattern. Species that usually inhabit leaf litter or rocky substrates have a cryptic coloration pattern consisting of brownish/yellowish ground color covered by dark blotches, saddles, stripes and other patterns, as seen in miniature aspredinids, astroblepids, loricariids, pseudopimelodids, scoloplacids, some bagrids, and akysids. On the other hand, many miniaturized siluriforms exhibit distinct degrees of body transparency: from the entirely diaphanous species of the trichomycterid genera *Tridens* Eigenmann and Eigenmann, 1889, *Tridensimilis* Eigenmann and Eigenmann, 1889, *Sarcoglanis* Myers and Weitzman, 1966, and *Malacoglanis* Myers and Weitzman, 1966 to catfishes with translucent bodies but with discrete pigmentation patterns, such as species of *Ammoglanis* Costa, 1994, *Microcambeva* Costa and Bockmann, 1994, and the African amphiliid genus *Zaireichthys*. Reduced body pigmentation is a common feature of meiofaunal inhabitants, not only vertebrates, and is considered an important adaptation to an interstitial lifestyle, assuring camouflage in sandy habitats (Giere 1993). Regardless of the evolutionary constraints imposed by the miniaturization process and its morphological effects, miniaturized catfishes exhibit a striking degree of morphological diversity (Figure 5).

Under the light of all morphological evidence of the effects of miniaturization and more recent works that adopt DNA-based analysis, a question that was raised in the dawn of the studies of miniature fishes and persists until today is: how many times miniaturization occurred within a given lineage? This core question in the study of the evolution of small fishes was first raised by Weitzman and Vari (1988) and approached by subsequent authors for distinct groups of fishes. By the time their idea was published, there were only 26 species of miniature siluriforms (and, ever since, some of them

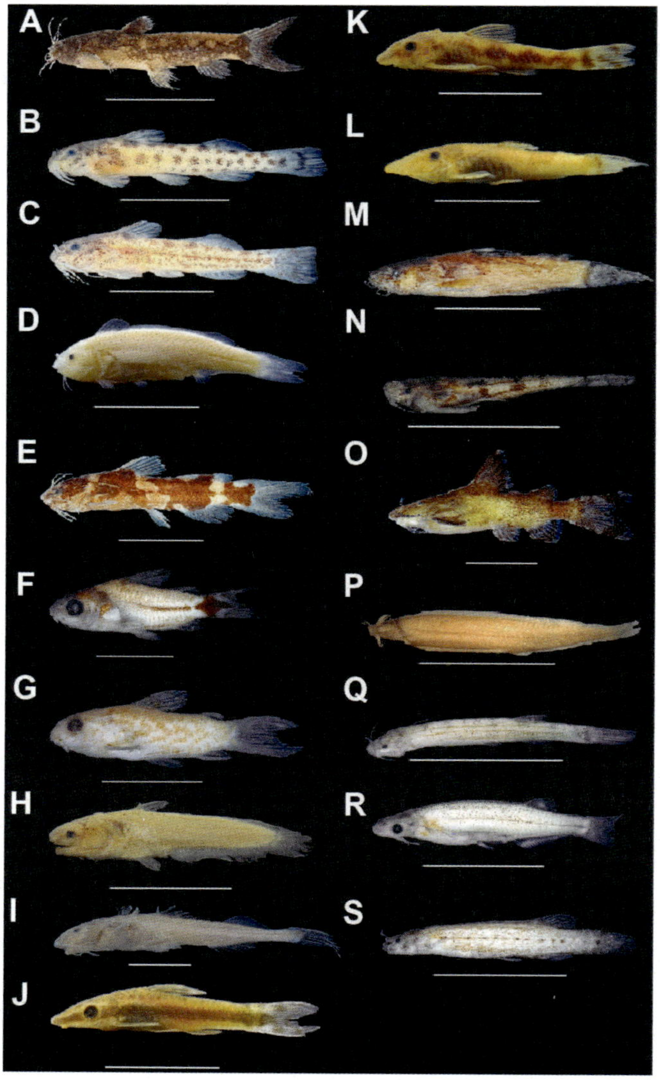

Figure 5. Morphological diversity of miniature catfishes. (A) *Parakysis hystriculus*, MZB 10985, 27.6 mm SL. (B) *Zaireichthys kunenensis*, SAIAB PBNAM03, 23.7 mm SL. (C) *Z. pallidus*, SAIAB 189968, 24.1 mm SL. (D) *Gelanoglanis varii*, MCP 49170, 17.7 mm SL. (E) *Nanobagrus lemniscatus*, ZRC 51981, 21.4 mm SL. (F) *Gastrodermus hastatus*, UFRJ 12284, 15.7 mm SL. (G) *G. pauciradiatus*, UFRJ 9153, 18.5 mm SL. (H) *Denticetopsis sauli*, ANSP 161432, 16.5 mm SL. (I) *Horiomyzon retopinnatus*, LIRP 10001, 25.0 mm SL. (J) *Otocinclus tapirape*, MCP 45581, 23.1 mm SL. (K) *Hisonotus bocaiuva*, MCP 36874, 22.3 mm SL. (L) *Rhinotocinclus halbothi*, MCP 48030, 19.4 mm SL. (M) *Microglanis oliveirai*, UFRJ 12214, 20.7 mm SL. (N) *Scoloplax dicra*, UFRJ 12217, 13.6 mm SL. (O) *Pseudolaguvia muricata*, UMMZ 248803, 25.6 mm SL. (P) *Pygidianops cuao*, AMNH 232978, 18.5 mm SL. (Q) *Ammoglanis obliquus*, UFRJ 12448, 13.7 mm SL. (R) *Tridentopsis pearsoni*, UFRJ 12815, 15 mm SL. (S) *Potamoglanis johnsoni*, UFRJ 12283, 15.6 mm SL. Scale bar = 10 mm.

were synonymized or removed from subsequent lists, as in the case of *Gastrodermus pauciradiatus*), phylogenetic hypothesis for the distinct groups of the order were very scarce, and the only available works relied on morphological data. Given this scenario, they considered that miniaturization occurred 12 times within Neotropical Siluriformes: in the ancestor of Scoloplacidae, independently in each of the trichomycterid subfamilies (Trichomycterinae, Sarcoglanidinae, Glanapteryginae, Vandelliinae, and Tridentinae); twice within Callichthyidae, in the genera *Aspidoras* and (based on species now allocated in the genus *Gastrodermus* Cope, 1878) and *Corydoras* Lacépède, 1803; twice in Aspredinidae, in the genera *Hoplomyzon* Myers, 1942 and *Dupouyichthys* Schultz, 1944; and once each within Pimelodidae and Loricariidae. Clearly, given the increased number of recognized miniature catfishes (hence, a higher number of available characters and terminals for phylogenetic analysis) and the inclusion of Afrotropical and Asian miniature species in evolutionary studies, we can easily assume that a much higher number of miniaturization events occurred in the Siluriformes. The widespread use of molecular techniques for assessing phylogenetic relationships from the early 2000s until today also challenges traditional phylogenetic hypotheses for most groups – and it represents a major advance, especially for inferring relationships of miniatures with closely related lineages. Morphological convergence is a common effect within miniaturized species: extreme reduction in body size might lead to similar reductions, losses, and structural simplifications that occurred, in fact, independently (i.e., homoplasies). Thus, adopting DNA sequences as the primary source for inferring phylogenetic relationships, whether in a total evidence matrix or using morphological characters only for ancestral state reconstruction, reduced the use and weight of homoplasies that were eventually misidentified as homologous characters, an error that led to the erroneous assumptions that miniature species are more closely related to each other than to other lineages. For example, within trichomycterids, not only has the number of miniature species increased from 14 in 1988 to 46 today, but also many advances in phylogenetic studies questioned traditional hypotheses. Weitzman and Vari (1988) listed the miniature trichomycterines *Trichomycterus hasemani* (Eigenmann, 1914), *T. johnsoni* (Fowler, 1932), and *T. santaeritae* (Eigenmann, 1918) as more closely related to each other than to other trichomycterine species, considering a single miniaturization event for the subfamily. Later, de Pinna (1989) suggested that *T. santaeritae* would be related to *Scleronema* Eigenmann, 1917, whereas *T. hasemani* and *T. johnsoni* would be related to the Tridentinae. More recently, Dutra et al. (2012) coined the term "*Trichomycterus hasemani* group" to illustrate a close relationship between *T. hasemani*, *T. johnsoni* and the newly discovered *T. anhanga*, considering a single miniaturization event for these three species and another for *T. santaeritae*. However, five years later Henschel et al. (2018), based on a molecular phylogeny, recovered the "*T. hasemani* group" as a new genus of the Tridentinae, naming it *Potamoglanis* Henschel, Mattos, Katz and Costa, 2018. In the subsequent year, Costa et al. (2019) redescribed the enigmatic *T. santaeritae* and discovered that it is not miniature. This discovery was followed by the molecular phylogeny by Ochoa et al. (2020), which placed *Potamoglanis* as sister to a clade composed of Amazonian sarcoglanidines and glanapterygines, being the genus now considered *incertae sedis* in Trichomycteridae. Another striking example of how molecular-based phylogenies are challenging traditional relationship hypotheses of the evolution of small-sized taxa was the description of the subfamily Microcambevinae, which comprises the genera *Microcambeva* (with six miniature species) and *Listrura* de Pinna, 1988 that were previously allocated in the Sarcoglanidinae and Glanapteryginae, respectively. These are just examples of how new taxa and new methods impact hitherto accepted hypotheses and, as a consequence, affect the understanding of how many times miniaturization occurred and its impacts on morphology and ecology.

Even though deep morphological assessments on reductions, losses, and simplifications of miniature catfish species are scarce, anatomy is more or less present in species descriptions. Unfortunately, that is not the case for ecological data. Information on habitat is usually common in taxonomic works (within "ecological data"), although the content *per se* is not the same. While some papers present a detailed description of the habitat, including substrate type, watercourse depth and width, water type, and rainfall seasonality, most papers are limited to a gross analysis of substrate type, and others do not even present any habitat preference information. As discussed before, labeling substrates as "sand" or "rocks," for example, masks finer properties of habitats, such as interstice size, water flow, and others. Clearer and deeper habitat analysis requires funding for proper equipment, trained staff, multidisciplinary collaboration, and consistency throughout distinct fieldwork in collecting the same data. As illustrated in Tables 1–4, substrate type data for neo- and Afrotropical species are widely available, which is not the case for Asian species, hampering comparative habitat preference studies. The lack of information on water type and flow is another major obstacle for this type of analysis. Roberts (1972) provided the first wide assessment of comparative environmental data for the fishes in the Amazon and Congo basins, contemplating factors such as water type, rainfall regime, tides, nutrient availability, and seasonal fluctuations in water level and considered a small-body size as an adaptive response to predation and food resource exploitation. If more data on neo- and Afrotropical and Asian species were available, a wider comparative study encompassing miniatures from these three biogeographic realms would be feasible. In that sense, more comprehensive data on miniatures' dietary preferences are also lacking – most detailed ecological information is only available for Neotropical species (Tables 1–4; Figure 2), but also in terms of other variables, such as ontogenetic or seasonal changes on diet, taxonomic identification of prey, quantification of classes of food items and others. Overall, the trophic ecology of fishes is understudied, but in the case of miniature species, the scenario is worsened by their small sizes (thus being prone to destruction during dissection) and, in many cases, the low number of specimens available in scientific collections. Also, identifying microscopic (e.g., diatoms, protists, and bacteria) and liquid (e.g., mucus, in the case of parasitic species) food items require additional microscopy steps and preparations and are not usually made unless in the case of ecology-focused studies, not in taxonomic descriptions.

Conclusions

The miniaturization concepts of Weitzman and Vari (1988) and Bragança and Burns (*in press*) share the assumption that to be considered miniatures, small-sized fish species should exhibit paedomorphic and/or reductive morphological characters. However, their ideas on the definition of size differ markedly: the first adopts an arbitrary cutoff size of 26 mm SL for miniatures, whereas the second considers a reduction in size when compared to closely related lineages; thus, species larger than 26 mm SL can also be considered miniatures. Adopting both concepts from a global perspective (i.e., extending beyond the Neotropics) increases our understanding of the diversity of miniaturized siluriforms and the effect of this evolutionary process in species' anatomy and ecology; we now consider a total of 135 miniature catfish species, when compared to the total of 81 of the last two lists (Toledo-Piza et al. 2014; Bragança et al. 2023).

In this chapter, we presented the current state of the art of the knowledge of minute members of Siluriformes and also pointed out the flaws and future directions for studies dedicated to unveiling their underestimated and unknown world. There are still many miniature species to be described and even more to discover about their relationships, dietary habits, habitat preferences, and anatomies.

Acknowledgments

We thank Jonathan Baskin for comments on the manuscript, Marcelo Britto for information on miniature species of *Aspidoras*, and Dione Seripierri for making available scientific literature. The authors thank the American Museum of Natural History (AMNH Ichthyology Department through Radford Arrindell and Scott Schaefer) for providing photographs of *Pygidianops cuao*; the Academy of Natural Sciences of Drexel University through Kyle R Luckenbill for photographs of *Denticetopsis sauli*; Heok Hee Ng for photographing *Nanobagrus lemniscatus*, *Parakysis hystriculus*, and *Pseudolaguvia muricata* and for sharing important literature; the NRF-SAIAB ichthyological collection for photographing specimens of *Zaireichthys*; Flávio Bockmann (FFCLRP/USP) and A. Esguícero (USP) for photographing *Horiomyzon retropinnatus*; Roberto Reis (PUCRS) for photographing *Gelanoglanis varii*, *Otocinclus tapirape*, *Hisonotus bocaiuva*, and *Rhinotocinclus halbothi*; and Wilson Costa (UFRJ) for making specimens of *Gastrodermus hastatus*, *G. pauciradiatus*, *Microglanis oliveirai*, *Scoloplax dicra*, *Ammoglanis obliquus*, *Tridentopsis pearsoni*, and *Potamoglanis johnsoni* available for photography.

References

Adriaens, D., J.N. Baskin and H. Coppens. 2010. Evolutionary morphology of trichomycterid catfishes: about hanging on and digging in. pp. 337–362. *In*: Nelson, J.S., H.-P. Schultze and M.V.H. Wilson (eds.). Origin and Phylogenetic Interrelationships of Teleosts. Verlag Dr. Friedrich Pfeil, München, Germany.

Alexander, R.M. 1966. Structure and function in the catfish. J. Zool. 1481: 88–152.

Baskin, J.N., T.M. Zaret and F. Mago-Leccia. 1980. Feeding of reportedly parasitic catfishes Trichomycteridae and Cetopsidae in the Rio Portuguesa Basin, Venezuela. Biotropica: 182–186.

Betancur-R, R., R.E. Broughton, E.O. Wiley., K. Carpenter, J.A. López, C. Li et al. 2013. The tree of life and a new classification of bony fishes. PLOS Currents Tree of Life. doi: 10.1371/currents.tol.53ba26640df0ccaee75bb165c8c26288.

Betancur-R, R., E.O. Wiley, G. Arratia, A. Acero, N. Bailly, M. Miya et al. 2017. Phylogenetic classification of bony fishes. BMC Evol. Biol. 17: 1–40. https://doi.org/10.1186/s12862-017-0958-3

Boaden, P.J.S. 1968. Water movement—a dominant factor in interstitial ecology. Sarsia 341: 125–136.

Bonato, K.O., P.C. Silva and L.R. Malabarba. 2018. Unrevealing parasitic trophic interactions—a molecular approach for fluid-feeding fishes. Front. Ecol. Evol. 6: 1–8.

Bonato, K.O., P.C. Silva, F.R. Carvalho and L.R. Malabarba. 2022. Trophic interactions of vampire catfishes Siluriformes: Vandelliinae revealed by metabarcoding analysis of stomach contents. Freshw. Biol. 673: 542–548.

Brandl, S.J., C.H. Goatley, D.R. Bellwood and L. Tornabene. 2018. The hidden half: ecology and evolution of cryptobenthic fishes on coral reefs. Biol. Rev. 934: 1846–1873.

Bragança, P.H.N., J. van der Zee, P.S. Lobel and U. Schliewen. 2023. Osteology and redescription of the miniature lampeye species of *Congopanchax* Poll, 1971 with an overview of miniature African freshwater fishes. Spixiana 46: 91–118.

Braganca, P.H.N. and M.D. Burns. *in press*. Body size evolution in Neotropical freshwater fishes: Patterns and processes of reduction and miniaturization. *In*: Albert, J.S. and R.E. Reis (eds.). Ecology and Evolution of Amazonian Fishes. Springer.

Britto, M.R., F.C.T. Lima and C.R. Moreira. 2002. *Aspidoras velites*, a new catfish from the upper rio Araguaia basin, Brazil (Teleostei: Siluriformes: Callichthyidae). Proc. Biol. Soc. Wash. 115: 727–736.

Britto, M.R., F.C.T. Lima and A.C.A. Santos. 2005. A new *Aspidoras* (Siluriformes: Callichthyidae) from rio Paraguaçu basin, Chapada Diamantina, Bahia, Brazil. Neotrop Ichthyol 3: 473–479.

Britz, R. and C.J. Ferraris, Jr. 2003. A new species of the Asian catfish genus *Pseudolaguvia* from Myanmar (Teleostei: Ostariophysi: Siluriformes: Erethistidae). Zootaxa 388: 1–8.

Britz, R., A. Ali and R. Raghavan. 2013. *Pseudolaguvia lapillicola*, a new species of catfish from peninsular India (Teleostei: Sisoridae). Ichthyol. Explor. Freshw. 23: 289–295.

Calegari, B.B., R.E. Reis and R.P. Vari. 2014. Miniature catfishes of the genus *Gelanoglanis* (Siluriformes: Auchenipteridae): monophyly and the description of a new species from the upper rio Tapajós basin, Brazil. Neotrop Ichthyol. 12: 699–706.

Calegari, B.B. and R.E. Reis. 2016. New species of the miniature genus *Gelanoglanis* (Siluriformes: Auchenipteridae) from the Tocantins River basin (Brazil) and osteological description of *G. nanonocticolus*. J. Fish Biol. 90(5): 1702–1716.

Carvalho, M. and A. Datovo. 2012. A new species of cascudinho of the genus *Hisonotus* (Siluriformes: Loricariidae: Hypoptopomatinae) from the upper Rio Tapajós basin, Brazil. Copeia 2012: 266–275.

Carvalho, T.P., J.G. Lundberg, J.N. Baskin, J.P. Friel and R.E. Reis. 2016. A new species of the blind and miniature genus *Micromyzon* Friel and Lundberg, 1996 (Silurifomes: Aspredinidae) from the Orinoco River: describing catfish diversity using high-resolution computed tomography. Proc. Acad. Nat. Sci. Philadelphia 165: 37–53.

Carvalho, T.P., R.E. Reis and J.P. Friel. 2017. A new species of *Hoplomyzon* (Siluriformes: Aspredinidae) from Maracaibo Basin, Venezuela: osteological description using high-resolution computed microtomography of a miniature species. Neotrop. Ichthyol. 15: 1–16.

Carvalho, T.P. and R.E. Reis. 2020. A new miniature species of *Acanthobunocephalus* (Silurifomes: Aspredinidae) from the lower Purus River basin, Amazon Basin, Brazil. Copeia 108: 347–357.

Castro, Í.S. and W.B. Wosiacki. 2017. *Ituglanis compactus*, a new species of catfish (Siluriformes: Trichomycteridae) from the rio Jari drainage, lower Amazon, Brazil. Zootaxa 4244: 207–218.

Conway, K.W. and T. Moritz. 2006. *Barboides britzi*, a new species of miniature cyprinid from Benin Ostariophysi: Cyprinidae, with a neotype designation for *B. gracilis*. Ichthyol. Explor. Freshw. 17: 73–84.

Costa, W.J.E.M. and F.A. Bockmann. 1994. A new genus and species of Sarcoglanidinae (Siluriformes: Trichomycteridae) from southeastern Brazil, with a re-examination of subfamilial phylogeny. J. Nat. Hist. 28: 715–730.

Costa, W.J.E.M. and P.Y. Le Bail. 1999. *Fluviphylax palikur*: a new poeciliid from the Rio Oiapoque Basin, Northern Brazil (Cyprinodontiformes: Cyprinodontoidei) with comments on miniaturization in *Fluviphylax* and other neotropical freshwater fishes. Copeia 1999: 1027–1034.

Costa, W.J.E.M., A.M. Katz, J.L.O. Mattos and F.S. Rangel-Pereira. 2019. Two new species of miniature psammophilic sarcoglanidine catfishes of the genus *Microcambeva* from the Atlantic Forest of eastern Brazil (Siluriformes: Trichomycteridae). J. Nat. Hist. 53: 1837–1851.

Costa, W.J.E.M., J.L.O. Mattos and A.C.A. Santos. 2019. *Ammoglanis multidentatus*, a new miniature sanddwelling sarcoglanidine catfish with unique osteological features from northeastern Brazil (Siluriformes: Trichomycteridae). Vertebr. Zool. 70: 61–67.

Costa, W.J.E.M., A.M. Katz, J.L.O. Mattos, P.F. Amorim, B.O. Mesquita, P.J. Vilardo et al. 2019. Historical review and redescription of three poorly known species of the catfish genus *Trichomycterus* from south–eastern Brazil (Siluriformes: Trichomycteridae). J. Nat. Hist. 53(47–48): 2905–2928.

Costa, W.J.E.M., E. Henschel and A.M. Katz. 2020. Multigene phylogeny reveals convergent evolution in small interstitial catfishes from the Amazon and Atlantic forests (Siluriformes: Trichomycteridae). Zool. Scr. 492: 159–173.

Dagosta, F.C.P. and M.C.C. de Pinna. 2021. Two new catfish species of typically Amazonian lineages in the Upper Rio Paraguay (Aspredinidae: Hoplomyzontinae and Trichomycteridae: Vandelliinae), with a biogeographic discussion. Pap. Avulsos Zool. 61: e20216147 [1–23].

Datovo, A. and M.I. Landim. 2005. *Ituglanis macunaima*, a new catfish from rio Araguaia basin, Brazil (Siluriformes, Trichomycteridae). Neotrop Ichthyol 3: 455–464.

Dutra, G.M., W.B. Wosiacki and M.C.C. de Pinna. 2012. *Trichomycterus anhanga*, a new species of miniature catfish related to *T. hasemani* and *T. johnsoni* (Siluriformes: Trichomycteridae) from the Amazon basin, Brazil. Neotrop. Ichthyol. 10: 225–231.

Eccles, D.H., D. Tweddle and P.H. Skelton. 2011. Eight new species in the dwarf catfish genus *Zaireichthys* (Siluriformes: Amphiliidae). Smithiana 13: 3–28.

Eigenmann, C.H. and R.S. Eigenmann. 1888. Preliminary notes on South American Nematognathi. I. Proc. Calif. Acad. Sci. 2: 119–172.

Eigenmann, C.H. 1918. The Pygidiidae, a family of South American catfishes. Mem. Carnegie Mus. 7: 259–398.

Ferraris, C.J., Jr. and F. Mago-Leccia. 1989. A new genus and species of pimelodid catfish from the Río Negro and Río Orinoco drainages of Venezuela (Siluriformes: Pimelodidae). Copeia 1989: 166–171.

Ferraris, C.J., Jr. 1996. Denticetopsis, a new genus of South American whale catfish (Siluriformes: Cetopsidae, Cetopsinae), with two new species. Proceedings of the California Academy of Sciences 49(6): 161–170.

Ferraris, C.J., Jr. and R. Britz. 2005. A diminutive new species of *Glyptothorax* (Siluriformes: Sisoridae) from the upper Irrawaddy River basin, Myanmar, with comments on sisorid and erethistid phylogenetic relationships. Ichthyol. Explor. Freshw. 16: 375–383.

Fricke, R., W.N. Eschmeyer and R. van der Laan (eds.). 2023. Eschmeyer's catalog of fishes: Genera, species, references. (http://researcharchive.calacademy.org/research/ichthyology/catalog/fishcatmain.asp). Electronic version accessed 12 Nov 2023.

Garavello, J.C. 1988. Three new species of *Parotocinclus* Eigenmann & Eigenmann, 1889 with comments on their geographical distribution (Pisces, Loricariidae). Naturalia (São Paulo) 13: 117–128.

Giere, O. 1993. Meiobenthology: The Microscopic Fauna in Aquatic Sediments. Springer–Verlag, Berlin, Germany.

Giller, P.S. and B. Malmqvist. 1998. The Biology of Streams and Rivers. Oxford University Press, London, UK.

Hanken, J. and D.B. Wake. 1993. Miniaturization of body size: organismal consequences and evolutionary significance. Annu. Rev. Ecol. Evol. Syst. 241: 501–519.

Henschel, E. 2016. A new catfish species of the *Trichomycterus hasemani* group (Siluriformes: Trichomycteridae), from the Branco river basin, northern Brazil. Vertebr. Zool. 66: 117–123.

Henschel, E., J.L.O. Mattos, A.M. Katz and W.J.E.M. Costa. 2018. Position of enigmatic miniature trichomycterid catfishes inferred from molecular data (Siluriformes). Zool. Scr. 471: 44–53.

Henschel, E., N.K. Lujan and J.N. Baskin. 2020a. *Ammoglanis natgeorum*, a new miniature pencil catfish (Siluriformes: Trichomycteridae) from the lower Atabapo River, Amazonas, Venezuela. J. Fish Biol. 97: 1481–1490 [1–10].

Henschel, E., P.H.N. Bragança, F. Rangel-Pereira and W.J.E.M. Costa. 2020b. A new psammophilic species of the catfish genus *Ammoglanis* (Siluriformes, Trichomycteridae) from the Amazon River basin, northern Brazil. Zoosyst. Evol. 96: 67–72.

Henschel, E., A.M. Katz and W.J.E.M. Costa. 2021a. A new candiru of the genus *Paracanthopoma* (Siluriformes: Trichomycteridae) from the Araguaia River basin, central Brazil. J Fish Biol. 996: 1990–1997.

Henschel, E., M.J. Bernt, J.N. Baskin, R.E. Schmidt and N.K. Lujan. 2021b. Osteology-focused redescription and description of the blood-feeding candirus *Paracanthopoma parva* and *Paravandellia alleynei* sp.n. (Trichomycteridae: Vandelliinae). J. Fish Biol. 100: 161–174.

Henschel, E., W.M. Ohara and W.J.E.M. Costa. 2023. Two new miniature translucent catfish species of the rare genus *Tridens* (Siluriformes: Trichomycteridae) from the Madeira River basin, northern Brazil. J. Fish Biol. 102: 1–17.

Hora, S.L. 1921. On some new or rare species of fish from the eastern Himalayas. Rec. Indian Museum: 22: 731–744.

IUCN. 2012. IUCN Red List Categories and Criteria: Version 3.1. Second edition. Gland, Switzerland and Cambridge, UK.

Kottelat, M. and C. Vidthayanon. 1993. *Boraras micros*, a new genus and species of minute freshwater fish from Thailand (Teleostei: Cyprinidae). Ichthyol. Explor. Freshw. 4: 161–176.

Lehmann A., P.N.K. Lujan and R.E. Reis. 2022. A new species of armored catfish (Loricariidae: Hypoptopomatinae) syntopic and superficially similar to *Parotocinclus collinsae*, from the Potaro River basin, Guyana. Ichthyol. Herpetol. 110: 69–76.

Leitão, R.P., J. Zuanon., S. Villéger, S.E. Williams, C. Baraloto, C. Fortunel et al. 2016. Rare species contribute disproportionately to the functional structure of species assemblages. Proc. R. Soc. B. 283: 20160084. http://doi.org/10.1098/rspb.2016.0084.

Lokeshwor, Y. and P.D. Marak. 2022. *Pseudolaguvia meghalayaensis*, a new diminutive sisorid catfish (Teleostei: Siluriformes) from northeastern India. Zootaxa 5175: 367–376.

Lundberg, J.G., A.H. Bornbusch and F. Mago-Leccia. 1991. *Gladioglanis conquistador* n. sp. from Ecuador with diagnoses of the subfamilies Rhamdiinae Bleeker and Pseudopimelodinae n. subf. (Siluriformes: Pimelodidae). Copeia 1991: 190–209.

Mahapatra, B.K. and S. Kar. 2015. *Hara nareshi*, a new species of catfish (Pisces: Erethistidae) from the Barak River system of Assam, India. Rec. Zool. Surv. India 115: 31–35.

McCauley, M., D.P. German, N.K. Lujan and C.R. Jackson. 2020. Gut microbiomes of sympatric Amazonian wood-eating catfishes Loricariidae reflect host identity and little role in wood digestion. Ecol. Evol. 1014: 7117–7128.

Medeiros, L.S. de, C.R. Moreira, M.C.C. de Pinna and S.M.Q. Lima. 2020. A new catfish species of *Microcambeva* Costa & Bockmann 1994 (Siluriformes: Trichomycteridae) from a coastal basin in Rio de Janeiro State, southeastern Brazil. Zootaxa 4895: 111–123.

Medeiros, L.S. De, L.M. Sarmento-Soares and S.M.Q. Lima. 2021. A new psammophilous catfish of the genus *Microcambeva* (Teleostei: Trichomycteridae) from the Rio Doce basin, southeastern Brazil. Ichthyol. Explor. Freshw. 31: 1–13.

Myers, G.S. 1958 Trends in the evolution of teleostean fishes. Biol. Bull. 7: 27–30.

Ng, H.H. 1996. *Akysis heterurus*, a new species of catfish (Teleostei: Akysidae) from eastern Sumatra, with notes on *Akysis hendricksoni* Alfred. Raffles Bull. Zool. 44: 3–10.

Ng, H.H. and M. Kottelat. 1998a. The catfish genus *Akysis* Bleeker (Teleostei: Akysidae) in Indochina, with descriptions of six new species. J. Nat. Hist. 32: 1057–1097.

Ng, H.H. and M. Kottelat. 1998b. *Hyalobagrus*, a new genus of miniature bagrid catfish from southeast Asia (Teleostei: Siluriformes). Ichthyol. Explor. Freshw. 9: 335–346.

Ng, H.H. and H.H. Tan. 1999. The fishes of the Endau drainage, Peninsular Malaysia with descriptions of two new species of catfishes (Teleostei: Akysidae, Bagridae). Zoological Studies 38: 350–366.

Ng, H.H. and M. Kottelat 2000. A review of the genus *Amblyceps* (Osteichthyes: Amblycipitidae) in Indochina, with descriptions of five new species. Ichthyol. Explor. Freshw. 11: 335–348.

Ng, H.H. and M. Kottelat 2003. *Parakysis notialis*, a new species of akysid catfish from Borneo (Siluriformes: Akysidae). Ichthyol. Res. 50: 48–51.

Ng, H.H. 2005. Two new species of *Pseudolaguvia* (Teleostei: Erethistidae) from Bangladesh. Zootaxa 1044: 35–47.

Ng, H.H. and W.J. Rainboth. 2005. Four new species of *Akysis* (Teleostei: Siluriformes: Akysidae) from mainland southeast Asia, with comments on *A. similis*. Raffles Bull. Zool. Suppl. 13: 33–42.

Ng, H.H. 2006. *Pseudolaguvia ferula*, a new species of sisoroid catfish (Teleostei: Erethistidae) from India. Zootaxa 1229: 59–68.

Ng, H.H. 2007. *Akysis pulvinatus*, a new species of catfish (Siluriformes: Akysidae) from southern Thailand. Zootaxa 1608: 51–58.

Ng, H.H. and M. Kottelat. 2007. A review of the catfish genus *Hara*, with the description of four new species (Siluriformes: Erethistidae). Rev. Suisse Zool. 114: 471–505.

Ng, H.H. and K.K.P. Lim, 2008. Catalogue of the siluriform (Teleostei: Ostariophysi) types in the Raffles Museum of Biodiversity Research, National University of Singapore. Raffles Museum of Biodiversity Research: ii+ 1–74.

Ng, H.H. 2008. A new species of *Nanobagrus* (Teleostei: Bagridae) from southern Borneo. Copeia 2008: 93–98.

Ng, H.H. 2009a. *Akysis fontaneus*, a new species of catfish (Siluriformes: Akysidae) from Sumatra. Zootaxa 1973: 41–50.

Ng, H.H. 2009b. *Parakysis hystriculus*, a new species of catfish (Siluriformes: Akysidae) from Sumatra. Zootaxa 1977: 39–48.

Ng, H.H. 2009c. Two new species of *Pseudolaguvia*, sisorid catfishes (Teleostei: Siluriformes) from northeastern India. Ichthyol. Explor. Freshw. 20: 277–288.

Ng, H.H. 2010. A new species of minature catfish from the Malay Peninsula (Teleostei: Bagridae: Nanobagrus). Zootaxa 2677: 60–68.

Ng, H.H. 2011. *Akysis bilustris*, a new species of catfish from scuthern Laos (Siluriformes: Akysidae). Zootaxa 3066: 61–68.

Ng, H.H. and L. Tamang. 2012. *Pseudolaguvia viriosa*, a new catfish (Teleostei: Sisoridae) from northeastern India. Zootaxa 3522: 81–88.

Ng, H.H., C.J. Ferraris, Jr. and D.A. Neely. 2012. The catfish genus *Erethistoides* (Siluriformes: Sisoridae) in Myanmar, with descriptions of three new species. Zootaxa 3254: 55–68.

Ng, H.H. and K.W. Conway. 2013. *Pseudolaguvia assula*, a new species of crypto-benthic sisorid catfish from central Nepal (Teleostei: Sisoridae). Ichthyol. Explor. Freshw. 24: 179–185.

Ng, H.H., Lalramliana and S. Lalronunga. 2016. A new diminutive sisorid catfish (Actinopterygii: Siluriformes) from northeastern India. Zootaxa 4105: 546–556.

Ng, H.H. 2018. Under an assumed name: the identity of *Nanobagrus fuscus* (Popta, 1904) clarified (Actinopterygii: Siluriformes). Vertebr. Zool. 68: 39–45.

Ochoa, L.E., A. Datovo, C. DoNascimiento, F.F. Roxo, M.H. Sabaj, J. Chang et al. 2020. Phylogenomic analysis of trichomycterid catfishes Teleostei: Siluriformes inferred from ultraconserved elements. Sci. Rep. 10: 2697.

Perkin, J.S., C.G. Montaña, E.J. Nogueira, B.B. Brandão, G.M. Mattox and K.W. Conway. 2022. Estimated richness and environmental correlates of miniature fish assemblages in the rio Jacundá, Brazil. Neotrop. Ichthyol. 20: 1–16.

de Pinna, M.C.C. 1989. A new sarcoglanidine catfish, phylogeny of its subfamily, and an appraisal of the phyletic status of the Trichomycterinae (Teleostei, Trichomycteridae). Am. Mus. Novit. 2950: 1–39.

de Pinna, M.C.C. and W.B. Wosiacki. 2003. Family Trichomycteridae. pp. 270–290. *In*: Reis, R.E., S.O. Kullander and C.J. Ferraris, Jr. (eds.). Check List of the Freshwater Fishes of South and Central America. Edipucrs, Porto Alegre.

de Pinna, M.C.C. and F.C.P. Dagosta. 2022. A taxonomic review of the vampire catfish genus *Paracanthopoma* Giltay, 1935 (Siluriformes, Trichomycteridae), with descriptions of nine new species and a revised diagnosis of the genus. Pap. Avulsos Zool. 62: 1–90.

Rayamajhi, A., M. Arunachalam and Usharamalakshmi. 2016. *Pseudolaguvia nepalensis*, a new catfish (Teleostei: Siluriformes: Erethistidae) from central Nepal. Int. J. Zool. Stud. 4: 46–60.

Reis, R.E. 2003. Callichthyidae (Armored catfishes). pp. 291–309. *In*: Reis, R.E., S.O. Kullander and C.J. Ferraris, Jr. (eds.). Check List of the Freshwater Fishes of South and Central America. Edipucrs, Porto Alegre.

Ribeiro-Silva, L.R., G.S.C. Silva, P.C. Venere, H.P. da Silva and F.F. Roxo. 2020. Description of a new species of *Hisonotus* (Loricariidae: Siluriformes) from rio Araguaia basin. Zootaxa 4860: 553–562.

Roberts, T.R. 1972. Ecology of the fishes in the Amazon and Congo basins. Bull. Mus. Comp. Zool. Harvard Univ. 143: 117–147

Roberts, T.R. 2003. Systematics and osteology of Leptoglaninae, a new subfamily of the African catfish family Amphiliidae, with descriptions of three new genera and six new species. Proc. Calif. Acad. Sci. 54: 81–132.

Rocha, M.S., R.R. de Oliveira and L.H. Rapp Py-Daniel. 2008. A new species of *Gladioglanis* Ferraris & Mago-Leccia from rio Aripuanã, Amazonas, Brazil (Siluriformes: Heptapteridae). Neotrop. Ichthyol. 6: 433–438.

Roxo, F.F., G.S.C. Silva, C. Oliveira and C.H. Zawadzki. 2013. *Hisonotus bocaiuva*, a new species from the rio São Francisco basin, Brazil (Teleostei: Loricariidae). Ichthyol. Explor. Freshw. 23: 319–326.

Roxo, F.F., N.K. Lujan, V.A. Tagliacollo, B.T. Waltz, G.S. Silva, C. Oliveira et al. 2017. Shift from slow–to fast–water habitats accelerates lineage and phenotype evolution in a clade of Neotropical suckermouth catfishes Loricariidae: Hypoptopomatinae. PLoS One 12: e0178240.

Roxo, F.F., F.L. Messias and G.S.C. Silva. 2019. A new species of *Parotocinclus* (Loricariidae: Hypoptopomatinae) from Rio Tocantins basin, Brazil. Zootaxa 4646: 346–356.

Ruiz, W.B.G. and O.A. Shibatta. 2010. A new species of *Microglanis* (Siluriformes, Pseudopimelodidae) from lower Rio Tocantins basin, Pará, Brazil, with description of superficial neuromasts and pores of lateral line system. Zootaxa 2632: 53–66.

Ruiz, W.B.G. and O.A. Shibatta. 2011. Two new species of *Microglanis* (Siluriformes: Pseudopimelodidae) from the upper-middle rio Araguaia basin, central Brazil. Neotrop. Ichthyol. 9: 697–707.

Schultz, L.P. 1944. The catfishes of Venezuela, with descriptions of thirty-eight new forms. Proc. U. S. Nat. Mus. 94(3172): 173–338.

Shibatta, O.A. 2016. A new species of bumblebee catfish of the genus *Microglanis* (Siluriformes: Pseudopimelodidae) from the upper rio Paraguay basin, Brazil. Neotrop. Ichthyol. 14: e160031.

Silva, G.S.C., F.F. Roxo, C.S. Souza and C. Oliveira. 2018. Two new species of *Corumbataia* (Hypoptopomatinae: Loricariidae) from Rio Corrente, upper Rio Paraná basin, Brazil. Zootaxa 4483: 317–330.

Skelton, P.H. and E.R. Swartz. 2011. Walking the tightrope: trends in African freshwater systematic ichthyology. J. Fish Biol. 79: 1413–1435.

Tamang, L. and B. Sinha. 2014. Two new species of the South Asian catfish genus *Pseudolaguvia* from northeastern India (Teleostei: Sisoridae). Zootaxa 3887: 37–54.

Tan, H.H. and H.H. Ng. 1999. The catfishes (Teleostei: Siluriformes) of central Sumatra. J. Nat. Hist. 34: 267–303.

Tilak, R. and A. Husain. 1975. A new sisorid catfish, *Laguvia ribeiroi kapuri* (Sisoridae: Siluriformes) from Uttar Pradesh. J. Inland Fishery Soc. India 6: 1–5.

Toledo-Piza, M., G.M. Mattox and R. Britz. 2014. *Priocharax nanus*, a new miniature characid from the rio Negro, Amazon basin Ostariophysi: Characiformes, with an updated list of miniature Neotropical freshwater fishes. Neotrop. Ichthyol. 12: 229–246.

Vari, R.P. and C.J. Ferraris Jr. 2003. Family Cetopsidae. pp. 257–260. *In*: Reis, R.E., S.O. Kullander and C.J. Ferraris Jr (eds.). Checklist of the Freshwater Fishes of South and Central America. Edipucrs, Porto Alegre, Brazil.

Vari, R.P., C.J. Ferraris Jr, and M.C.C. de Pinna. 2005. The Neotropical whale catfishes (Siluriformes: Cetopsidae: Cetopsinae), a revisionary study. Neotrop. Ichthyol. 3: 127–238.

Vidthayanon, C. 2002. Peat Swamp Fishes of Thailand. Office of Environmental Policy and Planning, Bangkok, Thailand.

Vijayakrishnan, B., J. Praveenraj and A. Mishra. 2023. *Pseudolaguvia permaris*, a new catfish from the Eastern Ghats of India (Teleostei: Sisoridae). Zootaxa 5297: 271–281.

Weitzman, S.H. and R.P. Vari. 1988. Miniaturization in South American freshwater fishes; an overview and discussion. Proc. Biol. Soc. Wash. 101: 444–65.

Wosiacki, W.B., D.P. Coutinho and L.F. de Assis Montag. 2011. Description of a new species of sand-dwelling catfish of the genus *Stenolicmus* (Siluriformes; Trichomycteridae). Zootaxa 2752: 62–68.

Zuanon, J. and I. Sazima. 2005. Free meals on long-distance cruisers: the vampire fish rides giant catfishes in the Amazon. Biota Neotropica. 5: 109–114.

Chapter 10

Parasitic Catfishes

Carlos DoNascimiento[1,*] and *Mário de Pinna*[2]

The candiru species of the family Trichomycteridae (subfamilies Stegophilinae and Vandelliinae) have been traditionally categorized as parasites, considering their trophic dependence on other species, not causing the death of their victims (at least not immediately as a result of direct attacks). However, in a strict definition of parasitism, true parasites are those organisms that have at least one stage of their life cycle living at the expense of a single host (de Meeûs and Renaud 2002; Lafferty and Kuris 2002). This subtle distinction based on the number of hosts individuals exploited per generation provides a key element to define a subcategory of parasitism: micropredation (Lafferty and Kuris 2002; Poulin and Randhawa 2013), which fits best the feeding strategy implemented by parasitic catfishes. The impact of micropredators on host fitness is generally minimal, being dependent on the disproportionate relation of size between the micropredator and their prey (Lafferty and Kuris 2002; Poulin and Randhawa 2013), a relation that becomes apparent when the size of parasitic trichomycterids (largest known species being *Vandellia cirrhosa* with 17 cm SL) is compared to that of their reported preys that commonly exceeds 30 cm SL (mostly members of large characiforms as bryconids, colossomatins, and cynodontids, and doradids and pimelodids among catfishes, and even stingrays) (de Pinna and Wosiacki 2003; Fernández and Schaefer 2009; de Pinna and Dagosta 2022). This disparate size relation can exceed a 46-fold length difference as verified in *Paracanthopoma parva* specimens (27.1–31.7 mm SL) found attached to a large specimen of *Zungaro zungaro* (125 cm TL) (Zuanon and Sazima 2005). The discrepancy will be far larger if measured in body mass.

Parasitic trichomycterids belong to the subfamilies Stegophilinae (Figure 1A) and Vandelliinae (Figure 1B) (Baskin 1973; de Pinna 1998). Members of those subfamilies show two markedly different feeding habits: Stegophilinae species (with the single exception of *Pareiodon*, which specializes in carrion feeding) feed on mucus and scales from other fishes (Winemiller and Yan 1989; Bonato et al. 2018), while Vandelliinae species are obligatory hematophagous (at least as adults) (Zuanon and Sazima 2004; de Pinna and Dagosta 2022). Reports on the feeding behavior of candiru trichomycterids in nature are fragmentary and often anecdotic (Ihering 1930; Guimarães 1935), with a few detailed descriptions (Kelley and Atz 1964; Machado and Sazima 1983), or remarks included in taxonomic essays (Miranda-Ribeiro 1946; de Pinna and Britski 1991; de Pinna and Dagosta 2022). Recently, molecular techniques have been incorporated into studying the diet of parasitic trichomycterids. Bonato et al. (2018), through metabarcoding analyses of the gut content of

[1] Grupo de Ictiología, Instituto de Biología, Universidad de Antioquia, Calle 67 # 53-108, Medellín, 050010, Colombia.
[2] Museu de Zoologia, Universidade de São Paulo, 481 Avenida Nazaré, 04263-000 São Paulo, Brazil.
Email: pinna@ib.usp.br
* Corresponding author: tepuglanis@hotmail.com

Figure 1. (A) *Schultzichthys gracilis* (Stegophilinae). (B) *Vandellia cirrhosa* (Vandelliinae): Guayabero River, Orinoco basin, Colombia; specimens not represented at the same scale; photographs by J. E. García-Melo (Proyecto CaVFish Colombia: https://cavfish.unibague.edu.co/catalogo). (C) *Megalocentor echthrus* (arrows) riding on *Inia geoffrensis*. (D) on *Sotalia fluviatilis*: Amazon basin, Colombia; photographs by F. Trujillo (Fundación Omacha: https://omacha. org/).

two stegophiline species (*Homodiaetus anisitsi* and *Pseudostegophilus maculatus*, both categorized therein as fluid-feeding), determined the precise taxonomic composition at the species level of their respective prey and the predominance of mucus over scales in the diet of both species.

Stegophilines comprise perhaps the most diverse group of trichomycterids in terms of morphological variation in their body plan, which is implicitly reflected in the large number of genera recognized in the family. However, regarding their buccal anatomy, except for *Pareiodon* with a unique mouth structure, all remaining stegophilines have a relatively conservative mouth morphology with a set of fixed features characteristic to the subfamily (coronoid process of dentary horizontally oriented, wing-like posterior projection at lateral tip of premaxilla, and mouth opening forming a crescent-shaped disk) and that have been proposed as synapomorphies of Stegophilinae (DoNascimiento 2015). On the other hand, vandelliines are more conservative in their *bauplan*, which has led to frequent misidentifications in ichthyological collections and even misallocation of recently described species. This issue is especially true for the genera *Paracanthopoma* and *Paravandellia*. Despite these challenges, these species nonetheless display extraordinary variability in mouth-associated structures. As de Pinna and Dagosta (2022) report for *Paracanthopoma*, a broad repertoire of feeding behavior is expected when considering the widely differing jaw bones, associated musculature, and tooth armature of this single genus. A positive demonstration of this premise is the video documentation of the oral functioning of a specimen belonging to an undescribed genus of relatively large size (≥ 100 mm SL) from the Amazon basin, which inflicts a clean transverse cut in the superficial layers of the flesh of its victim and introduces its head completely through the open wound, then attaching firmly with the use of their teeth and odontodes. One short video distributed on social media shows one specimen firmly attached to the back of a boy, leaving a bleeding triangular-shaped wound when the catfish is forcibly dislodged. A second video, also posted on social platforms, shows another specimen slicing the flesh of the dorsum (close to the dorsal fin) of a captured specimen of *Pseudoplatystoma* on a boat, introducing its head in the flesh with alternating oscillations of its head.

Another remarkable biological interaction observed in parasitic catfishes is phoresis (i.e., being temporarily transported by another organism of a different species), which has been recorded both in vandelliines (Zuanon and Sazima 2005) and stegophilines (Araújo-Wang et al. 2019), though differing considerably in its nature and selected hosts. In the case of vandelliines, this behavior has been recorded only in *Paracanthopoma parva*, which was observed firmly attached to the base of dorsal, caudal, and pectoral fins of *Zungaro zungaro* (Zuanon and Sazima 2005) and in *Pseudoplatystoma* sp. (both pimelodid hosts), with attachment sites near the caudal and anal fins, leaving clear red wounds when forcefully removed (de Pinna and Dagosta 2022). Indirect evidence (chunks of flesh clutched in teeth) found in preserved specimens of *Paracanthopoma truculenta* also suggests that at least this species attaches to the body of their hosts in the same way (de Pinna and Dagosta 2022). Zuanon and Sazima (2005) interpreted this behavior in *P. parva* as just a strategy of dispersal and protection from predators (dismissing the alternative of feeding). However, de Pinna and Dagosta (2022) extensively discuss the specialized mode of feeding (active oral sucking) that also occurs while riding their hosts, providing arguments that at least the large species of *Paracanthopoma* (*P. carrapata*, *P. parva*, and *P. truculenta*) can obtain blood and other body fluids from the systemic circulation of the soft tissues in their attachment sites on the host. This capability is enabled by a characteristic mouth morphology, supplied with sucker-like thick flattened lips, having superficial integumentary plicae, thick jaw bones, heavy jaw musculature, and strong tooth armature. Also, observations of the extended periods of permanency were recorded in some cases (up to nine hours, such as in the case of *Pseudoplatystoma* sp.).

Stegophilines, on the other hand, have been recorded riding on freshwater dolphins (*Inia geoffrensis*), where a variable number of specimens (from 1 to > 230 per dolphin individual) of *Ochmacanthus* sp. were observed in the Tocantins River (Araújo-Wang et al. 2019). This candiru-dolphin interaction was interpreted as most likely phoretic since any tissue harm as scars was not reported on the body surface of dolphins (although we were able to find what appear to be traces of skin damage in one of the enlarged photographs), nor any exhibition of negative reaction, irritation or behavior directed to remove candirus from their bodies, even under high loads of specimens (> 230) (Araújo-Wang et al. 2019). Recent photographic evidence (Figure 1C and D) facilitated by Dr F. Trujillo (Fundación Omacha: https://omacha.org/), while conducting ecological studies on populations of river dolphins in the Colombian Amazon, show specimens of *Megalocentor echthrus* attached to both *Inia geoffrensis* and *Sotalia fluviatilis*, indicating that this kind of phoresis in stegophilines does not seem to be uncommon.

From the textual descriptions and available photographic record of the riding events of *Paracanthopoma parva*, *Ochmacanthus*, and *Megalocentor echthrus*, it seems that the main structure used for attachment to their hosts is simply the mouth, although details are still missing: how parasitic catfishes initiate their riding, whether they use other morphological structure such as odontodes to secure at initial anchoring, or if they feed on their hosts (at least in the case of *Paracanthopoma*). Particularly interesting would be to reveal if some unique structures, such as the single hypertrophied opercular odontode of *Megalocentor*, are also used as anchoring devices to secure its first attachment to fast swimmers like river dolphins. Its possible sister group, *Apomatoceros*, has a unique spiny anal fin (without parallel among siluriforms) and nothing is known about its functional significance (whether it displays phoretic association and whether their spiny anal-fin rays are involved in anchoring or providing a fulcrum point for leverage during feeding). The natural history of parasitic catfishes is a promising and fascinating area of research in view of the wide spectrum of morphological variation shown in the body plan of stegophilines and in the buccal structure of vandelliines.

A contentious aspect repeatedly brought up by a prolific body of literature and now innumerable posts that flood the internet is the alleged invasion of human's natural orifices (especially, but not exclusively, the male urethra) by hematophagous candirus. These episodes are supposed to happen when bathers urinate in rivers populated by candirus in the Amazon basin (Gudger 1930; Herman 1973; Spotte 2002). It has been proposed that candirus are attracted by a stream of urine (either underwater or even above the surface of the river). However, at least a chemical affinity to human urine, as well as to other potential chemical attractants (ammonia, amino acids, and fresh fish slime), did not elicit any search response in a series of trials conducted with specimens of *Vandellia*, kept in aquaria in field locations in the Amazon basin (Spotte et al. 2001). Despite the abundant folklore around the subject, scientifically documented cases of such attacks on humans are limited to a single medical report by Dr. A. Samad in Manaus, who performed surgery to remove a large *Vandellia cirrhosa* from the urogenital tract of a male patient.

A key aspect underlying trophic specializations shown by the parasitic catfishes are the unique morphological innovations acquired by stegophilines and vandelliines along their evolutionary history. In the following sections, we provide a historical account of the contributions that have built our current understanding of the phylogenetic relationships of Stegophilinae and Vandelliinae, as well as a complete review of the synapomorphies supporting the monophyly of Tridentinae (as the unquestionable sister group of both parasitic subfamilies) and Vandellinae. The reader is directed to DoNascimiento (2015), where the monophyly of Stegophilinae and included genera is extensively reviewed.

Monophyly of Tridentinae, Stegophilinae, and Vandelliinae

Early recognition of Stegophilinae and Vandelliinae as closest relatives was formalized by Günther (1864) with its group Stegophilina that included the stegophiline and vandelliine species known at the time (*Stegophilus insidiosus*, *Vandellia cirrhosa*, and *V. plazaii*). Eigenman (1918) and Myers (1944) also recognized the evolutionary proximity of stegophilines and vandelliines, further adding Tridentinae as the closest to those two. Baskin (1973) was the first author of the cladistic era to recognize these three subfamilies as a Hennigian monophyletic group, which he called Vandelliinae group (for a detailed account of this pioneering work as one of the first explicit implementations of the foundations of the phylogenetic systematics to address the relationships of a group of Neotropical fishes; see de Pinna 2016). The group was then supported by five synapomorphies, of which only two (lacrimal-antorbital bone absent and eye large) are still corroborated (though homoplastic) today (DoNascimiento 2013, 2015). One of the synapomorphies proposed by Baskin (1973), i.e., parasitic feeding habits, was simultaneously proposed both for the Vandelliinae group and for the sister group relationship of Stegophilinae and Vandelliinae, based on the uncertainty about the feeding habits of Tridentinae. In relation to the two remaining synapomorphies, the lateral opening of the Weberian capsule at the end of a neck-shaped constriction and unossified third hypobranchial; the first one has a homoplastic distribution, being found in the clades *Tridens* + *Tridensimilis*, *Haemomaster* + *Stegophilus*, *Schultzichthys* + *Pareiodon* group and in an undescribed vandelliine genus; while the second one is uninformative for trichomycterids since a cartilaginous hypobranchial 3 represents the plesiomorphic condition, being present in *Copionodon*, *Trichogenes*, *Tridens*, *Tridentopsis*, Stegophilinae, and at least *Glanapteryx* within Glanapteryginae (DoNascimiento 2013).

De Pinna (1998) proposed two new synapomorphies for the Vandelliinae group of Baskin: mesethmoidal cornu with ventral process and mandibular teeth S-shaped. Interestingly, Baskin (1973) proposed these two characters in support of the alternative hypotheses of Stegophilinae plus Vandelliinae (supported for the first of those two characters) and Stegophilinae with Tridentinae (second character). However, more extensive comparative morphological analyses have revealed the presence of the two apomorphic conditions both in tridentines and vandelliines, thus corroborating the proposition of de Pinna (1998) as synapomorphies for the whole Vandelliinae group.

Datovo and Bockmann (2010) corroborated the traditional hypothesis of monophyly of the Vandelliinae group and proposed five additional unique synapomorphies: (1) A2 and A3 sections of the *adductor mandibulae* completely separated from each other; (2) the origin of A3 section of the *adductor mandibulae* limited to the anterior region of the suspensorium, but not reaching the autopalatine; (3) insertion of A3 section of the *adductor mandibulae* on the lower jaw extending anteriorly beyond Meckel's cartilage (reversed in Tridentinae); (4) the origin of *levator arcus palatini* wider than its insertion; and (5) the origin of the primary section of the *dilatator operculi* expanded anteriorly beyond the olfactory foramen.

In addition to the above-discussed characters, the following synapomorphies were also proposed by DoNascimiento (2013) to support the clade consisting of Tridentinae, Stegophilinae, and Vandelliinae: (1) anterior end of vomer wide (vs. pointed); (2) the presence of uncinate process on posterior margin of lateral end of epibranchial 1 (vs. absent), (3) supraorbital sensory pore s3 located inside of the internal rim of posterior naris and not visible externally [vs. exposed on the dorsal surface of the head and variably placed between the posterior naris and eye; this character is described and illustrated in *P. phaneronema* by Díaz del Basto et al. (2018)], and (4) rictal barbel equal or shorter than one-third of maxillary barbel length (vs. equal or longer than half of the maxillary barbel length).

Fernández and Schaefer (2009) conducted the first molecular phylogenetic analysis focused on Trichomycteridae (although with a clear emphasis on candirus) using multilocus data and obtained two contrasting hypotheses: on the one hand, the Vandelliinae group was only recovered in their maximum likelihood analysis but with low support; while in the maximum parsimony analysis, the only tridentine genus analyzed (*Tridens*) was placed as the sister group of the subfamilies Glanapteryginae and Sarcoglanidinae, receiving relatively robust support.

Subsequently, Ochoa et al. (2017), in a more taxonomically encompassing molecular analysis of Trichomycteridae, recovered a baffling position of Vandelliinae as more closely related to Glanapteryginae and the sarcoglanidine *Stauroglanis*, widely diverging from the traditional arrangement, i.e., allied with tridentines and stegophilines. This result was promptly dismissed as genomic data (ultraconserved elements – UCEs) were incorporated into a new phylogenetic analysis of Trichomycteridae (Ochoa et al. 2020), confirming once again the monophyly of the Vandellinae group in agreement with previous morphological hypotheses (Baskin 1973; de Pinna 1998; DoNascimiento 2015). The recent multilocus assessment by Costa et al. (2020) examined the phylogenetic relationships of the so-called TSVSG clade, which comprises subfamilies Tridentinae, Stegophilinae, Vandelliinae, Sarcoglanidinae, and Glanapteryginae. This study proposed Microcambevinae as a new trichomycterid subfamily and also recovers the sister relationship of Tridentinae to the pair of Stegophilinae plus Vandelliinae, adding to the growing consensus between morphology and DNA to the hypothesis of monophyly of the Vandellinae group of Baskin (1973).

Phylogenetic Relationships of Parasitic Trichomyterids

Contrary to the broad consensus on the monophyly of the Vandelliinae group, the interrelationships of their three constituting subfamilies have not converged to a single solution. Baskin (1973) pointed out that the internal relationships of his Vandelliinae group were problematic, given the conflicting character evidence supporting, on the one hand, the closest relationships between Stegophilinae and Vandelliinae, and on the other, between Stegophilinae and Tridentinae. The first hypothesis was selected by the author, based on five synapomorphies, namely, (1) parasitic feeding habits, (2) reduced maxillary and rictal barbels (although their presence in *Tridens* and *Tridensimilis* was noticed), (3) gill opening restricted to a small lateral opening (noticed as being homoplastically present in *Miuroglanis*), (4) branchiostegal membrane lacking free posterior margin (also homoplastic in *Miuroglanis*), and (5) ventral process at the end of the mesethmoidal cornu (actually a synapomorphy of the whole Vandelliinae group). In parallel, Stegophilinae would share four putative synapomorphies with Tridentinae: (1) teeth of the dentary and premaxilla arranged in separate rows, (2) teeth anterior to the premaxilla implanted directly on the upper lip, (3) S-shaped premaxillary teeth in lateral view (also discussed above), and (4) mouth width greater than the width of the neurocranium. In this sense, the sister group relationship between Stegophilinae and Vandelliinae was favored just by an additional synapomorphy.

De Pinna (1998) also recognized the sister group relationship of Stegophilinae and Vandelliinae but listed only three synapomorphies: (1) reduction of maxillary and rictal barbels, (2) presence of median premaxilla, and (3) semiparasitic feeding habits. Datovo and Bockmann (2010) pointed out that parasitic feeding habits are a questionable synapomorphy of Stegophilinae and Vandelliinae. Given the similar morphology and arrangement of the mandibular teeth in tridentines and stegophilines, as well as some observations of the behavior of tridentines in the field and laboratory (Baskin 1973; Adriaens et al. 2010), it was suggested that tridentines have feeding habits similar to those of stegophilines; thus, the semiparasitic condition would constitute a synapomorphy of the entire Vandelliinae group (Datovo and Bockmann 2010) instead.

Fernández and Schaefer (2009) obtained a sister group relationship of Stegophilinae and Vandelliinae in both parsimony (with robust support) and maximum likelihood analyses. On the contrary, Datovo and Bockmann (2010) found two synapomorphies: (1) the origin of section A3 of the *adductor mandibulae* distinctively narrow, and (2) section A3 of the same muscle with both muscular and tendinous origin, favoring the relationship of tridentines with stegophilines, in contrast to the traditional hypothesis, supported by only one synapomorphy (section A3 of the *adductor mandibulae* with insertion in the lower jaw extending beyond Meckel's cartilage), which was optimized as a synapomorphy of the Vandelliinae group, and subsequently reversed in Tridentinae. However, those authors recognized the relatively weak support for both hypotheses, recommending efforts to resolve the internal relationships of each subfamily, especially Tridentinae and Stegophilinae, where most homoplasies occur.

Aside from the unexpected position recovered for Vandelliinae by Ochoa et al. (2017), later phylogenetic analyses with multilocus (Costa et al. 2020) and genomic data (Ochoa et al. 2020) show broad corroboration of the hypothesis based on morphology, i.e., the sister group relationship of the parasitic subfamilies.

The monophyly of the Stegophilinae plus Vandelliinae clade was supported in DoNascimiento (2013) by five exclusive synapomorphies, among which two confirmed synapomorphies of Baskin (1973) (branchial opening restricted and branchiostegal membrane joined to isthmus), as well as the one proposed by de Pinna (1998): median premaxilla. The two remaining synapomorphies are: (1) hypobranchial foramen absent (vs. present) and (2) sensory pore s2 anterior to posterior naris (vs. medial).

Synapomorphies of Tridentinae

The monophyletic status of Tridentinae was formerly proposed by Baskin (1973) on the basis of eight putative synapomorphies:

1) dorsal process at the anterior end of hyomandibula (Figure 2A);
2) maxillary bone small (homoplastic in *Potamoglanis*, except *P. anhanga*) (Henschel et al. 2018; Figure 2A);
3) ventral process of opercle short (homoplastic at least in *Glanapteryx*);
4) eye exposed ventrally (homoplastic in the stegophilines *Apomatoceros* and *Haemomaster*);
5) expanded fontanel (homoplastic in *Potamoglanis*, in the vandelliines *Paracanthopoma* and *Paravandellia*, and *Ammoglanis*, except *A. amapaensis*);
6) opercular and interopercular odontodophores juxtaposed (this character has at least one exception, supposedly a reversal, in the recently described *Rhinotridens* (Datovo et al. 2023));
7) the origin of the dorsal fin just above or posterior to the origin of the anal fin (this character seems redundant and related to the following character);
8) 15 or more anal-fin rays.

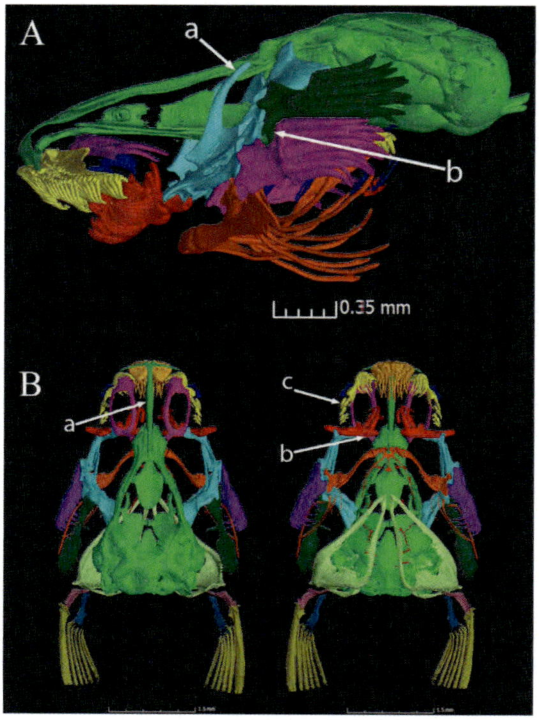

Figure 2. (A) Synapomorphies of Tridentinae (arrows): dorsal process at the anterior end of hyomandibula (a) and ventral process of opercle (b), seen in left lateral view of cranium of *Miuroglanis platycephalus* (ANSP 206004). (B) Synapomorphies of Vandelliinae (arrows): main body of mesethmoid slender (a); lower jaws free from each other and dentary teeth restricted to mesial portion of mandible; (b); scalpelloid teeth at distal end of premaxillae; (c); seen in dorsal (left) and ventral (right) views of cranium of *Paravandellia phaneronema*. CT-reconstructions provided by N. K. Lujan.

De Pinna (1998) listed the same synapomorphies provided by Baskin (1973) without modification, and Datovo et al. (2023) confirmed these characters as present in all previously known tridentine genera (except for character 3, absent in *Rhinotridens*). Given the homoplastic nature of characters 2, 3, 4, and 5, plus the presumably redundant nature of character 7, a single synapomorphy holds as exclusive supporting monophyly of Tridentinae (character 1).

Baskin (1973) did not have access to osteological material of *Miuroglanis*; however, based on the ventral exposure of the eye, juxtaposition of the opercular and interopercular odontodophores, and the presence of 15 or more rays in the anal fin, he confirmed its Tridentinae membership. Certainly, these conditions are derived from Trichomycteridae, and the last two are exclusively found in tridentines. Additionally, access to radiographs of the holotype of *M. platycephalus*, as well as C&S and CT-scan images of *Miuroglanis* specimens, allowed visualization of the dorsal process of the hyomandibula, as well as the reduction of the ventral process of the opercle (Figure 2A).

Synapomorphies of Vandelliinae

Baskin (1973) proposed the monophyly of vandelliines, based on 11 synapomorphies. However, two of those (vomer in the middle of the upper jaw and vomer toothed) are the result of the erroneous identification of the median premaxilla, indeed a neomorphic structure present exclusively in stegophillines and vandelliines (de Pinna and Bristki 1991). The remaining nine synapomorphies are: (1) scalpelloid teeth on distal end of premaxilla, enclosed in a skin pocket; (2) dentaries lacking medial articulation; (3) process on the posterolateral edge of pterotic directed posteriorly; (4) fifth ceratobranchial and associated teeth absent; (5) hypobranchials 2 and 3 absent (missing hypobranchial 2 homoplastic in *Tridens*); (6) fourth pharyngobranchial unossified (actually this bone is absent in vandelliines, as well as in stegophilines, *Potamoglanis* and at least *Glanapteryx* within glanapterygines); (7) upper pharyngeal tooth plate edentulous (the bone is absent in *Paracanthopoma*; see Dagosta and de Pinna 2022, at least two undescribed species of *Paravandellia* and *Tridens*); (8) blood-feeding on other vertebrates; and (9) mesocoracoid absent (homoplastic in *Tridentopsis* and *Glanapteryx*). Schmidt (1993) once again listed these 11 synapomorphies as valid characters supporting Vandelliinae, despite being aware of the arguments put forward by de Pinna and Britski (1991) about the actual identity of the bone identified as the vomer by Baskin (Schmidt 1993). De Pinna (1998) only recognized nine synapomorphies of Baskin (1973), excluding those two involving the median premaxilla interpreted as a vomer. Datovo and Bockmann (2010) added five new homoplasy-free synapomorphies, which support the monophyly of the subfamily: (1) section A3 of the *adductor mandibulae* originating exclusively from the autopalatine; (2) presence of the *constrictor palatini* muscle; (3) presence of the *hyporostralis* muscle; (4) presence of the *epirostralis* muscle; and (5) fibers of the *protractor operculi* inserting into the base of some of the more anteroventral opercular odontodes. To this list, ten additional synapomorphies can be added from DoNascimiento (2013): (1) the entire main body of mesethmoid slender (vs. laterally broad and laminar) (Figure 2B); (2) vomer absent (vs. present); (3) dentary teeth restricted to mesial portion of the dentary (vs. variably extending along mandibular ramus); (4) ligament between proximal region of premaxilla and distal end of mesethmoid cornu (vs. absent); (5) ligament between the medial bifid process of premaxilla and autopalatine (vs. absent); (6) premaxilla ligamentously joined to the coronoid process (vs. absent); (7) pharyngobranchial 3 absent (vs. present); (8) parapophysis of first free vertebra equal or shorter than corresponding vertebral centrum width (vs. always longer); (9) posterolateral process of cleithrum wide (vs. slender, when present); (10) scapulocoracoid joined to cleithrum via cylindrical cartilaginous block (vs. both bones directly articulating). It is quite clear that this long list of morphological synapomorphies for vandelliines is in large part the result of their extreme trophic specialization as obligatory hematophages (at least as adults), a feeding strategy otherwise seen in only three species of desmodontine bats among gnathostome vertebrates (Kwon and Gardner 2007).

Vandelliinae monophyly has also been consistently confirmed by all molecular analyses to date (Fernández and Schaefer 2009; Ochoa et al. 2017, 2020; Costa et al. 2020). Although stegophiline monophyly has never been questioned, for a long time it remained a poorly supported hypothesis. This situation changed with Datovo and Bockmann (2010) and DoNascimiento (2015), who proposed one and three new characters for the group, respectively. Finally, it is interesting to highlight that stegophiline monophyly and the interrelationships of its genera match almost perfectly between morphological analysis under parsimony (DoNascimiento 2015) and model-based analyses using a vast genomic dataset of UCEs (Ochoa et al. 2020), across an extensive taxonomic sampling of the subfamily. This further supports the relevance of maintaining phylogenetically informative morphological variation in phylogenetic studies, today vastly dominated by molecular data, to have a multi-layered understanding of biotic evolution and the phenotypic responses along the evolutionary history of lineages.

Comments on the Biogeography of Tridentinae, Stegophilinae, and Vandellinae

In general terms, Tridentinae is essentially an Amazon basin taxon (8 of 10 known species), while Stegophilinae, and Vandelliinae are widely distributed in the main large cis-Andean basins of South America (Figure 3): Orinoco, rivers of the Guianas draining to the Atlantic (Essequibo, Corantijn, Nickerie, Coppename, Suriname, Commewijne, and Maroni/Marowijne rivers), Amazon, São Francisco, Paraná-Paraguay, as well as in a few small coastal rivers of southeastern Brazil (de Pinna and Wosiacki 2003; Wosiacki and de Pinna 2007; Le Bail et al. 2012; Mol et al. 2012; Lasso et al. 2004; DoNascimiento et al. 2017; Taphorn et al. 2022). In sharp contrast with their widespread presence in cis-Andean basins, only a single representative each for Tridentinae and Vandelliinae

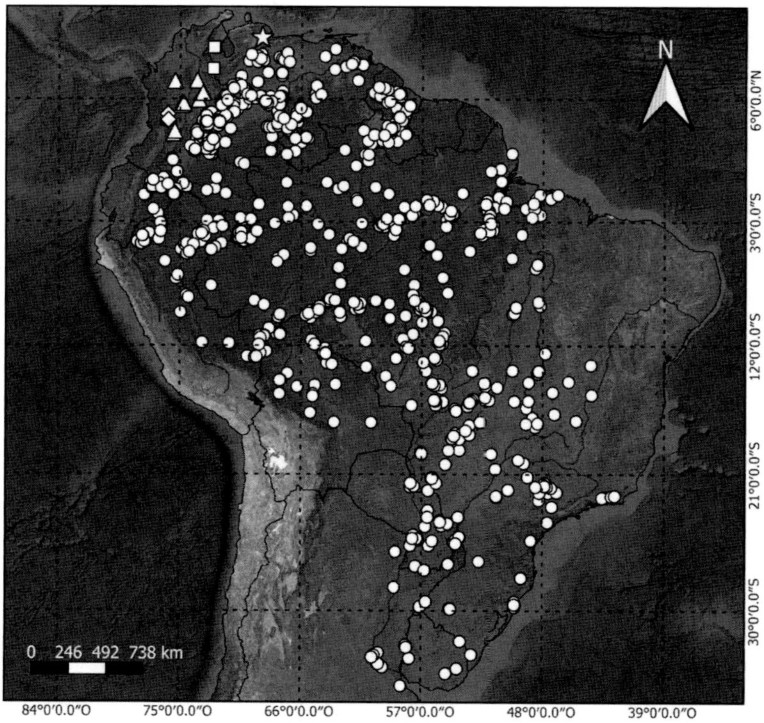

Figure 3. Geographic distribution of species of Stegophilinae, Tridentinae, and Vandellinae in South America. Diamond: undescribed stegophiline (Cauca River); triangles: *Paravandellia phaneronema* (Magdalena River basin); squares: *Tridensimilis venezuelae* (Maracaibo Lake basin); star: *Ochmacanthus alternus* (Tuy River); circles: cis-Andean records of stegophilines, tridentines, and vandelliines. Occurrence data from GBIF.org (September 15, 2023).

has been recorded in trans-Andean basins. For Tridentinae, *Tridensimilis venezuelae* is an endemic species of the Lake Maracaibo basin (Figure 3). According to morphological characters, the genus was recently confirmed as monotypic (Datovo et al. 2023) and hypothesized as sister group of *Tridens* (DoNascimiento 2013; Datovo et al. 2023). *Tridens* is found in the Amazon basin, but its presence in the Orinoco basin has also been confirmed by an undescribed species from the Manacacías River, a tributary of the Meta River, although erroneously identified as *Tridensimilis venezuelae* (Galvis et al. 2007). The last finding is especially relevant since the latter basin had a recent historical connection with Lake Maracaibo (Lundberg et al. 1998; Hardman and Lundberg 2006). However, *Tridens* representatives have not been recorded yet for the Orinoco River basin in Venezuela, despite extensive ichthyological sampling efforts in that country during the late 20th and first decade of 21st centuries, supplemented with exhaustive revision of trichomycterid material housed in the main ichthyological collections of Venezuela (MBUCV-V, MCNG, and MHNLS). Representatives of the entire Tridentinae remain non-existent from the Venezuelan side of the Orinoco or other tributaries on the Colombian side. *Tridens* and *Tridensimilis* are a distal clade in the tridentine phylogeny. It is reasonable to hypothesize that their divergence follow the interruption of the connection between the rivers that are currently part of the Lake Maracaibo basin and the proto-Orinoco, with the closure of the Maracaibo-Falcón outlet flowing north, an event that occurred approximately 8 million years ago (Lundberg et al. 1998; Hardman and Lundberg 2006).

On the other hand, the only other non-Amazonian tridentine is *Tridentopsis cahuali* from the Paraguay River drainage of Argentina (Azpelicueta 1990). The other two *Tridentopsis* species are essentially Amazonian (*T. pearsoni* from the Mamoré drainage and *T. tocantinsi* from the Tocantins basin). This distribution pattern involving Amazonian (but more specifically species from the Beni-Mamoré drainage) and Paraguayan elements fits a well-documented historical connection between both hydrographic systems (Carvalho and Albert 2011; Dagosta and de Pinna 2019, 2021).

The Vandelliinae is represented in the trans-Andean systems only by *Paravandellia phaneronema*, a species endemic to the Magdalena basin, distributed in both the Magdalena and Cauca rivers (Figure 3). That species was hypothesized as the sister group of a clade consisting of two undescribed cis-Andean species from the Amazon and Orinoco basins, respectively, although *Paravandellia oxyptera* and some other undescribed forms were missing from the analysis (DoNascimiento 2013). *Paravandellia* has been recently confirmed as the sister group of *Paracanthopoma* (de Pinna and Dagosta 2022), and both genera are the sister group of remaining Vandelliines (DoNascimiento 2013). Once again, a tenable hypothesis is an early divergence of *Paravandellia* and *Paracanthopoma* before the separation of the Magdalena from the proto-Orinoco by orogenesis of the northern Andes. The Magdalena basin, located west of the Western Cordillera of the Colombian Andes, was isolated long before the closure of the Maracaibo-Falcón outlet of the proto-Orinoco. The most recent event of isolation of the basins west of Lake Maracaibo is the uplift of the Sierra de Perijá, approximately 11.8 million years ago (Lundberg et al. 1998), thus being a working minimum age of differentiation of the genus *Paravandellia*.

Stegophilinae species occupy lowland rivers, and the most widely distributed species, such as those of the Amazon basin (e.g., *Apomatoceros alleni* and *Pareiodon microps*), inhabit the main channel of the Amazon River and its main tributaries. DoNascimiento (2013) reported the trans-Andean occurrence of an undescribed species known at the time from a single specimen collected in the Risaralda River, a tributary of the upper Cauca River drainage. Limited access to morphological information in that specimen prevented it from being quantitatively analyzed, and its tentative phylogenetic position was inferred from a few externally visible characters. The specimen in question has a wide mouth opening forming a crescent-shaped disk, a unique synapomorphy of the external morphology that diagnoses the Stegophilinae (Baskin 1973; de Pinna 1998; DoNascimiento 2015), and a pair of posterolateral lobes on the lower lip, a homoplastic condition but still informative for diagnosing the subfamily. The specimen has four rows of labial teeth uninterrupted medially, suggesting an aligned arrangement of median premaxillary teeth synapomorphic of a subclade comprising *Homodiaetus*, *Schultzichthys*, and members of the *Pareiodon* group (DoNascimiento 2013). On the other hand, the plesiomorphic condition of the distal end of the pelvic fin reaching the urogenital papilla suggests that this species is not part of the clade consisting of *Schultzichthys* plus the *Pareiodon* group. Besides, the presence of paired epiphyseal sensory pores (s6) refutes its affiliation with the *Pareiodon* group. The species then is hypothesized as probably related to *Homodiaetus*, based on a single external morphological character potentially informative, i.e., pelvic fins inserted on the anterior half of the standard length, the anterior position of the pelvic fin being diagnostic (although homoplastic) for *Homodiaetus* (DoNascimiento 2013). The Cauca species certainly has a closer phenetic similarity to species of *Homodiaetus* than to any other known genus of Stegophilinae. Additionally, some meristic characters that are taxonomically informative fit the range in *Homodiaetus* species. The trans-Andean species has four rows of labial teeth (2–5 rows in *Homodiaetus*), four rows of premaxillary teeth (three or four rows in *Homodiaetus*), five rows of dentary teeth (four to seven rows in *Homodiaetus*), and five or six opercular and interopercular odontodes (three to nine opercular odontodes and five to 12 interopercular odontodes in *Homodiaetus*). The size of the specimen (21.9 mm SL) also falls within the body size range of *Homodiaetus* species, which reaches a maximum recorded size of 46.1 mm SL (de Pinna and Wosiacki 2003). Based on the information discussed above, the trans-Andean stegophilline is either part of *Homodiaetus* or its closest relative. The recent availability of additional specimens of this species now allows access to its internal anatomy and DNA sequences, which are being prepared for publication elsewhere.

The Cauca stegophiline is restricted to a narrow section of the upper Cauca River (Figure 3), which contrasts with the more widespread presence of *Paravandellia phaneronema* along the entire Magdalena basin. Nonetheless, the presence of only three trans-Andean representatives of Tridentinae, Stegophilinae, and Vandelliinae reflects the depauperate condition already pointed out for the ichthyofauna of the Magdalena and Maracaibo basins (Lundberg et al. 1986, 1998;

Rodríguez-Olarte et al. 2011). This highlights their relictual biogeographic nature, as defined by Grandcolas et al. (2014), i.e., their current isolated and restricted geographic distribution contrasts to the vast cis-Andean area where the remaining tridentines, stegophilines, and vandelliines are present (Figure 3). This excludes an unlikely dispersal scenario posterior to isolation of the Magdalena and Maracaibo basins.

An interesting case of distribution in river systems peripherical to the Amazon core is illustrated by *Ochmacanthus alternus*, a widespread Orinoco species that is also present in the Tuy River (Figure 3), which is a small coastal river in north-central Venezuela that drains directly into the Caribbean Sea. A similar distribution pattern (Orinoco-Tuy) is exhibited by *Steindachnerina argentea*, a curimatid species that is also present in the Tocuyo River, another coastal river in north-western Venezuela and the western portion of the island of Trinidad (Vari 1991). It is plausible that the presence of both *O. alternus* and *S. argentea* in the Tuy River is the result of a secondary dispersal event, as proposed by Vari (1991). A fine-scale analysis of the historical biogeography (genus level) is impeded by two factors: first, limited knowledge of actual taxonomic diversity and geographic distribution of species, and second, incomplete taxonomic representation in available phylogenetic analyses. Efforts to remedy this situation are a priority in order to understand the evolutionary history of parasitic trichomycterids.

Non-Trichomycterid Catfish Trophic Parasites

A few instances of lepidophagy have been reported in catfishes other than Trichomycteridae. Contrary to the latter family, such cases are not associated with major monophyletic radiation but represent, for the most part, individual species adaptations. Also, contrary to trichomycterids, lepidophagy in other catfishes is not strict; instead, it is combined with other modes of feeding, such as carnivory. Still, some of those cases are associated with clear morphological and behavioral adaptations for scale-feeding and certainly qualify as independent instances of lepidophagy.

Scale feeding has been repeatedly reported in ariid catfishes from different taxa, life stages, and geographic regions: *Ariopsis felis* (Hoese 1966), *Neoarius berneyi* [Roberts (1978) as *Arius cleptolepis*], *Ariopsis seemanni* (Szelistowski 1989), *Arius solidus* and *A. utarus* (Coats 1991), *Genidens genidens*, *G. barbus*, and *Aspistor luniscutis* (Denadai et al. 2012). Some of the reported instances may represent results of necrophagy and scavenging [Tilney and Hecht (1990) in *Galeichthys feliceps* in South Africa], and no evidence of scales of target-item have been found in many ariid species (Blaber et al. 1994). However, a few are actually selective lepidophagous, especially as juveniles, with active stereotypic scale-eating behavior having been observed in juvenile *Ariopsis felis* in the Gulf of Mexico [Hoese (1966) as *Galeichthys felis*].

Pachypterus khavalchor (Horabagridae) from India is another demonstrably lepidophagous catfish, in this case combined with carnivory. It displays typical mouth parts adaptations for scale-eating, such as protruding teeth extending outside the oral cavity (Pise et al. 2022). Such morphology agrees with a feeding strategy of impact and dislodging of prey scales, with subsequent ingestion of fallen-off scales in midwater or on the substrate, a behavior reported in *P. khavalkhor* (Gosavi et al. 2018). This strategy is typical of lepidophagous characiforms (Sazima 1983) and contrasts with the rasping body-contact strategy of stegophiline trichomycterids. *Pachypterus khavalchor* is the only scale-eating fish species where the mechanism of scale digestion has been studied and thus constitutes a valuable model for future research on the structure and evolution of the digestive system in other lepidophagous fishes (Gosavi et al. 2018, 2019).

Brood Parasites

In the introductory section of this chapter, we emphasized the metabolic dependence to their respective heterospecific hosts, which is given to categorize an organism as a parasite (Price 1980; Gunn and Pitt 2012; Goater et al. 2014), and this conceptual framework has been extended to take also into account the provision of the parasite's habitat requirements by the same host (Lucius et al. 2017), and consequently, the ability of the parasite to evade the immune response of the host (Zelmer 1998). Interspecific brood parasitism is a reproductive strategy in which a parent of a given species leaves its eggs or early stages under the care of an unrelated species, imposing the costs of parental care, which also results in a variable loss of fitness to the foster parents (Wootton and Smith 2014; Di Liberto 2021). This biotic interaction is commonly treated as a specific form of social parasitism in a separate context from traditional trophic parasites (Lucius et al. 2017). Catfishes are again notable in this regard by including the only obligate non-avian vertebrate brood parasites, the cuckoo catfishes *Synodontis multipunctatus* and *S. grandiops*, Mochokidae from Lake Tanganyika (Sato 1986; Takahashi and Koblmüller 2020), and the Bombe, *Bathyclarias nyasensis*, Clariidae, from Lake Malaŵi (Stauffer and Loftus 2010).

In the case of the cuckoo catfishes, they primarily exploit maternally mouthbrooding Tropheini cichlids by laying and fertilizing their eggs simultaneously with cichlid spawning. Female mouthbrooders pick up their own eggs and the fertilized eggs of the cuckoo catfish. The catfish eggs hatch earlier than the host's eggs, and following yolk absorption, catfish fry eat the cichlid embryos inside the mouth of the parental cichlid, frequently eliminating the entire host progeny before completing embryo development. The cichlid remains brooding young catfishes until they leave as independent juveniles. Thus, the cuckoo catfish uses protection from female mouthbrooder cichlids and the provision of host embryos as the first exogenous food source (Sato 1986; Cohen et al. 2018; Reichard et al. 2023).

Cuckoo catfish is categorized as an intermediate generalist (i.e., lacks host specialization) that, under natural conditions, parasitizes at least six species of Tropheini and *Ophthalmotilapia nasuta* (Ectodini), all of them maternal mouthbrooders that spawn over a substrate (Brichard 1979; Sato 1986; Reichard et al. 2023). However, both *Synodontis multipunctatus* and *S. grandiops* also occasionally parasitize *Cyprichromis coloratus*, an open-water spawner cichlid (Cyprichromini), endemic of Lake Tanganyika (Takahashi and Koblmüller 2020). Laboratory essays have shown that the cuckoo catfish can successfully parasitize allopatric species from other major African rift valley lakes (Victoria and Malaŵi) and even South American mouthbrooding species such as *Geophagus steindachneri* (Wisenden 1999; Cohen et al. 2018), at rates higher than those observed for sympatric species interactions (Blažek et al. 2018; Cohen et al. 2018).

An extensive research program has been developed focused on various aspects of the behavior, reproductive biology, and evolutionary ecology of the cuckoo catfish (*S. multipunctatus*), especially given its ease of conducting experimental studies in laboratory conditions (Ochi et al. 2001; Blažek et al. 2018; Cohen et al. 2019; Polačik et al. 2019; Zimmerman et al. 2022, 2023; Reichard et al. 2023). However, as Takahashi and Koblmüller (2020) already point out, these studies have never considered the taxonomic uncertainties about the actual cuckoo catfish species involved, a problem that dates to the original behavioral report by Sato (1986). *Synodontis multipunctatus* was long considered a single species until Wright and Page (2006) described *S. grandiops*, which is a similar species that differ in the relative size of eye, pectoral-fin ray count, and body size, being a common species sympatric with *S. multipunctatus* in Lake Tanganyika. Koblmüller et al. (2006) analyzed the phylogenetic relationships and phylogeography of *Synodontis* in East Africa and estimated an age of about 2 mya for the most recent common ancestor of *S. multipunctatus*, coincident with the estimated age for the most recent common ancestor of the mouthbrooding haplochromines of Lake Tanganyika to which the current host-species of *S. multipunctatus* belong. These authors also found an accelerated rate of molecular evolution for that catfish species, which they attributed to be possibly a consequence of a dynamic co-evolutionary history between parasites and hosts.

The Bombe (*Bathyclarias nyasensis*) parasitizes *Bagrus meridionalis* (Kampango), which is a large substrate-spawning catfish that exhibits bi-parental care; both catfishes are endemic to Lake Malaŵi (Stauffer and Loftus 2010). In this case, Bombe juveniles are fed by both Kampango parents, congregating around the female vent to devour trophic eggs that are normally produced to feed Kampango young or swimming around the male opercular openings when Kampango male returns to the nest from foraging on surrounding sand substrates, carrying macroinvertebrates in the mouth to feed their young in the nest (Stauffer and Loftus 2010). Based on these observations, Stauffer and Loftus (2010) interpreted this interaction as an example of maladaptive, misdirected parental care.

Another instance of alloparental care, and possibly a parasitic relationship, implies another clariid catfish (*Dinotopterus cunningtoni*) from Lake Tanganyika. Ochi et al. (2001) found eggs and young of *D. cunningtoni* in nests of *Auchenoglanis occidentalis* (Auchenoglanididae) and observed that the host young of the latter disappeared before reaching independence from parental care. This was confirmed with observations in aquarium conditions of *D. cunningtoni* young preying over host young that had been collected from the same nest.

An interesting aspect of those examples of parasitic brooding is their environmental and geographic circumscription to the great lakes of Africa, where approximately 150 species of mouthbrooding cichlids are part of the assemblage of fish species of Lake Tanganyika (Koblmüller et al. 2008; Reichardt et al. 2023). This constitutes a widespread reproductive strategy potentially available for exploitation by the brood parasites, and three of the four examples of brooding parasites referred to above have been recorded from that single lake. The second relevant aspect is related to the phylogenetic independence in the occurrence of brood parasitism in catfishes since the families involved (Clariidae and Mochokidae) are distantly related (Diogo 2005; Sullivan et al. 2006; de Pinna et al. 2025). Also, *Dinotopterus* and *Bathyclarias* are not closely related within Clariidae (Agnèse and Teugels 2001, 2005; Devaere et al. 2007). This situation contrasts strikingly with the tight phylogenetic association in the trichomycterid parasites (sister group relationship of Stegophilinae and Vandeliinae).

Discussion and Conclusions

The clade formed by Tridentinae, Stegophilinae, and Vandeliinae is one of the most extraordinary radiations in the Siluriformes. Their feeding specializations and associated adaptations offer a unique opportunity for understanding the evolution of narrowly specialized ecologies. Most interestingly, they are a highly diverse clade in terms of number of genera and species. In fact, their taxonomic diversity is still in the process of being unraveled. Even less explored is their diversity of feeding habits. Given their pronounced diversity of mouth morphologies, it is to be expected that many fine-scale variations of blood-, scale-, and mucus-feeding exist, but yet undocumented. It is known that in vandelliines the blood-feeding oral adaptations change drastically in ontogeny. With such morphological modifications, associated feeding and behavioral changes are expected. Again, next to nothing is known in that area. Finally, the reproductive biology of parasitic catfishes is essentially unknown, with the only information available on the topic being reports of a few gravid females mentioned in taxonomic descriptions.

Without a closely related phylogenetically intermediate taxon, the only road to understanding the sequence of change in the evolution of parasitic catfishes is development. The study of ontogeny in a broad sense, i.e., including anatomy, ecology, behavior, and physiology, offers tremendous opportunities for understanding the sequence of modifications in the evolution and diversification of micropredatory trophic specializations.

Equally inciting possibilities exist in studying the physiology of parasitic catfishes, especially the hematophagous Vandelliinae. As is normal in most blood-feeding animals, some form of anti-coagulant is to be expected and, in this case, perhaps based on a biochemical basis entirely different from currently known ones in other hematophagous groups, given the entirely independent evolutionary history of parasitic catfishes. Also, some form of physiological pathway for iron metabolism is to be expected since excess iron is a ubiquitous problem in blood-feeding animals.

In sum, parasitic trichomycterids are a whole new field of study waiting to be developed, well beyond specialists' horizons. Vast possibilities for innovative discoveries exist in their natural history, behavior, ecology, physiology, development, and evolutionary biology. Such opportunities, however, depend on dedicated research efforts, which inevitably demand necessary resources. Hopefully, circumstances will permit the realization of such scientific potential before the unstoppable destruction of their natural habitats makes any such endeavors impossible.

Acknowledgments

We thank Jorge Enrique García-Melo (Universidad de Ibagué) and Fernando Trujillo (Fundación Omacha) for providing photographs of Figure 1, Nathan K. Lujan (Royal Ontario Museum) for CT-reconstructions of *Miuroglanis platycephalus* and *Paravandellia phaneronema*, used to compose Figure 2, and Karen Lineke Álvarez Álvarez (Universidad de Antioquia) for the map in Figure 3.

References

Adriaens, D., J.N. Baskin and H. Coppens. 2010. Evolutionary morphology of trichomycterid catfishes: about hanging on and digging in. pp. 337–362. *In*: Nelson, J.S., H.-P. Schultze and M.V.H. Wilson (eds.). Origin and Phylogenetic Interrelationships of Teleosts. Verlag Dr. Friedrich Pfeil, München.

Agnèse, J.F. and G.G. Teugels. 2001. The *Bathyclarias–Clarias* catfish species flock. A new model to understand rapid speciation in African Great Lakes. Compt. Rend. Acad. Sci. Paris Sciences de la vie 325: 683–688. https://doi.org/10.1016/s0764-4469(01)01348-8.

Agnèse, J.F. and G.G. Teugels. 2005. Molecular phylogeny of Clariidae (Teleostei, Siluriformes): implications for their body shape evolution, biogeography, and taxonomy. Mol. Phylogenet. Evol. 36: 546–553. https://doi.org/10.1016/j.ympev.2005.03.028.

Araújo-Wang, C., E.K. Schormans and J.Y. Wang. 2019. Ecological interaction of a "parasitic" candiru catfish and botos (*Inia geoffrensis*). Mar. Mam. Sci. 35: 1347–1354. https://doi.org/10.1111/mms.12593.

Azpelicueta, M.M. 1990. *Tridentopsis cahuali* n. sp. (Siluriformes, Trichomycteridae), a new miniature tridentine from Paraguay System, in Argentina. Rev. Suisse Zool. 97(4): 981–988.

Baskin, J.N. 1973. Structure and Relationships of the Trichomycteridae. Ph.D. Thesis, City University of New York, New York.

Blaber, S.J.M., D.T. Brewer and J.P. Salini. 1994. Diet and dentition in tropical ariid catfishes from Australia. Environ. Biol. Fishes 40: 159–174.

Blažek, R., M. Polačik, C. Smith, M. Honza, A. Meyer and M. Reichard. 2018. Success of cuckoo catfish brood parasitism reflects coevolutionary history and individual experience of their cichlid hosts. Sci. Adv. 4(5): eaar4380. https://doi.org/10.1126/sciadv.aar4380.

Brichard, P. 1979. Unusual behaviors in Lake Tanganyika cichlids. Buntbarsche Bull. 74: 10–12.

Bonato, K.O., P.C. Silva and L.R. Malabarba. 2018. Unrevealing parasitic trophic interactions - A molecular approach for fluid-feeding fishes. Front. Ecol. Evol. 6: 22. https://doi.org/10.3389/fevo.2018.00022.

Carvalho, T.P. and J.S. Albert. 2011. The Amazon-Paraguay divide, pp. 193–202. *In*: Albert, J.S. and R.E. Reis (eds.). Historical Biogeography of Neotropical Freshwater Fishes. University of California Press, Berkeley, Los Angeles, London.

Coates, D. 1991. Biology of fork-tailed catfishes from the Sepik River, Papua New Guinea. Env. Biol. Fish. 31: 55–74.

Cohen, M.S., M.B. Hawkins, J. Knox-Hayes, A.C. Vinton and A. Cruz. 2018. A laboratory study of host use by the cuckoo catfish *Synodontis multipunctatus*. Environ. Biol. Fish. 101: 1417–1425. https://doi.org/10.1007/s10641-018-0788-1.

Cohen, M.S., M.B. Hawkins, D.W. Stock and A. Cruz. 2019. Early life-history features associated with brood parasitism in the cuckoo catfish, *Synodontis multipunctatus* (Siluriformes: Mochokidae). Phil. Trans. R. Soc. B 374: 20180205. http://dx.doi.org/10.1098/rstb.2018.0205.

Costa, W.J.E.M., E. Henschel and A.M. Katz. 2020. Multigene phylogeny reveals convergent evolution in small interstitial catfishes from the Amazon and Atlantic forests (Siluriformes: Trichomycteridae). Zool. Scr. 49(2): 159–173. https://doi.org/10.1111/zsc.12403.

Dagosta, F.C.P. and M. de Pinna. 2019. The fishes of the Amazon: distribution and biogeographical patterns, with a comprehensive list of species. Bull. Am. Mus. Nat. Hist. 431(13): 1–163.

Dagosta, F.C.P. and M. de Pinna. 2021. Two new catfish species of typically Amazonian lineages in the Upper Rio Paraguay (Aspredinidae: Hoplomyzontinae and Trichomycteridae: Vandelliinae), with a biogeographic discussion. Pap. Avulsos Zool. 61(47): e20216147. http://doi.org/10.11606/1807-0205/2021.61.47.

Datovo, A. and F.A. Bockmann. 2010. Dorsolateral head muscles of the catfish families Nematogenyidae and Trichomycteridae (Siluriformes: Loricarioidei): comparative anatomy and phylogenetic analysis. Neotrop. Ichthyol. 8(2): 193–246.

Davoto, A., L. Ochoa, G. Vita, P. Presti, W.M. Ohara and M.C.C. de Pinna. 2023. A new genus and species of miniature tridentine catfish from the Amazon basin (Siluriformes: Trichomycteridae). Neotrop. Ichthyol. 21(3): e230076. https://doi.org/10.1590/1982-0224-2023-0076.

Denadai, M.R., E. Bessa, F.B. Santos, W.S. Fernandez, F.M.C. Santos, M.M. Feijó et al. 2012. Life history of three catfish species (Siluriformes: Ariidae) from southeastern Brazil. Biota Neotrop. 12: 74–83.

Devaere, S., G. Jansen, D. Adriaens and P. Weekers. 2007. Phylogeny of the African representatives of the catfish family Clariidae (Teleostei, Siluriformes) based on a combined analysis: independent evolution towards anguilliformity. J. Zool. Syst. Evol. Res. 45: 214–229. https://doi.org/10.1111/j.1439-0469.2006.00399.x.

Di Liberto, J.F. 2021. Interspecific Brood Parasitism. *In*: Vonk, J. and T. Shackelford. (eds.). Encyclopedia of Animal Cognition and Behavior. Springer, Cham. https://doi.org/10.1007/978-3-319-47829-6_1339-1.

Díaz del Basto, J.C., J.I. Mojica and H.-W. Koyro. 2018. External morphology of the fish parasite *Paravandellia phaneronema* (Miles 1943) (Siluriformes: Trichomycteridae), under the scanning electron microscope. Rev. Acad. Colomb. Cienc. Ex. Fis. Nat. 42(165): 323–329. https://doi.org/10.18257/raccefyn.646.

Diogo, R. 2005. Morphological Evolution, Adaptations, Homoplasies, Constraints, and Evolutionary Trends: Catfishes as a Case Study on General Phylogeny and Macroevolution. Science Publishers, Enfield.

DoNascimiento, C. 2013. Sistemática y relaciones filogenéticas de la subfamilia de bagres parásitos Stegophilinae (Siluriformes, Trichomycteridae). Ph.D. Thesis, Universidad Central de Venezuela, Caracas, Venezuela.

DoNascimiento, C. 2015. Monophyly of the subfamily of parasitic catfishes Stegophilinae (Siluriformes, Trichomycteridae) and phylogenetic diagnoses of its genera. Copeia 103(4): 933–960. https://doi.org/10.1643/CI-14-132.

DoNascimiento, C., E.E. Herrera-Collazos, G.A. Herrera-R., J.A. Maldonado-Ocampo, A. Ortega-Lara, J.S. Usma Oviedo et al. 2017. Checklist of the freshwater fishes of Colombia: a Darwin Core alternative to the updating problem. ZooKeys 708: 25–138.

Eigenmann, C.H. 1918. The Pygidiidae, a family of South American catfishes. Mem. Carnegie Mus. 7(5): 259–398.

Fernández, L. and S.A. Schaefer. 2009. Relationships among neotropical candirus (Trichomycteridae, Siluriformes) and the evolution of parasitism based on analysis of mitochondrial and nuclear gene sequences. Mol. Phylogenet. Evol. 52(2): 416–423. https://doi.org/10.1016/j.ympev.2009.02.016.

Galvis, G., J.I. Mojica, F. Provenzano, C.A. Lasso, D.C. Taphorn B., R. Royero et al. 2007. Peces de la Orinoquía Colombiana con Énfasis en Especies de Interés Ornamental. Instituto Colombiano de Desarrollo Rural (INCODER), Universidad Nacional de Colombia, Bogotá D.C.

Goater, T.M., C.P. Goater and G.W. Esch. 2014. Parasitism: The Diversity and Ecology of Animal Parasites. 2nd ed. Cambridge University Press, Cambridge.

Gosavi, S.M., S.S. Kharat, P. Kumkar and S.S. Navarange. 2018. Interplay between behavior, morphology and physiology supports lepidophagy in the catfish *Pachypterus khavalchor* (Siluriformes: Horabagridae). Zoology 126: 185–191.

Gosavi, S.M., C.R. Verma, S.S. Kharat, M. Pise and P. Kumkar. 2019. Structural adequacy of the digestive tract supports dual feeding habit in catfish *Pachypterus khavalchor* (Siluriformes: Horabagridae). Acta Histochem. 121(4): 437–449.

Grandcolas, P., R. Nattier and S. Trewick. 2014. Relict species: a relict concept? Trends Ecol. Evol. 29(12): 655–663. https://doi.org/10.1016/j.tree.2014.10.002.

Gunn, A. and S.J. Pitt. 2012. Parasitology: An Integrated Approach. Wiley-Blackwell, West Sussex.

Gudger, E.W. 1930. The Candirú. Paul B. Hoeber, New York.

Guimarães, J. 1935. Nota sobre a biologia do *Pseudostegophilus scarificator* Ihering. Revista de Indústria Animal 3: 273–275.

Günther, A. 1864. Catalogue of the fishes in the British Museum. Catalogue of the Physostomi, containing the families Siluridae, Characinidae, Haplochitonidae, Sternoptychidae, Scopelidae, Stomiatidae in the collection of the British Museum. Cat. Fishes v. 5: i–xxii + 1–455.

Hardman, M. and J.G. Lundberg. 2006. Molecular phylogeny and a chronology of diversification for "phractocephaline" catfishes (Siluriformes: Pimelodidae) based on mitochondrial DNA and nuclear recombination activating gene 2 sequences. Mol. Phylogenet. Evol. 40: 410–418.

Herman, J.R. 1973. Candirú: urinophilic catfish. Its gift to urology. Urology, 1(3): 265–267. https://doi.org/10.1016/0090-4295(73)90751-6.

Hoese, H.D. 1966. Ectoparasitism by juvenile sea catfish *Galeichthys felis*. Copeia 1966(4): 880–881.

Ihering, R. von. 1930. Notas ecológicas referentes a peixes d´água doce do estado de São Paulo e descrição de 4 espécies novas. Archivos do Instituto Biológico, São Paulo, 3: 93–103, Pl. 13.

Kelley, W.E. and J. Atz. 1964. A pygidiid catfish that can suck blood from goldfish. Copeia 1964: 702–704.

Koblmüller, S., C. Sturmbauer, E. Verheyen, A. Meyer and W. Salzburger. 2006. Mitochondrial phylogeny and phylogeography of East African squeaker catfishes (Siluriformes: *Synodontis*). BMC Evol. Biol. 6: 49. https://doi.org/10.1186/1471-2148-6-49.

Koblmüller, S., K.M. Sefc and C. Sturmbauer. 2008. The Lake Tanganyika cichlid species assemblage: recent advances in molecular phylogenetics. Hydrobiologia 615: 5–20. https://doi.org/10.1007/s10750-008-9552-4.

Kwon, M. and A.L. Gardner. 2007. Subfamily Desmodontinae. pp. 218–223. *In*: Gardner, A.L. (ed.). Mammals of South America, Volume 1: Marsupials, Xenarthrans, Shrews, and Bats. The University of Chicago Press, Chicago.

Lafferty, K.D. and A.M. Kuris. 2002. Trophic strategies, animal diversity and body size. Trends Ecol. Evol. 17(11): 507–513. https://doi.org/10.1016/S0169-5347(02)02615-0.

Lasso, C.A., D. Lew, D. Taphorn, C. DoNascimiento, O. Lasso-Alcalá, F. Provenzano et al. 2004. Biodiversidad ictiológica continental de Venezuela. Parte I. Lista de especies y distribución por cuencas. Mem. Fund. La Salle Cien. Nat. 159/160: 105–195.

Le Bail, P.-Y., R. Covain, M. Jégu, S. Fisch-Muller, R. Vigouroux and P. Keith. 2012. Updated checklist of the freshwater and estuarine fishes of French Guiana. Cybium 36(1): 293–319.

Lucius, R., B. Loos-Frank, R.P. Lane, R. Poulin, C.W. Roberts and R.K. Grencis. 2017. The Biology of Parasites. Wiley-VCH, Weinheim.

Lundberg, J.G., A. Machado-Allison and R.F. Kay. 1986. Miocene characid fishes from Colombia: evolutionary stasis and extirpation. Science 234: 208–209.

Lundberg, J.G., L.G. Marshall, J. Guerrero, B. Horton, L.R. Malabarba and F. Wesselingh. 1998. The stage for Neotropical fish diversifications: a history of tropical South American rivers. pp. 13–48. *In*: Malabarba, L.R., R.E. Reis, R.P. Vari, Z.M.S. Lucena and C.A.S. Lucena (eds.). Phylogeny and Classification of Neotropical Fishes. Edipucrs, Porto Alegre.

Machado, F.A. and I. Sazima. 1983. Comportamento alimentar do peixe hematófago *Branchioica bertonii* (Siluriformes, Trichomycteridae). Ciênc. Cult. 35: 344–348.

de Meeûs, T. and F. Renaud. 2002. Parasites within the new phylogeny of eukaryotes. Trends Parasitol. 18(6): 247–251. https://doi.org/10.1016/S1471-4922(02)02269-9.

Miranda Ribeiro, P. de. 1946. Notas para o estudo dos Pygidiidae Brasileiros (Pisces - Pygidiidae - Stegophilinae). Bol. Mus. Nac., Zool. (N. S.) 58: 1–20.

Mol, J.H., R.P. Vari, R. Covain, P.W. Willink and S. Fisch-Muller. 2012. Annotated checklist of the freshwater fishes of Suriname. Cybium 36(1): 263–292.

Myers, G.S. 1944. Two extraordinary new blind nematognath fishes from the Rio Negro, representing a new subfamily of Pygidiidae, with a rearrangement of the genera of the family and illustrations of some previously described genera and species from Venezuela and Brazil. Proc. Calif. Acad. Sci., Ser. 4, 23(40): 591–602.

Ochi, H., T. Onchi and Y. Yanagisawa. 2001. Alloparental care between catfishes in Lake Tanganyika. J. Fish Biol. 59: 1279–1286. https://doi.org/10.1006/jfbi.2001.1738.

Ochoa, L.E., F.F. Roxo, C. DoNascimiento, M.H. Sabaj, A. Datovo, M. Alfaro et al. 2017. Multilocus analysis of the catfish family Trichomycteridae (Teleostei: Ostariophysi: Siluriformes) supporting a monophyletic Trichomycterinae. Mol. Phylogenet. Evol. 115: 71–81. http://dx.doi.org/10.1016/j.ympev.2017.07.007.

Ochoa, L.E., A. Datovo, C. DoNascimiento, F.F. Roxo, M.H. Sabaj, J. Chang et al. 2020. Phylogenomic analysis of trichomycterid catfishes (Teleostei: Siluriformes) inferred from ultraconserved elements. Sci. Rep. 10: 2697. https://doi.org/10.1038/s41598-020-59519-w.

de Pinna, M.C.C. 1998. Phylogenetic relationships of Neotropical Siluriformes (Teleostei: Ostariophysi): historical overview and synthesis of hypotheses. pp. 279–330. *In*: Malabarba, L.R., R.E. Reis, R.P. Vari, Z.M. Lucena and C.A.S. Lucena (eds.). Phylogeny and Classification of Neotropical Fishes. Edipucrs, Porto Alegre.

de Pinna, M.C.C. and H.A. Britski. 1991. *Megalocentor*, a new genus of parasitic catfish from the Amazon basin: the sister group of *Apomatoceros* (Trichomycteridae: Stegophilinae). Ichthyol. Explor. Freshw. 2: 113–128.

de Pinna, M.C.C. and W.B. Wosiacki. 2003. Family Trichomycteridae (pencil or parasitic catfishes). pp. 270–290. *In*: Reis, R.E., S.O. Kullander and C.J. Ferraris (eds.). Check List of the Freshwater Fishes of South and Central America. Edipucrs, Porto Alegre.

de Pinna, M.C.C. 2016. The dawn of phylogenetic research on neotropical fishes: a commentary and introduction to Baskin (1973), with an overview of past progress on trichomycterid phylogenetics. Neotrop. Ichthyol. 14(2): e150127. http://doi.org/10.1590/1982-0224-20150127.

de Pinna, M. and F.C.P. Dagosta. 2022. A taxonomic review of the vampire catfish genus *Paracanthopoma* Giltay, 1935 (Siluriformes, Trichomycteridae), with descriptions of nine new species and a revised diagnosis of the genus. Pap. Avulsos Zool. 62: e202262072. http://doi.org/10.11606/1807-0205/2022.62.072.

de Pinna, M.C.C., L. Peixoto, V. Tagliacollo and M. Britto. 2025. Phylogenetic relationships and evolution of the major groups of Siluriformes. *In*: Arratia, G. and R.E. Reis (eds.). Catfishes, a Highly Diversified Group, Volume 2: Phylogeny and Evolution. CRC press, Boca Raton, FL.

Pise, M., M. Sachin, S.M. Gosavi, P.A. Gorule, C.R. Verma, S.S. Kharat et al. 2022. Osteological description of Indian lepidophagous catfish *Pachypterus khavalchor* (Siluriformes: Horabagridae) from the Western Ghats of India. J. Vertebr. Biol. 71: 1–13.

Polačik, M., M. Reichard, C. Smith and R. Blažek. 2019 Parasitic cuckoo catfish exploit parental responses to stray offspring. Phil. Trans. R. Soc. B 374: 20180412. http://dx.doi.org/10.1098/rstb.2018.0412.

Poulin, R. and H.S. Randhawa. 2015. Evolution of parasitism along convergent lines: from ecology to genomics. Parasitology 142: 6–15. https://doi.org/10.1017/S0031182013001674.

Price, P.W. 1980. Evolutionary Biology of Parasites. Monographs in Population Biology, 15. Princeton University Press, Princeton.

Reichard, M., S. Koblmüller, R. Blažek, H. Zimmermann, C. Katongo, A. Bryjová et al. 2023. Lack of host specialization despite selective host use in brood parasitic cuckoo catfish. Molecular Ecology 32: 6070–6082. https://doi.org/10.1111/mec.17173.

Roberts, T.R. 1978. An ichthyological survey of the Fly River in Papua New Guinea with descriptions of the new species. Smiths. Contr. Zool. 281: 1–72.

Rodríguez-Olarte, D., J.I. Mojica Corzo and D.C. Taphorn Baechle. 2011. Northern South America: Magdalena and Maracaibo basins, pp. 243–258. *In*: Albert, J.S. and R.E. Reis (eds.). Historical Biogeography of Neotropical Freshwater Fishes. University of California Press, Berkeley, Los Angeles, London.

Sato, T. 1986. A brood parasitic catfish of mouthbrooding cichlid fishes in Lake Tanganyika. Nature 323: 58–59.

Sazima, I. 1983. Scale-eating in characoids and other fishes. Environ. Biol. Fishes 9: 87–101.

Schmidt, R.E. 1993. Relationships and notes on the biology of *Paracanthopoma parva* (Pisces: Trichomycteridae). Ichthyol. Explor. Freshw. 4(2): 185–191.

Spotte, S., P. Petry and J.A.S. Zuanon. 2001. Experiments on the feeding behavior of the hematophagous candiru, *Vandellia* cf. *plazaii*. Environ. Biol. Fishes 60: 459–464.

Spotte, S. 2002. Candiru - Life and Legend of the Bloodsucking Catfishes. Creative Arts Book Company, Berkeley.

Stauffer, J.R. and W.F. Loftus. 2010. Brood parasitism of a bagrid catfish (*Bagrus meridionalis*) by a clariid catfish (*Bathyclarias nyasensis*) in Lake Malaŵi, Africa. Copeia 2010(1): 71–74.

Sullivan, J.P., J.G. Lundberg and M. Hardman. 2006. A phylogenetic analysis of the major groups of catfishes (Teleostei: Siluriformes) using rag1 and rag2 nuclear gene sequences. Mol. Phylogenet. Evol. 41: 636–662. https://doi.org/10.1016/j.ympev.2006.05.044.

Szelistowski, W.A. 1989. Scale-feeding in juvenile marine catfishes (Pisces: Ariidae). Copeia 1989(2): 517–519.

Takahashi, T. and S. Koblmüller. 2020. Brood parasitism of an open-water spawning cichlid by the cuckoo catfish. J. Fish. Biol. 96: 1538–1542. https://doi.org/10.1111/jfb.14350.

Taphorn, D.C., E. Liverpool, N.K. Lujan, C. DoNascimiento, D.D. Hemraj, W.G.R. Crampton et al. 2022. Annotated checklist of the primarily freshwater fishes of Guyana. Proc. Acad. Nat. Sci. Philadelphia 168: 1–95. https://doi.org/10.1635/053.168.0101.

Tilney, L. and T. Hecht. 1990. The food and feeding habits of two co-occurring marine catfish *Galeichthys feliceps* and *G. ater* (Osteichthyes: Ariidae) along the south-east coast of South Africa. J. Zool. 221(2): 171–193.

Vari, R.P. 1991. Systematics of the Neotropical characiform genus *Steindachnerina* Fowler (Pisces: Ostariophysi). Smiths. Contrib. Zool. 507: 1–118.

Wisenden, B.D. 1999. Alloparental care in fishes. Rev. Fish Biol. Fisher. 9: 45–70.

Wootton, R.J. and C. Smith. 2015. Reproductive Biology of Teleost Fishes. Wiley Blackwell, Chichester.

Wosiacki, W.B. and M.C.C. de Pinna. 2007. Família Trichomycteridae: Stegophilinae, pp. 72–73. *In*: Buckup, P.A., N.A. Menezes and M.S. Ghazzi (eds.). Catálogo das Espécies de Peixes de Água Doce do Brasil. Série Livros 23. Museu Nacional, Universidade Federal do Rio de Janeiro, Rio de Janeiro.

Wright, J.J. and L.M. Page. 2006. Taxonomic revision of Lake Tanganyikan *Synodontis* (Siluriformes: Mochokidae). Bull. Fla. Mus. Nat. Hist. 46: 99–154.

Zelmer, D.A. 1998. An evolutionary definition of parasitism. Int. J. Parasitol. 28: 531–533.

Zimmermann, H., R. Blažek, M. Polačik and M. Reichard. 2022. Individual experience as a key to success for the cuckoo catfish brood parasitism. Nat. Commun. 13: 1723. https://doi.org/10.1038/s41467-022-29417-y.

Zimmerman, H., D. Tolman and M. Reichard. 2023. Low incidence of cannibalism among brood parasitic cuckoo catfish embryos. Behav. Ecol. 34(4): 521–527. https://doi.org/10.1093/beheco/arad024.

Zuanon, J. and I. Sazima. 2005. Free meals on long-distance cruisers: the vampire fish rides giant catfishes in the Amazon. Biota Neotropica 5(1): 109–114.

Chapter 11

Whiskers Ahead

Catfish Migrations in Freshwater and Marine Environments

Jansen Zuanon[1,*] and *Gislene Torrente-Vilara*[2]

Fish migrations are well-known and conspicuous phenomena, where many individuals (often thousands of them) move through aquatic environments in search of habitats for reproduction, feeding, or seasonal refuge. In most cases, these migrations are highly visible in places with natural obstacles, such as waterfalls and rapids (or artificial ones like dams and fish ladders), that need to be overcome on their way to the destination habitat (Lucas and Baras 2001; Brönmark et al. 2014). However, migrations of catfishes (Siluriformes) tend to be much less conspicuous because most species have benthic or nektobenthic habits (both in fresh and saltwater), and the largest migrations occur in muddy rivers, where the visibility of these movements is hindered by low water transparency (often less than 30 cm). However, even though most catfish migrations do not constitute conspicuous movements, the seasonal occurrence of large schools of migratory catfish is well-known to local fishermen and riverside communities. These groups extensively exploit these resources through subsistence and commercial fishing (Figure 1).

In South America, catfishes support a significant portion of commercial fisheries in tropical floodplain river systems (Carolsfeld et al. 2004; Barthem and Goulding 1997; MPA 2013), especially in fisheries conducted in turbid rivers. Empirical knowledge passed down through generations by traditional communities has enabled fishermen to develop a refined understanding of species migrations and the development of specialized gear for successful fishing, such as the case of the pimelodid goliath catfishes in the Amazon basin (e.g., Goulding 1979, 1980; Smith 1979). Similarly, a deep knowledge of fish behavior, displacement routes and timing of seasonal movements has allowed the intensive commercial fishing of migratory pangasiid catfishes in Asian rivers (Baird et al. 2001, 2004; Hogan et al. 2004, 2007).

The current knowledge about freshwater fish migration in the Amazon basin was recently summarized by Duponchelle et al. (2021) and Herrera-R. et al. (2023). However, despite the important information derived from such studies and the included information on the migratory behavior of freshwater catfishes, comprehensive results for catfish migrations in estuarine and

[1] Rua Godofredo Fraga, 10, Apto. 65 A, Marapé, 11070-400 Santos, SP, Brazil.

[2] Laboratório de Física Aplicada, Materiais e Meio Ambiente - LaFAMA, Universidade Federal de São Paulo, Rua Dr. Carvalho de Mendonça, 144, 11070-100 Santos, SP, Brazil.

Email: gtvilara@unifesp.br

* Corresponding author: jzuanon3@gmail.com

Figure 1. (A) A group of fishermen capturing migrating goliath catfishes (mostly *Brachyplatystoma rousseauxii*, Pimelodidae) at the downstream portion of Teotônio waterfall in Madeira River, Brazilian Amazon. Photo by Michael Goulding. (B) Fishermen trying to capture large catfishes (Pimelodidae) with slingshots ("fisga:" a large hook attached to a rope and fitted to the tip of a long wooden pole) in a shallow stretch of the Teotônio waterfall, Madeira River, Brazil. Photo: G. Torrente-Vilara.

marine environments are still lacking. In fact, for most catfish species, there are no published studies about individual movement patterns or migration, and the criteria for determining whether a species is migratory is predominantly anecdotal. This is mostly based on visual records of movements across natural or artificial barriers, despite the existence of several methods to unravel fish migration (Boxes 1 and 2).

This chapter provides an overview of catfish migrations in various aquatic environments worldwide, with a stronger (non-intentional) focus on freshwater environments. To achieve this, we conducted an extensive (though not exhaustive) search of scientific articles, books, and gray literature (mainly unpublished dissertations and theses) to provide an overview of catfish migrations. We searched for literature records of catfish migration through Clarivate Analytics Web of Science,

BOX 1. Classical Methods to Study Fish Migration

Direct Observation (DO) – Some large rivers harbor remarkable geographic features that allow for the direct observation of migration (cat)fish schools: narrow falls and/or rapids. Historical catfish fisheries thrive in fishing villages near these features, where an impressive array of artisanal fishing devices have been developed based on an intimate knowledge of migratory, feeding, and other behavior of catfishes. Such favorable conditions made it possible to obtain highly valuable data on catfish migrations in South America (e.g., Gouding 1979, 1980 for pimelodid goliath catfishes of the genus *Brachyplatystoma*) and in Southeast Asia (e.g., Roberts and Baird 1995 for *Pangasius macronema* and *P. krempfi*). Despite the high-quality data and low cost of DO, the technique presents limitations in terms of spatial-temporal scale of observation, low level of individual information regarding the identity and biological characteristics of the catfish migrants, and low level of information regarding the involved migratory routes and its triggering factors.

Biological Monitoring (*BM*) – This is one of the most traditional methods used to infer presumed movements of fish populations along their life and in space and time. It is based on gathering cumulative information about sampled fish specimens and includes data regarding individual size, weight, sex, gonadal maturation, fat accumulation, stomach contents and degree of fullness. Examples of catfishes include the monthly catching of three ariid species in the southern Gulf of Mexico by Yáñez-Arancibia and Lara-Domínguez (1988); the four-year study of biological monitoring and seasonal movements of three ariid species at Patos Lagoon in southern Brazil by Araújo (1988); and a 33-year monitoring of Ariidae catfishes in the coastal region of West Africa by Simier et al. (2021). Despite the recovery of highly valuable ecological information about migratory fishes, BM presents several challenges and limitations: the data are logistically complex to obtain, progress slowly, usually have high economic costs, and have limited spatial and temporal coverage (Casas and Saborido-Rey 2023).

Mark Recapture (*MR*) – This method constitutes a classic and valuable approach in several ecological studies, including fish migration. It requires the capture and handling of each individual to receive a physical mark and the later recapture of marked individuals over time and space to detect population movements, estimate abundance and population demography. Some MR studies with migratory catfishes have produced excellent results in North America, such as those by Pellet et al. (1998) with *Ictalurus punctatus* and by Raabe and Hightower (2014) with *Pylodictis olivaris* and by Makrakis et al. (2007b) with the South American doradid *Pterodoras granulosus*. However, MR studies present some difficulties, with the main drawback being the usually low mark recovery rates due to tag shedding, unrecorded deaths, low capture rates, low tag return rates by artisanal and commercial fishing, or rapid dispersion of marked individuals to distant locations. These characteristics usually make MR methods expensive and frequently inefficient, especially in large, open aquatic systems such as the oceans. An example of an MR study with low yield is that of Godoy et al. (1979) with the pimelodid *Brachyplatystoma vaillantii*. The authors marked about 10,000 individuals at various locations along the Solimões-Amazonas River in the Brazilian Amazon, but only 63 tags were recovered (0.67%), which hindered solid conclusions about the species' movements and biological requirements.

Acoustic Telemetry (AT) – The principle of acoustic telemetry consists of the communication between an emitting tag (attached to a fish) and a receiving device (hydrophone) using an acoustic signal to determine the location of the tag (fish). The acoustic tags can be surgically attached externally to the fish's body or inserted into its abdominal cavity. Small coded tags (0.3 g) allow individual ID transmission for tracking group movements. They may include sensors for environmental data or changes in behavior or physiology (Crossing et al. 2017). The method can provide useful information about migration, individual behavior, habitat use, passage efficiency,

Box 1 contd. ...

...Box 1 contd.

predation, etc. Acoustic telemetry uses lower radio frequency signals (69–500 kHz) compared to radio telemetry (30–300 MHz), which both have advantages and disadvantages. Another unique advantage of acoustic telemetry is its capability to be used in salt and deep water (but may have a poor reading range in turbulent water systems). Acoustic telemetry includes both passive structures, where a network of static receivers detects animals moving through their environment, and active tracking, where an individual is followed by a boat with a mobile receiver. Examples of acoustic telemetry studies with catfishes include the work of Stoot et al. (2023) with the relation of environmental factors and the seasonal distribution of the estuarine ariid *Neoarius graeffei* in Northern New South Wales, Australia, and that of Hahn et al. (2019) to detect long-distance upstream movements of the large pimelodids *Phractocephalus hemioliopterus* and *Pseudoplatystoma punctifer* in Xingu River, Brazilian Amazon.

BOX 2. New Methods to Unravel Fish Migratory Patterns

Close Kin Mark Recapture (CKMR) – The method, proposed almost two decades ago (Nielsen et al. 2001; Skaug 2001), has recently re-emerged as a promising tool for estimating key demographic parameters and identifying kinship using genomic methods (Bravington et al. 2016). CKMR can be applied to estimate abundance, biomass, and other demographic parameters that are particularly challenging to infer for marine species with large, highly mobile, dispersed populations. This method has garnered considerable attention for its significant advantages over classic MR methods. CKMR can be applied to samples collected during a single sampling event and from samples obtained from dead (fished) specimens (Wacker et al. 2021). Furthermore, CKMR overcomes many challenges inherent to traditional MR modeling, as it does not suffer from effects that may occur after the initial capture, such as fishing trap avoidance or tag loss (Marcy-Quay et al. 2020). Advances in high-throughput sequencing technologies now enable rapid and accurate genotyping of a large number of samples at many *loci* to identify close kin accurately. Despite its potential, the successful application of the method remains very limited, as it has been used to assess only a few species (Delaval et al. 2022). The broad applicability of CKMR to fish populations has not yet been demonstrated, but it could potentially be important for migratory catfish.

Otoliths Microchemistry (OM) – This technique has been widely used in studies aiming to uncover fish migration routes, especially regarding diadromous fishes. Such studies use Inductively Coupled Plasma Mass Spectrometry (ICPMS), a powerful tool for quantifying isotope ratios, but is limited in the number of elements that can be sampled in a single ablation pass. Strontium (Sr) isotope ratios have proven useful in freshwater systems, but freshwater fish otoliths exhibit many other trace chemical elements, including manganese (Limburg et al. 2015), bromine (Limburg et al. 2010a), and selenium (Limburg et al. 2010b; Hauser et al. 2019b). More recently, oxygen has been used to infer environmental (climatic) changes throughout individuals' life history (Cavole et al. 2023), increasing the potential use of the technique. By comparing tissue isotope signatures with the isotopic landscape (isoscape), studies have unraveled migratory connectivity (Wassenaar and Hobson 2001; Rubenstein and Hobson 2004) and tracked the arrival of a species in a new environment (Vander Zanden et al. 2015). Regarding migratory species that use estuarine habitats during part of their life cycle, it can be difficult to quantify the time spent in these environments. Other potential limitations to the application of OM techniques are the high cost of microchemistry analysis and the need for enough isotopic variation among the aquatic habitats utilized by the

Box 2 contd. ...

...Box 2 contd.

species. Otolith microchemistry analyses have been successfully used in catfish migration research. Examples include Hegg et al. (2015), Hauser et al. (2019a, b), and Duponchelle et al. (2016), which studied isotopic ratios of strontium, selenium and calcium of Amazonian goliath pimelodids (*Brachyplatystoma* spp.). Strontium isotopes were also analyzed in *Zungaro jahu* by Avigliano et al. (2023) and in the Asian *Pangasius krempfii* by Hogan et al. (2007). Additionally, Kubota et al. (2015) and Tran et al. (2021) studied strontium and calcium isotopes in the ariid *Plicofollis tenuispinis*.

Crystalline Lenses (CL) – Similar to otoliths, the lenses of the eye's crystalline also function as archival tissues, revealing an individual's food sources by comparison with reference samples of $\delta^{13}C$ from primary producers, invertebrate primary consumers, fishes and other aquatic animals. It is also possible to detect the use of freshwater, estuarine, and marine environments throughout the life of fish (Wallace et al. 2014; Bell-Tilcock et al. 2021). Each layer (lamella) of the eye lens can mark a signature (isotopic ratio), and when analyzed in sequence, they constitute an isotopic representation of the primary carbon sources ($\delta^{13}C$) of an individual along its life, while $\delta^{15}N$ indicates its trophic level. This information can help track the spatial and temporal use of freshwater, estuarine, and marine environments throughout the lives of individuals. One limitation of the technique is the high cost of isotopic analysis due to the intricate structure of each fish's crystalline eyes, typically consisting of multiple layers of lenses. Additionally, isoscapes should cover a large spatial scale and encompass the habitats throughout the species' life cycle. As far as we could verify, there are no results in the literature regarding migratory catfishes.

Radio Frequency Identification (RFID) – It is a technology that has existed for decades but has been gaining new applications and more in-depth studies in recent years. RFID technology (a small passive tag inserted in the fish and one or more fixed and energized antennas) has enabled researchers to track individuals with precision and minimal invasiveness. When a fish marked with an RFID tag passes through an antenna connected to a reader, this information is sent to a hardware device (with specific software), which then forwards these details over the internet to a database. The collected data, such as movement patterns and social interactions, provide crucial information for the conservation of endangered species and the understanding of natural ecosystems. It is particularly valuable for monitoring migrations, as it allows the tracking of animals on a large scale and in real-time. For instance, RFID was used to track movements of the wels catfish *Silurus glanis* through fishways in the Meuse River in Belgium, with very good results (Ovidio et al. 2023). However, despite the obvious advantages of the technique, it is (sometimes) highly expensive, demands a large effort in capturing and marking individuals, and the antennas must be maintained energized for long periods, often in remote locations.

Scopus and Google Scholar, using several combinations of keywords that included "catfish(es)," "migration," "migratory," "freshwater," "saltwater," "marine," "estuarine," as well as more specific words such as "Ariidae" and "Plotosidae" when searching for records of marine and estuarine catfish migration. The recovered references were initially screened by title and abstract, during which we discarded references that did not contain explicit information on the occurrence and characteristics of catfish migratory behavior and/or events. We also checked the references cited in the recovered articles in search of additional information sources using a "snowball" approach.

Information regarding catfish species' maximum body size (Standard Length – SL, in centimeters) was obtained from FishBase (Froese and Pauly 2024). The obtained information was organized into tables containing details about the type of migration, estimated distance traveled, and the conservation status of the species, checked at IUCN (IUCN 2023) and ICMBio website (for the status in the Brazilian territory; ICMBIO 2024).

Throughout the chapter, we briefly introduce the concept of migration used in this study, the types of migration performed by catfishes, the main methods and techniques used in the study of fish migrations, the sensory mechanisms potentially involved in the migratory process, and the occurrence of migratory behavior among catfishes worldwide. We also present information on some ecosystem services provided by migratory catfishes and conclude with a brief assessment of the main threats to migratory catfishes and potential conservation policies aimed at this highly diverse group of fish.

Concept and Types of Migration and Their Occurrence Among Catfishes

Periodic and predictable movements between habitats, usually performed by most (or a large portion) of the population, constitute migrations (q.v. Northcote 1978). Moreover, migration differs from dispersal because individuals return to their initial habitat (Brönmark et al. 2014). Animal migration is motivated by three main functional aspects: (1) reproduction (spawning), (2) feeding, and (3) search for refuge (Northcote 1978, 1984). Regardless of the motivation, these movements result in the alternating use of two or more separate habitats, occurring periodically and predictably throughout the animal's life and involving a significant portion of the population. Migrations can occur at different spatial and temporal scales, ranging from daily foraging movements within a home range to annual long-distance displacements (or even a single migration along the animal's life).

Fish migration studies initially focused on movements (mostly of salmonids) between freshwater and marine environments (Myers 1938, 1949; McDowall 1988; Northcote 1997). The fundamental types of migration are defined by the kind of environment in which the population spends most of its adult life: (1) Oceanodromy – migrations that occur entirely in the sea and that seem uncommon among catfishes; (2) Potamodromy – migrations that occur entirely in freshwater and that are observed among several catfish families, such as Pimelodidae, Ictaluridae, some Pangasiidae and Siluridae; and (3) Diadromy – those occurring between freshwater and marine environments, occurring among the Ariidae, Pangasiidae, and Plotosidae.

Diadromy represents some of the most well-known and spectacular migrations in nature. There are three recognized forms of diadromy: anadromy, where adults migrate from a marine/estuarine environment to reproduce in freshwater; catadromy, where adults migrate from freshwater to reproduce in the marine/estuarine environment; and amphidromy, when such migrations are predictable between marine/estuarine and freshwater environments but are not related to reproduction (Myers 1949; McDowall 1988). Furthermore, migration between two freshwater habitats is known as potamodromy, and may involve two main movement directions in relation to the main river channel: longitudinal movements (upstream-downstream) and lateral movements (between the main river channel and floodplains). Estuarine environments are extremely important in the life cycle of many migratory species, both marine and freshwater, and they are especially important for catfish species of the families Ariidae, Pangasiidae, Plotosidae, Aspredinidae, and even for some species of Pimelodidae.

Natal homing, also known as philopatry or natal dispersal, is the migratory movement that returns an individual to the exact or nearly exact place of birth. It typically constitutes the longest effective dispersal movement that migratory animals can undertake during their lifetimes (Greenwood and Harley 1982; Paradis et al. 1998). Natal homing is observed in various animal groups, from invertebrates (e.g., insects) to vertebrates such as mammals, birds, reptiles, and fish (Weyer and Schmitt 2013; Rooker et al. 2008; Brothers and Lohmann 2015; Miller and Smallwood 1997; Hoffman and Forcada 2012). Natal homing is well documented in diadromous fish species that travel between freshwater and the sea. A classic example is the North American salmonids, which reproduce in freshwater, grow in the sea, and then return to their birthplace for reproduction (Mills 1986; Dittman and Quinn 1996). Natal homing has also been recently confirmed for the Amazon catfish *Brachyplatystoma rousseauxii* (Duponchelle et al. 2016; Hauser et al. 2019a, b), as well as for *Pseudoplatystoma corruscans* in the La Plata River basin in South America (Pereira et al. 2009) and *Ictalurus punctatus* in the Wisconsin River basin in North America (Pellett et al. 1998).

Diversity and Occurrence of Migratory Behavior in Catfishes

Siluriformes comprise a remarkable diversity of species, with 40 families and ca. 4,200 valid species (Fricke et al. January 2024). However, despite this high diversity, migratory behavior has been documented in only 32.5% (13 out of 40) of the catfish families. Natal homing behavior is an even rarer phenomenon, being recorded for only three catfish species so far (Table 1).

Nonetheless, short lateral migrations between habitats are quite common among freshwater fish, especially in the floodplain river systems of tropical environments (Junk et al. 1989). For example, our observations (JZ and GTV) in the Amazon indicate the seasonal use of floodable habitats during periods of rising and high water levels, as well as movement into river channels during the falling and dry phases of the hydrological cycle by several species of Cetopsidae, Heptapteridae and Auchenipteridae. These short and inconspicuous migrations likely result from unfavorable environmental conditions during the dry season, including abiotic factors such as low dissolved oxygen concentrations, high water temperatures, and elevated levels of decomposing organic matter, as well as biotic factors such as increased predation pressure, reduced food availability, and diminished shelter/refuge options against predators. A similar situation likely occurs in inland waters of tropical environments in Asia and Africa, where Siluriformes are highly diverse and abundant, inhabit aquatic systems with significant habitat heterogeneity, and experience strong seasonal variations in environmental conditions.

Among South American catfishes, the large pimelodids of the genus *Brachyplatystoma* (including *B. filamentosum*, *B. capapretum*, *B. rousseauxii*, *B. vaillantii*, *B. platynemum*, *B. tigrinum*, and *B. juruense* – Figure 2B and 2D) stand out as the most notable examples of long-distance migratory species that sustain important commercial fisheries throughout the Amazon basin (Petrere Jr. et al. 2004; Batista and Alves-Gomes 2006; Hegg et al. 2015; Duponchelle et al. 2016; Barthem et al. 2017; Hauser et al. 2018; Van Damme et al. 2019; Hauser et al. 2019b; Lima et al. 2020). The most remarkable case is that of the "dourada" *B. rousseauxii*, which reproduces in the headwaters of rivers in the western portion of the Amazon basin. Larvae and juveniles are carried by river currents to the Amazon River estuary, located approximately 5,000 km from the spawning area. The drift of the larvae has been estimated to take around 33–40 days to reach the estuary (Hauser et al. 2019b). After spending two or three years feeding and growing in that area, sub-adult individuals begin their return migration in presumably annual stages as they continue to grow and reach sexual maturity. This journey can last several years, during which the fish continue to grow and mature sexually. Captures of *B. rousseauxii* at the Teotônio waterfall in Madeira River (about 2/3 of the reproductive migration journey) have revealed pre-adult specimens in gonadal maturation, ranging in age from 4–6 years (Hauser et al. 2008, fig. 1). Upon completing this enormous journey, "douradas" will have traveled up to 11,600 km, making it the longest longitudinal migration of a potamodromous fish (Barthem et al. 2017).

In addition to *B. rousseauxii*, the "pintado" *Pseudoplatystoma corruscans* also exhibit natal homing migrations. This is supported by a genetic study of individuals from six rivers in the La Plata basin, which revealed the existence of six groups of individuals largely corresponding to the main branches on a population neighbor-joining tree. This study rejected the hypothesis of population panmixia and pointed out a strong tendency for individuals to return to their natal nursery regions for reproduction, supporting the hypothesis of homing behavior (Pereira et al. 2009).

Other medium and large-sized pimelodids are known to perform migration (Figure 2). These include the enormous "jaú" of the genus *Zungaro* (e.g., *Z. jahu*, Agostinho et al. 2003) (Figure 2C), the large benthic catfish *Platynematichthys notatus*, and the surubim catfishes of the genus *Pseudoplatystoma* (Goulding 1980; Makrakis et al. 2007a; Pereira et al. 2009; Hahn et al. 2019). Other examples are the carnivorous and scavenger catfishes of the genera *Calophysus*

Table 1. Diversity and occurrence of migratory and homing behavior among catfish families (Actinopterygii: Siluriformes), based on the literature reviewed in the present chapter (for details, please see Table 2). Taxonomic families are arranged in alphabetical order, and following Fricke et al. (2023). Migration type: POT = potamodromous; DIA = diadromous.

Catfish Families	Valid Genera	Valid Species	Migratory Behavior	Homing Behavior	Migration Type
Ailiidae	7	25	X		POT
Akysidae	5	59			
Amblycipitidae	4	47			
Amphiliidae	13	106			
Anchariidae	2	6			
Ariidae	34	157	X		DIA
Aspredinidae	13	49			
Astroblepidae	1	82			
Auchenipteridae	25	128	X		POT
Auchenoglanididae	3	21			
Austroglanididae	1	3			
Bagridae	18	229			
Callichthyidae	8	224			
Cetopsidae	5	44	X		POT
Chacidae	1	4			
Clariidae	16	120	X		POT
Claroteidae	9	65			
Cranoglanididae	1	3			
Diplomystidae	1	9			
Doradidae	31	96	X		POT
Heptapteridae	23	235			
Heteropneustidae	1	5			
Horabagridae	4	10			
Ictaluridae	7	51	X	X	POT
Kryptoglanidae	1	1			
Lacantuniidae	1	1			
Loricariidae	115	1,047	X		POT
Malapteruridae	2	19			
Mochokidae	9	228			
Nematogenyidae	1	1			
Pangasiidae	4	30	X		POT, DIA
Phreatobiidae	1	3			
Pimelodidae	30	116	X	X	POT
Plotosidae	10	42			DIA
Pseudopimelodidae	6	55			
Schilbeidae	5	35			
Scoloplacidae	1	6			
Siluridae	13	106	X		POT
Sisoridae	26	314	X		POT
Trichomycteridae	42	412	X		POT
TOTAL	**501**	**4,194**	**13**	**2**	

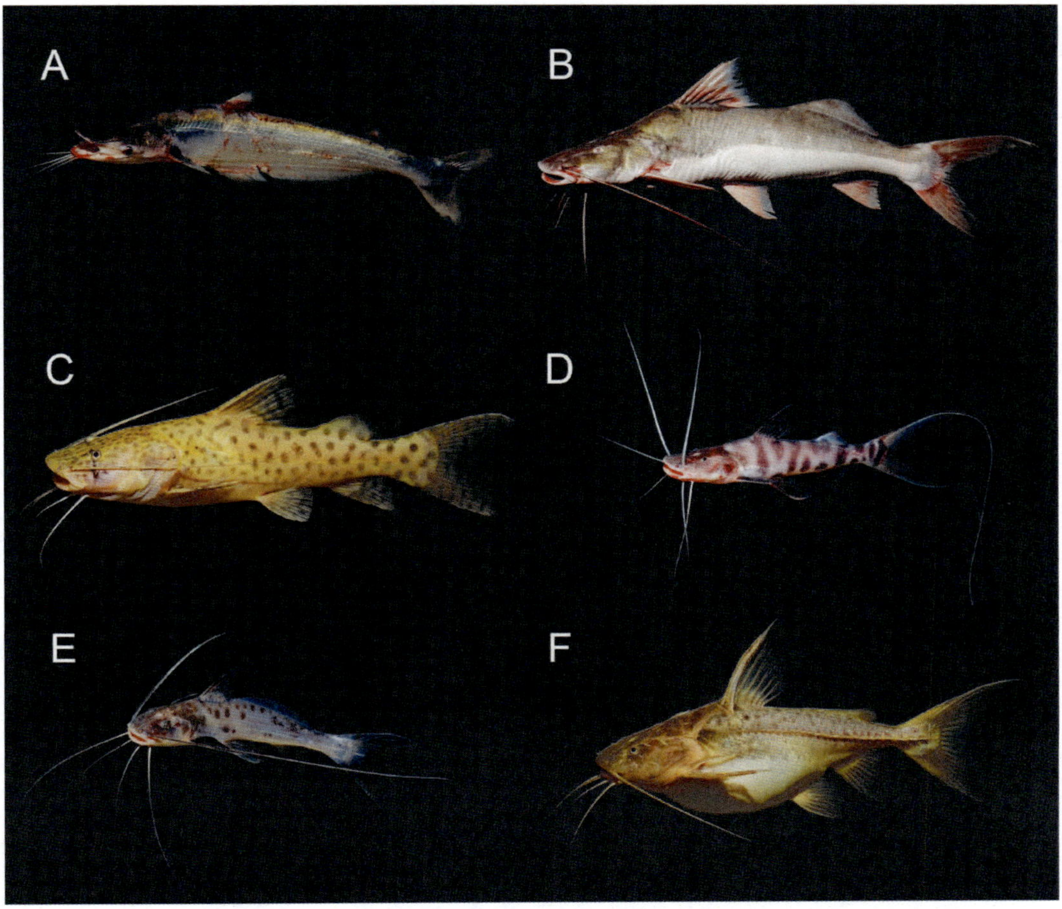

Figure 2. Some examples of South American migratory catfishes. (A) The pelagic and planktivorous *Hypophthalmus celiae*. (B) *Brachyplatystoma filamentosum*, a long-distance traveler of the Amazon. (C) the short-migrant "Jaú" catfish *Zungaro zungaro*. (D) Young individual of *Brachyplatystoma juruense*, another migratory goliath catfish. (E) The voracious scavenger and abundant pimelodid *Calophysus macropterus*. (F) The omnivorous migratory doradid *Pterodoras granulosus*. Photos by E. Ferreira and J. Zuanon.

(Figure 2E) and *Pinirampus*, as well as *Pimelodina flavipinnis* (Goulding 1980; Zapata and Usma 2013), the abundant and widely distributed mandi catfishes of the *Pimelodus* genus (e.g., Makrakis et al. 2007a), the gigantic and colorful "pirarara" *Phractocephalus hemioliopterus* (Hahn et al. 2019), and the shovelnose catfishes of the *Sorubim* genus (e.g., Carolsfeld et al. 2004; Makrakis et al. 2007a). Also noteworthy are the nektonic and planktivorous species of *Hypophthalmus* (known as "maparás" in Brazil) (Figure 2A), which form massive shoals in Amazonian rivers and lakes and are a significant commercial catch in the lower Amazon River in Brazil (Carvalho and Mérona 1986). In the São Francisco River, the enigmatic catfish known as "pirá-tamanduá," *Conorhynchos conirostris* (Figure 3A), which was tentatively classified among the Pimelodidae for many years (Lundberg et al. 2011), is known to undertake migrations in the São Francisco River, where it is endemic (Godinho and Godinho 2018).

Some South American catfish families with predominantly sedentary and territorial habits also harbor migratory species. For example, large doradid species undertake migrations, such as *Oxydoras niger* (Agostinho et al. 2007; Zapata and Usma 2013), *Pterodoras granulosus* (Makrakis et al. 2007a, b) (Figure 2F), and the enormous "bacu-pedra" *Lithodoras dorsalis* (Goulding 1980). The bacu-pedra uses the floodplains of muddy-water rivers in the Amazon basin to feed on fruits

Figure 3. Other examples of migratory catfishes. (A) The taxonomically enigmatic catfish "pirá-tamanduá" *Conorhynchos conirostris*, endemic to the São Francisco River in Brazil (Photo by Gilberto N. Salvador). (B) The African sharp tooth clariid *Clarias gariepinus* migrating upstream through a short waterfall (photo by Ben van der Waal).

and macroinvertebrates during the wet season and seeks refuge in the main river channels during the dry season. Little is known about the reproductive habits of the species, but young individuals are commonly recorded in the Amazon River estuary (Barthem et al. 1991), suggesting a pattern of upstream migratory reproduction and larval drift through the river channel to the estuary.

Other families of South American catfishes, typically with non-migratory habits, may include rare cases of migratory species. For example, the suckermouth catfishes (Loricariidae) encompass more than 1,000 species normally characterized by sedentary and territorial habits, spending most of their time inside or near shelters in clay banks, submerged logs, and rocky crevices. Surprisingly, only one species of loricariid, the black armored catfish *Rhinelepis aspera*, unquestionably performs migrations. This species inhabits the Paraná River system and has been observed migrating upstream and bypassing fish ladders (Agostinho et al. 1986; Makrakis et al. 2007a; Wagner et al. 2012). On the other hand, Auchenipteridae, Cetopsidae and Doradidae include some species with recorded short distance, seasonal lateral migrations between the river channel and floodplains (e.g., Van Damme et al. 2011; Zapata and Usma 2013). However, the true number of migratory species in several Amazonian catfish families can be much higher if the ecological niche conservation hypothesis is considered (q.v. Herrera-R. et al. 2023).

Trichomycteridae also hosts at least one long-distance migratory species, *Trichomycterus barbouri*, whose juveniles congregate in massive shoals and migrate up rivers in the Bolivian Amazon during the flood season in the Beni River basin (Miranda-Chumacero et al. 2015). There are also records of upstream migratory behavior in *T. largoperculatus* (and *T. pradensis*) in the Paraíba do Sul River basin in southeastern Brazil, but these movements likely cover relatively short distances. In both cases, these upstream migrations of young trichomycterids probably represent compensatory movements to downstream larval drift after adults have spawned in the headwaters of rivers where they permanently live after reaching sexual maturity (Miranda-Chumacero et al. 2015; Costa and Katz 2022). Finally, there is some genetic evidence of very small-scale migrations by the stream-dwelling trichomycterid *Trichogenes longipinnis* (Zamudio et al. 2009), but it is not certain whether these are actual migrations, passive drift of young fish, or simple population dispersal.

Among Asian catfishes, the main migratory species belong to Pangasiidae, with species of significant importance in commercial and subsistence fishing. These include *Pangasius krempfi* (Hogan et al. 2007; Tran et al. 2021; Duong et al. 2023), *P. conchophilus* (Baird et al. 2004), *P. macronema* (Baird et al. 2001), *P. bocourti* (So et al. 2006a), *Pangasianodon hypophthalmus* (So et al. 2006a, b; Soem et al. 2023), and the gigantic and highly threatened *Pangasianodon gigas* (Hogan et al. 2011). The large Sisoridae catfish *Bagarius bagarius* is also known to perform mid-distance migrations in rivers of Bangladesh (Saha et al. 2021). In the Mekong River basin, nocturnal migrations of both young and adult pangasiid species have been recorded, with adults migrating upstream to reproduce in the upper reaches of the basin; however, it is not entirely clear what motivates young pangasiids to migrate (Baird et al. 2004). The occurrence of long-distance longitudinal migrations of *P. krempfi* results in populations with no significant genetic structure (Duong et al. 2023), which has also been observed for *Pangasianodon hypophthalmus* and *Pangasius bocourti* (So et al. 2006a). Studies by Hogan et al. (2007) and Tran et al. (2021) demonstrated that *P. krempfi* spawn in the upper reaches of the Mekong River, and the larvae and juveniles drift downstream to the Mekong River estuary and areas in the South China Sea, where they spend a significant part of their lives until reaching sexual maturity. Afterward, they migrate more than 700 km upstream in the Mekong River to reproduce. These seasonal movements of large shoals of pangasiid catfishes are strongly threatened by the disruption of migratory routes by hydroelectric dams and by population depletion resulting from overfishing (Hogan et al. 2011; Soem et al. 2023). Nektonic catfishes of the genus *Clupisoma* (Ailiidae) are also known to perform potamodromous migrations in the Ganga-Brahmaputra River system and Indus River in the Indian subcontinent, where they are intensively exploited in commercial and subsistence fisheries (Riede 2004; Khan et al. 2014; Dorji and Tenzin 2023). However, there appear to be few published references (and difficult to access) regarding the life history of these species, hindering a more detailed analysis of their migratory behavior.

African rivers and lakes harbor several families of catfishes, most of which are endemic to the continent: Claroteidae, Austroglanididae, Bagridae, Schilbeidae, Malapteruridae, Mochokidae, Amphiliidae, Ariidae, and Clariidae. Despite this high taxonomic and phylogenetic diversity, published records of catfish migrations have been found only for species of the genus *Clarias* (e.g., *C. gariepinus*, *C. batrachus*) (Figure 3B), and these migrations seem to involve only short to medium-distance movements between habitats such as rivers, lakes, and swamps (Clay 1979; Lung'ayia 2022). It is highly likely that a much larger number of small African catfish species from different families undertake at least short-distance migrations between habitats for reasons such as reproduction, feeding, or in response to seasonal environmental changes that potentially affect food and shelter availability or produce harsh conditions. Unfortunately, the chronic lack of detailed ecological information about catfishes in African freshwaters seems far from being comprehensively addressed.

In Europe (Palearctic Region), the only native catfish family is Siluridae, and no conclusive records of migrations by these catfishes have been found. However, there seems to be downstream movement or dispersal of young individuals of the giant wels *Silurus glanis*. This movement is believed to be a way to share habitats with adults that prefer deep pools (25–30 m) along rivers (Cucherousset et al. 2018; Kuzishchin et al. 2018). In the Nearctic region (North America), there are records of short to medium-distance migrations by Ictaluridae catfishes, which house approximately 51 species (Berra 2001; Fricke et al. 2023; Table 1). The blue catfish *Ictalurus furcatus* shows site fidelity in freshwater environments, but part of the population exhibits downstream movements to higher-salinity stretches of tidal rivers that can facilitate the invasion of new basins along its non-native distribution (Bunch et al. 2021). Another species of ictalurid known to make migrations is the channel catfish *Ictalurus punctatus*. This catfish performs seasonal migrations between the Wisconsin and Mississippi rivers, where it occupies small home ranges during the summer, then migrates to the Mississippi River in autumn and back to the Wisconsin River in spring to spawn. Tagged individuals returned to occupy the same summer home sites they had used in previous years (Pellett et al. 1998). Interestingly, the size of the catfish influenced its degree of fidelity to summer home sites, with larger fish showing greater fidelity (Pellett et al. 1998). This seems to be the only case of natal homing migration performed by an Ictaluridae.

Evidence of site fidelity ("homing behavior" in the original article) was also obtained for the flathead catfish *Pylodictis olivaris* in the Oklahoma reservoir in North America by Hart and Summerfelt (1974). Individuals tagged with ultrasonic transmitters and released at different distances (1.3–2.7 km) from the capture sites tended to move back to the original capture areas, and some catfish reused some pools within their home range. The flathead catfish, a non-native species in the Little River, North Carolina, was also observed migrating upstream after removing three dams and a partial opening of a fourth dam (Raabe and Hightower 2014). The upstream migration allowed the catfish to access river stretches with a greater diversity of substrates that may favor reproduction, contributing to the dispersion of this non-native species along the basin. The flathead catfish's tolerance to the higher salinity of estuarine and marine environments has allowed its spreading to adjacent river basins (Bringolf et al. 2005; Montague and Shoup 2021).

In Australasia, catfishes of the families Ariidae and Plotosidae occur, with most species inhabiting marine and estuarine environments and not undertaking significant migratory movements. New Zealand harbors a single freshwater catfish species, *Ameiurus nebulosus* (Ictaluridae), that was intentionally introduced in the 1870s, but it apparently does not migrate (Paulin et al. 2001). In Australia and New Guinea, there are at least 43 species of Ariidae, with 18 species occurring in Australia, most of which (13) inhabit marine and estuarine environments; there are also 12 species of Plotosidae (q.v. Allen 1989). Australian freshwater catfishes perform short movements among habitats, both longitudinally (to spawn in upstream reaches) and laterally (to lagoons and floodplains), but do not make long migratory movements (Clunie and Kohen 2001). In the Murray-Darling River basin in Australia, there are only three catfish species: *Porochilus rendahli*, *Neosilurus hyrtlii*, and *Tandanus tandanus* (Plotosidae). *Porochilus rendahli* undertakes short migratory movements to lowland lagoons to spawn during the rainy season and migrates upstream at the onset of the dry season in search of refuge habitats, whereas *N. hyrtlii* migrates upstream to tributary streams, presumably for reproduction (Lintermans 2009).

Ariidae catfishes stand out in marine and estuarine environments, with approximately 157 species widely distributed in tropical, subtropical, and warm temperate seas (Berra 2001; Fricke et al. 2023; Table 1). Studies conducted with *Genidens barbus* by Avigliano and Volpedo (2015), Avigliano et al. (2017), and Avigliano et al. (2022), using strontium-calcium and barium isotopes ratios, demonstrated that this species is amphidromous, being born in freshwaters and migrating to estuarine and marine environments in southern South America (Uruguay, Argentina, and Brazil). Information regarding annual growth marks and annual Sr/Ca variations (Avigliano et al. 2022) observed in specimens of *G. barbus* indicate that the catfish migrates toward lower salinity environments during the austral spring; however, not all individuals present the same pattern of habitat use. Analyzing otolith core-to-edge Sr/Ca and Ba/Ca profiles, the authors found that some individuals can be considered freshwater residents, whereas others present three different types of amphidromy (Avigliano et al. 2017). Some individuals exhibited a Sr/Ca signature compatible with an initial freshwater use and moving (at least once) to a marine environment; other individuals apparently began their life in freshwater and then migrated to estuarine water; and other individuals started life in estuarine water and then moved to freshwater, marine or both (Avigliano et al. 2017). This reveals the diversity of the life cycle and complexity of the movements of juvenile *G. barbus*.

Such complexity of habitat use and movement patterns were also observed by Kubota et al. (2015), who studied a single individual of the marine ariid *Plicofollis tenuispinis* using high-resolution measurements of oxygen isotopes and Sr/Ca ratios in the otoliths. The study revealed that the catfish was born in the estuary of the Narmada River and, after nearly six months, migrated to the sea in the Gulf of Khambhat, Western India. The catfish lived in the sea for almost one year and would supposedly return to the estuary for spawning in its second year of life if not captured by a fisherman. In fact, Ariidae catfishes are considered euryhaline and eurythermal fishes that inhabit marine and estuarine waters worldwide. Yáñez-Arancíbia and Lara-Domínguez (1988) studied the habits of three ariids (*Arius felis*, *A. melanopus*, and *Bagre marinus*) in the Terminos Lagoon at the Gulf of Mexico and found three main life strategies in the occupation of the complex environment of fresh, estuarine, and marine waters: (1) spawning in freshwater (rivers and swamps) followed by a migration of juveniles toward the central estuarine system; (2) spawning in the estuary; and (3) spawning in the sea followed by migration of juveniles to nursery habitats in the fluvial-lagoon system. The authors proposed that ariid catfishes, which exhibit different areas for reproduction, growth, and feeding, generally occur in greater abundance, likely due to lower interspecific competition by temporal and spatial niche partition.

In Central America, Fox et al. (2021) studied the natural history and reproductive activity of two populations of the ariid *Cathorops tuyra*, one in the Santa Maria River that is connected to the sea and another that is land-locked in the Chagres River. The Santa Maria population was considered semi-anadromous since some individuals migrated to a marine environment to reproduce, whereas the Chagres River population completed the life cycle in freshwater. Despite some recognized fragilities of the study, the species can be considered facultative anadromous.

The life cycle of arid catfishes may be complex and involve movements in response to environmental factors and interactions with other ariid species. Araújo (1988) studied the spatial distribution and relative abundance of three ariids (*G. barbus*, *G. genidens* and *G. planifrons*) in the Patos Lagoon, Southern Brazil, based on bottom trawling catches between 1979 and 1983. The three catfish species cohabited both in the estuary and in the bay, but *G. genidens* segregates from *G. barbus* and *G. planifrons* and occupies mostly the interior of the Patos Lagoon, with only larger juveniles co-existing with the other two species at the estuary. The low sample size of *G. genidens* and *G. planifrons* precluded a better understanding of the life cycle of these species; however, the author presents a hypothesis for *G. barbus*. Young-of-the-year born in the austral summer at the estuary moved to more saline coastal waters in their first autumn and winter and later returned to intermediate areas and upper estuary during the following spring and summer. At the beginning of their second year of life (summer and autumn), they moved again to the lagoon's mouth but did not enter coastal waters. In the next summer, the now sub-adult (2+ years) catfish join the adults at the interior of Patos Lagoon and, after the spawning season, migrate to the sea to return to the estuary in the following winter for spawning.

Gomes and Araújo (2004) also studied the seasonal movements and reproductive activity of three arid catfishes (*Notarius luniscutis*, *Genidens barbus*, and *Bagre marinus*) in Sepetiba Bay, Southeastern Brazil. The authors found that the catfishes moved from the bay to the estuary of the Guarda River during the austral spring to early autumn to spawn, then returned to the bay from mid-late autumn to winter (April–August) where the individuals recovered their body condition, so characterizing short diadromous migrations.

In Australia, Stoot et al. (2023) used acoustic telemetry to study the movement patterns of the ariid *Neoarius graeffei* throughout an unregulated coastal river catchment, the Clarence River in northern New South Wales. The authors observed some distinctive movement patterns of *N. graeffei* within the river system, with individuals occupying areas at different distances from the river mouth

Table 2. A non-exhaustive list of catfish species with published information (selected references) regarding migration type, distance range, homing behavior, IUCN threat category (and the geographical scale of the analysis). IUCN threat categories (IUCN 2023). Migration type: POT = potamodromous; DIA = diadromous. Migration vector: LAT = lateral, between river channels and floodplains or lakes; LON = longitudinal, along the main river axis; HAB = among habitats of the same basin. Migration distance: L = long, hundreds to thousands of kilometers, usually along large rivers; M = mid, tens of kilometers, commonly among and through tributary rivers; S = short, a few kilometers, usually lateral migrations between habitats in rivers and floodplains or lakes. IUCN categories: CR = critically endangered, EN = endangered, VU = vulnerable, NT = near threatened, LC = least concern, DD = data deficient, NE = not evaluated.

Species	Migration Type	Migration Vector	Migration Distance	Homing Behavior	IUCN Category	IUCN Geographic Coverage	References for Migratory Behavior
Acanthodoras cataphractus	POT	LAT, HAB	S	No	LC	Global	Zapata and Usma 2013
Ageneiosus inermis	POT	LON, LAT	M	No	LC	Global	Van Damme et al. 2011; Zapata and Usma 2013
Ageneiosus pardalis	POT	LON, LAT	M	No	LC	Global	Zapata and Usma 2013
Ariopsis felis	DIA	LON, LAT	M	No	LC	Global	Yáñez-Arancíbia and Lara-Domínguez 1988; Walton et al. 2022
Auchenipterus nuchalis	POT	LAT	M	No	LC	Global	Fernandes 1997
Auchenipterus osteomystax	POT	LON	L	No	LC	Global	Rauber et al. 2021
Bagarius bagarius	DIA	LON, HAB	M	No	VU	Global	Baird et al. 2004; Saha et al. 2021
Bagre marinus	DIA	LON, LAT	M	No	LC	Global	Yáñez-Arancíbia and Lara-Domínguez 1988
Brachyplatystoma capapretum	DIA	LON, LAT	L	No	NT	Global	Huergo et al. 2011
Brachyplatystoma filamentosum	POT	LON	L	No	LC	Brazil	Agudelo-Córdoba et al. 2000; Barthem and Goulding 1997; Huergo et al. 2011 Barthem et al. 2017
Brachyplatystoma juruense	POT	LON	L	No	LC	Brazil	Barthem and Goulding 1997 Feng et al. 2020
Brachyplatystoma platynemum	POT	LON	L	No	LC	Brazil	Barthem and Goulding 1997 Feng et al. 2020
Brachyplatystoma rousseauxii	POT	LON	L	Yes	LC	Global	Barthem et al. 2017; Batista and Alves-Gomes 2006; Duponchelle et al. 2016; Hauser et al. 2019a; Hauser et al. 2019b
Brachyplatystoma vaillantii	POT	LON	L	No	LC	Brazil	Petrere Jr. et al. 2004
Calophysus macropterus	POT	LON, LAT	L	No	LC	Global	Barthem et al. 1991; Van Damme et al. 2011
Cathorops melanopus	DIA	LON, LAT	M	No	LC	Global	Yáñez-Arancíbia and Lara-Domínguez 1989
Cathorops tuyra	DIA	LAT, HAB	S	No	DD	Global	Fox et al. (2021)
Cetopsis plumbea	POT	LAT, HAB	S	No	LC	Global	Zapata and Usma 2013

Clarias batrachus	POT	LAT, HAB	M	No	LC	Global	Clay 1979
Clarias gariepinus	POT	LAT, HAB	M	No	LC	Global	Lung'ayia 2022; Clay 1979
Clupisoma garua	POT	LON	M	No	LC	Global	Riede 2004; Dorji and Tenzin 2023
Clupisoma naziri	POT	LON	M	No	NT	Global	Khan et al. 2014
Clupisoma sinense	POT	LON	M	No	LC	Global	Riede 2004
Conorhynchos conirostris	POT	LON	M	No	EN	Global	Godinho and Godinho 2018
Doraops zuloagai	POT	LAT, HAB	S	No	LC	Global	Zapata and Usma 2013
Genidens barbus	DIA	HAB	M	No	EN	Brazil	Araújo 1988; Avigliano and Volpedo 2015; Avigliano et al. 2017; Avigliano et al. 2022
Genidens genidens	DIA	HAB	M	No	LC	Global	Araújo 1988
Genidens planifrons	DIA	HAB	M	No	CR	Brazil	Araújo 1988
Hemisorubim platyrhynchos	POT	LON, LAT	M	No	LC	Global	Van Damme et al. 2011
Hypophthalmus edentates	POT	LON, LAT	M	No	LC	Brazil	Carvalho and Mérona 1986; Van Damme et al. 2011
Hypophthalus fimbriatus	POT	LON, LAT	M	No	LC	Brazil	Zapata and Usma 2013
Hypophthalmus oremaculatus	POT	LON, LAT	M	No	LC	Global	Makrakis et al. 2007a; Van Damme et al. 2011
Ictalurus furcatus	POT	HAB	M	No	LC	Global	Bunch et al. 2021
Ictalurus punctatus	POT	LON, HAB	M	Yes	LC	Global	Pellett et al. 1998
Leiarius longibarbis	POT	LON	M	No	DD	Global	Van Damme et al. 2011
Leiarius marmoratus	POT	LON	M	No	LC	Global	Van Damme et al. 2011; Doria et al. 2018
Leiarius pictus	POT	LON	M	No	LC	Global	Van Damme et al. 2011
Lithodoras dorsalis	POT	LON, HAB	L	No	LC	Brasil	Barthem et al. 1991; Arantes et al. 2019
Neoarius graeffei	POT	HAB	S	No	LC	Global	Stoot et al. 2023
Neosilurus hyrtlii	POT	LON	M	No	LC	Global	Lintermans 2009
Oxydoras niger	POT	LON, LAT	M	No	LC	Global	Agostinho et al. 2007; Zapata and Usma 2013
Pangasianodon gigas	POT	LON	L	No	CR	Global	Hogan et al. 2011
Pangasianodon hypophthalmus	POT	LON	L	No	EN	Global	So et al. 2006; So et al. 2006b; Soem et al. 2023
Pangasius bocourti	POT	LON	L	No	LC	Global	Baird et al. 2004; So et al. 2006

Table 2 contd. ...

...Table 2 contd.

Species	**Migration Type**	**Migration Vector**	**Migration Distance**	**Homing Behavior**	**IUCN Category**	**IUCN Geographic Coverage**	**References for Migratory Behavior**
Pangasius conchophilus	POT	LON	L	No	LC	Global	Baird et al. 2004
Pangasius krempfi	DIA	LON, HAB	L	No	VU	Global	Hogan et al. 2007; Tran et al. 2021; Duong et al. 2023
Pangasius macronema	POT	LON	L	No	LC	Global	Baird et al. 2001
Phractocephalus hemioliopterus	POT	LON, LAT	L	No	LC	Brazil	Hahn et al. 2019
Pimelodus blochii	POT	LON, LAT	L	No	LC	Brazil	Agostinho et al. 2007; Zapata and Usma 2013; Van Damme et al. 2011
Pimelodus grosskopfii	POT	LON	L	No	CR	Global	Zapata and Usma 2013
Pimelodus maculatus	POT	LON, LAT	L	No	LC	Global	Carolsfeld et al. 2004
Pimelodus ornatus	POT	LON, LAT	L	No	LC	Global	Makrakis et al. 2007a
Pimelodus pictus	POT	LON	S	No	LC	Brazil	Zapata and Usma 2013
Pinirampus pirinampu	POT	LON, LAT	L	No	LC	Global	Barthem et al. 1991; Van Damme et al. 2011; Zapata and Usma 2013
Platynematichthys notatus	POT	LON	L	No	LC	Global	Van Damme et al. 2011; Zapata and Usma 2013
Platysilurus malarmo	POT	LON	S	No	NE		Zapata and Usma 2013
Platysilurus mucosus	POT	LON	M	No	LC	Global	Van Damme et al. 2011
Platystomatichthys sturio	POT	LON	M	No	LC	Global	Van Damme et al. 2011
Plicofollis tenuispinis	DIA	HAB	M	No	DD	Golfo Pérsico	Kubota et al. 2015
Porochilus rendahli	POT	LON	M	No	LC	Global	Lintermans 2009
Pseudoplatystoma corruscans	POT	LON	L	Yes	VU	Brazil	Pereira et al. 2009
Pseudoplatystoma magdaleniatum	POT	LON	M	No	NE		Zapata and Usma 2013
Pseudoplatystoma metaense	POT	LON	L	No	NE		Zapata and Usma 2013
Pseudoplatystoma orinocoense	POT	LON	L	No	NE		Zapata and Usma 2013

Pseudoplatystoma punctifer	POT	LON	L	No	LC	Brazil	Barthem et al. 1991; Hahn et al. 2019; Duponchelle et al. 2021
Pseudoplatystoma tigrinum	POT	LON	L	No	LC	Brazil	Barthem et al. 1991; Duponchelle et al. 2021
Pterodoras granulosus	POT	LON, LAT	S	No	LC	Global	Bonetto et al. 1981; Makrakis et al. 2007b
Pterodoras rivasi	POT	LON, LAT	S	No	LC	Global	Zapata and Usma 2013
Pylodictis olivaris	POT	LON	M	No	LC	Global	Bringolf et al. 2005; Raabe and Hightower 2014; Montague and Shoup 2021
Rhinelepis aspera	POT	LON	L	No	LC	Brazil	Agostinho et al. 1986; Wagner et al. 2012
Silurus glanis	POT	HAB	S	No	LC	Global	Cucherousset et al. 2018; Kuzishchin et al. 2018
Sorubim cuspicaudus	POT	LON, LAT	M	No	NE		Zapata and Usma 2013
Sorubim elongatus	POT	LON, LAT	M	No	LC	Global	Van Damme et al. 2011
Sorubim lima	POT	LON, LAT	M	No	LC	Global	Van Damme et al. 2011; Zapata and Usma 2013
Sorubim maniradii	POT	LON, LAT	M	No	LC	Global	Van Damme et al. 2011
Sorubimichthys planiceps	POT	LON	L	No	LC	Global	Barthem et al. 1991; Van Damme et al. 2011; Zapata and Usma 2013
Trichogenes longipinnis	POT	HAB	S	No	NT	Brazil	Zamudio et al. 2009
Trichomycterus barbourin	POT	LON	S	No	NE	Bolivia	Miranda-Chumacero et al. 2015
Trichomycterus largoperculatus	POT	LON	S	No	NE	Brazil	Costa and Katz 2022
Trichomycterus pradensis	POT	LON	S	No	LC	Brazil	Costa and Katz 2022
Zungaro jahu	POT	LON	M	No	NT	Brazil	Alves et al. 2007; Avigliano et al. 2023
Zungaro zungaro	POT	LON	M	No	LC	Brazil	Barthem et al. 2017; Van Damme et al. 2011

along the year and influenced by water temperature and river height, including sex-specific responses to environmental factors. However, typical migration between marine and freshwater environments did not occur, and most individuals were considered residents. Based on these findings, the authors considered *N. graeffei* as "estuarine migrants," capable of completing their lifecycle within the estuary without using freshwater habitats.

A summary of published records of migratory behavior among catfish species is presented in Table 2. Although the literature search was not exhaustive, we believe the list of species and types of migratory behavior adequately represents the frequency of this behavior among catfish, its most common characteristics, and its distribution among different taxonomic families. However, our current knowledge of the world's catfish species is still far from complete, as several new species are described every year (Fricke et al. 2023). Considering that we know much less about the biology and ecology of fishes than their taxonomic diversity, we can expect to record several new cases of migratory behavior among catfishes in the near future.

Biological and Morphological Characteristics of Migratory Catfish

In addition to the ability to recognize environmental cues and navigate during migrations, fish must possess biological and morphological characteristics that increase their chances of success in finding their destination habitat and accomplishing various tasks such as reproduction, foraging, or seeking refuge. From a reproductive strategy perspective, most long-distance migratory catfish species are classified as r-strategists with seasonal or periodic reproduction (*sensu* Winemiller 1989; Winemiller and Rose 1992). Those species are characterized by producing a large number of tiny eggs (with little yolk) and no parental care. These traits are often associated with large body size, which allows for accommodating a larger volume of gonadal products (see Barneche et al. 2018) and the accumulation of significant energy reserves, usually in the form of intramuscular or cavity fat. Thus, the prevalence of medium to large-sized species among long-distance migratory catfish (Table 3) could be explained by the dual advantage offered by larger body sizes: higher fecundity (Barneche et al. 2018) and greater swimming efficiency during movements (Hugheny 1990; Roff 1991; Griffiths 2006).

Other morphological aspects of certain catfish species can also maximize energy use during migratory movements. These morphological features include a streamlined and laterally compressed body, a forked caudal fin, pectoral fins that can act as rudders, and an acuminate head and snout. These morphological characteristics are commonly observed in catfishes with nektobenthic or nektonic habits, such as the long-distance migratory pimelodids (e.g., *Brachyplatystoma rousseauxii* and *B. platynemum*), the planktivorous pimelodids of the genus *Hypophthalmus*, the invertivorous auchenipterids *Auchenipterus* spp., most Pangasiidae species, and the ailiids *Clupisoma* spp. However, several migratory catfishes do not possess conspicuously hydrodynamic bodies, such as most ariids, most *Clarias*, the large doradids of the genera *Lithodoras*, *Pterodoras*, and *Oxydoras*, most ictalurids, the heavily armed loricariid *Rhinelepis aspera*, the sisorid *Bagarius bagarius*, and pencil catfishes of the genus *Trichomycterus*. Therefore, despite the obvious advantages of a streamlined body for long migrations, the variety of morphological characteristics exhibited by migratory catfishes indicates that other anatomical, physiological, and behavioral traits allow for successful short- to mid-distance migrations in various environmental conditions.

As mentioned earlier, body size is positively correlated with migration distance, dispersal ability, and distribution range in fishes (e.g., Bernatchez and Dodson 1987; Hugheny 1990; Roff 1991; Griffiths 2006). This correlation is partially explained by the fact that larger fish have lower migratory costs per unit distance (Bernatchez and Dodson 1987) and can travel longer distances. However, a large size is not a mandatory characteristic for performing long-distance migrations. For instance, Miranda-Chumacero et al. (2015) described the migratory behavior of the "chipi chipi" catfish *Trichomycterus barbouri*. Despite its tiny size and a mean weight of only 0.3 g (or 0.0003 kg), these juvenile catfish form large shoals of thousands of individuals and migrate

Table 3. Body size and recorded (or estimated) traveled distances by migratory freshwater catfishes. Maximum body size obtained from FishBase (Froese and Pauly 2024).

Catfish Species	Family	Maximum Body Size (SL, cm)	Traveled Distance (km)	Reference
Brachyplatystoma rousseauxii	Pimelodidae	192	5,786	Barthem et al. 2017
Brachyplatystoma juruense	Pimelodidae	60	4,238	Barthem et al. 2017
Brachyplatystoma vaillantii	Pimelodidae	150	3,129	Barthem et al. 2017
Brachyplatystoma platynemum	Pimelodidae	100	> 3,000	Hauser et al. 2019
Pseudoplatystoma tigrinum	Pimelodidae	130	500–3,000	Zapata and Usma 2013
Pangasianodon gigas	Pangasiidae	300	> 1,200	Ngansiri et al. 2007
Pangasianodon hypophthalmus	Pangasiidae	130	> 1,200	So et al. 2006b
Brachyplatystoma tigrinum	Pimelodidae	60	> 1,000	Barthem et al. 2017
Pterodoras granulosus	Doradidae	90.1	1,054	Bonetto et al. 1981
Brachyplatystoma filamentosum	Pimelodidae	360	100–1,000	Barthem et al. 2017
Conorhynchos conirostris	Heptapteridae*	53.5	100–1,000	Godinho and Godinho 2018
Pangasius krempfi	Pangasiidae	120	> 700	Hogan et al. 2007
Pseudoplatystoma corruscans	Pimelodidae	182	100–500	Godinho et al. 2007
Zungaro zungaro	Pimelodidae	140	100–500	Zapata and Usma 2013
Trichomycterus barbouri	Trichomycteridae	3.7	370	Miranda-Chumacero et al. 2015
Phractocephalus hemioliopterus	Pimelodidae	135	347	Hahn et al. 2019
Pseudoplatystoma punctifer	Pimelodidae	105	164	Hahn et al. 2019
Zungaro jahu	Pimelodidae	140	< 100	Avigliano et al. 2023

* Provisional allocation; Please see de Pinna et al. (2025).

more than 300 km upstream along the Beni River in Bolivia. Comparing the migratory efforts of the "chipi chipi" and the "dourada" *B. rousseauxii* on a "gram for gram" basis, the effort required for the tiny "chipi chipi" is almost 20,000 times greater (Miranda-Chumacero et al. 2015). This exception to the relation between body size and migration distance suggests that other biological and ecological factors may influence the migration ability of catfishes.

Sensory Mechanisms Involved in Catfish Migration

Explaining the origin and evolution of migratory behavior is not easy, considering that migrations occur in various ways, in different habitats and animal groups (Dingle 2014). However, it seems clear that long-distance migrations play an important role in the structure of communities and ecosystem functioning (Gende et al. 2002). Migration comes with costs in terms of time, energy, and predation risk, which increase with the distance to be traveled and the number of obstacles along the way (Bernarchez and Dodson 1987). Thus, it is crucial for fish to have the ability to navigate precisely toward their destination habitat to minimize risks during the migratory process and maximize fitness.

Studies on fish migrations have predominantly focused on describing the migratory routes of species, with little attention given to the environmental factors that act as cues to trigger the migratory process at the right time (Dodson 1997). Different sensory mechanisms have been hypothesized in guiding fish during migration, such as the use of smell to follow chemical cues (cf. Northcote 1984), vision and the ability to navigate based on the position of the sun and the distinction of polarized light patterns (Quinn and Ogden 1984), and the use of biogenic magnetite crystals for orientation relative to the Earth's geomagnetic field (Walker et al. 1997; Quinn 1980), such as the crystals found in the olfactory lamellae of rainbow trout (*Oncorhynchus mykiss*) and in the lateral line of Atlantic salmon (*Salmo salar*) (Moore et al. 1990; Putman et al. 2013). In any case, a biological clock would be a basic requirement for the efficiency of geomagnetic mechanisms (Stabell 2012).

Olfaction, a highly developed sensory organ in catfishes, has been proposed as the primary sense involved in guiding fish migrations, according to two main hypotheses: (1) the "olfactory imprinting hypothesis" (Hasler and Wisby 1951), which suggests that unique water odor characteristics (resulting from interaction with local soil and vegetation) determine a site-specific odor imprinted in the young fish's olfactory memory as an environmental cue; and (2) the "pheromone hypothesis" (Nordeng 1971), where migrating adults would be guided by population-specific pheromones released by the young in the river or stream where they were born (Brönmark et al. 2014). Considering that a significant portion (or the majority) of Siluriformes species predominantly orient themselves using chemical cues (via olfaction and taste, through gustatory buds on the body and barbels) and that the main catfish migrations occur in fluvial environments, it is likely that olfaction and rheotaxis are the predominant mechanisms used for spatial orientation during movements between habitats. In natal homing migrations, the olfactory imprinting hypothesis proposed by Hasler and Wisby (1951) emerges as a probable primary homing mechanism guiding catfish to their birthplaces.

Migratory Catfishes and Ecosystem Services

A recent study by Pelicice et al. (2022) briefly described several ecosystem services provided by Neotropical freshwater fishes, including catfishes. These important services can be classified as provisioning, regulating, supporting, and cultural services. Among provisioning services, those related to fisheries stand out in function of the high participation of migratory catfishes in the landings in South America (e.g., Petrere Jr. et al. 2004) and southeastern Asian rivers (e.g., Baird et al. 2001). These services also include the supply of catfish parts and organs, such as dried swimbladder for the international trade of food and medicine, notably in Asia (e.g., Mitcheson et al. 2019), as well as the worldwide use of some migratory catfishes in aquaculture, both for food and for recreational fishing (e.g., Legendre and Pariselle 1998; Dunham and Elaswad 2018).

One regulating service provided by migratory catfishes is seed dispersal, which can play a significant role in maintaining riparian plant diversity and potentially contribute to the regeneration of disturbed riparian habitats (Correa et al. 2015). Some medium-sized and large migratory South American catfishes are known to frequently consume fruits and seeds, sometimes in substantial quantities, during the flood period in Amazonian rivers (Goulding 1980; Zuanon 1990; Correa et al. 2015). These include *Phractocephalus hemioliopterus*, *Pimelodus* spp., *Calophysus macropterus* (Pimelodidae), and the large thorny catfishes *Lithodoras dorsalis* and *Pterodoras granulosus* (Doradidae). However, several of these species are subject to high fishing pressure, which can deplete their populations, potentially impacting their role as seed dispersers in South American river-floodplain systems. Additionally, Goulding (1980) suggested a strong relationship between the migratory doradid *L. dorsalis* and some riparian trees (*Licania* spp., Chrysobalanaceae), highlighting the ecological importance of this interaction. Riparian plants that may benefit from this ecological interaction include palm trees such as *Astrocaryum jauari* and *Leopoldinia pulchra* (Arecaceae), pioneer trees like *Cecropia* spp. (Urticaceae), and abundant flooded forest species such as *Crataeva benthamii* (Capparaceae) and *Vitex cymosa* (Verbenaceae) (Goulding 1980; Zuanon 1990). Other regulating services include top-down control of fish communities by migratory catfishes, such as the goliath pimelodids in the Amazon basin (Barthem and Goulding 1997), and arguably also performed by other large-sized catfishes of the families Bagridae, Ictaluridae and Siluridae.

Supporting services provided by migratory catfishes include decomposition and nutrient cycling, probably performed by the migratory loricariid *Rhinelepis aspera* in the Paraná River basin. The synchronized movements of thousands of fish between habitats can indeed have profound impacts on ecosystem dynamics, including the transfer of nutrients and energy between different segments of river basins or between aquatic and terrestrial environments (Brodersen et al. 2008a; Brönmark et al. 2010). This phenomenon can also involve the transfer of nutrients between aquatic and terrestrial habitats mediated by large terrestrial carnivores (Gende et al. 2002). Longitudinal migrations performed by large Pimelodidae catfishes in South America and massive Pangasiidae schools in the rivers of South Asia and India likely move significant amounts of nutrients between estuaries and the headwaters of these continents' major rivers. However, the extent and importance of this ecosystem service have yet to be adequately evaluated.

Cultural services performed by migratory catfishes include recreational fishing, especially those related to trophy-sized large pimelodid catfishes of the genera *Brachyplatystoma*, *Zungaro* and *Pseudoplatystoma* in South American rivers, as well as *Bagarius*, *Pangasius* and *Pangasianodon* in Asian rivers, and large ictalurids in North American rivers and lakes. The abundance and timing of migrating catfish schools also generate scientific knowledge that informs about conservation needs and actions (e.g., Hogan 2011; Anderson et al. 2018; Duponchelle et al. 2021). Moreover, migratory catfishes are also involved in cultural practices of traditional communities worldwide, both in freshwater and marine realms, including specialized fishing techniques (Sant'Anna et al. 2020), food taboos (Goulding 1979; Smith 1979), cosmology, and existential foundations (Pelicice et al. 2022).

Threats to Migratory Catfishes

The most outstanding catfish migrations occur in freshwaters, making these siluriform species highly threatened. Freshwaters are among the most anthropogenically disturbed environments on Earth. The large shoals of migrating fishes are the main targets for commercial and industrial fisheries in tropical regions worldwide. Furthermore, the large size of most long-distance migratory catfishes makes them among the first to be removed from populations and assemblages. As large carnivorous fishes, they are considered top-quality protein sources and are among the most coveted fish for human consumption. In fact, stock depletion has been observed or presumed for several migratory catfishes in South America (Barthem et al. 1991; Petrere Jr. et al. 2004; Van Damme et al. 2019; Duponchelle et al. 2021) and Asia (Baird et al. 2004).

Besides the direct removal of individuals by fisheries, migration route disruptions stand out among the main threats to migratory catfishes. The proliferation of dam construction along the main river courses of tropical countries has caused habitat fragmentation and loss of connectivity among larval, juvenile, and adult habitats (Finner and Jenkins 2012; Anderson et al. 2018; Herrera-R. et al. 2020; Saha et al. 2021) by blocking access to migratory routes, spawning and foraging habitats (Pelicice et al. 2014; Duponchelle et al. 2021). Fish ladders, the most common system intended to allow the fish to surpass the barrier created by the dams, were initially designed to allow the passage of nektonic fishes, such as salmonids and characiforms, but are notoriously inefficient for catfishes (e.g., Godinho et al. 1991; Pelicice and Agostinho 2008). Fishways, such as that built around the Santo Antonio Hydroelectric Dam in the Madeira River in the Brazilian Amazon, apparently present a very low efficiency in mitigating the disruption of migratory routes of long-distance migratory goliath catfishes (Hahn et al. 2022). In order to perform adequately for migratory catfishes, such fishways should be built to mimic the hydrodynamic and bathymetric conditions present in their natural routes, which is rarely, if ever, done. Ironically (but tragically), when the fish transposition systems are efficient in attracting fish, they can turn into real ecological traps: individuals that manage to surpass the ladders may not find adequate habitats for foraging, spawning, or nursery habitats upstream or in the reservoirs, and are doomed to die without reproducing (Pelicice and Agostinho 2008).

Another strong negative effect of large dams on the fish fauna (and especially on channel-dwelling migratory fishes) is the reduction in the sediment load of turbid water rivers where catfishes are notoriously abundant. For instance, if the main large dams planned for the Amazon basin are constructed, the huge sediment retention would modify the sediment load downstream up to the plume of the Amazon River, causing environmental disturbances across the whole basin and affecting the sediment dynamics of the estuary and the coastal area of the Guianas (Latrubesse et al. 2017). Such mega environmental disturbance would disrupt the migratory route and the biological cycle of the most spectacular migratory catfish of the world, the dourada *B. rousseauxii*, with negative impacts in artisanal and commercial fisheries across five countries: Brazil, Bolivia, Peru, Colombia and Ecuador. Moreover, the mega dams planned to be constructed in the Andean Amazon are expected to result in a dramatic decrease in the biological productivity and fisheries yield across the floodplains of the Solimões/Amazon River, severely impacting the food security of traditional (indigenous and riverine) populations (Forsberg et al. 2017) and potentially compromising important biological interactions and ecological processes. Similar processes and negative impacts on catfish populations and fisheries also occur with migratory pangasiids in the Mekong River basin (Hogan 2011; Hogan et al. 2004).

The accelerated rate of deforestation in tropical regions of the world also produces strong negative effects on aquatic environments. The loss of vegetation cover disturbs local and regional rainfall patterns, causing unpredictable changes in the seasonal flood cycle. By altering the timing of the flood and dry periods, these changes disrupt the lateral migrations, blocking access to floodplains for spawning, feeding and shelter for adults, as well as to nursery habitats for larvae and juveniles. Changes in the annual flood pulse can also disrupt the movement patterns of mid- to long-distance migratory species by blurring environmental cues to start spawning and/or to move in and out of seasonal habitats, ultimately resulting in increased spawning failures and fish mortality.

The negative effects of deforestation can have synergistic effects with the ongoing climate change by further altering rainfall patterns and reducing the regularity and predictability of flood cycles, as well as by increasing water temperature and dissolved oxygen depletion. This can strongly affect the highly productive floodplain habitats of turbid water river systems, which carry heavy loads of decomposing organic matter, thus pushing the aquatic fauna toward dangerous physiological thresholds. Repeated events of low fish larvae survival in floodplain habitats would result in accumulated recruitment failures, with strong negative impacts on the ecological dynamics of river-floodplain fish assemblages, disruption of fish stocks and fisheries collapse in certain areas.

Water pollution also threatens fish worldwide, although not so strongly in huge basins like the Amazon due to the low human population density in most of the basin. However, in highly populated countries like India and China, even large rivers already show clear signs of environmental deterioration. Aside from the obvious direct negative effects of water pollution on fish, water quality changes can have strong deleterious effects on fish behavior by blurring chemical signals for habitat choice, by altering avoidance responses to polluted, low-quality habitats (and further exposing the fish to the pollutants), and by disturbing homing migrations by masking chemical signals that fish use to find their natal places and spawn (van Brummelen 1990; Jacquin et al. 2020; Deng and Demisse 2023). These effects can be especially deleterious to catfishes, which strongly rely on olfactory and other gustative and chemosensory structures for orientation, and especially to species that perform long homing migrations, such as the South American pimelodids *B. rousseauxii* and *P. corruscans* (Barthem et al. 2017; Batista and Alves-Gomes 2006; Duponchelle et al. 2016; Hauser et al. 2019a, b). Together with the negative effects of river damming, connectivity loss and habitat fragmentation, these disruptions to the homing behavior of long-distance migratory catfishes can result in changes to the genetic characteristics of subpopulations in headwater and tributary rivers and biodiversity losses, such as the case of *B. rousseauxii* in the Amazon basin (Batista and Alves-Gomes 2006; Barthem et al. 2017; Duponchelle et al. 2021).

Concluding Remarks

Apart from anecdotal information, empirical evidence of migratory behavior by catfishes is still scarce and stands out among the most challenging research in the function of the need for large-scale investigations over multiple watersheds to accurately assess spawning, feeding, or overwintering movements (Kwak et al. 2011). A recent systematic literature review by Herrera-R. et al. (2023) on migratory movements of Amazon freshwater fishes revealed that 223 species are known to migrate longitudinally or laterally in the rivers of the basin and remarks that the true number of migratory species is probably much higher. However, the authors also note the scarcity of results obtained through the application of robust methods (telemetry, otolith microchemistry, eye lenses isotopic signature; see Boxes 1 and 2) to confirm and clearly understand the migratory patterns of most species, for which the only evidence is based on secondary sources of information (and frequently anecdotal) (Herrera-R. et al. 2023). This situation probably also holds true for migratory catfishes in general, especially regarding less visible, small-sized species and those without commercial value and not exploited by fisheries.

Published evidence of catfish migrations in marine environments seems to be even scarcer than in freshwater, possibly because of the difficulty of detecting and recording migratory events in the huge and open oceans. On the contrary, the morphology of river systems (linear and dendritic pathways, with obstacles that must be surpassed to complete the migration) makes it easier to observe and understand migratory habits, chronology, and pathways. However, there is a strong bias toward riverine environments and large catfish species that perform longitudinal, long-distance migrations and are targeted by commercial fisheries (Table 3).

Of the 82 migratory catfish species listed in Table 2, 13 (~ 16%) are considered under some extinction risk or very close to it (Critically Endangered CR = 3, Endangered EN = 3, Vulnerable VU = 3, and Near Threatened NT = 4), resulting from different combinations of overexploitation, habitat loss, disruption of migratory routes, and water pollution. Considering the high ecological and economic importance of migratory catfishes worldwide, sound conservation actions are urgently needed. However, conservation policies for such multi-habitat, broad-range (and frequently transboundary) migratory catfishes are challenging. Concerted efforts for transnational policies must be made, aiming to ensure unimpeded migratory routes along river basins, such as the case of *Brachyplatystoma rousseauxii* whose distribution range encompasses five Amazonian countries. International regulations aiming to reach truly sustainable use of long-distance migratory catfishes targeted by commercial fisheries are also urgent, especially considering that these fishes use different areas/countries throughout their life cycle. This implies that geographically isolated conservation units, the most common conservation strategy used by the countries, cannot produce sound results. Formally protected large corridors on the geographic scale needed to conserve long-distance migratory catfishes seem progressively more difficult to implant since the dispute for spaces for production (especially agricultural use of floodplains) keeps growing, as well as the urban occupation of river margins. In this sense, strong diplomatic efforts aiming at the construction of cooperation treaties and transnational regulations regarding the conservation of critical habitats for migratory catfishes (especially spawning and nursery areas, sometimes far apart from each other, as well as wetlands in general) are mandatory.

It is also extremely important to stimulate and support scientific research on the basic biology of catfishes (feeding, spawning, growth, and longevity) and other aspects of their natural history (e.g., habitat use, foraging, and courtship behavior) in order to better know the ecological needs of migratory species and to identify key aspects of its conservation. Such directed research also has the potential to generate new scientific hypotheses regarding ecological relations involving migratory catfishes and their role in ecosystem processes, thus helping to produce more efficient conservation strategies and the maintenance of aquatic biodiversity (Ricklefs 2012).

Acknowledgments

JZ thanks Brazil's National Council for Scientific and Technological Development (CNPq) for a Research Productivity grant, the Amazonas State Foundation for Research Support (FAPEAM) for long-term financing, and the National Institute for Amazonian Research (INPA) for support along his career. GTV thanks the São Paulo Research Foundation (FAPESP, PROCESS #2016/07910-0 to GTV) and the Laboratory of Applied Physics, Materials and Environment (LaFAMA) of the Federal University of São Paulo (UNIFESP) for support along her career. We warmheartedly thank Kalebe Pinto for graciously editing Figure 2.

References

Agostinho, A.A., G. Barbieri, J.R. Verani and C.S. Agostinho. 1986. Ciclo reprodutivo e primeira maturação de *Rhinelepis aspera* (Agassiz 1829), (Teleostei-Loricariidae) no Rio Paranapanema. Revista Unimar 8(1): 17–27.

Agostinho, A.A., L.C. Gomes, H.I. Suzuki and H.F. Júlio. 2003. Migratory fishes of the upper Paraná River basin Brazil. pp. 19–98. *In*: Carolsfeld, J., B. Harvey, C. Ross and A. Baer (eds.). Migratory Fishes of South America: Biology, Fisheries and Conservation Status. Co-published by IDRC and the World Bank.

Agostinho, C.S., C.R. Pereira, R.J. de Oliveira, I.S. Freitas and E.E. Marques. 2007. Movements through a fish ladder: temporal patterns and motivations to move upstream. Neotrop. Ichthyol. 5(2): 161–167.

Agudelo-Córdoba, E.Y., Y. Salinas-Coy, C.L. Sánchez, D.L. Muñoz-Sosa, J.C. Alonso-González, M.E.A. Arteaga-Díaz et al. 2000. Bagres de la amazonía colombiana: un Recurso sin Fronteras. Instituto Amazónico de Investigaciones Científicas—SINCHI, Bogota.

Allen, G.R. 1989. Freshwater Fishes of Australia. T.F.H. Publications, Neptune City.

Alves, C.B.M., L.G.M da Silva and A.L. Godinho. 2007. Radiotelemetry of a female jaú, *Zungaro jahu* (Ihering, 1898) (Siluriformes: Pimelodidae), passed upstream of Funil Dam, rio Grande, Brazil. Neotrop. Ichthyol. 5(2): 229–232.

Anderson, E.P., C.N. Jenkins, S. Heilpern, J.A. Maldonado-Ocampo, F.M. Carvajal-Vallejos, A.C. Encalada et al. 2018. Fragmentation of Andes-to-Amazon connectivity by hydropower dams. Science Advances 4: eaao1642. DOI: 10.1126/sciadv.aao1642.

Arantes, C.C., K.O. Winemiller, A. Asher, L. Castello, L.L. Hess, M. Petrere Jr. et al. 2019. Floodplain land cover affects biomass distribution of fish functional diversity in the Amazon River. Scientific Rep. 9(1): 1–13. DOI: 10.1038/s41598-019-52243-0.

Araújo, F.G. 1988. Distribuição, abundância relativa e movimentos sazonais de bagres marinhos (Siluriformes, Ariidae) no Estuário da Lagoa dos Patos (RS), Brasil. Rev. Brasil. Zool. 5: 509–543.

Avigliano, E. and A.V. Volpedo. 2015. New records of anadromous catfish *Genidens barbus* (Lacépède, 1803) in the Paraná Delta (South America): evidence of extension in the migration corridor? Mar. Biodivers. Rec. 8: e23.

Avigliano, E., M. Leisen, R. Romero, B. Carvalho, G. Velasco, M. Vianna et al. 2017. Fluvio-marine travelers from South America: Cyclic amphidromy and freshwater residency, typical behaviors in *Genidens barbus* inferred by otolith chemistry. Fish. Res. 19: 184–194.

Avigliano, E., J. Pisonero, A. Méndez, A. Tombari and A.V. Volpedo. 2022. Habitat use of the amphidromous catfish *Genidens barbus*: first insights at its southern distribution limit, New Zealand J. Mar. Freshw. Res. 56(2): 284–290. DOI: 10.1080/00288330.2021.1879178.

Avigliano, E., M. Pouilly, F. Vargas, J. Penha, N. Silva, J. Bouchez et al. 2023. Environmental history of one of the largest freshwater Neotropical fish (*Zungaro jahu*) from otolith isotopic analysis. River Res. Applic. 2023: 1–10. DOI: 10.1002/rra.4120.

Baird, I., Z. Hogan, B. Phylaivanh and P. Moyle. 2001. A communal fishery for the migratory catfish *Pangasius macronema* in the Mekong River. Asian Fish. Sci. 14(1). DOI: 10.33997/J.AFS.2001.14.1.004.

Baird, I.A., M.S. Flaherty and B. Phylavanh. 2004. Mekong River Pangasiidae catfish migrations and the Khone Falls wing trap fishery in Southern Laos. Nat. Hist. Bull. Siam Soc. 52(1): 81–109.

Barneche, D.R., D.R. Robertson, C.R. White and D.J. Marshall. 2018. Fish reproductive-energy output increases disproportionately with body size. Science 360: 642–645. DOI: 10.1126/science.aao6868.

Barthem, R. and M. Goulding. 1997. Os bagres balizadores. Sociedade Civil Mamirauá, MCT – CNPq. Brasília.

Barthem, R.B., M.C.L.B. Ribeiro and M. Petrere Jr. 1991. Life strategies of some long-distance migratory catfish in relation to hydroelectric dams in the Amazon Basin. Biol. Conserv. 55(3): 339–45. DOI: 10.1016/0006-3207(91)90037-A.

Barthem, R.B., M. Goulding, R.G. Leite, C. Cañas, B. Forsberg, E. Venticinque et al. 2017. Goliath catfish spawning in the far western Amazon confirmed by the distribution of mature adults, drifting larvae and migrating juveniles. Sci. Rep. 7: 41784. DOI: 10.1038/srep41784.

Batista, J.S. and J.A. Alves-Gomes. 2006. Phylogeography of *Brachyplatystoma rousseauxii* (Siluriformes - Pimelodidae) in the Amazon Basin offers preliminary evidence for the first case of "homing" for an Amazonian migratory catfish. Gen. Mol. Res. 5(4): 723–740.

Bell-Tilcock, M., C.A. Jeffres, A.L. Rypel, T.R. Sommer, J.V.E. Katz, G. Whitman et al. 2021. Advancing diet reconstruction in fish eye lenses. Methods Ecol. Evol. 12(3): 449–457.

Bernarchez, L. and J.J. Dodson. 1987. Relationship between Bioenergetics and Behavior in Anadromous Fish Migrations. Can. J. Fish. Aquat. Sci. 44: 399–407.

Berra, T. 2001. Freshwater Fish Distribution. Academic Press, San Diego.

Bonetto, A.A., D. Roldán and M.C. Veron. 1981. Algunos aspectos estructurales y ecológicos de la ictiofauna del sistema del Iberá (Corrientes, Arg.). Ecosur 8: 79–89.

Bravington, M.V., H.J. Skaug and E.C. Anderson. 2016a. Close-kin mark recapture. Stat. Sci. 31: 259–274. DOI: 10.1214/16-STS552.

Bravington, M.V., P.M. Grewe and C.R. Davies. 2016b. Absolute abundance of southern bluefin tuna estimated by close-kin mark-recapture. Nat. Commun. 14(7): 13162. DOI: 10.1038/ncomms13162.

Bringolf, R.B., T.J. Kwak, W.G. Cope and M.S. Larimore. 2005. Salinity tolerance of Flathead Catfish: implications for dispersal of introduced populations. Trans. Amer. Fish. Soc. 134: 927–936. DOI: 10.1577/T04-195.1.

Brodersen, J., E. Ådahl, C. Brönmark and L.-A. Hansson. 2008. Ecosystem effects of partial fish migration in lakes. Oikos 117: 40–46. DOI:10.1111/j.2007. 0030-1299.16118.x.

Brönmark, C., J. Brodersen, B.B. Chapman, A. Nicolle, P.A. Nilsson, C. Skov et al. 2010. Regime shifts in shallow lakes: the importance of seasonal fish migration. Hydrobiologia 646: 91–100.

Brönmark, C., K. Hulthén, P.A. Nilsson, C. Skov, L.-A. Hansson, J. Brodersen et al. 2014. There and back again: migration in freshwater fishes. Can. J. Zool. 92: 467–479. DOI: 10.1139/cjz-2012-0277.

Brothers, J.R. and K.J. Lohmann. 2015. Evidence for geomagnetic imprinting and magnetic navigation in the natal homing of sea turtles. Curr. Biol. 25: 392–396. DOI: 10.1016/j. cub.2014.12.035.

van Brummelen, T.C. 1990. Chemicals affecting the spawning migration of anadromous fish by causing avoidance responses or orientational disability, with special reference to concentrations in the River Rhine. Unpubl. Techn. Rep. Univer. Utrecht, Depart. Exper. Zool., Section of Aquatic Toxicology. 76p. Available at https://edepot.wur.nl/326247.

Bunch, A.J., H. Hershey, G.R. Reger, F.J. Hoogakker, B.S. Donovan and T.M. Farmer. 2022. A Bayesian multistate approach to evaluate movements of an invasive freshwater estuarine-opportunist. Can. J. Fish. Aquat. Sci. 79(2): 213–222.

Carolsfeld, J., B. Harvey, C. Ross and A. Baer. 2004. Migratory Fishes of South America: Biology, Fisheries and Conservation Status. Ottawa: International Development Research Centre and the World Bank. http://hdl.handle.net/10986/14929 (available at http://hdl.handle.net/10986/14929).

Carvalho, J.L. and B. Mérona. 1986. Estudos sobre dois peixes migratórios do Baixo Tocantins, antes do fechamento da barragem de Tucuruí. Amazoniana 9(4): 595–607.

Casas, L. and F. Saborido-Rey. 2023. A review of an emerging tool to estimate population parameters: the close-kin mark-recapture method. Front. Mar. Sci. 10: 1087027. DOI: 10.3389/fmars.2023.1087027.

Cavole, L.M., K.E. Limburg, N.D. Gallo, A.G.V. Salvanes, A. Ramírez-Valdez, L.A. Levin et al. 2023. Otoliths of marine fishes record evidence of low oxygen, temperature and pH conditions of deep Oxygen Minimum Zones. Deep Sea Research Part I: Oceanographic Research Papers 191: e103941. DOI: 10.1016/j.dsr.2022.103941.

Clay, D. 1979. Sexual maturity and fecundity of the African catfish (*Clarias gariepinus*) with an observation on the spawning behaviour of the Nile catfish (*Clarias lazera*). Zool. J. Linn. Soc. 65: 351–365.

Clunie, P. and J. Kohen. 2001. Freshwater Catfish: A Resource Document. Volume 2. Final Report for Natural Resource Management Strategy Project R7002 to the Murray Darling Basin Commission. Heidelberg, Australia. 184 p.

Correa, S.B., R. Costa-Pereira, T. Fleming, M. Goulding and J.T. Anderson. 2015. Neotropical fish-fruit interactions: eco-evolutionary dynamics and conservation. Biol. Rev. 90(4): 1263–1278. DOI: 10.1111/brv.12153.

Costa, W.J.E.M. and A.M. Katz. 2022. A new catfish of the genus *Trichomycterus* from the Rio Paraíba do Sul Basin, south-eastern Brazil, a supposedly migrating species (Siluriformes, Trichomycteridae). Zoosyst. Evol. 98(1): 13–21. DOI: 10.3897/zse.98.72392.

Crossin, G.T., M.R. Heupel, C.M. Holbrook, N.E. Hussey, S.K. Lowerre-Barbieri, V.M. Nguyen et al. 2017. Acoustic telemetry and fisheries management. Ecol. Appl. 27(4): 1031–1049. DOI: 10.1002/eap.1533.

Cucherousset, J., P. Horky, O. Slavík, M. Ovidio, R. Arlinghaus, S. Boulêtreau et al. 2018. Ecology, behaviour and management of the European Catfish. Rev. Fish Biol. Fish. 28: 177–190. DOI: 10.1007/s11160-017-9507-9.

Delaval, A., V. Bendall, S.J. Hetherington, H.J. Skaug, M. Frost, C.S. Jones et al. 2022. Evaluating the suitability of close-kin mark-recapture as a demographic modelling tool for a critically endangered elasmobranch population. Evol. Appl. 16(2): 461–473. DOI: 10.1111/eva.13474.

Deng, G.T. and B.T. Demisse. 2023. Freshwater fish migration: fisheries management strategy insight. pp. 1–14. *In*: Perveen, K. and Farzana (eds.). Sustainable Wildlife Management. Environmental Sciences Series, IntechOpen. DOI: 10.5772/intechopen.110027.

Dingle, H. 2014. Migration: The Biology of Life on the Move. Oxford University Press, USA.

Dittman, A.H. and T.P. Quinn. 1996. Homing in Pacific salmon: mechanisms and ecological basis. J. Exp. Biol. 199(1): 83–91.

Dodson, J.J. 1997. Fish migration: an evolutionary perspective. pp. 10–36. *In*: Godin, J.G. (ed.). Behavioral Ecology of Teleost Fishes. Oxford University Press, Oxford.

Doria, C.R., F. Duponchelle, M.A.L. Lima, A. Garcia, F.M. Carvajal-Vallejos, C.C. Méndez et al. 2018. Review of fisheries resource use and status in the Madeira River Basin (Brazil, Bolivia, and Peru) before hydroelectric dam completion. Rev. Fisheries Sci. Aquacult. 26(4): 494–514. DOI: 10.1080/23308249.2018.1463511.

Dorji, S. and J. Tenzin. 2023. New distribution record of fish *Clupisoma garua* (Hamilton, 1822) (Siluriformes: Ailiidae) from the Sarpang District in southern central part of Bhutan. J. Threat. Taxa 15(9): 23920–23924. DOI: 10.11609/jott.8496.15.9.23920-23924.

Dunham, R.A. and A. Elaswa. 2018. Catfish Biology and Farming. Annu. Rev. Anim. Biosci 6: 305–325. DOI: 10.1146/annurev-animal-030117-014646.

Duong, T.Y., N-T.T. Nguyen, D.D. Tran, T.H. Le and S.A.M. Nor. 2023. Multiple genetic lineages of anadromous migratory Mekong catfish *Pangasius krempfi* revealed by mtDNA control region and cytochrome *b*. Ecol. Evol. 13: e9845. DOI: 10.1002/ece3.9845.

Duponchelle, F., M. Pouilly, C.P. Echeyran, M. Hauser, J.-F. Rennó, J. Panfili et al. 2016. Trans-Amazonian natal homing in giant catfish. J. Appl. Ecol. 53: 1511–1520. DOI: 10.1111/1365-2664.12665.

Duponchelle, F., V.J. Isaac, C. Doria, P.A. Van Damme, G.A. Herrera-R., E.P. Anderson et al. 2021. Conservation of migratory fishes in the Amazon basin. Aquatic Conserv: Mar. Freshw. Ecosyst. 2021: 1–19. DOI: 10.1002/aqc.3550.

Feng, D., R. Raoufi, E. Beighley, J.M. Melack, M. Goulding, R.B. Barthem et al. 2020. Future climate impacts on the hydrology of headwater streams in the Amazon River Basin: Implications for migratory goliath catfishes. Hydrological Processes 34(26): 5402–5416. DOI: 10.1002/hyp.13952.

Fernandes, C.C. 1997. Lateral migration of fishes in Amazon floodplains. Ecol. Freshwater Fish 6: 3644.

Finer, M. and C.N. Jenkins. 2012. Proliferation of Hydroelectric Dams in the Andean Amazon and Implications for Andes-Amazon Connectivity. PLoS ONE 7(4): e35126. DOI:10.1371/journal.pone.0035126.

Forsberg, B.R., J.M. Melack, T. Dunne, R.B. Barthem, M. Goulding, R C.D. Paiva et al. 2017. The potential impact of new Andean dams on Amazon fluvial ecosystems. PLoS ONE 12(8): e0182254. DOI: 10.1371/journal.pone.0182254.

Fox, J., A.M. Jiménez and M. Stange. 2021. Natural history study of an understudied sea catfish species from Panama (Siluriformes: Ariidae). Neotrop. Ichthyol. 19(4): e200105. DOI: 10.1590/1982-0224-2020-0105.

Fricke, R., W.N. Eschmeyer and R. van der Laan (eds.). 2023. Eschmeyer's catalog of fishes: genera, species, references. http://researcharchive.calacademy.org/research/ichthyology/catalog/fishcatmain.asp). Electronic version accessed 21 August 2023.

Froese, R. and D. Pauly (eds.). 2024. FishBase. World Wide Web electronic publication. www.fishbase.org [accessed 12 January 2024].

Gende, S.M., R.T. Edwards, M.F. Willson and M.S. Wipfli. 2002. Pacific Salmon in Aquatic and Terrestrial Ecosystems. BioScience 52(10): 917–928.

Godinho, A.L. and H.P. Godinho. 2018. Fisheries, life-history and conservation status of the catfish pirá *Conorhynchos conirostris* (Ostariophysi: Siluriformes) in Brazil. Neotrop. Ichthyol. 16(1): e170131. DOI: 10.1590/1982-0224-20170131.

Godinho, H.P., A.L. Godinho, P.S. Formagio and V.C. Torquato. 1991 Fish ladder efficiency in a Southeastern Brazilian river. Ciência e Cultura 43(1): 63–67.

Godinho, A.L., B. Kynard and H.P. Godinho. 2007. Migration and spawning of female surubim (*Pseudoplatystoma corruscans*, Pimelodidae) in the São Francisco River, Brazil. Environ. Biol. Fish. 80: 421–433. DOI 10.1007/s10641-006-9141-1.

Godoy, M.P., A.A. Pantoja and A.M.S. Monteiro. 1979. Marcação e migração da piramutaba *Brachyplatystoma vaillantii* (Val. 1840) na Bacia Amazônica (Pará e Amazonas) Brasil. (Pisces, Nematognathi, Pimelodidae). Bol. Facul. Ciências Agrárias do Pará 11: 1–21.

Gomes, I.D. and F.G. Araújo. 2004. Influences of the reproductive cycle on condition of marine catfishes (Siluriformes, Ariidae) in a coastal area at southeastern Brazil. Environ. Biol. Fishes 71: 341–351.

Goulding, M. 1979. Ecologia da pesca do Rio Madeira. INPA, Manaus.

Goulding, M. 1980. The fishes and the forest – Explorations in Amazonian natural history. University of California Press, Berkeley.

Greenwood, P.J. and P.H. Harvey. 1982. The natal and breeding dispersal of birds. Ann. Rev. Ecol. Sys. 13(1): 1–21.

Griffiths, D. 2006. Pattern and process in the ecological biogeography of European freshwater fish. J. Anim. Ecol. 75: 734–751. DOI: 10.1111/j.1365-2656.2006.01094.x.

Hahn, L., E.G. Martins, L.D. Nunes, L.F. da Câmara, L.S. Machado and D. Garrone-Neto. 2019. Biotelemetry reveals migratory behaviour of large catfish in the Xingu River, Eastern Amazon. Sci. Rep. 9: 8464. DOI: 10.1038/s41598-019-44869-x.

Hahn, L., E.G. Martins, L.D. Nunes, L.S. Machado, T.M. Lopes and L.F. da Câmara. 2022. Semi-natural fishway efficiency for goliath catfish (*Brachyplatystoma* spp.) in a large dam in the Amazon Basin. Hydrobiologia 849: 323–338. DOI: 10.1007/s10750-020-04438-0.

Hart, L.G. and R.C. Summerfelt. 1974. Homing behavior of flathead catfish, *Pylodictis olivaris* (Rafinesque), tagged with ultrasonic transmitters. Proc. Annu. Conf. Southeast. Game Fish Comm. 27: 520–531.

Hasler, A.D. and W.J. Wisby. 1951. Discrimination of stream odors by fishes and its relation to parent stream behaviour. Am. Nat. 85: 223–238. DOI:10.1086/281672.

Hauser, M., C.R.C. Doria, L.R.C. Melo, A.R. Santos, D.M. Ayala, L.D. Nogueira et al. 2018. Age and growth of the Amazonian migratory catfish *Brachyplatystoma rousseauxii* in the Madeira River basin before the construction of dams. Neotrop. Ichthyol. 16(1): e170130, DOI: 10.1590/1982-0224-20170130.

Hauser, M., C.R.C. Doria, R.V. Santos, A. García-Vasquez, M. Pouilly, C. Pécheyran et al. 2019a. Shedding light on the migratory patterns of the Amazonian goliath catfish, *Brachyplatystoma platynemum*, using otolith 87Sr/86Sr analyses. Aquatic Conserv. Mar. Freshw. Ecosyst. 29(3): 397–408. DOI: 10.1002/aqc.3046.

Hauser, M., F. Duponchelle, T.W. Hermann, K.E. Limburg, L. Castello, D.J. Stewart et al. 2019b. Unmasking continental natal homing in goliath catfish from the upper Amazon. Freshwater Biol. 65(2): 325–336. DOI: 10.1111/fwb.13427.

Hegg, J.C., T. Giarrizzo and B.P. Kennedy. 2015. Diverse early life-history strategies in migratory amazonian catfish: implications for conservation and management. PLoS ONE 10(7): e0129697. DOI:10.1371/journal.pone.0129697.

Herrera-R., G.A., T. Oberdorff, E.P. Anderson, S. Brosse, F.M. Carvajal-Vallejos, R. Frederico et al. 2020. The combined effects of climate change and river fragmentation on the distribution of Andean Amazon fishes. Glob. Chang. Biol. 26: 5509–5523.

Herrera-R., G.A., S.A. Heilpern, T.B.A. Couto, L. Victoria-Lacy, F. Duponchelle, S.B. Correa et al. 2023. A synthesis of the diversity of freshwater fish migrations in the Amazon basin. Fish and Fisheries 25(1): 114–133. DOI: 10.1111/faf.12795.

Hoffman, J.I. and J. Forcada. 2012. Extreme natal philopatry in female Antarctic fur seals (*Arctocephalus gazella*). Mamm. Biol. 77(1): 71–73.

Hogan, Z., I.G. Baird, R. Radtkek and M.J. Vander Zanden. 2007. Long distance migration and marine habitation in the tropical Asian catfish, *Pangasius krempfi*. J. Fish Biol. 71: 818–832. DOI: 10.1111/j.1095-8649.2007.01549.x.

Hogan, Z.S., P.B. Moyle, B. May, M.J. Vander Zanden and I.G. Baird. 2004. The Imperiled Giants of the Mekong. Amer. Sci. 92: 228–237.

Hogan, Z.S. 2011. Ecology and conservation of large-bodied freshwater catfish: a global perspective. Am. Fish. S. S. 77: 39–53.

Huergo, G.M., R.J. Filgueiras-Souza, J.S. Batista, K. Formiga-Aquino and J.A. Alves-Gomes. 2011. Molecular genetics as a tool for fisheries management in the Brazilian Amazon: Piraíba (*Brachyplatystoma filamentosum* and *Brachyplatystoma capapretum*) (Siluriformes: Pimelodidae) in white-water rivers. PANAMJAS 6(4): 280–289.

Hugueny, B. 1990. Geographic range of west African freshwater fishes: role of biological characteristics and stochastic processes. Acta Oecol. 11: 351–375.

ICMBio. 2024. Sistema de Avaliação do Risco de Extinção da Biodiversidade – SALVE. Available at https://salve.icmbio.gov.br/.

IUCN. 2023. The IUCN Red List of Threatened Species. Version 2024-1. https://www.iucnredlist.org. [Accessed on 30 August 2023].

Jacquin, L., Q. Petitjean, J. Côte, P. Laffaille and S. Jean. 2020. Effects of pollution on fish behavior, personality and cognition: some research perspectives. Front. Ecol. Evol. 8: 86. DOI: 10.3389/fevo.2020.00086.

Junk, W., P.B. Bayley and R.E. Sparks. 1989. The flood pulse concept in river-floodplain systems. pp. 110–127. *In*: Dodge, D.P. (ed.). Proceedings of the International Large River Symposium (LARS). Can. Spec. Publ. Fish. Aquat. Sci. 106.

Khan, U., Z. Hasan, M. Inayatullah and J. Arif. 2014. Feeding Habits of a Freshwater Catfish *Clupisoma naziri* (Pisces: Schilbidae) from Khyber Pakhtunkhwa Rivers Pakistan. Pakistan. J. Zool. 46(4): 1166–1169.

Kubota, K., Y. Yokoyama, Y. Kawakubo, A. Seki, S. Sakai, P. Ajithprasad et al. 2015. Migration history of an ariid Indian catfish reconstructed by otolith Sr/Ca and 18O micro-analysis. Geochem. J. 49: 469–480. DOI:10.2343/geochemj.2.0371.

Kuzishchin, K.V., M.A. Gruzdeva and D.S. Pavlov. 2018. Traits of biology of European wels catfish *Silurus glanis* from the Volga–Ahtuba water system, the Lower Volga. J. Ichthyol. 58: 833–844.

Kwak, T.J., M.T. Porath, P.H. Michaletz and V.H. Travnichek. 2011. Catfish Science: Status and Trends in the 21st Century. American Fisheries Society Symposium 77: 755–780.

Latrubesse, E.M., E.Y. Arima, T. Dunne, E. Park, V.R. Baker, F.M. d'Horta et al. 2017. Damming the rivers of the Amazon basin. Nature 546: 363–369. DOI: 10.1038/nature22333.

Legendre, M. and A. Pariselle. 1998. The biological diversity and aquaculture of clariid and pangasiid catfishes in south-east Asia. Proceedings of the mid-term workshop of the "Catfish Asia Project". Cantho, Vietnam, 11–15 May 1998. IRD, Toulouse, France.

Lima, M.A.L., A.R. Carvalho, M.A. Nunes, R. Angelini and C.R.C. Doria. 2020. Declining fisheries and increasing prices: The economic cost of tropical rivers impoundment. Fish. Res. 221: 105399.

Limburg, K.E., A. Lochet, D. Driscoll, D.S. Dale and R. Huang. 2010a. Selenium detected in fish otoliths: a novel tracer for a polluted lake? Environ. Biol. Fishes 89: 433–440. DOI: 10.1007/s10641-010-9671-4.

Limburg, K.E., H. Høie and D.S. Dale. 2010b. Bromine patterns in Norwegian coastal Cod otoliths—a possible marker for distinguishing stocks? Environ. Biol. Fishes 89: 427–432. DOI: 10.1007/s10641-010-9660-7.

Limburg, K.E., B.D. Walther, Z. Lu, G. Jackman, J. Mohan, Y. Walther et al. 2015. In search of the dead zone: Use of otoliths for tracking fish exposure to hypoxia. J. Mar. Syst. 141: 167–178. DOI: 10.1016/j.jmarsys.2014.02.014.

Lintermans, M. 2009. Fishes of the Murray-Darling Basin - An Introductory Guide. MDBC Publication 10/07, Canberra.

Lucas, M.C. and E. Baras. 2001. Migration of Freshwater Fishes. Blackwell Science, Oxford.

Lundberg, J.G., J.P. Sullivan and M. Hardman. 2011. Phylogenetics of the South American catfish family Pimelodidae (Teleostei: Siluriformes) using nuclear and mitochondrial gene sequences. Proc. Acad. Nat. Sci. Philadelphia 161: 153–189.

Lung'ayia, H.B.O. 2022. Some Observations on the African Catfish *Clarias gariepinas* (Burchell) in the Sondu-Miriu River of Lake Victoria, Kenya. E. Okemwa, E.O. Wakwabi and A. Getabu. (eds.). Proceedings of the Second EEC Regional Seminar on Recent Trends of Research on Lake Victoria Fisheries, Nairobi: ICIPE Science 105–114. (available at http://hdl.handle.net/1834/1159).

Makrakis, S., M.C. Makrakis, R.L. Wagner, J.H.P. Dias and L.C. Gomes. 2007a. Utilization of the fish ladder at the Engenheiro Sérgio Motta Dam, Brazil, by long distance migrating potamodromous species. Neotrop. Ichthyol. 5(2): 197–204.

Makrakis, M.C., L.E. Miranda, S. Makrakis, D.R. Fernandez, J.O. Garcia and J.H.P. Dias. 2007b. Movement patterns of armado, *Pterodoras granulosus*, in the Paraná River Basin. Ecol. Fresh. Fish 16: 410–416.

Marcy-Quay, B., S.A. Sethi, N.O. Therkildsen and C.E. Kraft. 2020. Expanding the feasibility of fish and wildlife assessments with close-kin mark–recapture. Ecosphere 11(10): e03259.

McDowall, R.M. 1988. Diadromy in Fishes: Migrations Between Freshwater and Marine Environments. Croom Helm, Christchurch.

Miller, K.E. and J.A. Smallwood. 1997. Natal dispersal and philopatry of southeastern American Kestrels in Florida. Wilson Bull. 109: 226–232.

Mills, D.H. 1986. Threats to the Atlantic salmon (*Salmo salar*) at various stages of its life-history. Environ. Cons. 13: 257–260. DOI: 10.1017/S0376892900036316.

Miranda-Chumacero, G., G. Álvarez, V. Luna, R.B. Wallace and L. Painter. 2015. First observations on annual massive upstream migration of juvenile catfish *Trichomycterus* in an Amazonian River. Environ. Biol. Fish. 98: 1913–1926. DOI: 10.1007/s10641-015-0407-3.

Mitcheson, Y.S., A. Wai-lun To, N.W. Wong, H.Y. Kwan and W.S. Bud. 2019. Emerging from the murk: threats, challenges and opportunities for the global swim bladder trade. Rev. Fish Biol. Fisheries 29: 809–835. DOI: 10.1007/s11160-019-09585-9.

Montague, G.F. and D.E. Shoup. 2021. Two Decades of Advancement in Flathead Catfish Research. 41(S1), Special Issue: Catfish 2020, Communicating Catfish Science: Proceedings of the Third International Catfish Symposium. October 2021. Pages S11–S26. DOI: 10.1002/nafm.10654.

Moore, A., S.M. Freake and L.M. Thomas. 1990. Magnetic particles in the lateral line of the Atlantic salmon (*Salmo salar* L.). Philos. Trans. R. Soc. B Biol. Sci. 329: 11–15. DOI: 10.1098/rstb.1990.0145.

MPA. 2013. Boletim estatístico de pesca e aquicultura do Brasil 2011. Brasília: Ministério da Pesca e Aquicultura.

Myers, G.S. 1938. Fresh Water Fishes and West Indian Zoogeography. Vol. 3465. US Government Printing Office.

Myers, G.S. 1949. Usage of anadromous, catadromous and allied terms for migratory fishes. Copeia 2: 89–97.

Ngamsiri, T., M. Nakajima, S. Sukmanomon, N. Sukumasavin, W. Kamonrat, U. Na-Nakorn et al. 2007. Genetic diversity of wild Mekong giant catfish *Pangasianodon gigas* collected from Thailand and Cambodia. Fish. Sci. 73: 792–799.

Nielsen, R., D.K. Mattila, P.J. Clapham and P.J. Palsbøll. 2001. Statistical approaches to paternity analysis in natural populations and applications to the north Atlantic humpback whale. Genetics 157: 1673–1682. DOI: 10.1093/genetics/157.4.1673.

Nordeng, H. 1971. Is the local orientation of anadromous fishes determined by pheromones? Nature 233: 411–413. DOI:10.1038/233411a0.

Northcote, T.G. 1978. Migratory strategies and production in freshwater fishes. pp. 326–359. *In*: Gerking, S.D. (ed.). Ecology of Freshwater Fish Production. Blackwell, Oxford.

Northcote, T.G. 1984. Mechanisms of fish migration in rivers. pp. 317–365. *In*: McCleave, J.D., G.P. Arnold, J.J. Dodson and W.H. Neill (eds.). Mechanisms of Migration in Rivers. Plenum Press, New York.

Northcote, T.G. 1997. Potamodromy in salmonidae: living and moving in the fast lane. N. Am. J. Fish. Manage. 17: 1029–1045. DOI: 10.1577/1548-8675.

Ovidio, M., A. Dierckx and J.-P. Benitez. 2023. Movement behaviour and fishway performance for endemic and exotic species in a large anthropized river. Limnologica 99: e126061. DOI: 10.1016/j.limno.2023.126061.

Paradis, E., S.R. Baillie, W.J. Sutherland and R.D. Gregory. 1998. Patterns of natal and breeding dispersal in birds. J. Anim. Ecol. 67(4): 518–536.

Paulin, C., C. Roberts, A. Stewart and P. McMillan. 2001. New Zealand Fish - A Complete Guide. Te Tapa Press, Wellington.

Pelicice, F.M and A.A. Agostinho. 2008. Fish-Passage Facilities as Ecological Traps in Large Neotropical Rivers. Cons. Biol. 22(1): 180–188.

Pelicice, F.M., P.S. Pompeu and A.A. Agostinho. 2014. Large reservoirs as ecological barriers to downstream movements of Neotropical migratory fish. Fish. Fish. 16(4): 697–715. DOI: 10.1111/faf.12089.

Pelicice, FM., A.A. Agostinho, V.M. Azevedo-Santos, E. Bessa, L. Casatti, D. Garrone-Neto et al. 2022. Ecosystem services generated by Neotropical freshwater fishes. Hydrobiologia 850: 2903–2926. DOI: 10.1007/s10750-022-04986-7.

Pellett, T.D., G.J. Van Dyck and J.V. Adams. 1998. Seasonal migration and homing of channel catfish in the lower Wisconsin River, Wisconsin, North Am. J. Fish. Manage. 18: 1, 85–95. DOI: 10.1577/1548-8675(1998)018<0085:SMAHOC>2.0.CO;2.

Pereira, L.H.G., F. Foresti and C. Oliveira. 2009. Genetic structure of the migratory catfish *Pseudoplatystoma corruscans* (Siluriformes: Pimelodidae) suggests homing behaviour. Ecol. Freshwater Fish 18: 215–225.

Petrere Jr., M., R.B. Barthem, E.A. Córdoba and B.C. Gómez. 2004. Review of the large catfish fisheries in the upper Amazon and the stock depletion of piraíba (*Brachyplatystoma filamentosum* Lichtenstein). Rev. Fish Biol. Fish. 14: 403–414. DOI: 10.1007/s11160-004-8362-7.

de Pinna, M.C.C., L. Peixoto, V. Tagliacollo and M. Britto. 2025. Phylogenetic relationships and evolution of the major groups of Siluriformes. *In*: Arratia, G. and R.E. Reis (eds.). Catfishes, a Highly Diversified Group, Volume 2: Phylogeny and Evolution. CRC press, Boca raton, FL.

Putman, N.F., K.J. Lohmann, E.M. Putman, T.P. Quinn, A.P. Klimley and D.L.G. Noakes. 2013. Evidence for geomagnetic imprinting as a homing mechanism in Pacific salmon. Curr. Biol. 23: 312–316. DOI: 10.1016/j.cub.2012.12.041.

Quinn, T.P. 1980. Evidence for celestial and magnetic compass orientation in lake migrating sockeye salmon fry. J. Comp. Physiol. 137: 243–248. DOI: 10.1007/BF00657119.

Quinn, T.P. and J.C. Ogden. 1984. Field evidence of compass orientation in migrating juvenile grunts (Haemulidae). J. Exp. Mar. Biol. Ecol. 81: 181–192. DOI: 10.1016/0022-0981(84)90005-4.

Raabe, J.K. and J.E. Hightower. 2014. Assessing distribution of migratory fishes and connectivity following complete and partial dam removals in a North Carolina River, North Am. J. Fish. Manage. 34: 5, 955–969. DOI: 10.1080/02755947.2014.938140.

Rauber, R.G., L. Strictar, L.C. Gomes, H.I. Suzuki and A.A. Agostinho. 2021. Spatial segregation in the reproductive activity of Neotropical fish species as an indicator of the migratory trait. J. Fish Biol. 98(3): 694–706. DOI: 10.1111/jfb.14614.

Ricklefs, R.E. 2012. Naturalists, natural history, and the nature of biological diversity. Amer. Nat. 179(4): 423–435.

Riede, K. 2004. Global register of migratory species - from global to regional scales. Final Report of the R&D-Projekt 808 05 081. Federal Agency for Nature Conservation, Bonn, Germany.

Roberts, T.R. and I.G. Baird. 1995. Traditional fisheries and fish ecology on the Mekong River at Rhone Waterfalls in Southern Laos. Nat. Hist. Bull. Siam Soc. 43: 219–262.

Roff, D.A. 1991. Life history consequences of bioenergetic and biomechanical constraints on migration. Amer. Zool. 31: 205–215.

Rooker, J.R., D.H. Secor, G. DeMetrio, R. Schloesser, B.A. Block and J.D. Neilson. 2008. Natal homing and connectivity in Atlantic blue fin tuna populations. Science 322: 742–744.

Rubenstein, D.R. and K.A. Hobson. 2004. From birds to butterflies: animal movement patterns and stable isotopes. Trends Ecol. Evol. 19(5): 256–263.

Saha, S., S. Nasren, D. Pandit and S. Mian, 2021. An overview of the biological features, distribution, and conservation of a critically endangered riverine catfish, *Bagarius bagarius* (Hamilton, 1822), in the natural waters of Bangladesh. Conservation 1: 350–367. DOI: 10.3390/conservation1040027.

Sant'Anna, I.R.A., R. Ferreira, E. Rodrigues, D.M. Pinto and C.R.C. Doria. 2020. Pescarias tradicionais da Cachoeira do Teotônio submersas pela usinas em Rondônia. Canoa Do Tempo 12(2): 229–248. DOI: 10.38047/rct.v12.n02.2020.d10.p.229.248.

Simier, M., O. Judicaël, F. Osse, O. Sadio and J.-M. Ecoutin. 2021. Biology and ecology of sea catfish (Ariidae) of estuarine, lagoon and coastal ecosystems in West Africa. J. Fish Biol. 2021: 1–15. DOI: 10.1111/jfb.14751.

Skaug, H.J. 2001. Allele-sharing methods for estimation of population size. Biometrics 57: 750–756. DOI: 10.1111/j.0006-341X.2001.00750.x.

Smith, N.J.H. 1979. A pesca no rio Amazonas. INPA, Manaus.

So, N., J.K.J. Van Houdt and F.A.M. Volckaert. 2006a. Genetic diversity and population history of the migratory catfishes *Pangasianodon hypophthalmus* and *Pangasius bocourti* in the Cambodian Mekong River. Fish. Sci. 72: 469–476.

So, N., G.E. Maes and F.A.M. Volckaert. 2006b. High genetic diversity in cryptic populations of the migratory sutchi catfish *Pangasianodon hypophthalmus* in the Mekong River. Heredity 96: 166–174.

Soem, S., Z.S. Hogan, B. Chan, S. Chhuoy, S. Uy, K. Pin et al. 2023. From staple food to scarce resource: the population status of an endangered striped catfish *Pangasianodon hypophthalmus* in the Mekong River, Cambodia. Sustainability 15: 9103. DOI: 10.3390/su15119103.

Stabell, O. 2012. Migration and navigation. pp. 82–95. *In*: Brönmark, C. and L.-A. Hansson (eds.). Chemical Ecology in Aquatic Systems. Oxford University Press, Oxford.

Stoot, L.J., G.L. Butler, Y. Niella, G.S. Doran, J.D. Thiem, M.D. Taylor et al. 2023. Environmental effects on the seasonal distribution of an estuarine species *Neoarius graeffei* in Northern New South Wales, Australia. Estuaries and Coasts 47: 229–243. DOI: 10.1007/s12237-023-01269-9.

Tran, N.T., M. Labonne, M-T. Chung, C-H. Wang, K-F. Huang, J-D. Durand et al. 2021. Natal origin and migration pathways of Mekong catfish (*Pangasius krempfi*) using strontium isotopes and trace element concentrations in environmental water and otoliths. PLoS ONE 16(6): e0252769. DOI: 10.1371/journal.pone.0252769.

Van Damme, P.A., F.M. Carvajal-Vallejos, J. Camacho, H. Muñoz and J. Coronel. 2011. Peces Migratorios de la Amazonía Boliviana. pp. 149–202. *In*: Van Damme, P.A., F.M. Carvajal-Vallejos and J. Molina Carpio (eds.). Los peces y delfines de la Amazonía boliviana: hábitats, potencialidades y amenazas. Edit. INIA, Cochabamba, Bolivia.

Van Damme, P.A., L. Córdova-Clavijo, C. Baigún, M. Hauser, C.R.C. Doria and F. Duponchelle. 2019. Upstream dam impacts on gilded catfish *Brachyplatystoma rousseauxii* (Siluriformes: Pimelodidae) in the Bolivian Amazon. Neotrop. Ichthyol. 17(4): e190118. DOI: 10.1590/1982-0224-20190118.

Vander Zanden, M.J., M.K. Jake, E.K. Clayton, C.T. Moody, C.T. Solomon and B.C. Weidel. 2015. Stable isotope turnover and half-life in animal tissues: a literature synthesis. PloS ONE 10(1): e0116182. DOI: 10.1371/journal.pone.0116182.

Wacker, S., H.J. Skaug, T. Forseth, Ø. Solem, E.M. Ulvan, P. Fiske et al. 2021. Considering sampling bias in close-kin mark–recapture abundance estimates of Atlantic salmon. Ecol. Evol. 11(9): 3917–3932.

Wagner, R.L., S. Makrakis, T. Castro-Santos, M.C. Makrakis, J.H.P. Dias and R.F. Belmont. 2012. Passage performance of long-distance upstream migrants at a large dam on the Paraná River and the compounding effects of entry and ascent. Neotrop. Ichthyol. 10(4): 785–795.

Walker, M.M., C.E. Diebel, C.V. Haugh, P.M. Pankhurst, J.C. Montgomery and C.R. Green. 1997. Structure and function of the vertebrate magnetic sense. Nature 390: 371–376. DOI: 10.1038/37057.

Wallace, A.A., D.J. Hollander and E.B. Peebles. 2014. Stable isotopes in fish eye lenses as potential recorders of trophic and geographic history. PLoS ONE 9(10): e108935.

Walton, B.W., C.F. Cotton, D.A. Gandy and M.M.P.B. Fuentes. 2022. Determining the influence of abiotic factors on spatial-temporal patterns of marine catfish (family: Ariidae) within the Apalachicola Bay System, Florida. Environ. Biol. Fish. 105: 369–383. DOI: 10.1007/s10641-022-01230-y.

Wassenaar, L.I. and K.A. Hobson. 2001. A stable-isotope approach to delineate geographical catchment areas of avian migration monitoring stations in North America. Environ. Sci. Technol. 35(9): 1845–1850.

Weyer, J. and T. Schmitt. 2013. Knowing the way home: Strong philopatry of a highly mobile insect species, *Brenthis ino*. J. Insect Cons. 17: 1197–1208. DOI: 10.1007/ s10841-013-9601-9.

Winemiller, K.O. 1989. Patterns of variation in life history among South American fishes in seasonal environments. Oecologia 81: 225–241.

Winemiller, K.O. and K.A. Rose. 1992. A patterns of life history diversification in North American fishes: implications for population regulation. Can. J. Fish. Aquat. Sci. 49: 2196–2218.

Yáñez-Arancíbia, A. and A.L. Lara-Domínguez. 1988. Ecology of three sea catfishes (Ariidae) in a tropical coastal ecosystem - Southern Gulf of Mexico. Mar. Ecol. Prog. Ser. 49: 215–230.

Zamudio, K.R., J.M. Robertson, L.M. Chan and I. Sazima. 2009. Population structure in the catfish *Trichogenes longipinnis*: drift offset by asymmetrical migration in a tiny geographic range. Biol. J. Linn. Soc. 97: 259–274.

Zapata, L.A. and J.S. Usma. 2013. Guía de las especies Migratorias de la Biodiversidad en Colombia. Peces. Vol. 2. Ministerio de Ambiente y Desarrollo Sostenible/WWF-Colombia. Bogotá, D.C. Colombia.

Zuanon, J.A.S. 1990. Aspectos da biologia, ecologia e pesca de grandes bagres (Pisces: Siluriformes, Siluroidei) na área da Ilha de Marchantaria, Rio Solimões, AM. Unpublished MSc. Thesis, INPA, Manaus. (Available at https://repositorio.inpa.gov.br/handle/1/38103).

Chapter 12

Diversity and Life Cycles of Subterranean Catfishes

Eleonora Trajano[1,*] and *Sandro Secutti*[2]

Subterranean habitats are networks of interconnected heterogeneous subsurface spaces filled with water or air, allowing the circulation of energy and matter, thus making them available for colonization by living organisms. Underground spaces large enough for human access are named caves (Ford and Williams 2007). This is clearly an anthropocentric definition. It is noteworthy that fishes and other subterranean organisms smaller than humans may use a larger extension of subterranean habitats, including spaces inaccessible to our species.

Subterranean habitats encompass spaces in a rock matrix, mainly carbonate rocks subject to karstification, i.e., the formation of terrains with distinct hydrology and landforms arising from a combination of rock solubility and fracture porosity, characterized by sinking streams, drainages, enclosed depressions, and steep walls (Ford and Williams 2007). Karst systems are functional units involving organized flow pathways forming a drainage unit with input-output zones and finite configurations (Aquilina et al. 2023). A karst aquifer is an open system with a boundary defined by the catchment limits and input, throughput and output flows, and mechanisms and controls (Ford and Williams 2007).

The main ecological filters restricting the effective colonization of the subterranean realm by epigean (surface) organisms, i.e., the establishment of subterranean self-sustained (source) populations, are a consequence of the permanent absence of light. These organisms must at least be able to orient themselves in darkness. In addition, subterranean habitats generally are characterized by the minimization of climatic fluctuations, providing refuge for survival during unfavorable times, such as dry phases of glacial cycles (Barr 1968; Trajano 1995, 2021).

Organisms regularly found in subterranean habitats have been classified into three categories of the Schiner-Racovitza system (Racovitza 1907). This classification was modified by Trajano (2012) to accommodate the source-sink population model. Source populations are populations with excess production that would continue to grow if isolated, in contrast with sink populations, that vanish if cut off from the source of migrants (Jones et al. 2004). **Trogloxenes** are individuals that regularly use subterranean habitats, usually as shelter, but are dependent on epigean (surface) resources to

[1] Instituto de Biociências, Universidade de São Paulo (retired), Rua Nobre Vieira, 364; 05587-180 São Paulo, SP, Brazil.
[2] Núcleo da Floresta - Pesquisa e Extensão em Fauna e Flora, Av. Marcos Penteado de Ulhôa Rodrigues, 4446; 06543-001 Santana de Parnaíba, SP, Brazil.
Email: sanoverkill@yahoo.com.br
* Corresponding author: etrajano@usp.br

complete their life cycles, and in this case, the source population is epigean. Bats that leave caves daily to feed are typical well-known trogloxenes. **Troglophiles** are subterranean source populations of epigean (surface) species connected by individuals commuting between these habitats, thus promoting genetic flow between the epigean and subterranean populations. Troglophilic populations that become genetically isolated in subterranean habitats, for instance, due to local extinction of epigean populations or topographic barriers, may differentiate into **troglobites**, i.e., exclusively subterranean species (Trajano and Carvalho 2017).

Subterranean organisms are herein defined as evolutionary units responding to subterranean selective regimes. Subterranean habitats provide resources (food, shelter, substrate, and climate) which affect survival/reproductive rates. In contrast, "accidentals" are organisms introduced into caves by mishap (such as being washed into caves or falling through upper openings) or when entering in search of a mild climate; their inability to properly orient themselves and to find food leads to their eventual demise (Trajano 2012; Trajano and Carvalho 2017).

Taxa preadapted (in the sense of exaptations, i.e., features that did not originally arise for their current use but rather were co-opted for new purposes; Buss et al. 1998) to subterranean life include nocturnal, cryptobiotic, predominantly chemo- and/or mechano-oriented animals, detritivores, opportunistic omnivores, or generalist carnivores able to cope with low nutrient availability. This is the case for most siluriforms, especially those living in streams, which are nocturnal, chemo-oriented generalist carnivores, frequently cryptobiotic. Therefore, it is no surprise to find such fishes in caves and other hypogean habitats in the condition of troglophiles or troglobites.

As a scientific hypothesis based on a negative trait, the troglobitic status for any subterranean population depends on corroboration of high and sufficient sampling efforts on the surface, which is rarely the case. Such a condition is usually inferred based on the presence of autapomorphies indicating differentiation under the particular subterranean selective regime. For this, the comparative method is fundamental.

Autapomorphies of troglobites that may be associated with subterranean life are called troglomorphisms. The most conspicuous troglomorphisms, generally used to recognize the troglobitic status, are regressive character states, mainly reduction of visual structures and dark pigmentation (melanin in fishes). Other regressive characteristics include a reduction of photonegative reactions and other avoidance behaviors, a loss of cryptobiotic habits, and a regression of circadian rhythmicity (Parzefall and Trajano 2010).

The most general explanation for character regression in both epigean and troglobitic organisms is the neutralistic model. When stabilizing selection for maintenance of visual structures and dark pigmentation of the skin is relaxed, these structures progressively regress due to the accumulation of neutral mutations. Other processes leading to troglomorphy have been proposed based on studies on Mexican cave tetras (*Astyanax*), such as pleiotropic effects due to the selection of some beneficial traits (see Wilkens and Strecker 2017 for a review). However, there is no convincing evidence that such processes are important for other troglobitic fishes, including siluriforms.

On the other hand, some character states are clearly adaptative, known as "constructive" traits. For instance, several vertebrate and invertebrate troglobites present enhanced non-visual sensory systems improving chemo-, mechano- and electroreception. Many constructive characters of troglobites are adaptations to food limitation, typical of most subterranean habitats, such as a decrease in metabolic rates, lipid storage in fat tissues, and a K-selected life cycle observed in organisms as diverse as planarians, crustaceans, beetles, spiders, fishes, and salamanders (Poulson 1963, 1966; Ginet and Decou 1977; Howarth and Moldovan 2018).

Diversity of Subterranean Catfishes

Distribution, Taxonomic and Phylogenetic Diversity

According to Proudlove (2023), roughly 1.7% (ca. 300 species) of all known freshwater fish species (ca. 18,000 species) have been reported as troglobites. Most freshwater fishes are ostariophysans belonging to three major orders: Cypriniformes with ca. 4,800 species (23 families) distributed in Africa, Eurasia, and Central Mexico in North America; Characiformes with ca. 1,290 spp. (24 families) distributed in the Neotropical Region and ca. 240 spp. (four families) in Africa; Siluriformes with ca. 4,217 species (ca. 40 families) worldwide, and ca. 2,690 spp. (16 families) in the Neotropical region (Fricke et al. 2024).

The majority of troglobitic fishes are Siluriformes, which are predominantly nocturnal, and Cypriniformes that include diurnal and nocturnal species. As expected, Characiformes, primarily diurnal, are under-represented in the subterranean biotope, with one highly specialized phreatobic species in Eastern Brazil and several troglobitic *Astyanax* populations in Central Mexico. Likewise, there is no troglobite among cichlids (Cichliformes), visually oriented fishes common in freshwaters around the world.

Families and genera with described troglobitic species throughout the world, based on Proudlove (op. cit.), are shown in Table 1. Among 246 described species of troglobitic fishes, 55 spp. (11 families) are siluriforms, the majority (41 spp.) from the Neotropical region. Several undescribed species have also been reported in publications, totaling 299 species of troglobitic fishes (90 siluriforms) known until May 2023 (Proudlove 2023).

These numbers are constantly changing, usually increasing due to the discovery and description of new species and the splitting of former nominal species after studies using more discriminating methods than the classic taxonomy, such as molecular techniques (e.g., *Trichomycterus* from Santander, Colombia; Flórez et al. 2021) and geometric morphometrics, revealing cryptic diversity. The occurrence of differentiating clades has been shown by geometric morphometrics for the Brazilian catfishes *Ancistrus cryptophthalmus* (at least two clades; Reis et al. 2006), *Rhamdiopsis krugi* (two clades; Bichuette et al. 2015), and *Trichomycterus dali* (three clades; Cordeiro 2014). Less frequently, this taxonomic diversity decreases due to the exclusion of species through synonymy or the discovery of source populations in epigean habitats, disproving the status of troglobites.

In accordance with what is observed on the surface, the largest taxonomic diversity (species richness) of troglobitic siluriforms is found in the Neotropical Region – six families and 41 described species versus four families and 14 described species in Asia, Africa, and North America. With 12 nominal species and nine undescribed ones, *Trichomyterus* is the richest genus with troglobitic species, followed by *Rhamdia* (eight nominal species plus six undescribed ones).

Table 1. Families and genera of siluriforms with described troglobitic species worldwide. In parenthesis: number of described species until May 2023. * Exclusively subterranean genera; ** *Isbrueckerichthys alipionis* cave form. Classification according to Fricke et al. (2024).

Family	Genera (Described spp.)	Country
Trichomycteridae	*Glaphyropoma* (1), *Ituglanis* (6), *Silvinichthys* (1), *Trichomycterus* (12)	Argentina, Bolivia, Brazil, Colombia, and Venezuela
Callichthyidae	*Aspidoras* (1)	Brazil
Loricariidae	*Ancistrus* (3), *Isbrueckerichthys* (1**)	Brazil and Venezuela
Astroblepidae	*Astroblepus* (2)	Ecuador and Peru
Amblycipitidae	*Xiurenbagrus* (1)	China
Siluridae	*Pterocryptis* (2)	Thailand and Vietnam
Kryptoglanidae	*Kryptoglanis* (1)	India
Heptapteridae	*Pimelodella* (2), *Rhamdia* (8), *Rhamdiopsis* (1)	Belize, Brazil, Mexico, Trinidad and Tobago, and Venezuela
Phreatobiidae	*Phreatobius** (3)	Bolivia and Brazil
Clariidae	*Clarias* (1), *Horaglanis** (4) *Uegitglanis** (1)	Namibia, India, and Somalia
Ictaluridae	*Prietella* (2)*, *Satan* (1)*, *Trogloglanis* (1)*	Mexico and USA
TOTAL	55	

However, this is a modest taxonomic diversity compared with China, where cypriniforms encompass 95 described troglobitic species (one-third of the total diversity of troglobitic fishes); the cyprinid genus *Sinocyclocheilus* alone includes 40 described species (Proudlove 2023). The vast extension of karst areas in China and the high degree of fragmentation may explain such impressive taxonomic diversity. However, there are only two known Chinese troglobitic siluriforms, *Xiurenbagrus dorsalis* and an undescribed *Pterocryptis*.

On the other hand, siluriforms present the highest phylogenetic diversity (a measure of diversity incorporating phylogenetic relationships among species) among the 11 fish orders with troglobitic species. For instance, the 55 described species of troglobitic catfishes are distributed in 21 genera and 11 families, whereas the 139 described troglobitic cypriniforms belong to 33 genera in only four families (Balitoridae, Cobitidae, Nemacheilidae, and Cyprinidae).

The Neotropical countries with the highest diversity of troglobitic fishes, in general, are Mexico, with representatives of six orders (Characiformes, Siluriformes, Ophidiiformes, Gobiiformes, Synbranchiformes, and Cyprinodontiformes), eight families and 12 species described, mostly *Rhamdia* (Heptapteridae) and Brazil, with three orders (Siluriformes, Gymnotiformes, and Characiformes), seven families and 22 described species, mostly trichomycterids (*Ituglanis* and *Trichomycterus*) and heptapterids (*Pimelodella*, *Rhamdia*, and *Rhamdiopsis*). This high diversity is due to former epigean diversity (the highest diversity provides more potential colonizers) and the extension of karst areas, where subterranean habitats are mostly developed and available.

In general, troglobites evolve from troglophilic populations genetically isolated in subterranean habitats (Barr 1968; Trajano 1995). Therefore, it is not surprising to find many troglophilic siluriforms in caves. As expected, most belong to genera with troglobitic derivatives (marked with *), for instance, *Trichomycterus** (Trichomycteridae), *Pimelodella**, *Rhamdia**, and *Imparfinis* (Heptapteridae), *Hypostomus* (Loricariidae) (Mattox et al. 2008). Their status as troglophiles was deduced based on the regular presence of all age/size classes throughout the cave during the year (Trajano and Carvalho 2017).

The recognition of troglophiles and trogloxenes and the distinction between them may be more difficult than for troglobites because these animals do not present distinctive morphological characteristics, and corroboration of their status is based on medium to long-term population studies. Fieldwork in caves is highly technical and time and effort-consuming, especially when focusing on biology and ecology. For this reason, such studies are rare, even for species intensively studied in the laboratory, such as the Mexican tetra characins *Astyanax*.

In general, there is little, if any, interest in troglophiles. However, for a good comprehension of subterranean evolution, they must be treated as potential ancestors of troglobites (Trajano and Carvalho 2017). Moreover, many troglophiles that are rare on the surface present high population densities in subterranean habitats, which opens the possibility of cave populations as sources of colonizers for epigean habitats, especially after long periods of adverse climate (Trajano and Carvalho 2017); therefore, such populations should be studied and protected.

Distribution and Habitat Diversity

Subterranean siluriforms are present in caves (in the human sense) and non-cave habitats, mainly in karst areas, and some species in other lithologies, such as quartzitic rocks. They occupy distinct habitats, from cave systems to aquifers. In addition, some species are found in subsurface interconnected spaces in lateritic soils (e.g., *Phreatobius cisternarum* in the Amazon delta and *Horaglanis* and *Kryptoglanis* in Kerala, India).

Many stream-dwelling troglobitic catfishes seem to be endemic to a single karst system (e.g., *Ancistrus formoso*, *A. galani*, *Aspidoras mephisto*, *Astroblepus riberae*, *Clarias cavernicola*, *Ituglanis bambui*, *I. epikarsticus*, *I. passensis*, *I. ramiroi*, *Pimelodella spelaea*, *Rhamdia enfurnada*, *R. guasarensis*, *R. laluchensis*, *R. macuspanensis*, *Trichomycterus donascimentoi*, *T. itacarambiensis*, *T. santanderensis*, *T. sketi*, *T. spectrum*, and *T. spelaeus*), or to different micro-basins in a relatively small area (e.g., *Pimelodella kronei*). On the other hand, phreatobic species may distribute through wide geographic areas (e.g., *Phreatobius sanguijuela* is known from localities ca. 290 km apart) (Trajano 2021). *Trogloglanis pattersoni*, and probably *Satan eurystomus* as well, are distributed in an area of 1,036 km^2 in the San Antonio Pool, Edwards Aquifer, USA (Longley and Karnei 1978).

A vertical zonation is observed in karst areas (Table 2) from permanent perched aquifers in the epikarst (heterogeneous interface between unconsolidated material and altered carbonate bedrock, capable of delaying or storing and locally rerouting vertical infiltration to the deeper regional phreatic zone; Jones et al. 2004) to the deep phreatic zone. This is the case of the ictalurids *Satan eurysthomus* and *Trogloglanis pattersoni*, captured in outlets of artesian wells, penetrating the Edwards Aquifer (Texas, USA) to depths between 400 and 580 m (Langecker and Longley 1993). No taxonomic correlation with habitat type is observed as illustrated by trichomycterids living in the epikarst, vadose tributaries, base-level streams, and the phreatic zone (Trajano and Bichuette 2010).

There is great variation in the ecological conditions of habitats occupied by subterranean fishes, from streams with well-oxygenated waters in riffles and rapids, alternating with slow-moving pools in open-channel passages to lentic environments, such as water bodies formed by the water table inside caves and inaccessible phreatic habitats (Trajano 2003).

Habitat use for individual species (Table 3) is inferred based on the presence of stable groups through time, indicating they are source populations adapted to a particular kind of environment. Sink populations, indicated by the occurrence of isolated or small unstable groups in transitory habitats (e.g., temporary streams and pools) or transitional habitats (between epigean and hypogean systems, the upper zone of natural and artificial wells), were not considered because the species did not evolve in these particular habitats.

In the case of source populations living in non-cave habitats, such as the epikarst and the phreatic zone connected to the surface through inaccessible fissures, samples have been obtained from sink populations, for instance, in rimstone pools (*Ituglanis epikarsticus*) and vadose streams (*I. bambui* and *Tichomycterus rubbioli*). The low recapture rates in the vadose tributary, where specimens of *I. bambui* have been found, and the occurrence of this species also in pools 4 km away indicate that its source population lives in the epikarst like *I. epikarsticus* (Trajano 2021; Bichuette and Trajano 2021).

Some species may occupy a special type of superficial subterranean habitat, the hypothelminorheic, constituted by small seeps, little more than small persistent spots in slight depressions in the ground, rich in organic matter and fed by subsurface water, underlain by a layer

Table 2. Habitats used by troglobitic siluriforms with examples, based on Trajano (2001). *Collected in artificial wells; ** S. Secutti, pers. obs.

Habitat Type		**Examples**
Epikarst		*Ituglanis bambui, I. epikarsticus*
Open channel stream passages	Vadose tributaries	*Ituglanis ramiroi* **, *Pimelodella spelaea*
	Base level streams	*Ancistrus cryptophthalmus, A. galani, Aspidoras mephisto, Glaphyropoma spinosum, Ituglanis passensis, Pimelodella kronei*, all *Rhamdia* spp., *Trichomycterus* spp.
Zone of Seasonal Oscillations of Water Table		
Phreatic zone	Connected to the surface through inaccessible fissures *	*Horaglanis* spp., *Phreatobius cisternarum, P. sanguijuela, Satan eurytomus, Silvinichthys bortayro, Trogloglanis pattersoni*
	Connected to the surface through caves, sinkholes, resurgences, flooded caves	*Ancistrus formoso, Astroblepus ribeira, Clarias cavernicola, Prietella phreatophila, P. lundbergi, Rhamdiopsis krugi, Trichomycterus dali, Uegitglanis zammaranoi*

Table 3. Intrapopulation variation in trogloglobitic catfishes regarding eye development and pigmentation degree. (A): variation encompassing the spectrum, from states close to the observed in epigean congeners to complete absence of dark pigmentation and eyes externally visible. (B): variation from reduced to absent eyes and pigmentation. * Undescribed species from Salitre Cave, Minas Gerais State, E Brazil.

Condition	Examples	Habitat
Population with slightly but significantly reduced eyes and pigmentation	*Ituglanis boticario*, *Pimelodella spelaea*, *Rhamdiopsis* sp.*	Base-level streams and vadose tributaries
Intrapopulation variability (A)	*Aspidoras mephisto*, *Pimelodella kronei*, *Rhamdia enfurnada*	Base-level streams and vadose tributaries
Intrapopulation variation (B)	*Ituglanis passensis*, *Rhamdia reddelli*, *Rhamdia zongolicensis*, *Trichomycterus chaberti*, *Trichomycterus santanderensis*	Base-level streams and vadose tributaries
Population homogeneously depigmented and anophthalmic	*Ancistrus formoso*, *Prietella phreatophila*, *Prietella lundbergi*, *Rhamdiopsis krugi*, *Satan eurystomus*, *Trogloglanis pattersoni*, *Uegitglanis zammaranoi*	Phreatic zone

of clay, and forming a kind of perched aquifer (Culver and Pipan 2019). This would be the case of *Horaglanis* and *Kryptoglanis* from Kerala (Raghavan et al. 2023; Binoy et al. 2012) and of undescribed *Phreatobius* species collected in thick layers of saturated leaf litter at the margins of creeks in the Amazon basin (E.Trajano pers. obs.).

Troglomorphic siluriforms are also known from non-subterranean aphotic habitats, such as the bottom of deep rivers with turbid waters, mainly the Amazon and Congo rivers. In the lower Congo, they include clariids (four genera), a claroteid and a mochokid; in the Amazon, they include loricariids (two genera), aspredinids (two genera), a trichomycterid, a cetopsid, and a pimelodid. Among these families, clariids, loricariids, trichomycterids, and aspredinids have troglobitic derivatives. In addition, most psamophilic troglomorphic catfishes, adapted to live in interstitial spaces on the beds of large rivers, are trichomycterids, belonging to Glanapteryginae, Sarcoglanidinae and Microcambevinae.

Phenotypical Diversity and Special Adaptations (Figure 1)

The evolution of troglomorphisms is a gradual process; therefore, some troglobites are expected to present intrapopulation variability in morphological, physiological, and behavioral traits. There is a correlation between the degree of troglomorphism and habitat (Table 3). Some fishes, usually stream-dwellers, have slightly reduced eyes and pigmentation, and only the comparative analysis allows for the detection of significant differences in relation to epigean congeners. On the other hand, phreatobitic species, in general, are homogeneously depigmented and anophthalmic.

All degrees of variation are observed between these extremes, frequently with a mosaic distribution of character states revealing independent differentiation of eyes and pigmentation (Table 3). In some species, such as *Rhamdia enfurnada* (NE Brazil), both characters show a normal distribution but with independent states in each individual: paler specimens may have relatively large eyes, dark specimens may exhibit greatly reduced eyes, and so on. In *Pimelodella kronei* (SE Brazil), pigmentation has a normal distribution, but 90 percent of the fish do not have externally visible eyes, and the remaining 10 percent have variable vestigial eyes. In these species, vestigial eyes are frequently asymmetrical in size, which is in accordance with neutralistic theories of the regressive evolution of eyes in cavefish (Secutti and Trajano 2021).

Figure 1. Brazilian troglobitic catfishes. Interspecific variability: (A) *Rhamdiopsis* from Campo Formoso, North Bahia. (B) *Pimelodella spelaea*; (C) *Rhamdia enfurnada* (Photographs courtesy D. Fenolio). Intrapopulation variability, *Pimelodella kronei*. (D1) Unpigmented individual (Courtesy I. Sazima); (D2) Pigmented individual (Courtesy D. Fenolio). Interpopulation variability, *Ancistrus cryptophthalmus*: (E1) Bezerra Cave (Courtesy A.L. Camargo); (E2) Passa Três Cave (S. Secutti); (E3) Angélica Cave (E. Trajano). (F) *Trichomycterus dali*: elongated nasal barbels, retention of the predorsal fold (Courtesy D. Fenolio). (G) *Rhamdiopsis krugi*, eggs seen by transparency (Courtesy A.L. Camargo). Reproduction and development in the laboratory: *Ancistrus cryptophthalmus*. (H1) Reproductive adults on rock dens; (H2) Egg cluster attached to den roof; (H3) Larvae (from Secutti and Trajano 2009). *Ituglanis passensis*: (I1) Adult female (Courtesy I. Sazima), (I2) Eggs and larvae, and (I2) Asymmetric development of eyes (photos from Secutti and Trajano 2021). (J1) *Trichomycterus itacarambiensis*, a pigmented individual with reduced eyes (Courtesy A.L. Camargo); (J2) *T. dali*, an unpigmented individual in frontal view (Courtesy A.L. Camargo).

Eye asymmetry was also reported for *Trichomycterus chaberti*: depending on the population, one eye or the other is affected (Pouilly and Miranda 2003). An extreme structural eye asymmetry was reported by Wilkens (2001) for Mexican troglobitic *Rhamdia*.

Different mechanisms may be involved in the reduction of dark pigmentation: decrease in the number/density of melanophores, decrease in the size of melanophores (usually associated with the former), DOPA (+) and DOPA (–) albinism, and lowered capacity of response to hormones and neurotransmitters involved in pigmentation control, when compared with epigean relatives (Grempel et al. 2020). Different genetic mechanisms result in continuous variation (polygenic bases for number of melanophores) versus discontinuous variation (for DOPA albinism).

Such mechanisms may act independently on the same population, resulting in color polymorphism, as observed in *Trichomycterus itacarambiensis* (E Brazil). Two color morphotypes, corresponding respectively to DOPA(–) and DOPA(+), are represented by individuals completely depigmented and variably pale but still clearly pigmented. Variably reduced eyes are present in both phenotypes, showing a normal distribution in relation to size, and externally visible eyes in depigmented fish are characteristically red colored (Trajano and Pinna 1996).

Interpopulation mosaics of character states are also observed and are well illustrated by four cave populations of the armored catfish *Ancistrus cryptophthalmus sensu lato* from Central Brazil. Different sequences, corresponding to transformation series, were obtained respectively for eyes, pigmentation, and body shape (studied using geometric morphometrics): **eyes**: epigean → Bezerra → Angélica → São Vicente I → Passa Três; **pigmentation**: epigean → Bezerra → Passa Três → São Vicente I and Angélica (with the largest variation, including individuals as pigmented as those from Bezerra Cave to the less pigmented among *A. cryptophthalmus*); **body shape**: epigean → Angélica + Bezerra → São Vicente I → Passa Três (1st axis) and epigean + Angélica + Bezerra + Passa Três → São Vicente I (2nd axis) (Reis et al. 2006). This is evidence of the independent differentiation of these populations.

There is a correlation between the degree of troglomorphism and habitat (Table 3). The most specialized troglobitic fishes (also invertebrates) live in the phreatic zone (e.g., *Ancistrus formoso*, *Prietella phreatophila*, *Rhamdiopsis krugi*, *Satan eurytomus*, *Trogloglanis pattersoni*, *Trichomycterus dali*, and *Uegitglanis zammaranoi*), whereas the less specialized ones, with intrapopulation variation, are usually stream dwellers (*Pimelodella kronei*, *P. spelaea*, *Rhamdia enfurnada*, and *Trichomycterus itacarambiensis*). A possible explanation arises from the transient condition (in terms of geological time) of caves within karst cycles. In general, troglobitic catfishes are derived from stream-dwellers. In the cessation phases of karst cycles, when large subterranean spaces disappear due to breakdowns or filling by chemical or clastic sediments, populations that have managed to adapt to the lentic conditions of the phreatic zone below may survive over a long geological period until the next cycle, frequently as relicts (Trajano 2021).

Species living in lotic habitats inside caves, in vadose and base-level streams, usually retain their generalized body morphology, as is the case of *Rhamdia* spp., *Pimelodella*, most *Trichomycterus* spp. and armored catfishes (*Ancistrus* and *Aspidoras*). Usually, size variation is also within the limits observed for epigean congeners. Among them, *Rhamdia enfurnada*, from NE Brazil, reaches the largest size (242 mm standard length – SL; E. Trajano unpublished. data) among troglobitic siluriforms. It shares with the synbranchid *Rakthamichthys digressus* the fourth position among the longest troglobitic fishes, after the synbranchids *Ophisternon candidum* (356 mm SL) and *O. infernale* (325 mm SL) and the cyprinid *Neolissochilus pnar* (329 mm SL) (Dahanukar et al. 2023).

On the other hand, effective colonization of small spaces is achieved through miniaturization (probably by paedomorphosis) and slender, elongated and flexible bodies with a reduced lateral line on the trunk, as observed in *Rhamdiopsis krugi*, *Ituglanis epikarsticus*, *I. bambui*, and *I. ramiroi* (Trajano 2021). Adaptations to the deep phreatic zone observed in *Satan eurysthomus* and *Trogloglanis pattersoni* include small body size, weak ossification, and reduced swim bladder (Langecker and Longley 1993).

In troglobitic catfishes, "sensorial compensation" for the unfeasibility of visual orientation is achieved through the elongation of barbels in the Mexican *R. reddelli*, *R. zongolicensis* and *R. laluchensis* (Weber et al. 2003), *Rhamdiopsis krugi* (Bockmann and Castro 2010) and *Rhamdiopsis* undescribed sp. from Toca do Gonçalo (NE Brazil) (F. Bockmann and E. Trajano, unpublished data), and *Trichomycterus dali* (SW Brazil) (Rizzato et al. 2011), as well as broad head laterosensory canals and pores in *Rhamdia guasarensis* (NW Venezuela) (DoNascimiento et al. 2004) and *Rhamdiopsis* from Toca do Gonçalo (Bockmann and Trajano, unpublished data).

Several phreatobic catfishes possess lipid storage in subcutaneous fat tissues, either in localized body parts, such as adipose fins (*Trichomycterus dali*) or all over the body, as in the case of *Rhamdiopsis krugi* and *Rhamdiopsis* from Toca do Gonçalo (Trajano 2021), *Satan* and *Trogloglanis pattersoni* (Langecker and Longley 1993).

Life Cycles: Reproduction, Early Development, Growth and Longevity

As an adaptation to food scarcity, many troglobites present a K-selected, or precocial lifestyle (*sensu* Balon 1981), characterized by delayed maturity, reduced proportion of the population breeding at any one time, less numerous but larger eggs, low growth rates, and longer life spans (Culver and Pipan 2019). We believe that the reduced fecundity per reproduction event (fewer eggs, decreased reproductive frequency) may be compensated by increased longevity (more reproductive events per individual) and increased probability of survival of the offspring, which hatch late at less vulnerable developmental stages. The first detailed and most comprehensive study on ecology and life cycles in amblyopsids was carried out by Poulson (1963, 1969), comparing epigean, troglophilic and troglobitic species.

Ideally, studies on the reproduction and life cycle of troglobites are based on fieldwork complemented by laboratory research. The latter provides opportunities for detailed observations and tests in a controlled environment, allowing for the isolation of variables in order to uncover cause-effect relationships. However, the biological meaning of laboratory data must be validated by field studies to exclude artifacts due to artificial conditions.

As aforementioned, fieldwork in caves is hard in general, technically difficult and time-consuming due to natural temporal and spatial variations (usually cyclical) and the impossibility of accessing the whole habitat of troglobitic organisms. For this reason, data on the biology of subterranean siluriforms are scanty and restricted to a few species.

Few Neotropical catfishes have been studied with a focus on biology and population ecology in nature, and even fewer in laboratory settings. Field studies on Brazilian stream-dwellers *Pimelodella kronei* (Trajano 1991), *Trichomyterus itacarambiensis* (Trajano 1997a, b), *Ancistrus cryptophthalmus* (Trajano and Bichuette 2007), and *Rhamdia enfurnada* (E. Trajano, unpublished data) yielded data on growth rates, average maximum length and longevity (calculated using von Bertalanffy formula), besides population size, density (in the accessible habitat), and individual movements (home ranges). Behavioral observations included the courtship behavior in the callichthyid *Aspidoras mephisto* (Tencatt and Bichuette 2017).

Development studies based on spontaneous reproduction in the laboratory were carried out for two Brazilian troglobitic siluriforms, the armored catfish *Ancistrus cryptophthalmus* (Secutti and Trajano 2009) and the trichomycterid *Ituglanis passensis* (Secutti and Trajano 2021). Data based on induced spawning in the Mexican *Rhamdia reddelli* and *R. zongolicensis* were described by Wilkens (2001).

Reproduction of troglobites in the laboratory is hampered by the paucity of data on their epigean relatives, which could guide the setup of proper conditions for reproduction, including physical and biotic variables. The only exceptions are loricariids (armored catfishes), several among them successfully bred in aquaria, such as *Hypostomus* spp. and *Liposarcus* spp. (Hypostominae), *Loricariichthys* spp. (Loricariinae), *Chaetostomus thomsoni*, *Hypancistrus zebra* and *Ancistrus* spp. (Loricariinae) (Secutti and Trajano 2009).

Reproductive Cycles

Studies on fish biology, including feeding and reproduction, are usually based on preserved samples. However, for most troglobites, sample sizes are limited by small population size and density and the general vulnerability of troglobites to disturbance (Trajano 2000, 2001). Hence, data on sex ratios and reproductive frequencies should ideally use less invasive methods, such as field observation on mature gonads seen by transparency (facilitated by the fish's pale skin).

In general, the size of the first individuals with mature gonads is half the maximum size recorded for the population. For instance, in *Ancistrus cryptophthalmus* and *Trichomycterus itacarambiensis*, the smallest females with developed gonads seen by transparency had, respectively, 40 mm and 43 mm SL, whereas maximum standard lengths recorded in the populations were around 80 mm and 83 mm SL. The same trend is observed for epigean species such as *Trichomycterus areolatus*: the minimum length of mature fishes was 56.7 mm for males and 51.1 for females, which corresponds to about half the maximum length that the species reaches (Manriquez et al. 1988).

Trajano (1987) compared the percentage of mature plus in maturation specimens in samples of the troglobitic *Pimelodella kronei* and epigean *P. transitoria* from the Alto Ribeira karst area, SE Brazil, captured during dry seasons: 43.5 percent in *P. kronei* from Areias Cave, 35 percent in *P. kronei* from Bombas Resurgence, and 94 percent in *P. transitoria*, indicating that the former reproduces less frequently than its epigean putative sister-species. In Areias Cave, a few mature individuals of *P. kronei* were observed throughout the year, which is consistent with the absence of a well-marked dry season in the area.

In tropical areas with pronounced seasonality, characterized by a well-marked dry season, difficulties to access most caves during rainy season, when the danger for speleologists to be caught by flashfloods is high, pose additional limitations to field studies. In such areas, biological and ecological data on subterranean organisms are mostly restricted to the dry season.

Seasonal reproduction, usually during the rainy season, has been reported for stream-dwelling troglobitic fishes living in these areas. Higher nutrient availability during this season of high waters would favor fat deposition and gonad maturation, with the production of yolk-rich eggs and the hatching of developed larvae.

A reproductive peak during the rainy season was recorded for *Trichomycterus itacarambiensis*, from Eastern Brazil, spawning mainly by the end of the rainy season, when the chances of the eggs and juveniles being washed away are decreased. It has been estimated that at least 50 percent of females in the population would reproduce every year (Trajano 1997a). Most females with developed gonads were pigmented; hence, it is possible that albino specimens reproduce less frequently or earlier during the rainy season. It is noteworthy that albinos present higher growth rates than pigmented individuals and seem to be more resistant to environmental stress (see below).

An extreme situation has been observed for *Pimelodella spelaea*, from the São Domingos karst area, Central Brazil. There was no specimen with developed gonads among 86 specimens captured during the dry season in 2000 (Trajano et al. 2004).

Catfishes we kept in the laboratory under nearly constant temperature (variation of 2–3°C along the year, as noted in many caves) were fed *ad libitum* once a week on average. Apparently, this feeding rate is higher than the observed in nature because all individuals were healthy, and no case of negative growth was recorded. Moreover, the proportion of individuals with developed gonads was well above the observed in caves. After a few weeks in the laboratory, all *P. kronei* specimens had mature gonads independently of the time of the year (Trajano 1987). This indicates that *P. kronei* is an opportunistic breeder adapting to a rather unpredictable environment, where rains may also occur during the "dry season" (May to September).

In contrast, in the laboratory, no specimen of *T. itacambiensis*, also fed *ad libitum*, showed developed gonads during the period corresponding with the dry season in the cave habitat (May–September). From November, some individuals had matured gonads, and the proportion of mature females (15.8 percent of the total number in captivity) was close to that observed in the cave habitat by the end of the rainy period (18 percent) (Trajano 1997a).

Spontaneous reproductive events observed in Passa Três Cave for *Ancistrus cryptophtalmus* (Secutti and Trajano 2009) and *Ituglanis passensis* (Trajano and Souza 1994; Secutti and Trajano 2021) were recorded at mid-wet season (January–February, i.e., summer peak). In the cave habitat, the proportion of mature females of *Ancistrus cryptophtalmus* was low in the dry season, varying from 1 to 7 percent per sampling occasion, whereas in Angélica Cave, no mature individual was found at the end of the wet season (March) (Trajano and Bichuette 2007). These data indicate that, for species living in highly seasonal and predictable habitats, endogenous factors regulate reproduction, which may be triggered by even small increases in water temperature.

Among phreatobic catfishes, most mature specimens of *Rhamdiopsis krugi* have been captured by the end of the wet season, but mature males were also found in the mid-dry season (Mendes 1995). This indicates prolonged reproductive cycles, including the dry season, with a peak at the beginning of the dry season. A prolonged reproductive season was also recorded in the laboratory for a population of *Trichomycterus* sp. 1 (*sensu* Cordeiro 2014) from Serra da Bodoquena, which lives in the upper phreatic zone. Three spontaneous reproductive events occurred from September to January, corresponding to the end of the dry season and the mid-wet season in the region (Cordeiro 2014). Such data indicate that phreatobic troglobites may present extended reproductive seasons, which do not necessarily follow the patterns observed for stream-dwellers.

Reproductive Behavior

Clear waters in many subterranean habitats, allied to the reduction of phobic reactions, especially photophobia, provide good opportunities for behavioral observation in the natural habitat. For this reason, there are several publications with data on spontaneous behavior (e.g., use of space, cryptobiotic habits, feeding behavior, and social interactions) and induced reactions to stimuli (luminous, mechanical, chemical) in troglobitic fishes (Parzefall and Trajano 2010). However, due to the relative rarity of reproductive events in many species, there are few published descriptions of reproductive behavior, including mating and parental care.

Courtship behavior in *Aspidoras mephisto* in the natural habitat was described in detail by Tencatt and Bichuette (2017): two individuals were observed close to each other when the smaller one (presumably male) started circling, the larger one (presumably female), gently touching it on the flanks, head and tail. A third individual approached, and the three individuals stopped parallel to each other. Then, the first male moved away, followed by the female. While the second male approached the couple, the first one went away, avoiding contact. The second male touched the flanks and tail of the female quickly, which also went away, avoiding mating. Although reproduction itself was not observed, according to R. McLure (pers. comm. in Tencatt and Bichuette 2017), the courtship behavior of *A. mephisto* would not differ from the observed in other corydoradines.

During pairing tests in the laboratory aiming to study the agonistic behavior of *Pimelodella kronei*, a few instances of putative reproductive behavior were observed; when similarly sized individuals were paired, a fully mature male bent sideways and convolved the other individual, gently squeezing its body in quick movements during ca. 30 sec (Trajano 1987).

In the summer of 1989/1990, some *Ituglanis passensis* specimens kept in the laboratory initiated reproductive behavior. Over several weeks, the male, already in the same den as the female, nervously pursued her away from the shelter, sometimes touching her anal region with its snout. This behavior continued even after the first juveniles appeared (February 1990), when both adults were transferred to another aquarium (Trajano and Souza 1994).

After several attempts to induce reproduction in the ictalurid *Prietella phreatophila*, through environmental manipulation and hormone administration, Hendrickson et al. (2001) eventually witnessed the reproductive behavior of a couple of individuals. Initially, they began associating much more closely than typical for this species, often staying in proximity or touching one another.

The presumed male had developed pronounced swelling of the dorsal head region, especially of the tissues covering the orbits. Large, cream-colored eggs were observed in the female. During eight days, the male flexed the posterior half of his body into a U-shaped arch and quivered as the female vigorously nuzzled, nipped, and pushed at his anal region with her mouth. This behavior lasted 18–23 seconds and was repeated at 10–30 minute intervals. Upon completing this sequence, the specimens were separated and rested motionless on the substrate. Actual reproduction was not observed, and although the female belly seemed less distended, no eggs were found.

Reproduction, particularly reproductive behavior, is a conservative trait. This is exemplified by the Mexican *Astyanax* characids. Although several troglomorphic populations are still interfertile with epigean congeners, originating viable hybrid populations (Wilkens and Strecker 2017), these cave populations correspond to suitable diagnosable species based on a series of autapomorphies (Trajano 2021). Therefore, it is probable that many troglobitic catfishes retain the behavior pattern of their epigean relatives.

Fecundity and Parental Care

Fecundity values of *Trichomycterus chaberti,* from Bolivia, showed an increased trend in the number of eggs from subterranean, headwaters, and valley populations. An opposite trend was observed for egg size, with significant differences between the two extreme groups (Pouilly and Miranda 2003). In contrast, egg size and fecundity in the Brazilian *Ituglanis passensis* were within the variation observed in epigean congeners. Egg laying in this species observed in the laboratory was parceled (Secutti and Trajano 2021), i.e., females have more than one spawning during the breeding season.

No differences in yolk content were detected for *Rhamdia reddelli* and *R. zongolicensis* when compared to epigean close relative *R. laticauda*. On the other hand, egg yolk content in *R. laticauda* is double that in *R. guatemalensis*, producing larger larvae with longer barbels. It is noteworthy that all troglobitic *Rhamdia* species belong to the *R. laticauda*-group, which would be preadapted to the subterranean life in opposition to *R. guatemalensis* (Wilkens 2001).

As mentioned, a group of *Ancistrus cryptophthalmus*, composed of one male and three females from Passa Três Cave, reproduced spontaneously twice in the laboratory in consecutive years. The eggs (diameter around 4 mm) were attached to the ceiling of a den formed by limestone rocks. This egg size is within the range recorded for several loricariids and approaches the observed for other ancistrines. As typical of many loricariids, the eggs were tended by the male, which stayed at the shelter entrance, defending the egg clutch, cleaning, and fanning the eggs with the fins for oxygenation (Secutti and Trajano 2009).

An interesting case of putative nursery cave, used for reproduction and growth, was found in *Trichomycterus* catfishes from Serra da Bodoquena, SW Brazil. A population formed basically by juveniles was found in a vadose cave, ca. 150 m above the regional base level. Low recapture rates, higher density of individuals near a phreatic crevice, and long-distance movements observed indicate that this is an open population with a high proportion of migrants. This indicates that this species, typically pheatobic, uses the vadose cave for safer reproduction, migrating vertically through crevices and cracks towards their definitive habitat (Cordeiro 2014). Segregation of habitat between juveniles and adults has been observed in other trichomycterids, such as *Trichomycterus areolatus*, *T. chiltoni* and *Bullockia maldonadoi* (Arratia 1983).

Development and Eye Regression

Since the 1950s, it has been known that troglobitic vertebrates with anophthalmic adults, such as the Mexican tetra characin *Astyanax*, and the East European salamander, *Proteus anguinus*, are born with relatively large eyes, which regress along ontogeny. The degeneration process in troglobitic *Astyanax* has been studied in great detail with a focus on development and genetics (Wilkens and Strecker 2017). Eye regression starts just after the larvae hatch, being achieved in a matter of days

(Wilkens 1988). Likewise, a fast eye regression, completed in less than one month, was described for the highly troglomorphic cyprinid *Phreatichthys andruzzi* from Somalia (Berti et al. 2001).

Early development has been studied in a few troglobitic catfishes. In these cases, larvae are born with eyes that regress along ontogeny. Eyed juveniles observed in the cave habitat also corroborate this phenomenon for siluriforms.

The Mexican *Rhamdia zongolicensis* and *R. reddelli* were the first troglobitic siluriforms studied with a focus on the regression of troglomorphic traits. Their eyes are smaller than those of the epigean relative *Rhamdia laticauda* from the beginning of ontogeny, but they present all single components. Gross differences are observed in the lenses, which are subject to variability. Because of the early ontogenetic existence of all single structures in *Rhamdia* and the structural left-right asymmetry between the eyes of many adults, it is quite possible that eye reduction is due mainly to the mutation of developmental genes and not structural ones (Wilkens 2001).

Eye development in Mexican *Rhamdia* catfishes proceeds much slower than in the intensively studied Mexican *Astyanax* (Wilkens 2001), considered the model for regressive evolution in subterranean fishes. The same is true for *Ancistrus cryptophthalmus* from Passa Três Cave (Figure 2) and *Ituglanis passensis*, indicating a pattern, at least for less specialized troglobitic catfishes.

The hatching larvae of *Ancistrus cryptophthalmus* from Passa Três cave are relatively large but present low growth rates from the beginning of the exogenous feeding stage. The average growth rates during the first years in the laboratory were estimated at 2.3 mm month^{-1} in the first six months of life, decreasing and keeping constant around 0.7–0.8 mm month^{-1} in the subsequent four semesters (Secutti and Trajano 2009). Low rates were also recorded in the natural habitat for individuals larger than 30 mm SL (probably one year old) during the dry season (Trajano and Souza 1994).

The eyes of adult *A. cryptophthalmus* are not visible externally. In specimens born in the laboratory, the eye regression started quite late during the ontogeny. By the end of the first semester of life (size around 26 mm total length – TL), a small increment in the eye was observed, followed by a long time of structural stasis (ca. 270 days); only then, the eye regression started, consisting of two phases. Initially, the only modification was the reduction of the dark pigmentation of the eyes. After several months (fish size ca. 45 mm TL, on average), the disorganization of the eye structure started: the typical circular shape changed into a horseshoe shape, and the eyes gradually sank into the ocular cavities, acquiring an elliptical shape. This second phase was clearly recognizable in juveniles ca. 380 days old. Concomitantly, folds protruded from the walls of the ocular cavity, eventually covering the regressing eyes; apparently, a centripetal growth of bones in the circumorbital region contributed to close (but not completely) the orbits. The orbits ended as small depressions covered by skin typical of the population from the Passa Três population (São Domingos kars area), as observed in 2-year-old individuals. Heterogeneous growth started in 100-days-old individuals (Secutti and Trajano 2009).

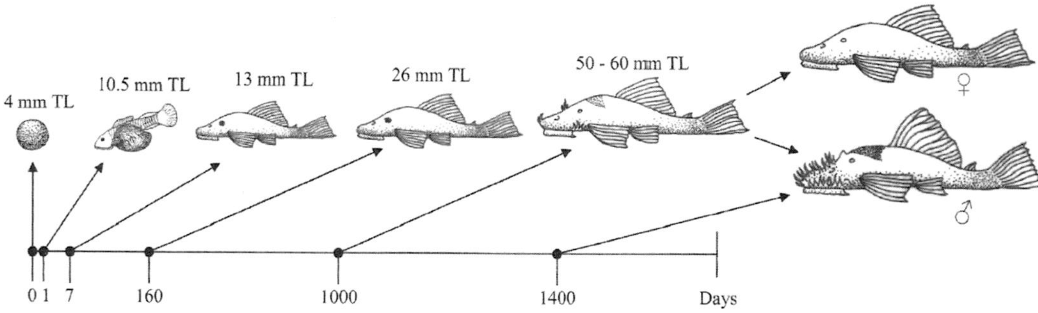

Figure 2. Summary of the life cycle of *Ancistrus cryptophthalmus* from Passa Três Cave, São Domingos karst area, Central Brazil (from Secutti and Trajano 2009).

Ituglanis passensis is characterized by reduced eyes, varying from vestigial black spots, symmetrical or not, to none externally visible. In laboratory-born specimens, eye regression started late, one year after birth, and following a pattern of stasis phases intercalated with slow growth and fluctuating asymmetric rates, i.e., variable degenerating rates affecting alternatively right and left eye of the same individual along ontogeny (Secutti and Trajano 2021).

The initial average growth rate was 1.24 mm month^{-1}, with significant individual variation. From the 40th day on, there was a considerable growth slowdown, but without perceptive eye regression. When the longest living juvenile reached ca. 15 mm SL (day 180), both eyes started presenting downward movements, as observed among fishes in general; low body growth rates persisted until day 370, followed by an acceleration until day 600, and finally deceleration. This configures an S-shaped growth curve in the first two years instead of a C-shaped one. In the second year, the increase in eye-size in relation to SL growth was slower. Furthermore, eyes became asymmetrical, with the right eye beginning to grow faster; this tendency reverted around the 550th day. Less than one month later, the left eye presented a small change in structure, losing the typical round-shape. This would be the result of fluctuating asymmetry in eye-size, observed from day 350 on, without dominance in size in either the left or the right eye (Secutti and Trajano 2021), contrary to the observed for *T. chaberti*, in which each population has a dominant one (Pouilly and Miranda 2003)

It is noteworthy that, in contrast with the stream-dwelling, moderately specialized troglobitic *Rhamdia* spp., *A. cryptophthalmus*, and *I. passensis*, an early eye-regression was observed for phreatobic *Trichomycterus* from Serra da Bodoquena, SW Brazil (Cordeiro and Trajano, unpublished data).

Growth Rates and Longevity

Mean longevity in the natural habitat may be estimated using growth rates obtained in MR studies. This has been done for some species; however, gathering field data may be limited by the need to return repeatedly to localities of difficult access and long hours of work under unsafe conditions. Examples with references are in the following paragraphs. Moreover, especially in the case of phreatobic species, only part of the habitat may be reached. In tropical regions, where seasons are not as well-delimited as in the temperate ones, examination of growth rings in otoliths and scales is not reliable. For these reasons, most data are based on fish survival in the laboratory under controlled conditions favoring such survival (e.g., exclusion of predators, *ad libitum* feeding). Such data must be regarded as potential maximum longevity.

In *Pimelodella kronei*, growth rates in the cave habitat for individuals with more than 80 mm SL were estimated at 1 mm month^{-1} between March and October; individuals captured at the cave entrance presented higher growth rates, as expected due to the higher food availability. For the same reason, growth rates of *P. kronei* catfish kept in a laboratory and fed *ad libitum* on average once a week were higher than deep inside the cave (Trajano 1991).

Mean longevities of epigean *Pimelodella* species estimated through MR techniques were 6.8 year for *P. transitoria* (Gerhard 1999) and 4–6.5 year for *P. pappenheimi* (Duboc 2003). A maximum longevity of 7 year in *P. transitoria* was estimated using otoliths (Yaguiu and Trajano, unpublished data). Therefore, longevity estimated at 10–15 year for *P. kronei* (Trajano1991) may be considered high when compared to congeners also studied in the field using MR techniques.

Lower growth rates have been recorded for *Trichomycterus itacarambiensis*, which lives in a cave situated in a highly seasonal area that is flooded during part of the rainy season. In opposition, *P. kronei* localities are not subject to a typical dry season. For this reason, *T. itacarambiensis* individuals move more and for longer distances than the latter, at least during the dry season. Mean growth rates for individuals with more than 30.0 mm SL, estimated in the dry season of 1994, were 0.06 mm month^{-1}; for the 30.0 to 45.0 mm size class (immature), these rates were three times higher (0.17 mm month^{-1}) and the rates for the set of the DOPA (–) individuals was also higher (0.09 mm month^{-1}) than those for the whole studied population. Maximum sizes recorded in the cave habitat were around 80 mm SL, and longevity was estimated at 7+ year (Trajano 1997b).

Adult *T. itacarambiensis* specimens fed *ad libitum* in the laboratory had grown, on average, 1.2 mm month^{-1} (varying from zero to 3.5 mm month^{-1}) after one year. As expected, smaller individuals, 30.0 to 40.0 mm SL, grew faster, 2.1 mm month^{-1} on average. Albino individuals seem to be more specialized to the cave life than pigmented ones: they grow faster in the cave habitat, live longer, at least in the laboratory (E. Trajano pers. obs.), and seem to be more resilient to environmental stress, such as progressive decrease of flash floods, which brings nutrients throughout the cave, observed in the last decades (Trajano et al. 2009).

In São Domingos karst area, central Brazil, also characterized by a seasonal climate, low growth rates were recorded for *Ancistrus cryptophthalmus* in the dry season, 0.5 mm month^{-1} on average; 0.7 mm month^{-1} for individuals smaller than 42 mm SL, and 0.35 mm month^{-1} for larger individuals, with cases of negative growth (Trajano and Bichuette 2007). Longevity was estimated at 8–10 year in the cave habitat and 20 year in the laboratory, with particularly slow growth in late life stages (Secutti and Trajano 2009). Growth rates of *Ituglanis passensis* varied from −0.35 to +1.62 mm month^{-1} (mean = 0.9 mm month^{-1}); negative growth, as well as the smallest increases in length, were recorded for larger individuals.

Most specimens of *Pimelodella spelaea*, also from the São Domingos area and examined in the dry season in 2000, were skinny, and some were severely underfed. Growth rates varied from −0.7 to +0.2 mm month^{-1} (mean = 0.17 mm month^{-1}), and all but one recaptured fish lost weight (differences in weight ranged from 0 to 20.7 g), consistently with the hypothesis of feeding stress during the dry season (Trajano et al. 2004).

Several epigean teleost fishes exhibit heterogeneous growth (i.e., significant variation in growth rates among individuals of the same age and from the same habitat), especially under confinement, such as in aquiculture. These differences would be mostly due to growth reduction in most individuals, which is referred to as growth suppression (Volpato and Fernandes 1994).

Heterogeneous growth with cases of negative growth (SL at the recapture lower than in the previous capture) was observed in the field and the laboratory for all studied troglobitic species, including larvae and juveniles, and also for the epigean *P. transitoria* (ca. 10% of recaptures; Gerhard 1999). This number was particularly high for species living in highly seasonal areas, indicating strong feeding stress during the dry season. Volpato and Fernandes (1994) suggested that heterogeneous growth is an adaptive strategy to optimize the survival of the population in a restricted space. This would be the case for fish in aquiculture and also in subterranean habitats. However, social control of growth alone does not explain heterogeneous growth because it has been observed for larvae and juveniles of *Ancistrus cryptophthalmus*, *Ituglanis passensis*, and *Trichomycterus* sp. 1 from Serra da Bodoquena, born in the laboratory and kept isolated, under the same conditions of space and food availability. Likewise, genetic differences do not account for variation in growth rates of larvae and juveniles because they shared at least one parent (*A. cryptophthalmus* and *I. passensis*) or both (*Trichomycterus* sp. 1).

High longevities compared to similarly sized epigean relatives have also been recorded under laboratory conditions. For instance, *Ituglanis passensis* and *I. bambui* individuals survived for 20 years in captivity (Bichuette and Trajano 2021). Observations on *Aspidoras mephisto* also indicate a high life expectancy: one individual lived for nine years in the laboratory, and another for seven years. Both were preserved for their species description; therefore, they could have survived longer. Since both were adults when captured, it is reasonable to infer a life span of 10+ years, at least in the laboratory (Tencatt and Bichuette 2017), which is high longevity for a small fish (maximum SL around 45 mm).

Particularly high longevity, up to 30 years, has been estimated for the highly specialized phreatobic *Rhamdiopsis* from Toca do Gonçalo (NE Brazil), based on the initial and final size of specimens kept in the laboratory for several years. Two of these individuals, respectively with 15–20 mm SL and around 40 mm SL when captured, died in the laboratory after 17–18 years, with

40.7 mm and SL 58.4 mm SL. They were fed with *Artemia* crustaceans *ad libitum*, on average once a week; such food intake is probably higher than in the highly oligotrophic natural habitat. Based on their initial size and growth rate in the laboratory, it is reasonable to suppose that the larger catfish was at least 10 years old when captured and the smaller one was five years old. Therefore, a lifespan of 25–30 years is a reasonable assumption (Trajano and Bockmann, unpublished data).

Conclusions

Among the eleven fish orders with troglobitic species, Siluriformes is the second in taxonomic diversity (90 described and undescribed species versus 140 for Cypriniformes) and the first in phylogenetic diversity (11 families and 20 genera versus four families and 33 genera in Cypriniformes). The highest diversity of troglobitic catfishes is observed in the Neotropical region, mainly Mexico and Brazil. Trichomycteridae and Heptapteridae are the richest families in troglobitic species (33 and 24 described and non-described species, respectively) and also in troglophilic populations. This is probably a consequence of the high epigean diversity of siluriforms in these regions as potential colonizers of subterranean habitats.

Subterranean fishes may occupy different habitats within the vertical zonation observed in karst areas, from permanent perched aquifers in the epikarst, to vadose and base-level streams to the upper and deep phreatic zone, presenting specific adaptations to life in these distinctive habitats. Most stream-dwelling troglobitic catfishes are endemic to a single karst system or different micro-basins in a relatively small area. On the other hand, phreatobic species tend to have a wider geographic distribution.

Troglobitic catfishes show broad intra- and interspecific variability in troglomorphisms, such as eye development and tegumentary dark pigmentation, from slightly but significantly reduced to very reduced or absent eyes and pigmentation, with all degrees of variation between these extremes, frequently with a mosaic distribution of character states and without taxonomic correlation. A correlation with habitat is observed for several traits: phreatobic species tend to exhibit a higher degree of troglomorphism than stream-dwellers, wide geographic distributions, prolonged reproductive seasons, higher longevity, and possibly an early eye regression.

Reproductive cycles have been recorded for several troglobitic catfishes, with peaks usually related to food availability, which is higher in the rainy season. In species living in highly seasonal and predictable habitats, endogenous factors would regulate reproduction, which may be triggered by even small increases in water temperature. No differences in reproductive behavior in relation to epigean relatives have been detected for the few studied troglobitic siluriforms.

Low growth rates from early ages and high longevity estimated for all studied catfishes are characteristic of K-selected, or precocial life cycles, as an adaptation to food-limited habitats, at least during part of the annual cycle. However, in the absence of comparable data for epigean close relatives, it is not possible to conclude unambiguously that these trends are troglomorphisms of the subterranean species or preadaptations (in the sense of exaptations) of the epigean ancestors that favored the colonization of subterranean habitats.

Heterogeneous growth, a common phenomenon among epigean fishes, was recorded for all studied troglobitic catfishes, including a significant number of instances of negative growth, also related to feeding stress. Therefore, body size is not a reliable indicator of fish age, and individual growth rates must be estimated on a case-by-case basis.

Late eye regression, associated with asymmetry in eye development and intra-population variability of troglomorphic traits, which frequently present a mosaic of character states, provides support for the Neutral Mutation Theory (accumulation of deleterious mutations under reduced selection pressure) to explain character regression in cave fishes (Secutti and Trajano 2021).

References

Aquilina, L., C. Stumpp, D. Tonina and J.M. Buffington. 2023. Hydrodynamics and geomorphology of groundwater environments. pp. 3–37. *In*: Malard, F., C. Griebler and S. Retaux (eds.). Groundwater Ecology and Evolution (second edition). Elsevier Academic Press, London.

Arratia, G. 1993. Preferencias de habitats de peces Siluriformes de aguas continentales de Chile (Fam. Diplomystidae and Trichomycteridae, Stud. Neotrop. Fauna Environm. 18: 1[illegible]7–237.

Balon, E.K. 1981. Saltatory processes and altricial to precocial forms in the ontogeny of fishes. Am. Zool. 21: 573–596.

Barr, T.C., Jr. 1968. Cave ecology and the evolution of the troglobites. Evol. Biol. 2: 35–102.

Berti, R., J.P. Durand, S. Becchi, R. Brizzi, N. Keller and G. Ruffat. 2001. Eye degeneration in the blind cave-dwelling fish *Phreatichthys andruzzii*. Can. J. Zool. 79: 1278–1285. https://doi.org/10.1139/z01-084.

Bichuette, M.E., B. Rantin, E. Zingst-Zaher and E. Trajano. 2015. Geometric morphometrics throws light on evolution of the subterranean catfish *Rhamdiopsis krugi* (Teleostei: Siluriformes: Heptapteridae) in eastern Brazil. Biol. J. Linn. Soc. 114: 136–151.

Bichuette, M.E. and E. Trajano. 2021. Monitoring Brazilian cavefish: ecology and conservation of four threatened catfish of genus *Ituglanis* (Siluriformes: Trichomycteridae) from Central Brazil. Diversity 13: 9. https:// doi.org/10.3390/d13020091.

Binoy, V.V., K.S. Roshan and V.B. Rakesh. 2012. Occurrence of *Kryptoglanis shajii*, an enigmatic subterranean-spring catfish (Siluriformes, Incertae sedis) in the channels of paddy fields. Current Science 102: 161.

Bockmann, F.A. and R.M.C. Castro. 2010. The blind catfish from the caves of Chapada Diamantina, Bahia, Brazil (Siluriformes: Heptapteridae): description, anatomy, phylogenetic relationships, natural history, and biogeography. Neotrop. Ichthyol. 8: 673–706.

Buss, D.M., M.G. Haselton, T.K. Shackelford, A.L. Bleske and J.C. Wakefield. 1998. Adaptations, exaptations, and spandrels. Am. Psychol. 53: 533.

Cordeiro, L.M. 2014. Distribuição, ecologia e filogeografia dos bagres troglóbios do gênero *Trichomycterus* na área cástica da Serra da Bodoquena, MS. Ph.D. Thesis, Universidade de São Paulo, São Paulo, Brazil. [Distribution, ecology and phylogeography of troglobitic catfishes, genus *Trichomycterus*, from the Serra da Bodoquena karst area, MS].

Culver, D.C. and T. Pipan. 2019. The Biology of Caves and Other Subterranean Habitats. Oxford University Press. Oxford, UK. Second edition.

Dahanukar, N., R.L. Sundar, D. Rangad, G. Proudlove and R. Raghavan. 2023. The world's largest cave fish from Meghalaya, Northeast India, is a new species, *Neolissochilus pnar* (Cyprinidae, Torinae). Vertebr. Zool. 73: 141–152. DOI 10.3897/vz.73.e101011.

DoNascimiento, C., F. Provenzano and J.G. Lundberg. 2004. *Rhamdia guasarensis* (Siluriformes: Heptapteridae), a new species of cave catfish from Sierra de Perijá, northwestern Venezuela. Proc. Biol. Soc. Wash. 117: 564–574.

Duboc, L.F. 2003. Ecologia de bagres heptapterídeos no Rio Morato, Guaraqueçaba – PR (Siluriformes: Heptapteridae). Ph.D. Thesis, Universidade de São Paulo, São Paulo, Brazil. [Ecology of heptapterid catfishes in Rio Morato, Guaraqueçaba - PR (Siluriformes: Heptapteridae).

Flórez, J.S., C.D. Cadena, C. DoNascimento and M. Torres. 2021. Repeated colonization of caves leads to phenotypic convergence in catfishes (Siluriformes: *Trichomycterus*) at a small geographical scale. Zool. J. Linn. Soc. 193: 772–788.

Ford, D. and P. Williams. 2007. Karst Hydrology and Geomorphology. John Wiley & Sons, Chichester, UK.

Fricke, R., W.N. Eschmeyer and J.D. Fong. 2024. Eschmeyer›s Catalog of Fishes: genera/species by family/subfamily. (http://researcharchive.calacademy.org/research/ichthyology/catalog/SpeciesByFamily.asp). Electronic version accessed Jan 2024.

Gerhard, R. 1999. Ecologia de populações e comportamento de quatro espécies de bagres Heptapterinae em riachos do Alto Vale do rio Ribeira, São Paulo (Teleostei: Siluriformes). M.Sc. Thesis, Universidade de São Paulo, São Paulo, Brazil. [Population ecology and behavior of four Heptapterinae catfish species in streams from the Upper Ribeira Valley, São Paulo].

Ginet, R. and V. Decou. 1977. Initiation à la biologie et à l'écologie souterraines. Jean Pierre Delarge, Paris, France.

Grempel, R.G., E. Trajano and M.A. Visconti. 2020. Regression of dark color in subterranean fishes involves multiple mechanisms: response to hormones and neurotransmitters. Neotrop. Ichthyol. 18: e190120. doi.org/10.1590/1982-0224-2019-0120.

Hendrickson, D.A., J.K. Krejca and J.M.R. Martinez. 2001. Mexican blindcats genus *Prietella* (Siluriformes: Ictaluridae): an overview of recent explorations. Environm. Biol. Fish. 62: 315–337.

Howarth, F.G. and O.T. Moldovan. 2018. The ecological classification of cave animals and their adaptations. pp. 41–67. *In*: Moldovan, O.T., L. Kovač and S. Halse (eds.). Cave Ecology. Ecological Studies 235. Springer Nature, Switzerland.

Jones, W.K., D.C. Culver and J.S. Herman (eds.). 2004. Epikarst. Karst Waters Inst. Spec. Publ. 9: 1–160.

Langecker, T.G. and G. Longley. 1993. Morphological adaptations of the Texas blind catfishes *Trogloglanis pattersoni* and *Satan eurystomus* (Siluriformes: Ictaluridae) to their underground environment. Copeia 1993: 976–986.

Longley, G. and H. Karnei, Jr. 1978. Status of *Trogloglanis pattersoni* Eigenmann, the Toothless Blindcat. Scientific Report, U.S. Fish and Wildlife Service, Albuquerque, USA.

Manriquez, A., L. Huaquín, M. Arellano and G. Arratia. 1988. Aspectos reproductivos de *Trichomycterus areolatus* Valenciennes, 1846 (Pisces: Teleostei: Siluriformes) en Rio Angostura, Chile. Stud. Neotrop. Fauna Environm. 23: 89–102. doi: 10.1080/01650528809360749.

Mattox, G.M.T., M.E. Bichuette, S. Secutti and E. Trajano. 2008. Surface and subterranean ichthyofauna in the Serra do Ramalho karst area, northeastern Brazil, with updated lists of Brazilian troglobitic and troglophilic fishes. Biota Neotropica 2008: 145–152.

Mendes, L.F. 1995. Ecologia populacional e comportamento de uma nova espécie de bagres cavernícolas da Chapada Diamantina, BA (Siluriformes, Pimelodidae). M.Sc. thesis, Universidade de São Paulo, São Paulo, Brazi. [Population ecology and behavior of a new species of cave catfishes from Chapada Diamantina, BA (Siluriformes, Pimelodidae)].

Parzefall, J. and E. Trajano. 2010. Behavioral patterns in subterranean fishes. pp. 81–114. *In*: Trajano, E., M.E. Bichuette and B.G. Kapoor (eds.). Biology of Subterranean Fishes. Science Publishers, Enfield, UK. https://doi.org/10.1201/ebk1578086702-c4.

Pouilly, M. and G. Miranda. 2003. Morphology and reproduction of the cavefish *Trichomycterus chaberti* and the related epigean *Trichomycterus* cf. *barbouri*. J. Fish Biol. 63: 490–505. doi:10.1046/j.1095-8649.2003.00171.x, a.

Poulson, T.L. 1963. Cave adaptation in amblyopsid fishes. Am. Midl. Natur. 70: 259–290.

Poulson, T.L. 1966. The Life of the Cave. McGraw-Hill, NY, USA.

Poulson, T.L. 1969. Population size, density, and regulation in cave fishes. Actes IVe Congrès Internationale de Spéologie Yugoslavia 4–5: 189–192.

Proudlove, G. 2023. Subterranean Fishes of the World. An account of the subterranean (hypogean) fishes with a bibliography from 1436. https://cavefishes.org.uk/checklist.php?type=cave.

Racovitza, E.G. 1907. Essai sur les problems biospéologiques. Archives de Zoologie Expérimentale et Générale 4(6): 371–400. *Fac símile In*: X. Bellés. 2004. Assaig sobre els Problems Bioespeleològics. Institut D´Estudis Catalans, Barcelona, Spain.

Raghavan, R., R.L. Sundar, C.P. Arjun, R. Britz and N. Dahanukar. 2023. Evolution in the dark: Unexpected genetic diversity and morphological stasis in the blind, aquifer-dwelling catfish *Horaglanis*. Vertebrate Zoology 73: 57–74 | DOI 10.3897/vz.73.e98367.

Reis, R.E., E. Trajano and E. Hingst-Zaher. 2006. Shape variation in surface and cave populations of the armoured catfish *Ancistrus* (Siluriformes: Loricariidae) from the São Domingos karst area, Upper Tocantins River, Brazil. J. Fish Biol. 68: 414–429. DOI 10.1111/j.0022-1112.2006.00891.x.

Rizzato, P.P., E.P.D. Costa Jr., E. Trajano and M.E. Bichuette. 2011. *Trichomycterus dali*: a new highly troglomorphic catfish (Silurifomes: Trichomycteridae) from Serra da Bodoquena, Mato Grosso do Sul State, Central Brazil. Neotrop. Ichthyol. 9: 477–491.

Secutti, S. and E. Trajano. 2009. Reproductive behavior, development and eye regression in the cave armored catfish, *Ancistrus cryptophthalmus* Reis, 1987 (Siluriformes: Loricariidae), breed in laboratory. Neotrop. Ichthyol. 2009: 479–490.

Secutti, S. and E. Trajano. 2021. Reproduction, development, asymmetry and late eye regression in the Brazilian cave catfish *Ituglanis passensis* (Siluriformes, Trichomycteridae): evidence contributing to the neutral mutation theory. Subt. Biol. 38: 91–112. doi: 10.3897/subtbiol.31.60691.

Tencatt, L.F.C. and M.E. Bichuette. 2017. *Aspidoras mephisto*, new species: The first troglobitic Callichthyidae (Teleostei: Siluriformes) from South America. PLoS ONE 12: e0171309. doi:10.1371/journal.pone.0171309.

Trajano, E. 1987. Biologia do bagre cavernícola, *Pimelodella kronei,* e de seu provável ancestral, *Pimelodella transitoria* (Siluriformes, Pimelodidae). Ph.D. Thesis, Universidade de São Paulo, São Paulo, Brasil. [Biology of the cave catfish, *Pimelodella kronei,* and its putative ancestor, *Pimelodella transitoria* (Siluriformes, Pimelodidae)].

Trajano, E. 1991. Population ecology of *Pimelodella kronei*, troglobitic catfish from Southeastern Brazil (Siluriformes, Pimelodidae). Environm. Biol. Fish. 30: 407–421.

Trajano, E. and A.M. Souza. 1994. The behavior of *Ancistrus cryptophthalmus*, an armoured blind catfish from caves of Central Brazil, with notes on syntopic *Trichomycterus* sp. (Siluriformes, Loricariidae, Trichomycteridae). Mém. Biospéol. 21: 151–159.

Trajano, E. 1995. Evolution of tropical troglobites: Applicability of the model of Quaternary climatic fluctuations. Mém. Biospéol. 22: 203–209.

Trajano, E. and M.C.C. de Pinna. 1996. New cave species of *Trichomycterus* from Eastern Brazil (Siluriformes, Trichomycteridae). Rev. fr. d'Aquariol. 23: 85–90.

Trajano, E. 1997a. Food and reproduction of *Trichomycterus itacarambiensis*, cave catfish from south-eastern Brazil. J. Fish Biol. 51: 53–63.

Trajano, E. 1997b. Population ecology of *Trichomycterus itacarambiensis*, a cave catfish from eastern Brazil (Siluriformes, Trichomycteridae). Environm. Biol. Fish. 50: 357–369.

Trajano, E. 2000. Cave faunas in the Atlantic tropical rain forest: Composition, ecology and conservation. Biotropica 32: 882–894.

Trajano, E. 2001. Ecology of subterranean fishes: an overview. Environm. Biol. Fish. 62: 133–160.

Trajano, E. 2003. Ecology and ethology of subterranean catfishes. pp. 601–635. *In*: Arratia, G., B.G. Kapoor, M. Chardon and R. Diogo (eds.). Catfishes, v. 2. Science Publishers, Enfield, UK.

Trajano, E., R.E. Reis and M.E. Bichuelle. 2004. *Pimelodella spelaea*: A new cave catfish from Central Brazil, with data on ecology and evolutionary considerations (Siluriformes: Heptapteridae). Copeia 2004: 315–325.

Trajano, E. and M.E. Bichuette. 2007. Population ecology of cave armoured catfish, *Ancistrus cryptophthalmus* Reis 1987, from central Brazil (Siluriformes: Loricariidae). Ecol. Freshw. Fish 16: 105–115.

Trajano, E., S. Secutti and M.E. Bichuette. 2009. Population decline in a Brazilian cave catfish, *Trichomycterus itacarambiensis* Trajano & Pinna, 1996 (Siluriformes): reduced flashflood as a probable cause. Speleobiol. Notes 1: 24–27.

Trajano, E. and M.E. Bichuette. 2010. Subterranean fishes of Brazil. pp. 331–355. *In*: Trajano, E., M.E. Bichuette and B.G. Kapoor (eds.). Biology of Subterranean Fishes. Science Publ., Enfield, UK.

Trajano, E. 2012. Ecological classification of subterranean organisms. *In*: White, W.B. and D.C. Culver (eds.). Encyclopedia of Caves. Academic Press, Waltham, USA.

Trajano, E. and M.R. Carvalho. 2017. Towards a biologically meaningful classification of subterranean organisms: a critical analysis of the Schiner-Racovitza system from a historical perspective, difficulties of its application and implications for conservation. Subt. Biol. 22: 1–26. DOI 638 10.3897/subtbiol.22.9759.

Trajano, E. 2021. Diversity of Brazilian troglobitic fishes: Models of colonization and differentiation in subterranean habitats. Diversity 13: 106. https://doi.org/10.3390/ d13030106.

Volpato, G.L. and M.O. Fernandes. 1994. Social control of growth in fish. Braz. J. Med. Biol. Res. 27: 797–810.

Weber, A., G. Allegrucci and V. Sbordoni. 2003. *Rhamdia laluchensis*, a new species of troglobitic catfish Siluriformes: Pimelodidade) from Chiapas, Mexico. Ichthyol. Explor. Freshw. 14: 273–280.

Wilkens, H. 1988. Evolution and genetics of epigean and cave *Astyanax fasciatus* (Characidae: Pisces). Evol. Biol. 23: 271–367.

Wilkens, H. 2001. Convergent adaptations to cave life in the *Rhamdia laticauda* catfish group (Pimelodidae, Teleostei). Environm. Biol. Fish. 62: 251–261.

Wilkens, H. and U. Strecker. 2017. Evolution in the Dark. Darwin's Loss without Selection. Springer, Berlin, Germany.

Chapter 13

Sound Production and Communication in Catfishes

Ingrid M. Kaatz,[1,*] *Phillip S. Lobel*[2] and *Aaron N. Rice*[3]

Catfishes are among the most diverse and highly vocal fish orders (Rice et al. 2022). Due to their widespread distribution, abundance and diversity in aquatic habitats around the world, combined with their amenability to vocalize in captive settings, catfishes have been among the intensively studied groups of taxa of fish for acoustic communication (e.g., Tavolga 1960; Fine et al. 1997; Ladich 2001). Catfishes are among the earliest vocal fishes discovered (Geoffroy St. Hilaire 1827; Dufossé 1874; Darwin 1938), predating the invention of the hydrophone. Thus, much of the early work on catfish vocalizations was focused on the underlying functional morphology responsible for sound production (Haddon 1881; Bridge and Haddon 1893; Sörenson 1894–1895). The development and availability of hydrophones led to the earliest underwater recordings of catfish sounds in marine habitats (e.g., Dobrin 1947; Fish 1948; Breder 1968; Fish and Mowbray 1970).

The study of sound production in catfishes has provided the opportunity to examine acoustic communication across hierarchical levels of biology and examine a wide range of studies. The literature review on catfish sound included 139 publications covering the following subjects: video or other sound recordings, human hearing and direct observations, morphological correlates of vocal mechanisms, physiology, and neurobiology. Fundamentally, studies in catfish sound production have established catfishes as an excellent taxonomic system for vocal behavior and mechanisms of acoustic communication.

A vocal catfish census to date shows that more than 150 species in 24 families produce sound based on the ability to produce observable or recordable sounds (Table 1). Additional taxa suggest evidence of sound production inferred solely from vocal morphology (Figure 1). This chapter will consider the morphological, behavioral and ecological adaptations of vocal catfishes.

[1] 40 Apple Valley Road, Stamford, CT 06903, USA.

[2] Department of Biology, Boston University, 5 Cummington St. Boston, MA 02215, USA.

[3] K. Lisa Yang Center for Conservation Bioacoustics Cornell Lab of Ornithology, Cornell University, 159 Sapsucker Woods Rd, Ithaca, New York 14850, USA.

Emails: plobel@bu.edu; arice@cornell.edu

* Corresponding author: imkaatz@yahoo.com

Table 1. Catfishes that have exhibited vocal behavior in the literature, with taxonomy reflecting currently recognized nomenclature (following Fricke et al. 2024). Species bolded are supported by results from I. Kaatz (unpublished).

Family	**Vocal Species** (Total Number of Species Within the Family)	**Relevant Citation** Current Examples
TRICHOMYCTERIDAE	***Bullockia maldonadoi***	G. Arratia pers. com.
CALLICHTHYIIDAE	(27) *Callichthys callichthys*; *Corydoras adolfoi*; *C. aeneus*; *C. agassizii*; *C. arcuatus*; *C. equus*; *C. gossei*; *C. imitator*; *C. melanotaenia*; *C. melini*; *C. metae*; *C. multiradiatus*; *C. napoensis*; *C. ornatus*; *C. paleatus*; *C. punctatus*; *C.* ***rabauti***; *C. reticulatus*; *C. robineae*; *C. simulates*; *C. sodalist*; *C.* ***sterbai***; *C.* ***sychri***; *C. virginiae*; *Dianema urostriatum*; *Megalechis thoracata*; ***Scleromystax barbatus***	Heyd and Pfeiffer 2000; Kaatz et al. 2010; Hadjiaghai and Ladich 2015
LORICARIIDAE	(11) *Dekeyseria picta*; *Hypostomus commersoni*; *Hypostomus derbyi*; ***Hypostomus plecostomus***; *Otocinclus affinis*; *Paraloricaria vetula*; *Pterygoplichthys gibbiceps*; *Pterygoplichthys pardalis*; *Ricola macrops*; *Rineloricaria uracantha*; *Panaqolus maccus*	Tellechea et al. 2013; Vallone et al. 2017; Webb McAdams et al. 2023
CHACIDAE	*Chaca chaca*	Kaatz et al. 2010
PLOTOSIDAE	*Plotosus lineatus*	Fish 1948; Burgess 1989
HORABAGRIDAE	*Horabagrus brachysoma*	Kaatz et al. 2010
BAGRIDAE	(13) ***Bagrichthys macracanthus***; *Hemibagrus nemurus*; *Leiocassis micropogon*; *Leiocassis poecilooptera*; *Mystus carcio*; *Mystus gulio*; *Mystus vittatus*; *Pseudomystus siamensis*; *Sperata aor*; *Sperata seenghala*; *Tachysurus brashnikowi*; *Tachysurus sinensis*; *Tachysurus vachellii*	Protasov 1965; Heyd and Pfeiffer 2000; Kaatz et al. 2010
SISORIDAE	*Sisor rhabdophorus*	Mahajan 1963; de Pinna 1996
PANGASIIDAE	*Pangasianodon hypophthalmus*	Kaatz et al. 2010; Parmentier et al. 2010
SILURIDAE	*Siluris glanis*	Dufossé 1874; Tiepelt 2005
ASPREDINIDAE	(2) *Bunocephalus coracoideus*; *Bunocephalus verrucosus*	Heyd and Pfeiffer 2000; Kaatz et al. 2010; Kaatz and Stewart 2012
AUCHENIPTERIDAE	(10) *Auchenipterichthys thoracatus*; *Liosomadoras morrowi*; *Liosomadoras oncinus*; ***Spinipterus moijiri***; *Tatia aulopygia*; *Tocantinsia piresi*; *Trachelyichthys exilis*; *Trachelyichthys exilis*; *Trachelyopterus galeatus*; *Trachycorystes* sp.	Birindelli and Zuanon 2012; Birindelli et al. 2012; Kaatz and Stewart 2012; Birindelli 2014
DORADIDAE	(27) *Acanthodoras cataphractus*; *Acanthodoras spinosissimus*; *Agamyxis albomaculatus*; *Agamyxis pectinifrons*; *Amblydoras affinis*; *Astrodoras asterifrons*; *Doras micropoeus*; *Hassar orestis*; *Hemidoras stenopeltis*; *Leptodoras acipenserinus*; ***Lithodoras dorsalis***; *Megalodoras uranoscopus*; *Nemadoras humeralis*; *Nemadoras trimaculatus*; *Hemidoras morrisi*; *Orinocodoras eigenmanni*; *Ossancora punctata*; *Oxydoras kneri*; *Oxydoras niger*; *Physopyxis* sp.; *Platydoras armatulus*; *Platydoras costatus*; *Platydoras hancockii*; *Pterodoras granulosus*; *Rhynchodoras xingui*; *Tenellus leporhinus*; ***Trachydoras brevis***	Birindelli 2014; Knight and Ladich 2014; Boyle et al. 2015a; van der Sleen et al. 2020
HEPTAPTERIDAE	(3) *Brachyrhamdia meesi*; *Pimelodella cristata*; *Rhamdia quelen*	Schachner and Schaller 1981; Heyd and Pfeiffer 2000; Parmentier et al. 2010

Table 1 contd. ...

...Table 1 contd.

Family	Vocal Species (Total Number of Species Within the Family)	Relevant Citation Current Examples
PIMELODIDAE	(6) *Iheringichthys labrosus*; ***Leiarius marmoratus***; *Pimelodus blochii*; *Pimelodus ornatus*; *Pimelodus pictus*; *Phractocephalus hemioliopterus*	Tellechea et al. 2011; Rocha 2012; Ladich and Maiditsch 2020
PSEUDOPIMELODIDAE	(3) ***Lophiosilurus apurensis***; *Batrochoglanis raninus*; *Microglanis iheringi*	Kaatz et al. 2010; Birindelli and Shibata 2011; Kaatz and Stewart 2012
CLARIIDAE	(2) *Clarias batrachus*; *C. gariepinus*	Goel 1966; Baron et al. 2005; Jawad et al. 2021
HETEROPNEUSTIDAE	*Heteropneustes fossilis*	Agrawal and Sharma 1965; Kaatz et al. 2010
ARIIDAE	(9) *Ariopsis felis*; *Ariopsis seemanni*; *Arius arius*; *Arius maculatus*; *Bagre marinus*; *Bagre panamensis*; *Cathorops melanopus*; *Netuma thalassina*; *Potamarius nelson*	Hughes et al. 2014; Wang et al. 2017; Souza et al. 2023
ICTALURIDAE	(5) *Ameiurus natalis*; *Ameiurus nebulosus*; *Ictalurus furcatus*; *I. punctatus*; *Notutus flavus*	Mohajer et al. 2015; Albersheim-Carter 2015; Lecorchick et al. 2023
MALAPTERURIDAE	(2) *Malapterurus beninensis*; *M. electricus*	Fish 1948; Howes 1985; Boyle et al. 2015b
MOCHOKIDAE	(25) *Atopochilus* sp.; *Microsynodontis cf. batesii*; *Mochokiella paynei*; *Synodontis angelicus*; ***S. brichardi***; *S. decorus*; *S. eupterus*; *S. flavitaeniatus*; *S. gambiensis*; *S. grandiops*; *S. marmoratus*; *S. membranaceus*; *S. nigrita*; *S. nigriventris*; *S. njassae*; *S. notatus*; *S. obesus*; *S. ocellifer*; *S. pardalis*; *S. petricola*; *S. pleurops*; *S. robbianus*; *S. schall*; *S. schoutedeni*; *S. sorex*	Kéver et al. 2020a, b; Jawad et al. 2021
AUCHENOGLANIDIDAE	(3) ***Auchenoglanis occidentalis***; ***Notoglanidium macrostoma***; *Parauchenoglanis ngamensis*	Skelton 1993; Kaatz 1999; Kaatz et al. 2010
SCHILBEIDAE	*Schilbe intermedius*	van der Waal 2000

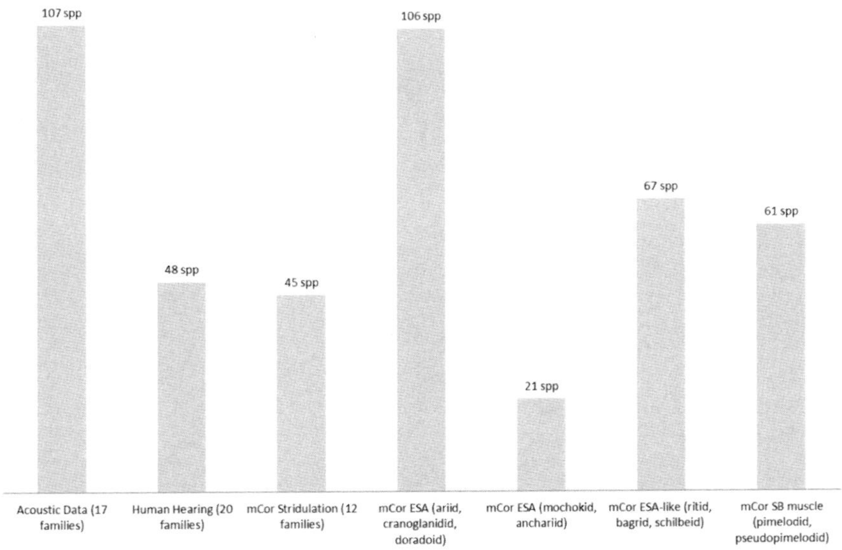

Figure 1. Evidence for sound production in catfishes representing the number of species (#): acoustic data analysis, direct observations lacking data analysis and morphological inference from vocal structures for species that lack evidence of vocal behavior. mCor = structures correlated with mechanism components present in known vocal catfishes; SB = swimbladder mechanism; STR = stridulation mechanism; ESA-like = potential vocal muscle and bones connected with the swimbladder that is not interpreted as derived from elastic spring mechanism homologous structures.

Call Characteristics

Catfish sounds have been observed and described qualitatively without any specialized recording equipment and through the use of hydrophones to record calls underwater or in the air. For sounds observed by humans listening, different catfish sounds were described and distinguished in the literature through onomatopoeia. Dobrin (1947) described the sounds of *Ariopsis felis* as drumming

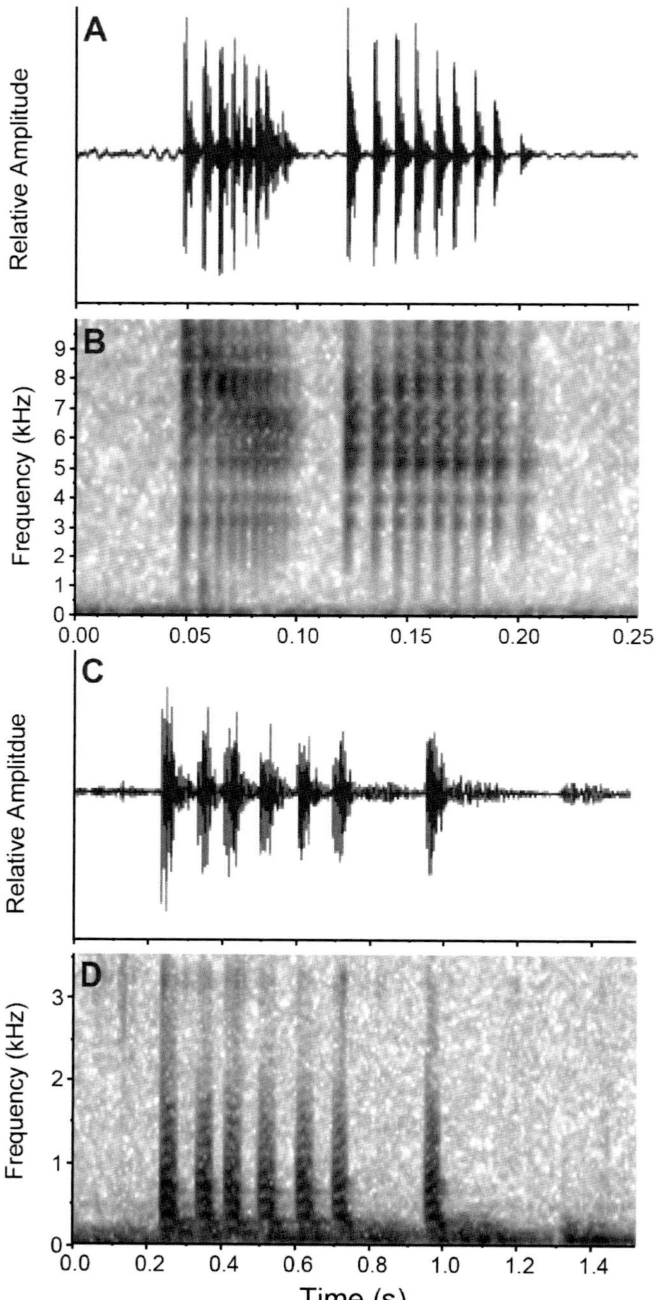

Figure 2. *Liosomadoras oncinus* social context sounds in undisturbed aquarium populations (from Kaatz 1999). (A) Waveform and (B) spectrogram of pectoral fin stridulation sounds. (C) Waveform and (D) spectrogram male courtship call consisting of multiple short elements produced in series.

consisting of evenly spaced beats. Doradid and auchenipterid swimbladder sounds audibly differ to a human listener (hum, grunt, growl, pop, toot, buzz, and drill) and differ in frequency (Kaatz and Stewart 2012). Stridulation sounds are often described as grating, creaking, or squeaking (Kaatz et al. 2010).

A call is a patterned sound of one or more elements (Odom et al. 2021). The majority of catfish calls contain multiple pulses in a series (Pruzsinszky and Ladich 1998; Tellechea et al. 2011; Kaatz and Stewart 2012; Knight and Ladich 2014). Event trains produced in disturbance with inter-event intervals ≤ 1,000 ms (Boyle et al. 2014, 2015a) could be treated as calls, although the higher variability of intersound intervals suggests a sound series.

Catfishes have been reported to produce one to four different types of sounds per species in studies, indicating a potential range of different functions for vocalizing, although the maximum number of sounds for any repertoire is likely incompletely known. The maximum catfish acoustic repertoire reported thus far includes four types in callichthyids and three in doradids, mochokids, pimelodids, ariids, and malapterurids.

These sounds fall into different sound type categories: pulsed and continuous. Pulsed sounds contain a discrete series of pulses. Catfish stridulation disturbance sounds are commonly pulsed broadband frequency, ranging well over 1,000 Hz up to the limited detection of 16,000 Hz (Heyd and Pfeiffer 2000). Disturbance stridulation sound patterns are commonly produced in sweep pairs, triples, or groups, and isolated sweeps also occur. Free-moving catfishes in aquarium social groups produced pulsed stridulation sounds (Kaatz 1999; Figure 2A). Waveform patterns of stridulation sweeps vary across species in maximum pulse number per sweep (Kaatz 1999). An exception is *Synodontis membranaceus*, which produces sounds with zero or few pulse numbers, sometimes with harmonic and frequency-modulated components (Kaatz et al. 2010). Another exception is the ictalurid *Ictalurus furcatus*, which produces a slider sound that lacks a pulsed pattern in a part of the call (Mohajer et al. 2015). Stridulation sounds detected from aquarium populations of free-moving individuals of *Corydoras* are broadband and pulsed (Kaatz et al. 2017).

Continuous pulse series (CPS) sounds consist of individual pulses lacking pulse off-times (no silent gaps between consecutive pulses). Catfish disturbance swimbladder sounds are commonly CPS sounds and typically low frequency (< 500 Hz dominant frequency) with tonal or harmonic components. Catfish calls can consist of short CPS swimbladder sounds produced in trains. Examples of catfish swimbladder disturbance calls that consist of a series of short CPS pulses include *Megalodoras uranoscopus* (Ladich and Maidtisch 2014) and *Pimelodus pictus* (Ladich and Maiditsch 2020). Male swimbladder courtship trumpet calls of *Liosomadoras* spp. consist of a series of short CPS toot sounds (Figure 2C,D; Kaatz 1999).

Call Temporal Traits

The frequency sensitivity of the catfishes' auditory system aligns with the frequency ranges of the sound characteristics they are known to produce (Ladich 2023). Catfish swimbladder disturbance sounds are well described and differ within families in duration for both Doradidae and Pimelodidae (Figure 3). Differences in call types within species of *Acanthadoras* and *Agamyxis* have also been found in these data sets (Figure 3). The duration of disturbance calls (recorded in air or underwater) for 14 species in six families ranges from < 15 ms to approximately 4 s (Pruzsinszky and Ladich 1998; Kaatz and Lobel 2001; Tellechea et al. 2013; Boyle et al. 2014).

For catfishes, a few social sounds have been acoustically described. The duration of agonistic calls for four species in four families ranges from approximately 65 to 500 ms. The male courtship-associated sound of *C. paleatus* was longer in duration (mean ± SE), 1,248 ± 128 ms, than the non-reproductive encounter sound, 836 ± 112 ms (Pruzsinszky and Ladich 1999). By comparison, the male courtship sound of *Arius arius* was a shorter sound, approximately 100 ms (Vijayyakumar et al. 2012).

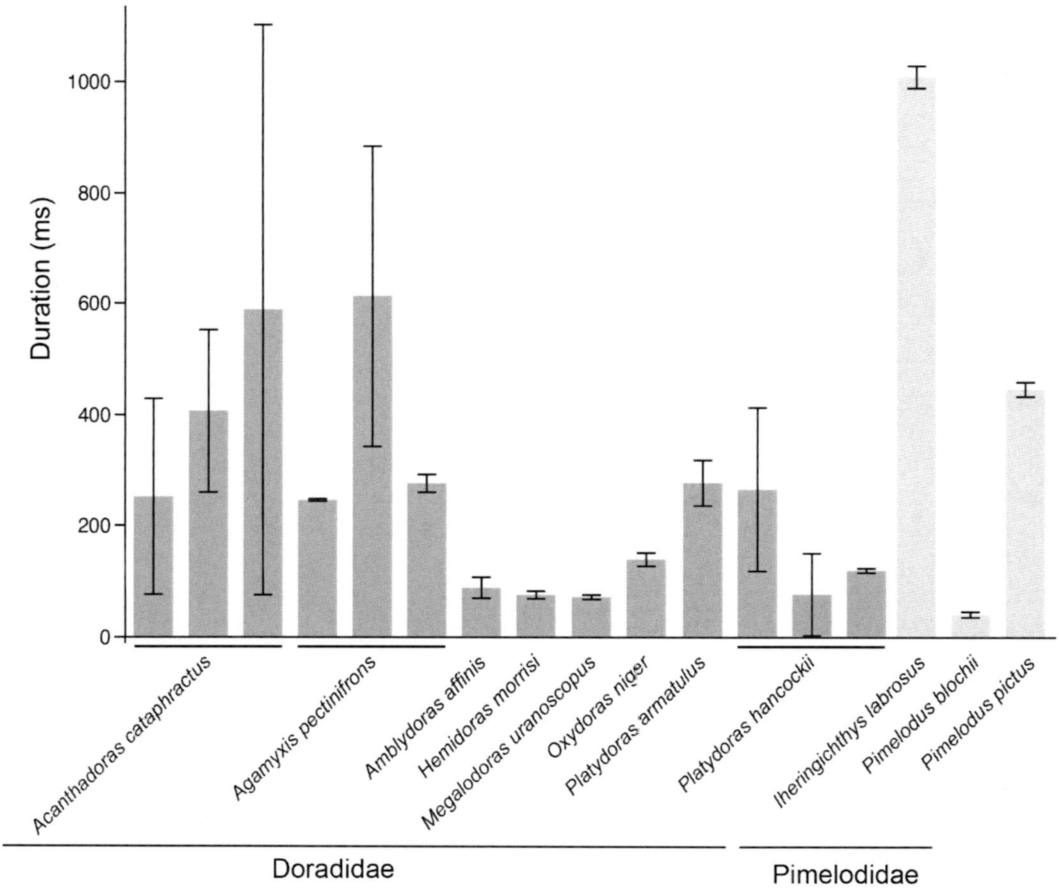

Figure 3. Neotropical catfish disturbance calls (mean ± SD or S.E.M.; based on Ladich 1997; Papes and Ladich 2011; Tellechea 2011; Knight and Ladich 2014). Multiple plots indicate different call types per species following Boyle et al. (2015a).

What is the duration of extended catfish vocal displays? Doradid *Oxydoras niger* disturbance sounds were elicited for 22 minutes, with a sound production rate of 6–9 sounds/min (Kastberger 1977). Knight and Ladich (2014) found a maximum of 52 disturbance sounds/minute among doradid species. The call rate for non-reproductive aquarium populations of *Corydoras* species was 1–2 sounds/h compared to 72–439 sounds/h for reproductive condition populations (Kaatz and Lobel 1999). Other temporal call parameters have not been widely described for interspecies comparisons: pulse number, pulse duration, interpulse interval, pulse period, and call period.

Spectrographic Characteristics

Catfish stridulatory sounds have higher frequencies than swimbladder sounds, with stridulatory sounds having peak frequencies above 1,000 Hz (from seven species in four families), while swimbladder-generated disturbance sounds have peak frequencies between 100–400 Hz (based on 31 species in six families). Doradids, mochokids and ariids produce frequency-modulated sounds (Tavolga 1960; Ladich 1997; Vijayakumar et al. 2012). Harmonics or tonal call components are narrowband uniform frequency traces present in the spectrogram. These sounds are produced by doradid, ariid, pimelodid and mochokid families in the disturbance context (15 species), in field recordings of an ariid (Mazlan et al. 2008), and in social contexts by one callichthyid, one malapterurid and three ariids.

Individual Differences and Ontogeny

Individual call differences within species can inform mate choice and may represent fitness parameters selected. During courtship, males are often the vocal entity, and females are silent listeners, as in *Corydoras paleatus* (Pruzsinszky and Ladich 1998) and *Liosomadoras* species (Kaatz 1999). Males and females of *Megalechis thoracata* produced different sounds in agonism. Female crackle calls were shorter in duration and higher in dominant frequency than male barks (Hadjiaghai and Ladich 2015). In *Arius arius*, male aggressive call peak frequency was greater than for females. There were no sex differences for disturbance vocalizations in a pimelodid (Tellechea et al. 2011).

Catfish sounds also change through ontogeny. Calls of juvenile *Synodontis schoutedeni* differ with size (Lechner et al. 2010). Call duration and amplitude increased, and dominant frequency decreased with size. It is proposed that smaller individuals hear higher frequencies to listen to other vocal juveniles. Juvenile *C. paleatus* produced shorter and higher frequency disturbance sounds but were silent in social contexts (Pruzsinszky and Ladich 1998). Ariid juveniles produced more sounds than adults (Vijayakumar et al. 2012).

Species Call Differences

Catfish calls show differences in temporal and frequency characteristics among taxa (Table 2). Pfeiffer and Eisenberg (1965) observed temporal differences between mochokid and doradid stridulation sounds. Dominant frequency differed for some doradoids' swimbladder sounds for similarly sized individuals (Kaatz and Stewart 2012). Doradid swimbladder sounds differed among species (Boyle et al. 2015a). New studies continue to support the finding of species differences in disturbance calls of catfishes (Lecorchick et al. 2023; Webb and Smith 2023).

Table 2. Statistically significant differences in the calls of catfish species.

Taxa Represented **Family (# Species Examined)**	**Acoustic Characteristics** * = significant statistical differences	**Literature**
Pimelodidae, Doradidae, Mochokidae (2 spp. per family)	Sound Duration*, Pulse Duration between families, Pulse Period* between families, Fundamental Frequency*, Frequency Modulation*	Ladich 1997
Auchenipteridae, Doradidae, Aspredinidae, Pimelodidae, Pseudopimelodidae, Mochokidae (29 spp.)	Dominant Frequency*	Kaatz and Stewart 2012
Mochokidae (5 spp)	Sound Duration NS, #Sounds/train, Inter-Event Interval*, Event Rate*	Boyle et al. 2014
Doradidae (3 spp.)	Sound Duration, Dominant Frequency*, #Amplitude Modulations, Amplitude Modulation Rates*	Boyle et al. 2015a
Doradidae, Mochokidae (5 spp.)	Sound Duration*, "Pausen Dauer" = Intersound*	Pfeiffer and Eisenberg 1965
Ariidae, Aspredinidae, Bagridae, Callichthyidae, Doradidae, Loricariidae, Mochokidae, Pimelodidae (11 spp.)	Sound Duration *, "Pausen Dauer"*	Heyd and Pfeiffer 2000
Loricariidae (2 spp.)	Sound Duration*, Dominant Frequency*	Webb 2011
Mochokidae (6 spp.)	#Jerks, Pulse Number/Jerk, Pulse Duration, Pulse Interval, Pulse Period, Dominant Frequency (overlap, no species statistics)	Parmentier et al. 2010
Mochokidae (2 spp.)	Pulse Period*, Peak Frequency*	Kéver et al. 2020b
Ictaluridae (4 spp.)	Sound Duration, Pulse Number*, Peak Time, Center Frequency, Bandwidth	Lecorchick et al. 2023

Sound Producing Mechanisms

Catfish Acoustic Mechanism Analytical Tools

Fish acoustic mechanism function has been studied extensively (e.g., Demski et al. 1973; Parmentier et al. 2017). Techniques applied to understanding the vocal mechanisms of catfishes include muscle morphology (Birindelli and Shibatta 2011), high-speed video (Ghahramani et al. 2014), electron microscopy (Kaatz et al. 2010), computerized tomography (Boyle et al. 2015a), and cellular techniques (e.g., recordings, tracing motor neuron pathways, and immunohistochemistry) (Ladich and Fine 1994; Albersheim-Carter et al. 2015). The two common mechanism types are swimbladder-associated muscles and stridulation.

Swimbladder Mechanisms

Acoustic compressors (Bridge and Haddon 1893), constrictors, or drumming extrinsic muscles attached directly to the swimbladder are found in pseudopimelodids (Birindelli and Shibatta 2011) and pimelodids (Lundberg and Akama 2005). A tensor tripodus muscle that also attaches to the swimbladder may be involved in sensory dampening (Ladich and Fine 1994). Catfish swimbladder sounds can also be generated by a pair of protractor muscles arising from the postcranium and inserted on the elastic spring apparatus (ESA), whose bony elements attach to the swimbladder (Ladich 2001). Doradoid ESA Müllerian ramus shapes vary from the widest disk-shaped (Figure 4A) to reduced plug-like (Figure 4B) (Kaatz and Stewart 2012). Auchenipterids belonging to the genus *Ageneiosus* have the most reduced ESA, and tested species have been proven to be silent (Kaatz et al. 2010). The Müllerian ramus in ariids is more shelf-like in shape (Kailola 2004), raising the question of how ESA structural variation may influence call traits.

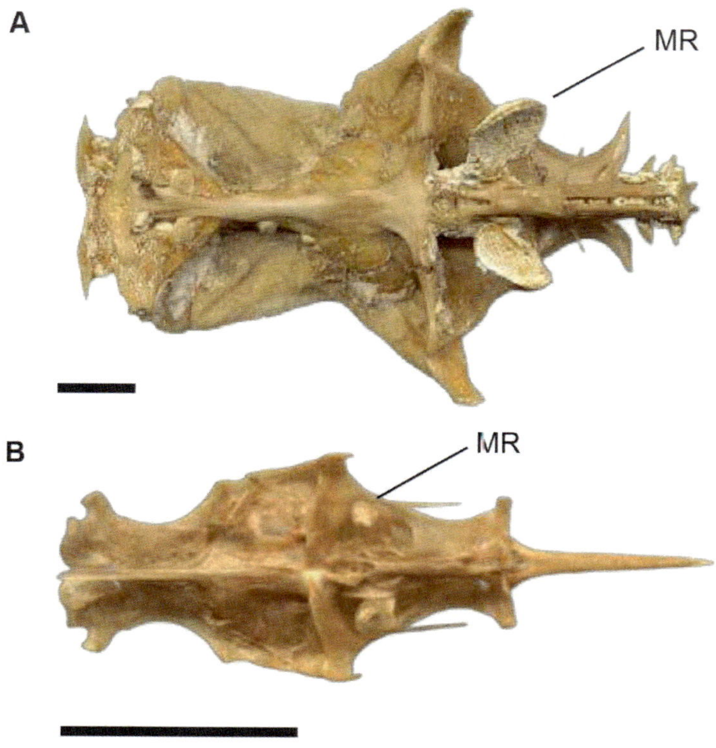

Figure 4. Elastic spring apparatus Müllerian ramus shape variation in doradoids (scale bar = 1 cm). (A) Disk-shaped, in the auchenipterid *Tocantinsia piresi*. (B) Plug-like, in the doradid *Ossancora punctata*. MR = Müllerian ramus.

Stridulation Mechanisms

Pectoral spine stridulation is produced by fin sweeping (primary spine abduction) forward and/or backward (primary spine adduction) (Parmentier et al. 2010). Functional morphology studies identify ictalurid pectoral stridulation as a stick-slip mechanism (Mohajer et al. 2015). *Synodontis* pectoral stridulation is described as a brake shoe model for the motion of the process inside the pectoral girdle fossa (Parmentier et al. 2010). Microscopic bony ridges are present on the dorsal process of the pectoral fin spine proximal region and contact the inner surface groove of the cleithrum to produce sounds (Figure 5). Vocal ridges have been documented with microscopy (de Pinna 1996; Fine et al. 1997; Kaatz et al. 2010; Parmentier et al. 2010; Webb McAdams and Smith 2023). Stridulation sounds produced by teeth have been documented for four loricariid species (Tellechea et al. 2013), a heteropneustid *Heteropneustes fossilis* (Agrawal and Sharma 1965) and the mochokid *Synodontis gambiensis* (Freytag 1968).

The most uncommon stridulation mechanism is the use of the anterior interspinous bones in proximity to the dorsal spine. This stridulation apparatus has only been found fully developed in three museum specimens of sisorids (de Pinna 1996). Mahajan (1963) demonstrated stridulation behavior in a fourth species *Sisor rhabdophorus*.

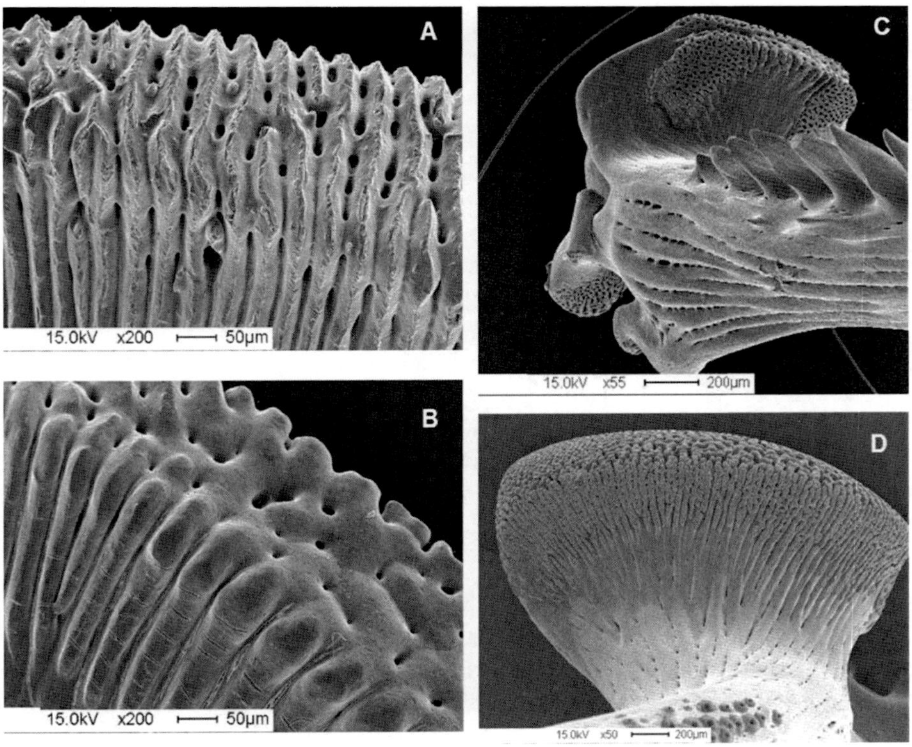

Figure 5. SEM surface morphology of the pectoral spine dorsal process of certain vocal catfishes. (A) Bladed ridges in *Corydoras rabauti*. (B) Rounded ridges in *Corydoras arcuatus*. (C) Edge knobs in *Microglanis iherengi*. (D) Raised convoluted ridges in *Dianema urostriata*.

Novel and Speculative Mechanisms

The electric catfish *Malapterurus beninensis* is vocal, but the ESA was not conclusively confirmed as the conventional mechanism for all sound types produced (Boyle et al. 2015b). A novel mechanism involving gas movement within the swimbladder system is conjectured. The presence of an ESA in anchariids is the only evidence for the family and remains untested for vocal ability (Parmentier and

Diogo 2006). The ESA lacking in the genus *Helicophagus* is present in other pangasiids (Sörenson 1894–1895), but its efficacy in sound production is unknown. The cranoglanidid *Cranoglanis bouderius* is described as having an ESA and a protractor muscle associated with a Muellerian ramus (Diogo et al. 2002); however, no sounds have been reported. The aspredinid *Bunocephalus verrucosus* has a transparent, thin-walled swimbladder with no obvious means of producing its swimbladder-like low-frequency humming sound (Kaatz and Stewart 2012).

ESA-like mechanisms do not have muscles attached to a Müllerian ramus but provide muscle swimbladder connections and present additional morphological correlates for identifying vocal taxa (Figure 1). Parmentier and Diogo (2006) reviewed the presence of a protractor post-temporalis hypothesized vocal mechanism in the Bagridae for *Bagrichthys*, *Hemibagrus*, and *Bagrus*. This ESA-like mechanism was described in Diogo et al. (1999) for *Bagrus docmac*. Complex motions of the post-temporalis muscle could generate swimbladder sounds. Similar extrinsic swimbladder-associated muscles identified in species descriptions by Mo (1991) added a new vocal family, Ritidae, and additional vocal species in the Bagridae and Schilbeidae (Figure 1). Two adult specimens of *Tachysurus sinensis* produced low-frequency sounds, which could be the first evidence of a swimbladder mechanism in this family (Kaatz 1999). Weakly audible and low-frequency sounds (modal dominant frequency 100 Hz) were also recorded in disturbance from *Notoglanidium* cf. *macrostoma* (Kaatz 1999) and might be evidence for this mechanism.

Vocal Mechanism Morphology Inference

The total count of vocal catfishes would increase from a total of more than 150 to at least 450 species if we included morphological correlates to identify vocal species for all mechanism types for species not known to produce sounds (Table 1; Figure 1). Morphological inference from museum specimens allows us to hypothesize the extent of vocalization behavior further. Sörenson (1894–1895) proposed the earliest use of morphology in identifying vocal fishes. For swimbladder mechanisms, the presence of associated muscles identifies a potentially vocal fish. Vocal morphology is the sole evidence for four catfish families (Diplomystidae, Arratia 2003; Ritidae, Mo 1991; Anchariidae, Parmentier and Diogo 2006; Cranoglanididae, Diogo et al. 2002) and vocal behavior supports 24. Thirteen catfish families lack evidence of vocal behavior or morphology: Nematogenyidae (G. Arratia pers. com), Scoloplacidae, Astroblepidae, Cetopsidae (Kaatz et al. 2010), Ailiidae, Akysidae, Amblycipitidae, Kryptoglanidae, Phreatobiidae, Austroglanididae, Lacantuniidae, Amphiliidae, Claroteidae. The presence of non-vocal and vocal species within catfish families (Kaatz et al. 2010) also provides an experimental landscape for studying the evolution of acoustic communication, placing a priority on knowing the vocal versus silent status of taxa within clades.

Vocal Species Inference for Swimbladder Mechanisms

Morphological correlates of catfish swimbladder mechanism add more vocal species to the total catfish vocal count (Figure 1). Pimelodids have extrinsic acoustic muscles with origins in the cranium and insertion directly on the swimbladder (Stewart and Pavlik 1985; Stewart 1986; Ladich 2001; Lundberg and Akama 2005; Buitrago-Suarez and Burr 2007). A taxonomic survey of pimelodids found constrictor muscles in most species (n = 88%, n = 57); seven species lacked swimbladder muscles (Rocha 2012).

How reliable is the presence of an ESA in identifying a vocal fish? Malapterurids (Boyle et al. 2015b) have an ESA-like structure that is silent. ESAs are present in mochokids (Vigliotta 2008), and some produce swimbladder sounds (Kaatz and Stewart 2012). However, no swimbladder sounds are produced by the ESA muscles of *Synodontis nigriventris* (Boyle et al. 2014). Therefore, ESA presence alone cannot identify an acoustic species in this family. Protractor muscles can produce both electric and vocal signals or only one or the other (Kéver et al. 2020a, b). Conditions specific to swimbladder silence in *S. nigriventris* remain to be further examined and may be found in differences in protractor motor neurons.

In other families, the ESA seems to be a reliable indicator of vocal ability. Most ariids examined have an ESA (Kailola 2004), and the family is vocal. No doradoid species with an ESA has been found to be silent (Kaatz and Stewart 2012). The sound-generating ESA protractor muscles of the vocal driftwood catfishes *Liosomadoras* (Birindelli and Zuanon 2012) are substantial. Even the smaller ESA in *Hassar acipenserinus* produces a buzzing sound. Therefore, it is reasonable to hypothesize that the presence of an ESA protractor muscle is a valid morphological correlate of vocal doradoids (Birindelli 2014).

Vocal Species Inference for Stridulation Morphology

The catfish pectoral spine morphological correlate of stridulation in disturbance proposed by Kaatz et al. (2010) is a well-developed dorsal process with surface-bladed or rounded ridges, edge knobs or raised convoluted ridges (Kaatz et al. 2010; Figure 5). Otherwise, undocumented vocal catfishes that have these pectoral stridulation morphological correlates of vocal behavior add to the vocal catfish count (Figure 1). Social sounds have been recorded in *Megalechis thoracata* (Hadjiaghai and Ladich 2015), whose dorsal process has a convoluted ridge surface. *Corydoras* species whose social sounds were recorded have bladed or rounded ridges (Kaatz and Lobel 1999). Teugels (1986) illustrated pectoral spine ridged dorsal processes for clariids for several subgenera not tested for vocal ability. Equally well-developed dorsal processes with prominent ridges are present in illustrations of pectoral spines for several *Clarias* species (Skelton and Teugels 1991; Teugels et al. 2001; Devaere et al. 2007a).

Silent Morphological Correlates

The lack of swimbladder-associated muscles in a species is indicative of silence in that mechanism. The auchenoglanidid *Auchenoglanis biscutatus* lacks compressor swimbladder muscles (Bridge and Haddon 1893). Pseudopimelodid swimbladder muscles were not found in two of the six examined genera (Birindelli and Shibatta 2011). Swimbladder vocal muscles common among pimelodids are absent in *Pimelodus sapo* and *Pimelodus pulcher* (Bridge and Haddon 1893); *Calophysus macropterus*, *Hypophthalmus edentatus*, *Megalonema platycephalum*, *Megalonema amaxanthum*, *Megalonema platanum*, *Pimelodina flavipinnis*, and *Luciopimelodus pati* (Rocha 2012; Stewart 1986). The lack of a compressor muscle is plesiomorphic for genera in the Rhamdiinae (Lundberg et al. 1991). Reversals between vocal and silent character states are suggested by the reduction of the swimbladder, as seen in the rheophilic pimelodid *Cetopsorhamdia phantasia* (Stewart 1985). Stridulation silent auchenipterid *Ageneiosus* species are derived in the phylogeny (Birindelli et al. 2012) with vocal genera such as *Tatia* at the base. *Ageneisosis* spp. bony encapsulation of the swimbladder and a reduced fragile ESA suggest they also lack swimbladder vocal ability.

Are there structural designs associated with the loss or absence of a stridulation mechanism? The primary pectoral spine and its dorsal process are reduced in catfishes that were silent during disturbance (Kaatz et al. 2010). Electric catfishes have short, soft fin rays and do not stridulate (Boyle et al. 2015b). The cranoglanidid *Cranoglanis bouderius* has a reduced dorsal condyle (Diogo et al. 2002), which could limit the ability to make sounds, but its vocal behavior has not been tested. de Pinna (1996) proposes atrophy of dorsal spine-associated vertebral vocal elements as an indicator of silent sisorid species. Pectoral stridulation dorsal process surfaces of species silent in disturbance included smooth, irregular vacuolated, beaded, flat-topped convoluted ridges or cob-webbed structures in *Hara jerdoni*, *Denticetopsis praecox* and *Ageneiosus* sp. (Kaatz et al. 2010).

Vocal Mechanism Evolution

Swimbladder Mechanism Evolution

Morphological systematics that places diplomystids as basal Siluriformes leads to the hypothesis that swimbladder silence could be the early vocal condition for the order (Diogo and Peng 2010).

Swimbladder mechanisms are found throughout the order Siluriformes, but their presence and absence can vary within families at the genus level (Parmentier and Diogo 2006). The ESA is widespread but not homologous across catfish families (Curran 1989: Auchenipteridae; Howes 1985: Malapteruridae; Kailola 2004: Ariidae; Vigliotta 2008: Mochokidae). Birindelli (2014) interpreted the doradoid ESA as an apomorphy.

All doradid subfamilies include swimbladder vocal genera; five of six are supported by vocal behavior and morphology, and Wertheimerinae genera only by the presence of ESA muscles (Sabaj and Arce H. 2022). Among doradids, there is a reduction in the size of ESA elements located in crown genera (Kaatz and Stewart 2012). *Hassar* and *Leptodoras* produce sounds with their smaller ESA for species examined, but their sounds are less audible to a human bystander. Auchenipterid swimbladder vocal genera *Tocantinsia*, *Liosomadoras*, and *Trachelyicthys* are basal with other vocal genera in higher branches (Curran 1989; Birindelli et al. 2012). *Centromochlus* is basal and has muscles associated with the ESA (Diogo et al. 2003).

Swimbladder acoustic morphology is described in phylogenetic studies for pimelodids (Rocha 2012) and pseudopimelodids (Birindelli and Shibatta 2011). Pimelodid swimbladder drumming muscles are considered a synapomorphy for the Pimelodinae (Diogo 2005), which includes vocalizing *Pimelodus* species. Vocal genera occur in the tree at all levels (Rocha 2012). Pseudopimelodids with swimbladder constrictor muscles include *Lophiosilurus*, *Cephalosilurus*, and *Pseudopimelodus* (Birindelli and Shibatta 2011). Diogo et al. (2004) illustrated vocal muscle presence in *Batrochoglanis raninus*. Vocal swimbladder sounds are present in taxa throughout both tree hypotheses (Birindelli and Shibatta 2011).

The ariid catfish vocal apparatus of *Galeichthys* lacks ESA muscles connecting the anterolateral arm of the Müllerian ramus to the neurocranium and anterior vertebrae (Kailola 2004). The ESA muscle lacking ariid genus *Galeichthys* is in the earlier branching Galeichthyinae, while species with fully developed ESA muscles belong to genera throughout the derived Ariinae (Betancur-R. 2009).

Stridulation Mechanism Evolution

A molecular phylogeny that places callichthyids and loricariids as branching earlier than diplomystids suggests stridulation as an early evolving character in the Siluriformes (e.g., Kappas et al. 2016). The catfish fossil record of bone morphology is extensive (Gayet and Meunier 2003). Pectoral spine stridulation morphology is preserved in the fossil record and present in Late Cretaceous catfishes and their pectoral spine dorsal process straie (Alves et al. 2019). Examples of species with fossilized striae extending from the pectoral spine base out to the edge of the dorsal process include the fossils identified as *Diplomystes* from Bolivia (Gayet and Meunier 1998; see Arratia 2025), the loricariid *Hypostomus commersoni* (Vallone et al. 2017), and an extinct clariid *Heterobranchus austriacus* (Kovalchuk and Ferraris 2016).

Both callichthyid subfamilies include vocal genera (Reis 1998). The Callichthyinae are vocal throughout the phylogeny. The Corydoradinae basal genus *Aspidoras* has stridulation striae present that supports a vocal status, while *Corydoras* are vocal as derived taxa. Based on Britto's (2003) phylogenetic hypothesis Callichthyinae is less derived than the Corydoradinae, suggesting vocal behavior as an ancestral state.

For other less well-sampled families, some stridulating *Clarias* species are both early branching and some derived in the clariid tree, while the reduced pectoral spines of *Channallabes* are also found throughout the phylogeny but might not be able to produce sounds (Devaere et al. 2007a, b). Ictalurid stridulation sounds have been recorded in three of the seven genera, vocal *Ictalurus* and *Ameiurus* branch earlier in the phylogeny (Arce-H. et al. 2017), with derived genus *Noturus* also vocal (Lecorchick et al. 2023).

Dual Vocal Mechanisms

Catfish stridulation and swimbladder mechanisms produce divergent call types (Ladich 1997). This presents the possibility of communicating with a broader vocal repertoire and targeting a wider listening audience. Catfish families with swimbladder vocal muscle and stridulation mechanisms present in species with established vocalization include (# species 50): auchenipterid (7); doradid (24); heptapterid (1); pimelodid (4); pseudopimelodid (1); ariid (5); and mochokid (8).

In mochokids, dual mechanisms are present in species of the genera *Mochokiella* and *Microsynodontis*, which are early branching genera in the family, as well as among species of the derived genus *Synodontis* (Pinton et al. 2013; Schedel et al. 2022). Two ariid *Arius* species in derived positions produce stridulation and swimbladder sounds. Catfishes with dual vocal mechanisms among pimelodids (Rocha 2012) include *Pimelodus pictus* and are widespread throughout the doradid (e.g., *Acanthadoras cataphractus*) and auchenipterid (e.g., *Auchenipterichthys thoracatus*) families (Kaatz et al. 2010; Kaatz and Stewart 2012).

Behavioral Contexts for Communication

Fish acoustic signals originate from noise-producing actions such as feeding, fin movements or modification of buoyancy mechanisms (Parmentier et al. 2017). Some sounds recorded from catfishes might be noise, such as air expulsion (Dufossé 1874; Kaatz 1999). Feeding competition sounds produced by derived or simple mechanisms could be considered agonistic signals (Fish and Mowbray 1970). The context of a silurid species' field foraging associated sound recordings remains unknown (Tiepelt 2005) but could be a candidate for feeding competition calls. Documentation of vocal catfish species through acoustic analysis shows sounds produced in typical fish acoustic behavioral contexts, including startle (seven species), schooling (three species), agonism (including territoriality and aggressive interactions; 22 species), reproduction (courtship, spawning; 15 species), predator-prey interactions (three species), and disturbance (either hand-held or restrained; 94 species). Fish vocal communication signals function to defend or compete for resources, to attract and choose mates, or to avoid predators (Kaatz et al. 2017). The behavioral categories of catfish calls are examined below.

Startle, Schooling and Chorusing

Some catfish sounds are produced in contexts where exact functions have not been fully investigated. Startle sounds of *Corydoras* species (Kaatz and Lobel 1999) were produced by stridulation by free-moving individuals and have pulsed patterns like other known signals in the genus. They occur when individuals appear to be alarmed, and they are widespread in the documentation of other vocal fishes (Fish and Mowbray 1970). Vocalizing within schools or aggregations has been detected in marine catfishes. In aquaria, *Bagre marinus* individuals within groups produced yelp and sob sounds (Tavolga 1960). Sounds produced by schools of *Ariopsis felis* (Tavolga 1960) and *Ariopsis seemani* in aquaria have also been recorded and described (Schmidtke et al. 2013). Passive acoustic recordings in the ocean also detect swimbladder sounds (Breder 1968; Tavolga 1960). *Bagre marinus* and *A. felis* produce regularly patterned and distinctive swimbladder sounds in the field described as a percolator, yelp and short grunts (Tavolga 1960). *Ariopsis felis* choruses lasted as little as nine minutes to as long as four hours (Breder 1968). Baltz and Campos (1996) detected sounds in sciaenid drum choruses they hypothesized to be *Bagre panamensis* from their field surveys. Another ariid, *Arius maculatus* produced sounds in tropical mangrove coastal habitats (Mazlan et al. 2008).

Social Contexts

Agonism is the second most common context for vocalization in catfishes. *Corydoras* spp. creaks are produced during chases and claps during pre-chases (Kaatz and Lobel 1999). Sounds were also detected during intra- and inter-specific aggression in a clariid hybrid (Baron et al. 2005).

Male and female *Megalechis thoracata* produce agonistic sounds in contests (Hadjiaghai and Ladich 2015). Species in several families produced calls during conspecific and heterospecific agonistic interactions in aquarium populations: chase, lateral display, head-butting and displacing each other out of cover sites (Kaatz 1999). Playback of intruder sounds in *Ameiurus nebulosus* appeased the territorial residents (Rigley and Muir 1979). Playback of artificial sounds to *Rhamdia quelen* elicited aggressive reactions (Schachner and Schaller 1981). Males of the callichthyid *Callichthys callichthys* that build bubble nests grunt when guarding eggs (Burgess 1989). Head-butting and nudging in vocal ariids was proposed by Tavolga (1971) as the presence of a dominance hierarchy. Signal differences could influence dominance relationships. Larger catfishes produce louder or lower frequency sounds (Pruzsinszky and Ladich 1998; Lechner et al. 2010; Knight and Ladich 2014).

The third most common context for signaling in catfishes is reproduction. Catfish males vocalize during courtship in the Ariidae (*Arius arius*, Vijayakumar et al. 2012) and Callichthyidae (e.g., *Corydoras paleatus*, Pruzsinszky and Ladich 1998; Kaatz 1999). Human listening detected reproductive sounds in a bagrid (*Mystus vittatus*; Heyd and Pfeiffer 2001), a mochokid (*Synodontis schall*, Abu-Gideiri and Nasr 1973), a loricariid (*Rineloricaria uracantha*, Moodie and Powers 1962), a heteropneustid (*Heteropneustes fossilis*, Agrawal and Sharma 1965) and a callichthyid (*Callichthys callichthys*, Burgess 1989). Hydrophone listening in the field revealed rapid pectoral fin knocking by a courting male, mouth rasping from the courting pair by a loricariid (Moodie and Power 1982). Further evidence of reproductive sounds comes from the change in catfish vocalizations during dry to wet season treatments in aquarium populations (Kaatz 1999). Sound abundance increased, and call types changed in mochokid, doradid and auchenipterid species (Kaatz 1999).

Disturbance Behaviors

Disturbance is the most common behavioral context for vocal catfishes. Of vocal families, only the Siluridae and Trichomycteridae lack disturbance sounds. Localization of a conspecific stridulation disturbance sound was demonstrated through training of a loricariid (Smith et al. 2009), raising the possibility that these sounds could be social signals. Disturbance sounds produced by different mechanisms in different numbers in different habitats, in the air or underwater, could target different listeners. Sounds produced in these different habitats differ in acoustic parameters (Ladich 1997; Knight and Ladich 2014). The authors proposed stridulation is a distress call and may target different predators, while swimbladder sounds are intraspecific signals.

Predator-Prey Interactions

Encounters between natural predators and prey are directly associated with fish vocalization (Ladich 2022). A heteropneustid makes pectoral stridulation sounds when chasing prey (Agrawal and Sharma 1965). Predator presence releases disturbance sounds from catfishes that are prey targets; however, bass (Bosher et al. 2006) and piranha (Markl 1968) predators were not deterred. Vocalizing can make catfishes vulnerable as prey targets. Dolphin predators engaged in head-turning responses to fish playbacks that included *Ariopsis felis* sounds (Gannon et al. 2005), indicating passive listening by predators. Sharks similarly approached *A. felis* sounds in field playbacks in response to distress and normal pulsed sounds (Banner 1972). Playbacks of *A. felis* sounds reduced the foraging behavior of prey crab species (Hughes et al. 2014).

Abiotic Influences on Calls

Propagation Distance and Amplitude

Early field explorers noted the propagation distance of catfish sounds in the air: *Pterodoras maculatus* 30 m (Sörenson 1894–1895); 3 m for *Clarias batrachus* (Goel 1966); 5 m for grunting

Synodontis schall (Abu-Gideiri and Nasr 1973). Catfish sounds have been recorded underwater at less than 1 m for 78/85 species in 15 families, leaving maximum propagation distance unknown for most. The callichthyid *Megalechis thoracata* produced some sounds within one to two body lengths of the listener, suggesting close communication with conspecifics (Hadjighai and Ladich 2015). *Doras* sp. swimbladder sounds were spectrographically similar between tested distances at 10, 30, 60 and 100 cm (Kastberger 1977), indicating a longer call range.

Catfish sounds can propagate further than 1 m. *Ariopsis felis* sounds were recorded 3 m from the hydrophone (Fish and Mowbray 1970). The ictalurid *Ictalurus punctatus* (Mohajer et al. 2015), loricariids *Hypostomus commersoni, H. derbyi, Paraloricaria vetula, Ricola macrops* (Tellechea et al. 2013), and pimelodid *Iheringichthys labrosus* (Tellechea et al. 2011) had sounds recorded at 1 m. These catfishes may be broadcasting sounds rather than attempting acoustic crypsis or whispering to nearby conspecifics.

Catfish call amplitude is described at known distances underwater for nine species in four families with varying methods. Comparable unweighted sound pressure levels for five species of doradids (Knight and Ladich 2014) underwater near the caller ranged from mean 107–131 dB re 1 μPa or swimbladder to 127–139 dB re 1 μPa for stridulation sounds. Social sounds near the caller for a callichthyid ranged from 101–137 dB re 1 μPa (Hadjiaghai and Ladich 2015). Sounds recorded at 1 m for *Ictalurus furcatus* disturbance stridulation were 122–145 dB re 1 μPa (Ghahramani et al. 2014). Lower amplitude sounds might indicate a conspecific listener at close range, while louder sounds that propagate several body lengths from the caller may indicate a greater broadcast range. Catfish with amplitude data are neither among the loudest nor quietest fishes (Kaatz et al. 2017).

Does Temperature Influence Sound Production?

Environmental factors influencing calling patterns in fishes include the lunar cycle, freshwater habitat flooding, and tidal or temperature changes (Kaatz et al. 2017). Ocean water temperature limits the calling periods and intensity in the field for *Ariopsis felis* (Breder 1968). Winter months were too cold for their calling, and high-temperature peaks limited calling intensity. Papes and Ladich (2011) found significant changes in hearing and dominant frequency and temporal call traits at different temperatures for *Platydoras armatulus*. Call amplitude, pulse number and maximum pulse period did not vary with temperature changes. Recording conditions have implications for comparing sounds recorded in different laboratory or field environments or at different seasons or depths. In a pimelodid *Pimelodus pictus* significant effects of temperature were also discovered (Ladich and Maiditsch 2020). Calls were shorter at higher temperatures, and the fundamental frequency increased. Calling activity and dominant frequency, which are key vocal traits in behavior, were unaffected.

Diel Patterns

Many vocal fishes, especially catfishes, call nocturnally (Kaatz 1999; Kaatz et al. 2017). *Corydoras* sounds were observed and recorded primarily at night (Kaatz and Lobel 1999). *Rhamdia quelen* was found to call in darkness treatments in aquaria (Schachner and Schaller 1981). *Malapterurus beninensis* was active and vocal at night (Boyle et al. 2015b). For two species of auchenipterids (*Liosomadoras morrowi* and *Trachelyopterus galeatus*) under an aquarium light cycle that included direct sunlight to simulate varying light regimes, acoustic activity was abundant at night and early morning (Kaatz 1999). In field populations, ariid *Ariopsis felis* calling began at twilight and continued into the full darkness of night (Breder 1968). The exception was the ariid *A. seemani*, which produced short swimbladder sounds in aquarium groups during the day with greater vocal activity in the morning (Heyd and Pfeiffer 2000).

Seasonal Patterns

Catfish calling patterns exhibit seasonality. Breder (1968) observed seasonal patterns of sounds in the field in Florida in marine bay waters. *Ariopsis felis* exhibited no calling changes with lunar or tidal patterns. Percolator sounds were seasonal, most common in May, lessening in the summer and greatest intensity in the fall, tracked to follow temperature changes. Catfish species in several families produced different types of calls when maintained under wet season conditions in aquaria (Kaatz 1999; Kaatz and Lobel 1999), suggesting seasonal calling patterns.

Catfish Ecoacoustics

What is the potential for catfish voices in soundscapes? Surveying natural habitats acoustically is a burgeoning study area that includes fish sounds (Lindseth and Lobel 2018; Hawkins 2022). Globally, the Palearctic has one documented acoustic catfish, *Siluris glanis*. Sounds of vocal freshwater catfishes in Asian families are known but have not been widely described. Asian marine ariids are vocally active, and their sounds are detected in the field (Mazlan et al. 2008). Other ariids are found across biogeographic regions in marine habitats and significantly contribute to the soundscape (Tavolga 1960). Australasian marine catfishes include one vocal stridulating species *Plotosus lineatus*. Ethiopean mochokids in the genus *Synodontis* are highly vocal, producing both swimbladder and stridulation sounds (Table 1). The freshwater habitats of two biogeographic regions host the best-described vocal catfishes, the Nearctic and the Neotropics.

Ictalurids are the only freshwater vocal catfishes of the Nearctic that have been documented. Anderson et al. (2008) detected sounds in the field similar to stridulation sounds recorded in disturbance for *Ameiurus nebulosus*. Other ictalurid species producing disturbance stridulation sounds could be adding sounds to the soundscape (Lecorchick et al. 2023).

Catfishes are speciose and diverse in Neotropical freshwater ecosystems and could be major contributors to these soundscapes. Their sounds should be distinguishable based on mechanism types and across some taxa. Stridulation sounds of catfishes range from several hundred Hz to > 10 kHz (Kaatz et al. 2010; Heyd and Pfeiffer 2000). Swimbladder sounds do not overlap with stridulation sounds at the highest frequencies and differ among species (Figure 3). The frequency of swimbladder sound frequency differs across taxa. Disturbance sound dominant frequency is lowest in aspredinids (133 $\pm$ 35 Hz) and pseudopimelodids (108 $\pm$ 10 Hz) (Kaatz and Stewart 2012). Pimelodid dominant frequency examples are higher and include 220 $\pm$ 46 Hz for *Iheringichthys labrosus* (Tellechea et al. 2011) and 310 $\pm$ 49 Hz for *Phractocephalus hemioliopterus* (Kaatz and Stewart 2012). Dominant frequency varies among species of doradids and auchenipterids (Kaatz and Stewart 2012). The mean dominant frequency of doradid *Pterodoras granulosus* growls is lower than *Platydoras* spp. sounds. Among auchenipterids, *Tatia aulopygia* sounds have a higher mean dominant frequency than *Trachelyichthys exilis* growls. Neotropical catfish sounds also differ in temporal characteristics, and these acoustic traits could be used to distinguish taxa (Table 2) if their sounds are detected in Neotropical waters.

Future Directions

Morphological correlates of vocal fish behavior have greatly contributed to the identification of new vocal species. Vocal structures have been included in revisions, phylogenetic studies and the description of new species: pimelodids (Azpelicueta and Britski 2012) and doradids (Diogo and Chardon 2004; Sabaj and Birindelli 2008; Sousa and Birindelli 2011; Birindelli et al. 2011). We need more systematic work combining vocal behavior and morphology in phylogenies, combining the work of ichthyologists working with living fishes and museum specimens. Broadly surveying families currently known as silent could contribute significantly to understanding evolutionary trends of vocal behavior.

Documenting sounds associated with specific social behaviors is important to understanding communication. Catfishes identified as silent tested only in the disturbance context (Kaatz et al. 2010) might be vocal in other contexts. Health, age, sex, sexual maturity, and environmental factors like temperature during recording could also influence the vocal behavior of observed fish. Catfish vocal studies could include questions asking if there is geographical variation in calls and if species in sympatry have acoustically displaced calls. These questions have been confirmed for other vocal fishes in the wild. Further investigation of how sounds may vary by individual, sex, size, and reproductive condition is ideally suited to laboratory aquarium studies (Kaatz and Lobel 1999; Webb McAdams et al. 2023).

The question remaining for vocal catfishes studied in captivity is: do they exhibit these vocal behaviors in native habitats? Monitoring such populations is a good starting point for field passive acoustics. Most catfishes identified as vocal in laboratory work lack field data and require acoustic monitoring of undisturbed populations. Field sampling could involve a netted pen with an underwater camera or passive acoustic monitoring with a hydrophone.

References

Abu-Gideiri, J.B. and D. Nasr. 1973. Sound production by *Synodontis schall* (Bloch-Schneider). Hydrobiologia 43: 415–428.

Agrawal, V.P. and R.S. Sharma. 1965. Sound producing organs of an Indian catfish *Heteropneustes fossilis*. Ann. Mag. Nat. Hist. 8: 339–344.

Albersheim-Carter, A. Blubaum, I.H. Ballagh, K. Missaghi, E.R. Siuda, G. McMurray et al. 2016. Testing the evolutionary conservation of vocal motorneurons in vertebrates. Resp. Physiol. Neurobiol. 224: 2–10.

Alves, Y.M., L.P. Bergqvist and P.M. Brito. 2019. The dorsal and pectoral fin spines of catfishes (Ostariophysi: Siluriformes) from the Bauru Group (Late Cretaceous), Brazil: A comparative and critical analysis. J. S. Am. Earth Sci. 92: 32–40.

Anderson, K.A., R.A. Rountree and F. Juanes. 2008. Soniferous fishes in the Hudson River. Tran. Am. Fish. Soc. 137: 616–626.

Arce-H., M., J.G. Lundberg and M.A. O'Leary. 2017. Phylogeny of the North American catfish family Ictaluridae (Teleostei: Siluriformes) combining morphology, genes and fossils. Cladistics 33: 406–428.

Arratia, G. 2003. The Siluriform postcranial skeleton—An overview. pp. 121–157. *In*: Arratia, G., B.G. Kapoor, M. Chardon and R. Diogo (eds.). Catfishes, Vol. 1. Science Publishers, Enfield, NH.

Arratia, G. 2025. Siluriforms and the South American fossil record: their significance and the origin and evolutionary radiation of the group. *In*: Arratia, G. and R.E. Reis (eds.). Catfishes, a Highly Diversified Group. Volume 2: Evolution and Phylogeny. CRC Press, Boca Raton, FL.

Azpelicueta, M.M. and H.A. Britski. 2012. *Iheringichthys syi* (Siluriformes: Pimelodidae), a new pimelodid species from upper Paraná basin, Brazil. Neotrop. Ichthyol. 10: 45–52.

Baltz, D.M. and J.A. Compos. 1996. Hydrophone identification and characterization of *Cynoscion squamipinnis* (Perciformes: Sciaenidae) spawning sites in the Guld of Nicoya, Costa Rica. Rev. Biol. Trop. 44: 743–751.

Banner, A. 1972. Use of sound in predation by young lemon sharks, *Negaprion brevirostris* (Poey). Bull. Mar. Sci. 22: 251–283.

Baron, V.D., A.A. Orlov and D.E. El'yashev. 2005. Investigations of electric and acoustic activity of some representatives of the ichthyofauna of South Vietnam. J. Ichthyol. 45: S271–S279.

Betancur-R R. 2009. Molecular phylogenetics and evolutionary history of ariid catfishes revisited: a comprehensive sampling. BMC Evol. Biol. 9: 175.

Birindelli, J.L.O., D.F. Fayal and W.B. Wosiacki. 2011. Taxonomic revision of the thorny catfish genus *Hassar* (Siluriformes: Doradidae). Neotrop. Ichthyol. 9: 515–542.

Birindelli, J.L.O. and O.A. Shibatta. 2011. Morphology of the gas bladder in bumblebee catfishes (Siluriformes, Pseudopimelodidae). J. Morphol. 272: 890–896.

Birindelli, J.L.O., A. Akama and H.A. Britski. 2012. Comparative morphology of the gas bladder in driftwood catfishes (Siluriformes: Auchenipteridae). J. Morphol. 273: 651–660.

Birindelli, J.L.O. and J. Zuanon. 2012. Systematics of the jaguar catfish genus *Liosomadoras* Fowler, 1940 (Auchenipteridae: Siluriformes). Neotrop. Ichthyl. 10: 1–11.

Birindelli, J.L.O. 2014. Phylogenetic relationships of the South American Doradoidea (Ostariophysi: Siluriformes). Neotrop. Ichthyol. 12: 451–564.

Bosher, B.T., S.H. Newton and M.L. Fine. 2006. The spine of the Channel catfish *Ictalurus punctatus*, as an anti-predator adaptation: An experimental study. Ethology 112: 188–195.

Boyle, K., O. Colleye and E. Parmentier. 2014. Sound production to electric discharge: sonic muscles evolution in progress in *Synodontis* spp. catfishes (Mochokidae). Proc. R. Soc. B 281: 20141197.

Boyle, K.S., S. Riepe, G. Bolen and E. Parmentier. 2015a. Variation in swim bladder drumming sounds from three doradid catfish species with similar sonic morphologies. J. Exp. Biol. 218: 2881–2891.

Boyle, K.S., G. Bolen and E. Parmentier. 2015b. Agonistic sounds and swim bladder morphology in a malapterurid electric catfish. J. Zool. 296: 249–260.

Breder, C.M., Jr. 1968. Seasonal and diurnal occurrences of fish sounds in a small Florida bay. Bull. Am. Mus. Nat. Hist. 138: 327–378.

Bridge, T.W. and A.C. Haddon. 1893. III. Contributions to the anatomy of fishes—II. The airbladder and Weberian ossicles in the siluroid fishes. Phil. Trans. R. Soc. B 184: 65–334.

Britto, M.R. 2003. Phylogeny of the subfamily Corydoradinae Hoedeman, 1952 (Siluriformes: Callichthyidae), with a definition of its genera. Proc. Acad. Nat. Sci. Phila. 153: 119–154.

Buitrago-Suarez, U.A. and B.M. Burr. 2007. Taxonomy of the catfish genus *Pseudoplatystoma* Bleeker (Siluriformes: Pimelodidae) with recognition of eight species. Zootaxa 1512: 1–38.

Burgess, W.E. 1989. An Atlas of Freshwater and Marine Catfishes: A preliminary survey of the Siluriformes. Neptune City, New Jersey: Tropical Fish Hobbyist Publications, Inc.

Curran, D.J. 1989. Phylogenetic relationships among the catfish genera of the family Auchenipteridae (Teleostei: Siluroidea). Copeia 1989: 408–419.

Darwin, C. 1838. Voyage of the Beagle. Herausgeber: Verlag.

Demski, L.S., J.W. Gerald and A.N. Popper. 1973. Central and peripheral mechanisms of teleost sound production. Amer. Zool. 13: 1141–1167.

Devaere, S., D. Adriaens and W. Verraes. 2007a. Survey of the anguilliform Clariidae (Teleostei, Siluriformes) of Gabon and Republic of Congo, with description of two new species and key to the African clariid genera. Belg. J. Zool. 137: 99–109.

Devaere, S., G. Jansen, D. Adrieaens and P. Weekers. 2007b. Phylogeny of the African representatives of the catfish family Clariidae (Teleostei, Siluriformes) based on a combined analysis: independen evolution towards anguilliformity. J. Zool. Syst. Evol. Res. 45: 214–229.

Diogo, R., P. Vandewalle and M. Chardon. 1999. Morphological description of the cephalic region of *Bagrus docmac*, with a reflection on Bagridae (Teleostei: Siluriformes) autapomorphies. Neth. J. Zool. 49: 207–232.

Diogo, R., M. Chardon and P. Vandewalle. 2002. Osteology and myology of the cephalic region and pectoral girdle of the Chinese catfish *Cranoglanis bouderius*, with a discussion on the autapomorphies and phylogenetic relationships of the Cranoglanididae. J. Morphol. 253: 229–242.

Diogo, R., M. Chardon and P. Vandewalle. 2003. Osteology and myology of the cephalic region and pectoral girdle of *Centromochlus heckelii*, comparison with other auchenipterids, and comments on the synapomorphies and phylogenetic relationships of the Auchenipteridae (Teleostei: Siluriformes). Anim. Biol. 53: 397–416.

Diogo, R. and M. Chardon. 2004. On the osteology and myology of the cephalic region and pectoral girdle of *Franciscodoras marmoratus* (Lütken 1874), comparison with other doradids, and comments on the synapomorphies and phylogenetic relationships of the Doradidae (Teleostei: Siluriformes). Anim. Biol. 54: 175–193.

Diogo, R., M. Chardon and P. Vandewalle. 2004. Osteology and myology of the cephalic region and pectoral girdle of *Batrochoglanis raninus*, with a discussion on the synapomorphies and phylogenetic relationships of the Pseudopimelodinae and Pimelodidae (Teleostei: Siluriformes). Anim. Biol. 54: 261–280.

Diogo, R. 2005. Osteology and myology of the cephalic region and pectoral girdle of *Pimelodus blochii*, comparison with other pimelodines, and comments on the synapomorphies and phylogenetic relationships of the Pimelodinae (Ostariophysi: Siluriformes). Eur. J. Morphol. 42: 115–126.

Diogo, R. and Z. Peng. 2010. State of the art of Siluriform higher-level phylogeny. pp. 465–515. *In*: Grande, T., F.J. Poyato-Ariza and R. Diogo (eds.). Gonorynchiformes and Ostariophysan Relationships. CRC Press: Boca Raton.

Dobrin, M.B. 1947. Measurement of underwater noise produced by marine life. Science 105: 19–23.

Dufossé, L. 1874. Recherches sur les bruits et les sons expressifs que font entendu les poissons d'Europe et sur les organes producteurs de ces phenomenes acoustiques ainsi que sur les appareils de L'Audition de plusieurs de ces animaux. Deuxieme partie. Ann Sci. Nat. Zool. Paléontol. 20 (Cinquième Série): 1–134.

Fine, M.L., P.J. Friel, D. McElroy, C.B. King, K.E. Loesser and S. Newton. 1997. Pectoral spine locking and sound production in the channel catfish, *Ictalurus punctatus*. Copeia 1997: 777–790.

Fish, M.P. 1948. Sonic Fishes of the Pacific. Woods Hole Oceanographic Institution, Woods Hole, MA.

Fish, M.P. and W.H. Mowbray. 1970. Sounds of the Western North Atlantic Fishes: A Reference File of Biological Underwater Sounds. The Johns Hopkins Press: Baltimore.

Fricke, R., W.N. Eschmeyer and J.D. Fong. 2024. Eschmeyer's Catalog of Fishes: Species by family/Subfamily. Electronic version accessed March 8, 2024. Available at: http://researcharchive.calacademy.org/research/ichthyology/catalog/SpeciesByFamily.asp.

Freytag, G. 1968. Ergebnisse zur marinen bioakustik. Protok. Fisch. Tech. 11: 252–352.

Gannon, D.P., N.B. Barros, D.P. Nowocek, A.J. Read, D.M. Waples and R.S. Wells. 2005. Prey detection by bottlenose dolphins (*Tursiops truncatus*): an experimental test of the passive listening hypothesis. Anim. Behav. 69: 709–720.

Gayet, M. and F.J. Meunier. 1998. Maastrichtian to early late paleocene freshwater Osteichthys of Bolivia: additions and comments. pp. 85–144. *In*: Malabarba, L.R., R.E. Reis, R.P. Vari, Z.M.S. Lucena and C.A.S. Lucena (eds.). Phylogeny and Classification of Neotropical Fishes. Editora Universitaria EDIPUCRS, Porto Alegre, Brasil.

Gayet, M. and F.J. Meunier. 2003. Paleontology and paleobiology of catfishes. Catfishes, vol. 2. pp. 491–524. *In*: Arratia, G., B.G. Kapoor, M. Chardon and R. Diogo (eds.). Catfishes, Vol. 1. Science Publishers, Enfield, NH.

Geoffroy Saint-Hilaire, I. 1827. Poissons du Nil, de la mer Rouge et de la Méditerranée. pp 141-338. Description de l'Egypte ou recueil des observations et des recherches qui ont été faites en Égypte pendant l'expedition de l'Armée français, publié par les ordres de sa Majesté-L'Empereur Napoléon le Grand. Imprimerie Impériale, Paris.

Ghahramani, Z., Y. Mohajer and M.L. Fine. 2014. Developmental variation in sound production in water and air in the blue catfish *Ictalurus furcatus*. J. Exp. Biol. 217: 4244–4251.

Goel, H.C. 1966. Sound production in *Clarias batrachus* (Linnaeus). Copeia 1966: 622–24.

Haddon, A.C. 1881. The stridulating apparatus of *Callomystax gagata*. J. Anat. Physiol. 15: 321–326.

Hadjiaghai, O. and F. Ladich. 2015. Sex-specific differences in agonistic behaviour, sound production and auditory sensitivity in the callichthyid armoured catfish *Megalechis thoracata.* PLoS One 10: e0121219.

Hawkins, A.D. 2022. The use of underwater active and passive acoustics to locate and study fishes. Int. J. Aquat. Fish. Sci. 8: 80–86.

Heyd, A. and W. Pfeiffer. 2000. Ueber die Lauterzeugung der Welse (Siluroidei, Ostariophysi, Teleostei) und Ihren Zusammenhang mit der Phylogenie und der Schreckreaktion. Rev. Suisse Zool. 107: 165–211.

Howes, G.J. 1985. The phylogenetic relationships of the electric catfish family Malapteruridae (Teleostei: Siluroidei). J. Nat. Hist. 19: 37–67.

Hughes, A.R., D.A. Mann and D.L. Kimbro. 2014. Predatory fish sounds can alter crab foraging behaviour and influence bivalve abundance. Proc. R. Soc. B 281: 20140715.

Jawad, L.A., M.M.S. Farrag and J.M. Park. 2021. Interspecific and intraspecific differences in pectoral-fins spine morphology in Nile River and Lake Nasser catfishes, Siluriformes. Proc. Zool. Inst. RAS 325: 308–322.

Kaatz, I.M. 1999. The Behavioral and Morphological Diversity of Acoustic Communication Systems in a Clade of Tropical Catfishes (Pisces, Siluriformes). Syracuse, SUNY ESF. Ph.D. Thesis, SUNY College of Environmental Science and Forestry, Syracuse, NY.

Kaatz, I.M. and P.S. Lobel. 1999. Acoustic behavior and reproduction in five species of *Corydoras* catfishes (Callichthyidae). Biol. Bull. 197: 241–242.

Kaatz, I.M., D.J. Stewart, A.N. Rice and P.S. Lobel. 2010. Differences in pectoral fin spine morphology between vocal and silent clades of catfishes (Order Siluriformes): ecomorphological implications. Curr. Zool. 56: 73–89.

Kaatz, I.M. and D.J. Stewart. 2012. Bioacoustic variation of swimbladder disturbance sounds in neotropical doradoid catfishes (Siluriformes: Doradidae, Auchenipteridae): Potential morphological correlates. Curr. Zool. 58: 171–188.

Kaatz, I.M., A.N. Rice and P.S. Lobel. 2017. How fishes use sound: Quiet to loud and simple to complex signaling. pp. 1–12. *In*: Farrell, A.P. (ed.). Reference Model in Life Science: Encyclopedia of Fish Physiology: From Genome to Environment. Elsevier: Amsterdam.

Kailola, P.J. 2004. A phylogenetic exploration of the catfish family Ariidae (Otophysi: Siluriformes). The Beagle, Rec. Mus. Art Gall. N. Terr. 20: 87–166.

Kappas, I., S. Vittas, C.N. Pantzartzi, E. Drosopoulou and Z.G. Scouras. 2016. A time-calibrated mitogenome phylogeny of catfish (Teleostei: Siluriformes). PLOS ONE 11: e0166988.

Kastberger, G. 1977. Der trommelapparat der Doradiden (Siluriformes, Pisces). Zool. Jahrb. Allg. Zool. 81: 281–309.

Kéver, L., A.H. Bass, E. Parmentier and B.P. Chagnaud. 2020a. Neuroanatomical and neurophysiological mechanisms of acoustic and weakly electric signaling in synodontid catfish. J. Comp. Neurol. 528: 2602–2619.

Kéver, L., E. Parmentier, A.H. Bass and B.P. Chagnaud. 2020b. Morphological diversity of acoustic and electric communication systems of mochokid catfish. J. Comp. Neurol. 529: 1787–1809.

Knight, L. and F. Ladich. 2014. Distress sounds of thorny catfishes emitted underwater and in air: characteristics and potential significance. J. Exp. Biol. 217: 4068–4078.

Kovalchuk, O. and C.J. Ferraris. 2016. Late Cenozoic catfishes of Southeastern Europe with inference to their taxonomy and palaeogeography. Palaeont. Electr. 19.3.34A: 1–17.

Ladich, F. and M.L. Fine. 1994. Localization of swimbladder and pectoral motoneurons involved in sound production in pimelodid catfish. Brain Behav. Evol. 44: 86–100.

Ladich, F. 1997. Comparative analysis of swimbladder (drumming) and pectoral (stridulation) sounds in three families of catfishes. Bioacoustics 8: 185–208.

Ladich, F. 2001. Sound-generating and -detecting motor system in catfishes: design of swimbladder muscles in doradids and pimelodid. Anat. Rec. 263: 297–306.

Ladich, F. and I.P. Maiditsch. 2020. Temperature affects sound production in fish with two sets of sonic organs: The pictus cat. Comp. Biochem. Physiol. A 240: 110589.

Ladich, F. 2022. Shut up or shout loudly: Predation threat and sound production in fishes. Fish Fish. 23: 227–238.

Ladich, F. 2023. Hearing in catfishes: 200 years of research. Fish Fish. 24: 618–634.

Lechner, W., L.E. Wysocki and F. Ladich. 2010. Ontogenetic development of auditory sensitivity and sound production in the squeaker catfish, *Synodontis schoutedeni*. BMC Biol. 8: 10.

Lecorchick, J.R.L., A. Flecker and A.N. Rice. 2023. Acoustic variation in ictalurid catfishes. Cornell Undergrad. Res. J. 2: 41–53.

Lindseth, A. and P. Lobel. 2018. Underwater soundscape monitoring and fish bioacoustics: A review. Fishes 3: 36.

Lundberg, J.G., A.H. Bornbusch and F. Mago-Leccia. 1991. *Glandioglanis conquistador* N. Sp. from Ecuador with diagnoses of the Subfamily Rhambdiinae Bleeker and Pseudopimelodinae N. Subf. (Siluriformes: Pimelodidae). Copeia 1991: 190–209.

Lundberg, J.G. and A. Akama. 2005. *Brachyplatystoma capapretum*: a new species of goliath catfish from the Amazon Basin, with a reclassification of allied catfishes (Silutiformes: Pimelodidae). Copeia 2005: 492–516.

Mahajan, C.L. 1963. Sound producing apparatus in an Indian catfish *Sisor rhabdophorus* Hamilton. Zool. J. Linn. Soc. 44: 721–724.

Markl, H. 1968. Das schutzverhalten eines welses (*Hassar orestis* Steindacher) gegen angriffe von pirhanas (*Serrasalmus nattereri* Kner). Z Tierpsychol. 26: 385–389.

Mazlan, A.G., S. Abdullah, M.G. Shariman and A. Arshad. 2008. On the biology and bioacoustic characteristics of spotted catfish *Arius maculatus* (Thunberg 1792) from the Malaysian Estuary. Res. J. Fish. Hydrobiol. 3: 63–70.

Mo, T. 1991. Anatomy, relationships and systematics of the Bagridae (Teleostei: Siluroirdei) with a hypothesis of silluroid phylogeny. Theses Zool. 17: 1–216.

Mohajer, Y., Z. Ghahramani and M.L. Fine. 2015. Pectoral sound generation in the blue catfish *Ictalurus furcatus*. J. Comp. Physiol. A 201: 305–15.

Moodie, G.E.E. and M. Powers. 1982. The reproductive biology of an armored catfish, *Loricaria uracantha*, from Central America. Environ. Biol. Fish. 7: 143–148.

Odom, K.J., M. Araya-Salas, J.L. Morano, R.A. Ligon, G.M. Leighton, C.C. Taff et al. 2021. Comparative bioacoustics: A roadmap for quantifying and comparing animal sounds across diverse taxa. Biol. Rev. 96: 1135–1159.

Papes, S. and F. Ladich. 2011. Effects of temperature on sound production and auditory abilities in the striped raphael catfish *Platydoras armatulus* (Family Doradidae). PLOS ONE 6: e26479.

Parmentier, E. and R. Diogo. 2006. Evolutionary trends of swimbladder sound mechanisms in some teleost fishes. pp. 45–70. *In*: Ladich, F., S. Collin, P. Moller and B. Kapoor (eds.). Communication in Fishes Vol. 1. Science Publishers: Enfield, NH.

Parmentier, E., G. Fabri, I.M. Kaatz, N. Decloux, S. Planes and P. Vandewalle. 2010. Functional study of the pectoral spine stridulation mechanism in different mochokid catfishes. J. Exp. Biol. 213: 1107–1114.

Parmentier, E., R. Diogo and M. Fine. 2017. Multiple exaptations leading to fish sound production. Fish Fish. 18: 958–966.

Pfeiffer, W. and J.F. Eisenberg. 1965. Die lauterzeugung der dornwelse (Doradidae) und der fiederbartwelse (Mochokidae). Z. Morph. Oekol. Tiere 54: 669–679.

de Pinna, M.C.C. 1996. A phylogenetic analysis of the Asian catfish families Sisoridae, Akysidae, and Amlycipitidae with a hypothesis on the relationships of the Neotropical Aspredinidae (Teleostei, Ostariophysi). Fieldiana: Zool. 84: 1–83.

Pinton, A., J-F. Agnèse, D. Paugy and O. Otero. 2013. A large-scale phylogeny of *Synodontis* (Mochokidae, Siluriformes) reveals the influence of geological events on continental diversity during the Cenozoic. Mol. Phylogenet. Evol. 66: 1027–1040.

Protasov, V.R. 1965. Bioacoustics of Fishes. Nauka, Moscow.

Pruzsinszky, I. and F. Ladich. 1998. Sound production and reproductive behavior of the armored catfish *Corydoras paleatus* (Callichthyidae). Environ. Biol. Fish. 53: 183–191.

Reis, R.E. 1998. Anatomy and phylogenetic analysis of the neotropical callichthyid catfishes (Ostariophysi, Siluriformes). Zool. J. Linn. Soc. 124: 105–168.

Rice, A.N., S.C. Farina, A.J. Makowski, I.M. Kaatz, P.S. Lobel, W.E. Bemis et al. 2022. Evolutionary patterns in sound production across fishes. Ichthyol. Herpetol. 110: 1–12.

Rigley, L. and J. Muir. 1979. The role of sound production by the brown bullhead *Ictalurus nebulosus*. Proc. Penn Acad. Sci. 53: 143–134.

Rocha, M.S. 2012. Systematica da familia Pimelodidae Swainson, 1838 (Teleostei: Siluriformes). Ph.D. Dissertation, Instituto Nacional de Pesquisas da Amazonia- INPA, Manaus, Amazonas.

Sabaj, M.H. and J.L.O. Birindelli. 2008. Taxonomic revision of extant *Doras* Lacepède, 1803 (Siluriformes: Doradidae) with descriptions of three new species. Proc. Acad. Nat. Sci. Phila. 157: 189–233.

Sabaj, M.H. and M. Arce H. 2022. Towards a complete classification of the Neotropical thorny catfishes (Siluriformes: Doradidae). Neotrop. Ichthyol. 19: e210064.

Schachner, G. and F. Schaller. 1981. Schallerzeugung und schallreaktionen beim antennenwels (Mandim) *Rhamdia sebae sebae* Val. Zool. Beitr. 27: 375–392.

Schedel, F.D.B., A. Chakona, B.L. Sidlauskas, M.O. Popoola, N.U. Wingi, D. Neumann et al. 2022. New phylogenetic insights into the African catfish families Mochokidae and Austroglanididae. J. Fish Biol. 100: 1171–1186.

Schmidtke, D., J. Schulz, J. Hartung and K-H. Esser. 2013. Structure and possible functions of constant-frequency calls in *Ariopsis seemanni* (Osteichthyes, Ariidae). PLOS ONE 8: e64864.

Skelton, P.H. and G.G. Teugels. 1991. A review of the clariid catfishes (Siluroidei, Clariidae) occurring in southern Africa. Rev. Hydrobiol. Trop. 24: 241–260.

Skelton, P.H. 1993. A Complete Guide to the Freshwater Fishes of Southern Africa: Southern Book Publishers, Halfway House, South Africa.

Sleen, P. van der, A. Lugo Carvajal, J. Zuanon and M. Holmgren. 2020. Cats singing in the dark? Spawning aggregations of sound-producing fish in Amazonian floodplain forests. Environ. Biol. Fish. 103: 1265–1267.

Smith, M.E., P.C. Stewart, A.L. Webb and B.D. Rogers. 2009. Sound production and localization in loricariid catfishes. J. Acoust. Soc. Am. 125: 2487.

Sörenson, W. 1894–1895. Are extrinsic muscles of the air-bladder in some Siluroidae and the "elastic spring"' apparatus of others subordinate to the voluntary production of sounds? What is according to our present knowledge, the function of the Weberian ossicles? J. Anat. Physiol. 29: 109–139; 205–229; 399–423; 518–552.

Sousa, L. and J.L.O. Birindelli. 2011. Taxonomic revision of the genus *Scorpiodoras* (Siluriformes: Doradidae) with resurrection of *Scorpiodoras calderonensis* and description of a new species. Copeia 2011: 121–140.

Souza, P.M., Z. Olsen and S.J. Brandl. 2023. Paired passive acoustic and gillnet sampling reveal the utility of bioacoustics for monitoring fish populations in a turbid estuary. ICES J. Mar. Sci. 80: 1240–1255.

Stewart, D.J. 1985. A new species of *Cetopsorhamdia* (Pisces: Pimelodidae) from the Rio Napo basin of Eastern Ecuador. Copeia 1985: 339–344.

Stewart, D.J. and M.J. Pavlik. 1985. Revision of *Cheirocerus* (Pisce: Pimelodidae) from tropical freshwaters of South America. Copeia 1985: 356–367.

Stewart, D.J. 1986. Revision of *Pimelodina* and description of a new genus and species from the Peruvian Amazon (Pisces: Pimelodidae). Copeia 1986: 653–672.

Tavolga, W.N. 1960. Sound production and underwater communication in fishes. pp. 93–136. *In*: Lanyon, W.E. and W.N. Tavolga (eds.). Animal Sounds and Communication. American Institute of Biological Sciences: Washington, DC.

Tavolga, W.N. 1971. Acoustic orientation in the sea catfish, *Galeichthys felis*. Ann. NY Acad. Sci. 188: 80–97.

Tellechea, J.S., F.T-d Mello, I. Gonzalez-Bergonzoni and N. Vidal. 2011. Sound production and pectoral spine locking in a Neotropical catfish (*Iheringichthys labrosus*, Pimelodidae). Neotrop. Ichthyol. 9: 889–894.

Tellechea, J.S., F. Teixeira-De-Mello, I. Gonzalez-Bergonzoni and N. Vidal. 2013. Sound production in four species of the Loricariidae family. Braz. J. Biol. 73: 679–680.

Teugels, G.G. 1986. A systematic revision of the African species of the genus *Clarias* (Pisces: Clariidae). Ann. Mus. Roy. Afr. Cent. Sci. 247: 1–199.

Teugels, G.G., Sudarto and L. Pouyaud. 2001. Description of a new *Clarias* species from Southeast Asia based on morphological and genetical evidence (Siluriformes, Clariidae). Cybium 25: 81–92.

Tiepelt, H. 2005. *Siluris glanis* recordings. FishBase. https://www.fishbase.se/physiology/FishSoundsSummary.php?autoctr=172.

Vallone, E.R., R.I. Vezzosi and A.L. Cione. 2017. First fossil fish (Teleostei, Siluriformes) from the late Pleistocene of Santa Fe Province, Argentina. Alcheringa 41: 369–377.

van der Waal, B.C.W. 2000. Fish as a resource in a rural river catchment in the Northern Province, South Africa. Afr. J. Aquat. Sci. 25: 56–70.

Vigliotta, T.R. 2008. A phylogenetic study of the African catfish family Mochokidae (Osteichthyes, Ostariophysi, Siluriformes), with a key to its genera. Proc Acad. Nat. Sci. Phila. 157: 73–136.

Vijayakumar, P., N. Veerappan and T. Balasubramanian. 2012. Acoustic behavior of marine catfish *Arius arius* (Hamilton). Int. J. Env. Biol. 2: 187–194.

Wang, Z-T., D.P. Nowacek, T. Akamatsu, K-X. Wang, J-C. Liu, G-Q. Duan et al. 2017. Diversity of fish sound types in the Pearl River Estuary, China. PeerJ 5: e3924.

Webb, A.L. 2011. Sound Production in Two Loricariid Catfishes. M.Sc. Thesis, Department of Biology, Western Kentucky University, Bowling Green, KY.

Webb McAdams, A.L. and M.E. Smith. 2023. The relationship between body size and stridulation production in loricariid catfishes. J. Acoust. Soc. Am. 154: 3672–3683.

Wysocki, L.E. and F. Ladich. 2003. The representation of conspecific sounds in the auditory brainstem of teleost fishes. J. Exp. Biol. 206: 2229–2240.

Chapter 14

Insights Into Catfish Respiratory Mechanisms

Giacomo Zaccone,[1,*] *José Manuel Icardo*[2] and *Alessio Alesci*[3]

The catfish family Clariidae (order Siluriformes) comprises at least 14 genera and 116 species of freshwater, air-breathing fish. In addition, the family Heteropneustidae (genus *Heteropneustes*, two species) is often considered to be a subfamily of the Clariidae or, even to form together with the clariids, the superfamily Clarioidea (Sullivan et al. 2006). The species of the genus *Clarias* are widely distributed throughout Southeast Asia and continental Africa, whereas those of the genus *Heteropneustes* appear to be restricted to southern Asia. Both *Clarias* and *Heteropneustes* possess air-breathing organs (ABOs) and exhibit air-breathing capabilities, although several exceptions may occur (Graham 1997).

The ABOs in clariids include paired suprabranchial chambers that are lined by a modified respiratory epithelium, the suprabranchial chamber membrane. The suprabranchial chambers contain the so-named labyrinthine organs that appear as tree-like extensions of the gill arches 2 and 4. The labyrinthine organs show a large surface area and are covered by respiratory epithelium. In addition, the fusion of the epibranchial filaments of each gill arch results in the formation of flattened structures known as gill fans or plates. Gill fans are also covered by respiratory epithelium, separate the branchial from the suprabranchial cavity, and control ABO ventilation and air retention. Numerous studies have focused on the structure of the clariid ABOs over the years (Singh and Hughes 1971; Singh 1982; Dutta and Munshi 1985; Maina and Maloiy 1986; Graham 1997; Maina 2018; Maina et al. 2022). The existence of interspecific variations in the degree of ABO development has been related to variable dependence on aerial oxygen. Thus, several species, such as *Clarias mossambicus* have been described as obligate air breathers, while others, like *Clarias gariepinus*, are facultative air breathers (Graham 1997). Of note, obligate air breathers die when forced to be submersed, while facultative air breathers use the ABOs as supplementary/main source of oxygen when the levels of O_2 decrease or the need for O_2 uptake increases. It must be underscored that, in all the species of *Clarias*, the ABOs coexist with well-developed gills and that the capability to extract oxygen from both water and air depends on many variables, such as the morphological development of the ABOs, the O_2 content of the water and/or the capability for hypoxia acclimation

[1] Department of Veterinary Sciences, University of Messina, 98168 Messina, Italy.
[2] Department of Anatomy and Cell Biology, Faculty of Medicine, University of Cantabria, 39011 Santander, Spain.
[3] Department of Chemical, Biological, Pharmaceutical and Environmental Sciences, University of Messina, 98166 Messina, Italy.
Emails: jose.icardo@unican.es; aalesci@unime.it
* Corresponding author: zacconegiacomo@gmail.com

(Graham 1997). Most other determinants of hypoxia tolerance (such as the capability to extract O_2 from the air, oxygen affinity for hemoglobin, or the thickness of the barrier exchange) have not been explored, or the data only refer to a few clariid species (Graham 1997; Maina 2018).

The members of the family Heteropneustidae also exhibit ABOs. However, they differ from those found in *Clarias*. The species most studied, *Heteropneustes fossilis*, contains similar suprabranchial chambers and gill fans. However, *H. fossilis* lacks labyrinthine organs. Instead, the suprabranchial chambers are connected to a pair of tubular structures, the air sacs, that extend back into the body, one on each side of the spine, embedded in the epaxial muscles (Rawat et al. 2022). Well-developed gill fans regulate air sac ventilation (Olson et al. 1990; Munshi 1993; Graham 1997).

This chapter aims to review the structural characteristics of the gills and ABOs in two species, the North African sharp tooth catfish *Clarias gariepinus* and the Asian stinging catfish *H. fossilis*. We have tried to establish similarities and differences in organ composition, including the structure of the exchange barrier. In addition, the ultrastructure and immunochemical distribution of the cells involved in oxygen sensing and immune defense will be analyzed. Gills and ABOs interface with the external environment, are the site of O_2 uptake and O_2-sensing regulation and constitute a first-order defense barrier against external pathogens and toxic insults.

Morphological Features of Gills and ABOs – Comparison Between *Clarias* and *Heteropneustes*

The Gills

The gills of the two species analyzed here exhibit the typical teleost configuration with gill arches giving rise to paired gill filaments (Figure 1A) that support the respiratory lamellae (Figure 1B). The filaments arise almost free from the arches, allowing water to flow directly from the buccal cavity to the opercular chambers. The filament epithelium is covered by pavement, squamous cells characterized by the presence of surface microridges arranged in fingerprint whorls (Figure 1C). The gill epithelium also contains light and dark chloride cells and mucous (goblet) cells (Maina 2018; Maina et al. 2022) (Figure 1D). Chloride cells (ionocytes) are characterized by the presence of a large number of mitochondria and a well-developed tubular system. Chloride cells have long been implicated in ion transport and acid-base regulation and may be joined to each other and pavement cells by tight junctions (Olson 2000). On the whole, *Clarias* and *Heteropneustes* show many similarities in gross morphology. However, they also show important structural and ultrastructural differences.

In *Clarias gariepinus*, the gill filament contains two types of mast cells (Figure 1D). Typical mast cells exhibiting large, very dense, rounded granules are located on the apical side of the epithelium. These cells are globular in the lamellae but show a pyramidal shape in the interlamellar space (Figure 1D). Here, the cell apex often reaches the epithelial surface, and the granules may extend tube-like prolongations that appear to serve as conduits for secretion (Figure 1E). In fact, dense material similar to that contained in the mast cell granules appears stuck to the surface microridges (Figure 1E, inset). The second mast cell type is located in the basal side of the epithelium, contains a very large number of small granules, and appears often associated with granular leukocytes (Maina et al. 2002). Leukocytes show numerous, small rod-shaped granules with a dense core (Figure 1F). They lack junctional specializations and often show a bilobed nucleus and amoeboid cytoplasmic prolongations (Maina et al. 2022). These cells have tentatively been identified as eosinophil leukocytes. Notably, eosinophils show a very variable granular morphology in fish (Icardo 2019).

By contrast, the gill epithelium in *Heteropneustes fossilis* (Figure 1G) contains a very low number of mast cells, always with large cytoplasmic granules, and lacks granular leukocytes. However, the epithelium contains numerous cells that could be identified as neuroendocrine (NEC) cells. NECs are characterized by a clear cytoplasm and the presence of small, membrane-bound dense granules (Figure 1H). NECs appear either isolated or forming small groups and are present

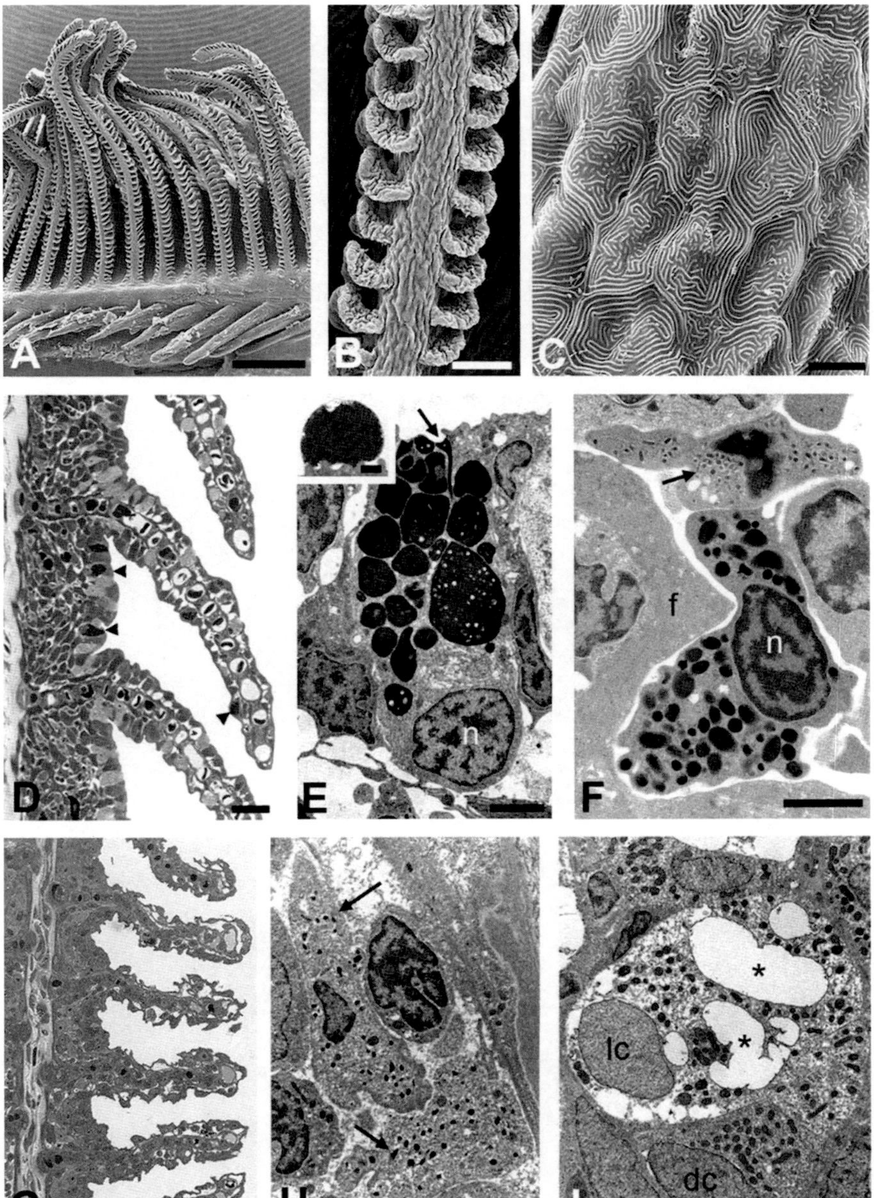

Figure 1. Gill structure. (A) SEM of *Heteropneustes fossilis*. Filaments arise free from the gill rack. Bar: 500 µm. (B) SEM of *Heteropneustes fossilis*. The filaments show regularly arranged lamellae. Bar: 50 µm. (C) SEM. *Heteropneustes fossilis*. Filament surface. Pavement cells show microridges in whorl patterns. Bar: 5 µm. (D) *Clarias gariepinus*. Semithin section. Detail of lamellae with the vascular space. Arrowheads indicate mast cells. Bar: 20 µm. (E) *Clarias gariepinus*. TEM. Detail of pyramidal mast cell with large granules. A tube-like extension abuts the epithelial surface. n, nucleus. Bar: 2 µm. Inset of E: *Clarias gariepinus*. TEM. A seemingly secretory drop is stuck to surface microridges. Bar: 500 nm. (F) *Clarias gariepinus*. TEM. Mast cells with numerous, small granules co-localize with eosinophils (arrow). n, mast cell nucleus. (f) filament-rich cells. Bar: 2 µm. (G) *Heteropneustes fossilis*. Semithin section. Filament and lamellae. Mast cells are not apparent. Dark points correspond to erythrocytes. Bar: 20 µm. (H) *Heteropneustes fossilis*. TEM. Neuroendocrine cells. Bar: 2 µm. (I) *Heteropneustes fossilis*. TEM. Light (lc) and dark (dc) chloride cells. Asterisks indicate a large apical cavity. Bar: 3 µm.

in the epithelium of filaments and lamellae. On the other hand, light and dark chloride cells appear to be more numerous than in *C. gariepinus*. Also, many light chloride cells in *H. fossilis* show large apical cavities (Figure 1I). These cavities are much less evident in *C. gariepinus*.

ABOs in Clarias gariepinus

As stated above, the ABOs in *C. gariepinus* include the suprabranchial membrane, the labyrinthine organ, and the gill fans. *Heteropneustes fossilis* also has a suprabranchial membrane and gill fans but lacks the labyrinthine organ. All of these ABOs share a common structure. They exhibit respiratory islets and intervascular, non-respiratory areas (Figure 2A). In the two areas, the surface is covered by pavement cells that exhibit microridges arranged in whorls (Maina 2018; Maina et al. 2022). The respiratory epithelium is traversed by capillaries that orientate transversely to the epithelial surface (Figure 2A and inset of A). These capillaries show protruding endothelial cells and are surrounded by cells rich in microfilaments (Figure 2B). Microfilament-rich cells are joined by desmosomes,

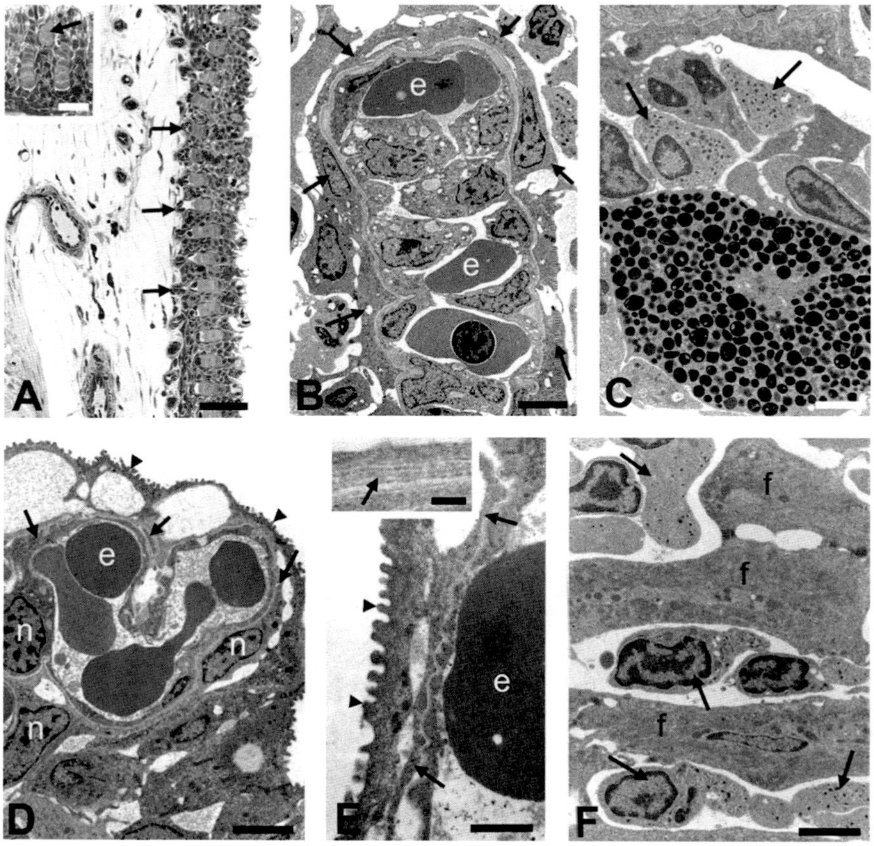

Figure 2. Structure of ABOs in *Clarias gariepinus*. (A) Gill fan. Semithin section. The respiratory epithelium is traversed by capillaries (arrows) arranged in parallel. Bar: 40 µm. Inset of A: Labyrinthine organ. Detail of two capillaries and a goblet cell (arrow). Bar: 20 µm. (B) Suprabranchial membrane. TEM. The capillary in the image is surrounded by a basement membrane and ensheathed by pillar-like cells (arrows). Endothelial cells protrude into the capillary lumen. e, erythrocyte. Bar: 3 µm. (C) Suprabranchial membrane. TEM. A large mast cell with numerous granules co-localizes with eosinophils (arrows). Bar: 2 µm. (D) Suprabranchial membrane. TEM. The distal tip of the capillary contains erythrocytes (e) and is surrounded by pillar-like cells (arrows). n, pillar-like cell nucleus. Arrowheads, pavement cells. Note large spaces between pavement and pillar-like cells. Bar: 3 µm. (E) Suprabranchial membrane. TEM. The space between pavement cells (arrowheads) and pillar-like cells (arrows) is narrow in this area. e, erythrocyte. Bar: 1 µm. Inset of E: Labyrinthine organ. In this area, only an extremely thin cytoplasmic rim (arrow) of pillar-like cells separates the capillary wall and the pavement cells. Bar: 200 m. (F) Gill fan. In the non-respiratory areas, a flattened network of microfilament-rich cells (f) accommodates numerous eosinophils (arrows). Bar: 2 µm.

appear closely apposed to the capillary basement membrane, and ensheath the capillaries, isolating them from the surrounding tissue. In addition, microfilament-rich cells form a continuous network accommodating several cell types. A recent study in *C. gariepinus* has shown the presence within this network of goblet cells, chloride cells, eosinophils, and mast cells (Maina et al. 2022). Mast cells are restricted to the basal half of the epithelium and are frequently seen in contact with eosinophils (Figure 2C).

The distal tip of the capillaries develops a complex exchange barrier at the epithelial surface. In most cases, the capillary tip is covered by the cytoplasm of the microfilament-rich cells, which is thus interposed between the thick basement membrane of the capillaries and the pavement epithelium (Figure 2D). The exchange barrier is further thickened by the presence of spaces of variable width located under the pavement epithelium. These spaces develop despite the presence of tight junctions between the microfilament-rich cells and pavement cells. This complex structural arrangement suggests that the ABO potential for efficient gas exchange is low. Only occasionally does the exchange barrier become thinner with a great reduction of the distance between capillaries and pavement cells (Figure 2E). However, even in the thinnest areas, a cytoplasmic rim appears always to be forming part of the exchange barrier (Figure 2E, inset).

The non-respiratory areas of the ABOs show a flat epithelium that contains, except for the capillaries, similar cell types and structural organization to that observed in the respiratory areas. The epithelium is also formed by a network of microfilament-rich cells that accommodates other cell types (Figure 2F). The main difference is probably the flattening of the network. The absence of capillaries and the flat arrangement explain the height differences between the two areas. In addition, the number of mast cells appears to have mostly been reduced, whereas the number of eosinophils appears to be higher than in the respiratory areas.

The Air Sacs

The air sacs of *Heteropneustes fossilis* are tubular structures that extend along the dorsal trunk, organize into concentric layers (Figure 3A), and show a roughly circular cross-section (Figure 3A, inset). The luminal surface of the air sacs exhibits respiratory islets and non-respiratory areas covered by pavement cells with microridges (Figure 3B and inset). Noticeably, this surface is similar to that shown by all the ABOs in *C. gariepinus* despite the large differences in gross structure (Maina 2018; Maina et al. 2022).

The epithelium of the respiratory areas contains capillaries that course transversely to reach the epithelial surface (Figure 3C). However, the transverse arrangement is less apparent in the air sac than in the ABOs of *C. gariepinus*. Also, capillaries may course under the epithelial surface instead of abutting directly (Figure 3C). Ultrastructural observations indicate that the capillaries are surrounded by a basement membrane and ensheathed by a layer of pillar-like cells (Figure 3D). These two elements appear to be less developed than in *C. gariepinus*. Nonetheless, the layer formed by pillar-like cells is maintained at the epithelial surface, where the exchange barrier develops. This barrier shows a close apposition of the capillary wall and pavement layer with only a thin cytoplasmic rim of pillar-like cells interposed (Figure 3E). This indicates a higher potential for respiratory interchange than in the ABOs of *C. gariepinus*. Spaces between the pavement epithelium and the layer of pillar-like cells, which are very apparent in the ABOs of *C. gariepinus*, are only occasionally observed in the air sac.

Similar to what occurs in the ABOs of *C. gariepinus*, the air sac epithelium is formed by a network of microfilament-rich cells joined by desmosomes. Goblet cells (see Figure 3C) and NECs (Figure 3F) accommodate within this network. Most NECs are located in the basal half of the epithelium. The epithelium also contains a very small number of typical mast cells. Chloride cells could not be recognized. In addition, the air sac contains a special type of granulocyte that shows regular, rounded granules (Figure 3G). Each of these granules is packed with small, rounded vesicles that could be filled with a dense material or show a clear appearance. However, despite

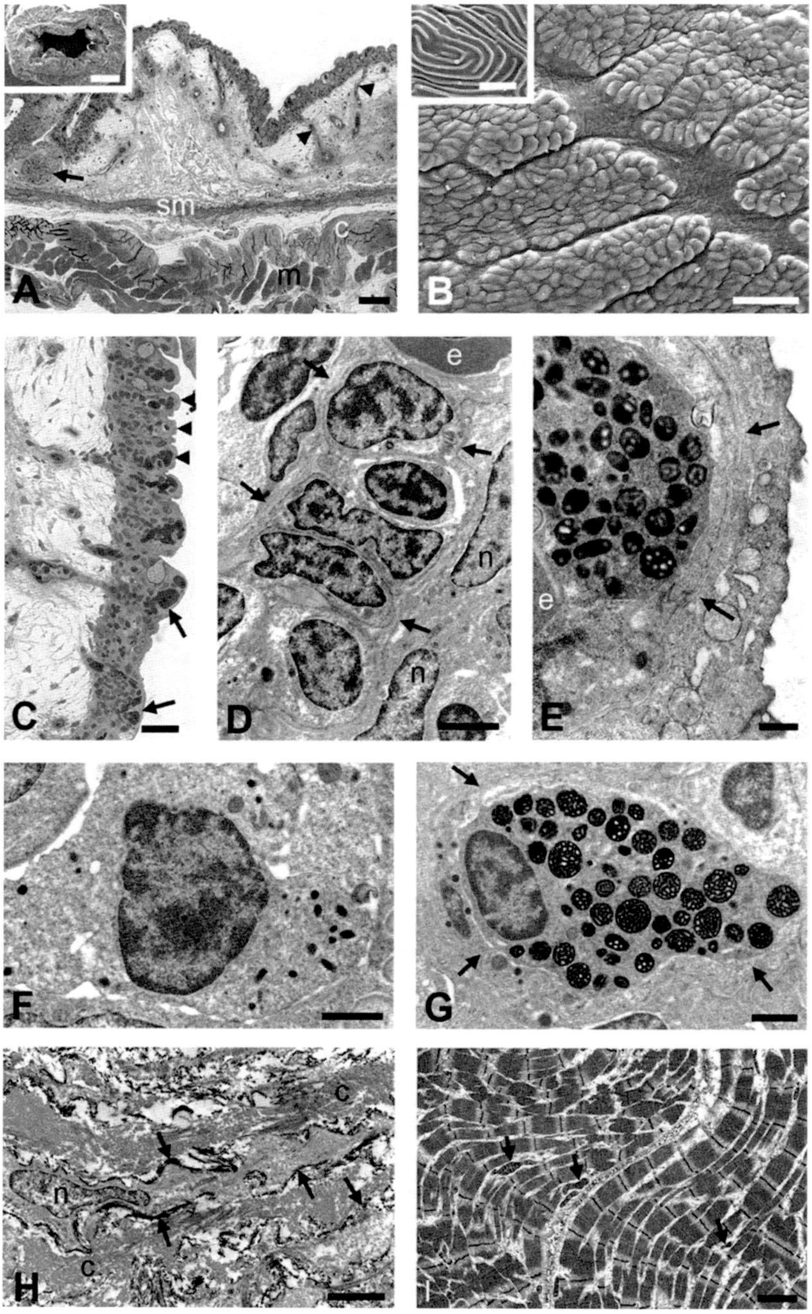

Figure 3. *Heteropneustes fossilis*. Air sac. (A) Semithin section. The wall is organized into layers. The thick subepithelium contains a nerve (arrow) and several capillaries (arrowheads) that reach the respiratory epithelium. c, collagen. m, skeletal muscle. sm, smooth muscle. Bar: 50 μm. Inset of A: SEM. Cross-section of the air sac. Bar: 500 μm. (B) SEM. The respiratory areas are elevated due to the underlying capillaries. Bar: 100 μm. Inset of B: SEM. Detail of cell surface microridges. Bar: 2 μm. (C) Semithin section. Several capillaries (arrowheads) traverse the epithelium in parallel, while others (arrows) run under the epithelium. Large goblet cells are apparent. Bar: 20 μm. (D) TEM. The capillary wall is surrounded by the basement membrane and pillar-like cells (arrows). e, erythrocyte. n, pillar cell nucleus. Bar: 2 μm. (E) TEM. The exchange barrier includes a thin rim of pillar-like cell cytoplasm (arrows). The capillary lumen contains an erythrocyte (e) and a granulocyte. Bar: 500 μm. (F) TEM. Detail of neuroepithelial cells with membrane-bound granules. Bar: 1 μm. (G) TEM. Granulocyte in a capillary (arrows) exhibits rounded granules with dense and empty vesicles. Bar: 1 μm. (H) TEM. Elongated smooth muscle cells are surrounded by collagen (c) and decorated with elastin fibers (arrows). n, cell nucleus. Bar: 2 μm. (I) TEM. Two skeletal muscle cells show aligned myofibrils and scarce mitochondria (arrows). Bar: 2 μm.

a previous assertion, these cells appear exclusive of the vascular system and are not observed in the tissue (Figure 3E and G). This suggests that they are a particular type of granular leukocyte, probably basophils or eosinophils. The empty appearance of many of the small vesicles could be due to a process of piecemeal degranulation that is typical of eosinophils (Simon et al. 2020). However, other cells involved in immune responses, such as basophils and mast cells, use similar ways of secretion. The reason these cells were not observed in the gills of the same *H. fossilis* specimens is unknown.

Under the luminal epithelium, the air sacs show a thick connective layer rich in vascular elements and nerves, a thin, smooth muscle layer, and a thick external layer formed by skeletal (cucullaris) muscle embedded in dense collagen (Figure 3A). In the smooth muscle layer, most smooth muscle cells appear to be oriented circumferentially, are surrounded by collagen and are often decorated with elastin material (Figure 3H). The skeletal muscle shows the typical myofibrils, is richly innervated with cholinergic nerve fibers (Zaccone et al. 2018), and contains a seemingly small number of mitochondria (Figure 3I).

The present observations add to previous structural descriptions of the gills and ABOs of both *C. gariepinus* (Maina 2018; Maina et al. 2022) and *H. fossilis* (Lauriano et al. 2021). We highlight the existence in the ABO epithelia of microfilament-rich cells connected by desmosomes that form a structural network that houses mast cells, granular leukocytes, and monocytes. In addition, the capillaries that traverse the respiratory epithelium of the ABOs are entirely covered by microfilament-rich cells. These cells resemble the pillar cells that define the vascular compartment of the gill lamellae (Olson et al. 1994). As a result, they ought to be categorized as pillar cells. The ABO capillaries have previously been reported to be covered by pilaster or pillar-like cells in *Clarias* and other bimodal breathers (Hughes and Munshi 1973; Maina and Maloiy 1986; Munshi and Hughes 1991; Maina et al. 2022). Thus, pillar cells appear to be crucial to the structural organization and maintenance of the respiratory epithelium. The fact that the gills and ABOs share structural similarities supports the notion that the latter are gill derivatives (Hughes and Munshi 1973; Olson et al. 1994; Abd-El Maksoud et al. 2008). It should be underscored that the air sacs of *H. fossilis* are organized into concentric tissue layers and lack chloride cells and any cartilaginous structure. However, they share the general organization of the respiratory epithelium.

Immunology of Mucosal Barrier in Catfish

The mucosal barriers of catfish constitute the first line of defense against pathogen invasion. Simultaneously, mucosae carry out a diverse array of critical physiological processes such as nutrient adsorption, osmoregulation, waste excretion and environmental sensing. Catfish depend more heavily on mucosal barriers than terrestrial vertebrates since they are continuously interacting with the aquatic microbiota (Peatman et al. 2015). Here we describe aspects of innate and adaptative immunoregulatory molecules in the primary mucosal tissues (gills and ABOs) of two catfish species (see Lauriano et al. 2021; Maina et al. 2022).

Immune Cells and Neuroepithelial Cell Systems in the Gills and Air Sac of Heteropneustes fossilis – Immunocytochemical Markers

Heteropneustes fossilis belongs to one of the most diversified and well-adapted groups of vertebrates that may be found in marshy settings. In order to accommodate the combined functions of gills and ABOs, structural changes of the respiratory organs and their circulatory supply are necessary to shift from aquatic to aerial respiration (Dutta and Munshi 1985). The expression of acetylcholine (ACh) via its nicotinic and muscarinic receptors in the gills and ABOs of the species studied—possibly a site for O_2 sensing—could be strictly correlated with the response to hypoxia due to the ACh stimulatory effects on sensory nerve fibers and cardiorespiratory reflexes (Burleson and Milson 1995a, b). However, the cellular mechanisms implicated in this response are still unclear. NECs and mucous cells in the gill filaments and the lamellae of the respiratory air sac express ACh. A cholinergic route in the gills and ABOs of *H. fossilis* may contribute to the hyperventilatory response to hypoxia via the nicotinic and muscarinic receptors (Figure 4A) that are present in VAChT-positive NECs (Zaccone et al. 2020) and in cholinergic mucous cells (Burleson and Milson 1995; Regan et al. 2011). GABA R1 is present in the gill neurons of several teleost species, and cholinergic neurons have been described in the filament core of trout and mudskippers (Porteus et al. 2013; Zaccone et al. 2017, 2019). ACh has also been demonstrated in the dome apical caps of the epithelial cells of the gill lamellae of *Pantodon bucholzi* (Capillo et al. 2021).

A co-localization of the antimicrobial peptide Piscidin 1 (Pis1) and ACh occurs in fish gill NECs and goblet cells, as reported by Lauriano et al. (2021). A substantial body of evidence indicates that goblet cells are implicated in the innate and adaptive immune responses that take place at the mucosal surfaces in vertebrates (Knoop and Newberry 2018). In the human lung, goblet cells can synthesize antimicrobials and mucins that have been related to the development of host defense (Widdicombe and Wine 2015; Whitsett 2018). The mucous release is mediated, at least partially, by prostaglandins and ACh (Knoop and Newberry 2018). In the digestive tract, the immune functions of sentinel goblet cells have recently come to light (Whitsett 2018). These additional functions may also be operative in fish (Alesci et al. 2022a, 2023). Further research will be needed to ascertain whether all the fish goblet cells are implicated in immune responses or whether these responses are unique to a subset of goblet cells located in the airway epithelium of air-breathing fish coping with hypoxic environments. In addition, the release of ACh from goblet cells in gills and ABOs, the interaction of these cells with the nervous system, and the significance of the interaction, if any, with the local immune system are promising areas of research.

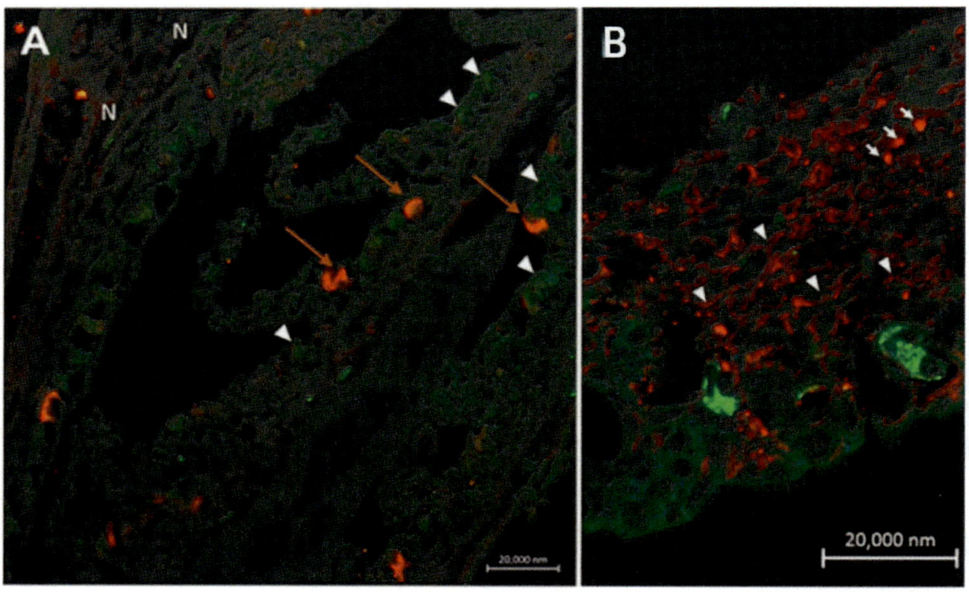

Figure 4. Immunocytochemistry of gills and air-breathing organs. (A) Gill filaments of *Heteropneustes fossilis* labeled with antibodies against 5-HT and the alpha 7 subunit of the acetylcholine nicotinic receptor (alpha 7 nAChR). Neuroepithelial cells (NECs) located within the multi-layered epithelium of the gill filament and gill lamellae coexpress the two markers, as shown by red arrows. (B) Superbranchial chamber of *Clarias gariepinus* immunostained with antibodies against Piscidin1 and 5 HT. Several Pis1 positive mast cells (arrows) rest close to a network of Pis1 positive nerve fibers. Note also the colocalization of eosinophils (green) and mast cells (red). Scale bars: 20 μm.

We discovered in the gills of *Heteropneustes fossilis* a NEC population that co-expressed GAD67 and 5-HT (Lauriano et al. 2021). In addition, serotonergic and cholinergic NEC populations have been described in the gills of *Kryptolebias marmoratus* (Regan et al. 2011). It is currently unknown whether GABA release by NECS may regulate mucus synthesis and secretion by goblet cells. Recent research (Kobayashi and Tata 2018; Barrios et al. 2019) has shown that GABA secretion induces mucous discharge for activating mucous cells. Our research suggests that the neurocircuitry that innervates NECs in the gill and ABO epithelia may regulate NEC secretion (Lauriano et al. 2021). Although typically thought of as O_2 sensors in the mammalian lung, NECs also have a role in controlling goblet cell hyperplasia (Kobayashi and Tata 2018). It is becoming increasingly clear that goblet cells, besides secreting mucus, play a significant role in mucosal immunity. The immunohistochemical localization of Pis1 in the goblet cells of gills and ABOs is consistent with reports of Piscidin expression in mast cells, rodlet cells, and other epithelial cell types in the filaments and lamellae of teleost gills of a wide taxonomic range of fish species. In the Atlantic cod, Pis1 immunoreactivity is detected in the gas gland cells, parasympathetic ganglia of the swim bladder wall, mucosal epithelium of the proximal intestine, and cartilage cells of the gill filaments (Ruangsri et al. 2012a). Nerve bundles in the submucosal layer of the air sac and the nerve plexuses surrounding the vascular walls of the gill efferent filament arteries also show a significant Pis1 immunoreactivity in *H. fossilis*. The NECs of the gills and air sac are also Piscidin 1-positive in *H. fossilis* (Lauriano et al. 2021). In addition, the neural tissues of the Atlantic cod are positive for Pis1 (Ruangsri et al. 2012a), suggesting that Piscidin 1 may be neurogenic. This finding raises the possibility of interactions between the immune and nervous systems. Intersystem communication is also suggested by the association of eosinophil granulocytes with NECs and NEC clusters in *H. fossilis*. Pis1 generated in NECs may also have immunological functions since GABA-producing NECs may collaborate with other cell types, such as Pis1-positive mucous cells, to establish an adaptive immune defense network. Piscidin 1 is widely distributed in fish epithelia, both in immune-related and non-immune tissues (Ruangsri et al. 2012b). The possible role of Pis1 as a cytoprotective molecule that may help to maintain tissue homeostasis points to yet another potential function for this ubiquitous peptide (Ruangsri et al. 2012b). Finally, since both NECs and sensory neurons contain neuropeptides, the release of piscidins by NECs (Figure 4B) may be a component of a sophisticated network that establishes interactions between the nervous and immune systems (Zaccone et al. 2006; Aresti Sanz and El Aidy 2019). The neuropeptides released by sensory neurons stimulate immune cells to proliferate and migrate. The majority of the known piscine antimicrobial peptides (AMPs) are present in mucous secretions (Fernandes et al. 2010). As previously described, fish AMPS (i.e., piscidins) are species-specific and support the innate immune system in fish skin and several other organs. Pis1 immunoreactivity is present in the goblet cells in the gills and the air sac of the catfish *H. fossilis*. In addition to their function as hypoxia-sensitive chemoreceptors, NECs are candidates for potential immunomodulatory effects by releasing chemicals that induce mucous production and activate the mucosal immune system. In turn, goblet cells can modulate the activity of the mucosal immune system in response to environmental variables.

Immune Cells in the Gills and ABOs of Clarias gariepinus

This section deals with the presence and function of immune cells and with the immune regulatory signaling, in the respiratory organs of the catfish *Clarias gariepinus*. Immune cells include lymphocytes, granular leukocytes and mast cells. However, it must be underscored that proper identification of the granulocyte types in fish remains conflictive. Granulocytes can be identified primarily by hematological stains or by the ultrastructural resemblance to mammalian granulocytes (Ainsworth 1992; Icardo 2019). For instance, carps show all three types of granulocytes (neutrophils, eosinophils, and basophils), although the basophils resemble mammalian mast cells (Ainsworth 1992). However, the number, shape, and content of individual granules vary widely even between related species. Furthermore, fish eosinophils and basophils appear to be almost species-specific (Icardo 2019).

Eosinophils

Serotonin (5-HT) is a central nervous system signaling mediator. Wan cells, such as mast cells and T lymphocytes, can synthesize and release 5-HT (Roumier et al. 2019). This affects immune regulation through various 5-HT receptors of the 5-HT1, 5-HT2, 5-HT3, 5-HT4, and 5-HT7 classes that are expressed by immune cells of both the innate and adaptative immune cell systems. Most of the body's 5-HT circulates in the bloodstream, which is absorbed by the blood platelets and released upon platelet activation. In recent years, 5-HT has been linked to a variety of immunoregulatory processes, including the activation of T cells and the recruitment of granular leucocytes (Herr et al. 2017). The primary cell type attracted to the lung during inflammation is the eosinophil leukocyte (Kang et al. 2013). Previous research has shown that 5-HT controls the migration and recruitment of eosinophils as well as the function of the specific signaling molecules involved in this process (Kang et al. 2013). In *Carassius auratus*, inflammatory responses to parasite infections are mediated by eosinophils (Fujimaki and Isoda 1990; Zinkl et al. 1991) Fish have cellular and humoral innate components at mucosal surfaces that provide for innate immunity (Gomez et al. 2013; Salinas 2015). Antimicrobial peptides are expressed in the skin, gill, and gut of fish (Gomez et al. 2013) and are essential components of the gill mucus and mucosal immunity (Capillo et al. 2021). Eosinophils, mast cells, and basophils are the primary effector cells of innate immunity and are crucial to the processes of defense against bacterial, viral, and parasite infections (Metcalfe 2008; Kita 2013; Lauriano et al. 2019). However, immune cells differ in terms of growth, maturation, and location within the tissues (Metcalfe 2008). Mast cells are among the first cells to perceive danger signals coming from the external environment because of their advantageous location at the major body surfaces, such as the skin and the respiratory and gastrointestinal epithelia (Metcalfe 2008). Activated mast cells help to coordinate the early stages of the innate immune response, stimulate the recruitment of eosinophils and basophils and take part at the beginning of the regulatory and adaptative immune responses (Metcalfe 2008). Following chemotactic stimuli, eosinophils are attracted to the tissues (Zinkl et al. 1991; Simon et al. 2020). As nearly all immune cells express 5-HT receptors, the evidence for 5-HT significance in the development of immune responses is clear. 5-HT regulates cytokine release and cell activation by stimulating monocytes and lymphocytes.

Mast Cells

Many disorders, including infections, induce the simultaneous activation of mast cells, eosinophils, and basophils. The chemicals that are secreted from, or expressed on the surface of, activated immune cells have mostly been studied in mammalian tissues (Metcalfe 2008). The presence of cell-specific surface molecules indicates specific cell involvement in a particular illness or defense mechanism and is correlated with a particular phenotype in tissue repair processes and adaptive immunological responses. By expressing complement receptors, eosinophils can detect infections and stimulate the innate immune response (Simon et al. 2020). Activated eosinophils release several soluble mediators, including cytokines, chemokines, growth factors, and bioactive lipids (Travers and Rothenberg 2015). Tyrosine hydroxylase is expressed by eosinophils, which also regulate several homeostatic processes in adipose, mucosal, and bone marrow tissues (Jung 2016). By releasing the granule content, eosinophils induce mast cells to release a variety of soluble mediators such as cytokines, histamine, and serotonin (Piliponsky et al. 2019). Eosinophils also produce nitric oxide that inhibits viral multiplication (Phipps et al. 2007). Since surface receptors bind neurotransmitters and neurochemicals at the site of neuron release, the expression of receptors on immune cells, including eosinophils and mast cells, suggests close interactions between these cells and the nervous system. Fish have a lot of mast cells in the gills, gut, and skin, where they operate very similarly to the mammalian mast cells (Piliponsky et al. 2019; Alesci et al. 2022b, c, d; Zaccone et al. 2022; Lauriano et al. 2023). While the structure, internal contents, and cell surface receptors of eosinophils have been defined in a wide range of vertebrate species, this matter has received much less attention in fish. The coexistence of mast cells and eosinophils in both the epithelium and subepithelium in the gills and ABOs of *Clarias* is a notable characteristic. This finding agrees with previous reports on the gills of the butterfly fish *Pantodon bucholzi* (Capillo et al. 2021) and on mammalian vertebrates where the two types of immune cells appear in homeostatic settings (Ratcliffe 2016). In *Clarias* and *Pantodon*, the use of a battery of antibodies against 5-HT, vesicular acetylcholine transporter (VAChT), neuronal nitric oxide synthase, and Piscidin 1 reveals the co-localization of eosinophils and mast cells (Figure 4B) as well as the presence of contacts between the immune cells and nerves (Maina et al. 2022). The increase of mast cells in the subepithelium and the connections with eosinophils may be related to the eosinophil production of chemoattractants like 5-HT. Nitric oxide (NO), ACh, and 5-HT are released by mast cells. According to Gilchrist et al. (2004), NO plays a crucial role in controlling the generation of chemokines and cytokines by mast cells. Mast cells may even have a secretory role, as shown by the formation of tube-like structures in the interlamellar epithelium of *C. gariepinus* (Maina et al. 2022). While this is surprising, cells evolve with many biological roles, and this phenotypic characteristic may be congruent with the local environment (Metcalfe 2008). On the other hand, we have provided evidence of a close relationship between Piscidin 1-positive mast cells and Pis1-positive nerve fibers (Figure 4B). In the ABOs of *C. gariepinus* (Maina et al. 2022). This demonstration opens new avenues for research. Forsythe and Bienenstock (2012) noted the importance of mast cell-nerve functional units in physiology and pathology. According to recent research on the cholinergic system of immune cells, ACh synthesized by immune cells plays a critical role in controlling immune functions by activating signals that modulate the synthesis of cytokines (Fujii et al. 2017). ACh and the choline acetyltransferase, the ACh-degrading enzymes AChE and BuChE, and the ACh muscarinic and nicotinic receptors (mAChRs and nAChRs) are the main components of the cholinergic system. Lipopolysaccharides activate the cholinergic system of immune cells and the synthesis of ACh. By influencing cytokine production, ACh produced by immune cells modulates in an autocrine/paracrine manner their own muscarinic and nicotinic receptors, thus contributing to modulating the cell's innate immune responses (Fujii et al. 2017). Recent research also points to a connection between the cholinergic system of immune cells and the neuroendocrine cells of the fish gills. For instance, the expression of mAChR and alpha-2 subunit of nAChR by NECs is similar to that observed in immune cells and at the sites of immune cell/nerve contacts (Zaccone et al. 2020; Capillo et al. 2021; Lauriano et al. 2021). This strongly suggests that NECs have cholinergic activity.

Co-Localization of Mast Cells and Eosinophils

We have described by immunohistochemical methods the localization of 5-HT, tyrosine hydroxylase (TH), and nNOS in fish eosinophils for the first time. In addition, eosinophils and mast cells were found to co-localize in the gills and ABOs of *C. gariepinus* (Maina et al. 2022). Eosinophils in the *Pantodon* gills were also positive for 5-HT (Capillo et al. 2021). According to Fillion et al. (1993), 5-HT is chemotactic for eosinophils and aids in eosinophil recruitment. Eosinophil functions can be modulated by mast cells and vice versa. The two cell types produce a wide variety of unique mediators. TH is produced by eosinophils (Jung 2016) and can be used to identify catecholaminergic neurons in the central and peripheral nervous systems of zebrafish (Chen and Zon 2009). Another neurotransmitter, dopamine, has also been found in a variety of immune cell types, including regulatory T cells that are CD4+ and CD25+, granulocytes, and T and B cells. Dopamine receptors in these cells modulate the proliferation and differentiation of lymphocytes (Levite 2008; Pacheco et al. 2014). The regulation of the activity of immune cells by neurotransmitters produced by the nervous system and that of nervous cells by immunologically derived neurotransmitters are examples of reciprocal neurotransmitter-mediated intersystem communications. In our study, the co-localization of mast cells and eosinophils suggests that these immune cells interact with each other. In particular, the expression of the antimicrobial immunomodulatory dopamine and nitric oxide in the mast cells positive to nNOS, TH, and Pis1 peptide should be associated with neural activation and coordination of the immune responses.

Conclusions

Clarias gariepinus and *Heteropneustes fossilis* are two catfish species that have developed the capability to breathe air through air-breathing organs (ABOs). Despite gross morphological differences, all the ABOs are endowed with a respiratory epithelium. However, the exchange barrier is thick and irregular, suggesting a low potential for efficient gas exchange. As an exception, the air sac of *H. fossilis* appears to be better suited for gas diffusion.

The respiratory epithelium of the catfish contains, among other cells, mucous cells, mast cells, eosinophils and neuroendocrine cells. These cells express a variety of neurotransmitters, neurotransmitter receptors and antimicrobial peptides such as ACh, 5-HT, VAChT, GABA receptors, nNOS and Piscidin 1. Many of these cells co-localize and are found in the vicinity of Piscidin 1 and VAChT-positive nerve fibers. Neuropeptides released by neural activity may act as signaling molecules able to modify the activity of immune cells that, in turn, could modulate neural activity. Cross talks between the neural and immune systems appear to regulate local inflammatory reactions and may also affect distant organs, such as similar epithelia in other parts of the body. These cross-talks may also modulate the activity of the central nervous system. The study of the neuroimmune communications that are likely to be operative in the respiratory organs of the catfish constitutes a promising avenue of research. Future studies will contribute to our understanding of mechanisms that regulate mucosal barrier integrity and mucosal immune response in the ABOs of the catfishes.

Acknowledgments

The authors are indebted to Dr. J.N. Maina for the generous supply of the *Clarias* tissues.

References

Abd-El Maksoud, A., M. Kassab, A. Sayed and M. Fayed. 2008. Anatomical, light and scanning electron microscopic studies on the air breathing dendritic organs of the sharp tooth catfish (*Clarias gariepinus*). J. Vet. Anat. 1: 29–37.

Ainsworth, A.J. 1992. Fish granulocytes: morphology, distribution, and function. Ann. Rev. Fish Dis. 2: 123–148.

Alesci, A., S. Pergolizzi, S. Savoca, A. Fumia, A. Mangano, M. Albano et al. 2022a. Detecting Intestinal goblet cells of the broadgilled hagfish *Eptatretus cirrhatus* (Forster, 1801): A confocal microscopy evaluation. Biology 11: 1366.

Alesci, A., M. Albano, S. Savoca, D.M. Mokhtar, A. Fumia, M. Aragona et al. 2022b. Confocal identification of immune molecules in skin club cells of zebrafish (*Danio rerio*, Hamilton 1882) and their possible role in immunity. Biology 11: 1653.

Alesci, A., G. Capillo, A. Fumia, E. Messina, M. Albano, M., Aragona et al. 2022c. Confocal characterization of intestinal dendritic cells from myxines to teleosts. Biology 11: 1045.

Alesci, A., S. Pergolizzi, G. Capillo, P. Lo Cascio and E.R. Lauriano. 2022d. Rodlet cells in kidney of goldfish (*Carassius auratus*, Linnaeus 1758): A light and confocal microscopy study. Acta Histochem. 124: 151876.

Alesci, A., S. Pergolizzi, A. Fumia, C. Calabrò, P. Lo Cascio and E.R. Lauriano. 2022e. Mast cells in goldfish (*Carassius auratus*) gut: Immunohistochemical characterization. Acta Zool. 104: 366–379.

Aresti Sanz, J. and S. El Aidy. 2019. Microbiota and gut neuropeptides: a dual action of antimicrobial activity and neuroimmune response. Psychopharmacology 236: 1597–1609.

Barrios, J., A.T. Kho, L. Aven, J.A. Mitchel, J.-A. Park, S.H. Randell et al. 2019. Pulmonary neuroendocrine cells secrete γ-aminobutyric acid to induce goblet cell hyperplasia in primate models. Am. J. Resp. Cell Mol. Biol. 60: 687–694.

Burleson, M.L. and W.K. Milsom. 1995b. Cardio-ventilatory control in rainbow trout: I. Reflex effects of exogenous neurochemicals. Resp. Physiol. 101: 289–299.

Capillo, G., G. Zaccone, C. Cupello, J.M.O. Fernandes, K. Viswanath et al. 2021. Expression of acetylcholine, its contribution to regulation of immune function and O_2 sensing and phylogenetic interpretations of the African butterfly fish *Pantodon buchholzi* (Osteoglossiformes, Pantodontidae). Fish Shellfish Immunol. 111: 189–200.

Chen, A.T. and L.I. Zon. 2009. Zebrafish blood stem cells. J. Cell. Biochem. 108: 35–42.

Dutta, H.M. and J.S.D. Munshi. 1985. Functional morphology of air-breathing fishes: a review. Proc. Anim. Sci. 94: 359–375.

Fernandes, J.M.O., J. Ruangsri and V. Kiron. 2010. Atlantic cod piscidin and its diversification through positive selection. PLos ONE 5: e9501.

Fillion, G., M.P. Fillion, I. Cloez-Tayarani, H. Sarhan, F. Haour and F. Bolanos. 1993. 5-HT and the immune system. pp. 329–335. *In*: Vanhoutte, P.M., P.R. Saxena, R. Paoletti, N. Brunello and A.S. Jackson (eds.). Serotonin. From Cell Biology to Pharmacology and Therapeutics. Springer, Dordrecht, Holland.

Forsythe, P. and J. Bienenstock. 2012. The mast cell-nerve functional unit: A key component of physiologic and patho-physiologic responses. pp. 196–221. *In*: Bienenstock, J. (ed.). Allergy and the Nervous System. S. Karger AG, Basel, Switzerland.

Fujii, T., M. Mashimo, Y. Moriwaki, H. Misawa, S. Ono, K. Horiguchi et al. 2017. Physiological functions of the cholinergic system in immune cells. J. Pharmacol. Sci. 134: 1–21.

Fujimaki, Y. and M. Isoda. 1990. Fine-structural study of leucocytes in the goldfish, *Carassius auratus*. J. Fish Biol. 36: 821–831.

Gilchrist, M., S.D. McCauley and A.D. Befus. 2004. Expression, localization and regulation of nitric oxide synthase (NOS) in the human mast cells lines: Effects on leukotriene production. Blood 104: 462–469.

Gomez, D., J.O. Sunyer and I. Salinas. 2013. The mucosal immune system of fish: the evolution of tolerating commensals while fighting pathogens. Fish Shellfish Immunol. 35: 1729–1739.

Graham, B.J. 1997. Air-Breathing Fishes. Elsevier, New York.

Herr, N., C. Bode and D. Duerschmied. 2017. The effects of serotonin in immune cells. Front. Cardiovasc. Med. 4: 48.

Hughes, G.M. and J.S.D. Munshi. 1973. Nature of the air-breathing organs of the Indian fishes Channa, Amphipnous, Clarias and Saccobranchus as shown by electron microscopy. J. Zool. 170: 245–270.

Icardo, J.M. 2019. Cardiovascular system and blood. pp. 207–229. *In*: Kirschbaum, F. and K. Formicki (eds.). The Histology of Fishes. CRC Press, Florida, USA.

Jung, Y. 2016. eosinophils and Type 2 cytokine signaling in macrophages support the biogenesis of cold-induced beige fat. J. Bacteriol. Virol. 46: 44–46.

Kang, B.N., S.G. Ha, N.S. Bahaie, M.R. Hosseinkhani, X.N. Ge, N.M. Blumenthal et al. 2013. Regulation of serotonin-induced trafficking and migration of eosinophils. PLoS ONE 8: e54840.

Kita, H. 2013. Eosinophils: multifunctional and distinctive properties. Int. Archiv. Allergy Immunol. 161(Suppl. 2): 3–9.

Knoop, K.A. and R.D. Newberry. 2018. Goblet cells: multifaceted players in immunity at mucosal surfaces. Mucosal Immunol. 11: 1551–1557.

Kobayashi, Y. and P.R. Tata. 2018. Pulmonary neuroendocrine cells: sensors and sentinels of the lung. Dev. Cell 45: 425–426.

Lauriano, E.R., S. Pergolizzi, M. Aragona, G. Montalbano, M.C. Guerrera, R. Crupi et al. 2019. Intestinal immunity of dogfish *Scyliorhinus canicula* spiral valve: a histochemical, immunohistochemical and confocal study. Fish Shellfish Immunol. 87: 490–498.

Lauriano, E.R., G. Capillo, J.M. Icardo, J.M.O. Fernandes, V. Kiron, M. Kuciel et al. 2021. Neuroepithelial cells (NECs) and mucous cells express a variety of neurotransmitters and neurotransmitter receptors in the gill and respiratory air-sac of the catfish *Heteropneustes fossilis* (Siluriformes, Heteropneustidae): a possible role in local immune defence. Zool. 148: 125958.

Lauriano, E.R., A. Alesci, M. Aragona, S. Pergolizzi, A. Miller, K. Zuwala et al. 2023. Immunohistochemistry of the gut-associated lymphoid tissue (GALT) in African bonytongue (*Heterotis niloticus*, Cuvier 1829). Int. J. Mol. Sci. 24: 2316.

Levite, M. 2008. Neurotransmitters activate T-cells and elicit crucial functions via neurotransmitter receptors. Curr. Op. Pharmacol. 8: 460–471.

Maina, J.N. and G.M.O. Maloiy. 1986. The morphology of the respiratory organs of the African air-breathing catfish (*Clarias mossambicus*): a light, electron and scanning microscopic study, with morphometric observations. J. Zool. 209: 421–445.

Maina, J.N. 2018. Functional morphology of the respiratory organs of the air-breathing fish with particular emphasis on the African catfishes, *Clarias mossambicus* and *C. gariepinus*. Acta Histochem. 120: 613–622.

Maina, J.N., J.M. Icardo, G. Zaccone, M. Aragona, E.R. Lauriano, A. Alesci et al. 2022. Immunohistochemical and ultrastructural study of the immune cell system and epithelial surfaces of the respiratory organs in the bimodally breathing African sharptooth catfish (*Clarias gariepinus* Burchell, 1822). Anat. Rec. 305: 3212–3229.

Metcalfe, D.D. 2008. Mast cells and mastocytosis. Blood 112: 946–956.

Munshi, J.S.D. and G.M. Hughes. 1991. Structure of the respiratory islets of accessory respiratory organs and their relationship with the gills in the climbing perch, *Anabas testudineus* (Teleostei, Perciformes). J. Morphol. 209: 241–256.

Munshi, J.S.D. 1993. Structure and function of the air-breathing organs of *Heteropneustes fossilis*. Adv. Fish Res. 1: 99–137.

Olson, K.R., J.S.D. Munshi, T.K. Ghosh and J. Ojha. 1990. Vascular organization of the head and respiratory organs of the air-breathing catfish, *Heteropneustes fossilis*. J. Morphol. 203: 165–179.

Olson, K.R., P.K. Roy, T.K. Ghosh and J.S.D. Munshi. 1994. Microcirculation of gills and accessory respiratory organs from the air-breathing snakehead fish, *Channa punctata*, *C. gachua*, and *C. marulius*. Anat. Rec. 238: 92–107.

Olson, K.R. 2000. Respiratory system. pp. 357–367. *In*: Ostrander, G.K. (ed.). The Laboratory Fish. Academic Press, New York, USA.

Pacheco, R., F. Contreras and M. Zouali. 2014. The dopaminergic system in autoimmune diseases. Front. Immunol. 5: 117.

Peatman, E., M. Lange, H. Zhao and B.H. Beck. 2015. Physiology and immunology of mucosal barriers in catfish (*Ictalurus spp.).* Tissue Barriers 3: 4, e1068907.

Phipps, S., C.E. Lam, S. Mahalingam, M. Newhouse, R. Ramirez, H.F. Rosenberg et al. 2007. Eosinophils contribute to innate antiviral immunity and promote clearance of respiratory syncytial virus. Blood 110: 1578–1586.

Piliponsky, A.M., M. Acharya and N.J. Shubin. 2019. Mast cells in viral, bacterial, and fungal infection immunity. Int. J. Mol. Sci. 20: 2851.

Porteus, C.S., D.L. Brink, E.H. Coolidge, A.Y. Fong and W.K. Milsom. 2013. Distribution of acetylcholine and catecholamines in fish gills and their potential roles in the hypoxic ventilatory response. Acta Histochem. 115: 158–169.

Ratcliffe, M.J.H. (ed.). 2016. Encyclopedia of Immunobiology. Vol. X. Academic Press, New York.

Rawat, A., R. Chaube and K.P. Joy. 2022. Air sac and gill vasotocin receptor gene expression in the air-breathing catfish *Heteropneustes fossilis* exposed to water and air deprivation conditions. Fish Physiol. Biochem. 48: 381–395.

Regan, K.S., M.G. Jonz and P.A. Wright. 2011. Neuroepithelial cells and the hypoxia emersion response in the amphibious fish *Kryptolebias marmoratus*. J. Exp. Biol. 214: 2560–2568.

Roumier, A., C. Bechade and L. Maroteaux. 2019. Serotonin and the immune system. pp. 181–196. *In*: Pilowsky, P.M. (ed.). Serotonin. Academic Press, New York, USA.

Ruangsri, J., J.M.O. Fernandes, J.H.W.M. Rombout and M.F. Brinchmann. 2012a. Ubiquitous presence of piscidin-1 in Atlantic cod as evidenced by immunolocalisation. BMC Vet. Res. 8: 46.

Ruangsri, J., S.A. Salger, C.M.A. Caipang, V. Kiron and J.M.O. Fernandes. 2012b. Differential expression and biological activity of two piscidin paralogues and a novel splice variant in Atlantic cod (*Gadus morhua* L.). Fish Shellfish Immunol. 32: 396–406.

Salinas, I. 2015. The mucosal immune system of teleost fish. Biology 4: 525–539.

Simon, H.-U., S. Yousefi, N. Germic, I.C. Arnold, A. Haczku, A.V. Karaulov et al. 2020. The cellular functions of eosinophils: Collegium Internationale Allergologicum (CIA) Update 2020. Int. Arch. Allergy Immunol. 181: 11–23.

Singh, B.N. and G.M. Hughes. 1971. Respiration of an air-breathing catfish *Clarias Batrachus* (Linn.). J. Exp. Biol. 55: 421–434.

Singh, B.N. 1982. Development of the air breathing organ in the catfish, *Clarias batrachus* (Linn.). Zool. Anz. 208: 100–111.

Sullivan, J.P., J.G. Lundberg and M. Hardman. 2006. A phylogenetic analysis of the major groups of catfishes (Teleostei: Siluriformes) using rag1 and rag2 nuclear gene sequences. Mol. Phyl. Evo. 41: 636–662.

Travers, J. and M.E. Rothenberg. 2015. Eosinophils in mucosal immune responses. Mucosal Immunol. 8: 464–475.

Whitsett, J.A. 2018. Airway epithelial differentiation and mucociliary clearance. Ann. Am. Thor. Soc. 15(Suppl. 3): S143–S148.

Widdicombe, J.H. and J.J. Wine. 2015. Airway gland structure and function. Physiol. Rev. 95: 1241–1319.

Zaccone, G., A. Mauceri and S. Fasulo. 2006. Neuropeptides and nitric oxide synthase in the gill and the air-breathing organs of fishes. J. Exp. Zool. 305A: 428–439.

Zaccone, G., E.R. Lauriano, M. Kuciel, G. Capillo, S. Pergolizzi, A. Alesci et al. 2017. Identification and distribution of neuronal nitric oxide synthase and neurochemical markers in the neuroepithelial cells of the gill and the skin in the giant mudskipper, *Periophthalmodon schlosseri*. Zoology 125: 41–52.

Zaccone, G., E.R. Lauriano, G. Capillo and M. Kuciel. 2018. Air-breathing in fish: Air-breathing organs and control of respiration. Acta Histochem. 120: 630–641.

Zaccone, G., J. Maina, A. Germana, G. Montalbano, G. Capillo, M. Aragona et al. 2019. First demonstration of the neuroepithelial cells and their chemical code in the accessory respiratory organ and the gill of the sharptooth catfish, *Clarias gariepinus*: a preliminary study. Acta Zool. 100: 160–166.

Zaccone, G., C. Cupello, G. Capillo, M. Kuciel, A.L.R. Nascimento, A. Gopesh et al. 2020. Expression of acetylcholine- and G protein coupled muscarinic receptor in the neuroepithelial cells (NECs) of the obligated air-breathing fish, *Arapaima gigas* (Arapaimatidae: Teleostei). Zool. 139: 125755.

Zaccone, G., G. Capillo, J.M.O. Fernandes, V. Kiron, E.R. Lauriano, A. Alesci et al. 2022. Expression of the antimicrobial peptide Piscidin 1 and neuropeptides in fish gill and skin: a potential participation in neuro-immune interaction. Mar. Drugs 20: 145.

Zinkl, J.G., W.T. Cox and C.S. Kono. 1991. Morphology and cytochemistry of leucocytes and thrombocytes of six species of fish. Comp. Haematol. Int. 1: 187–195.

Index

V

W

X

Made in the USA
Las Vegas, NV
16 September 2026

b17fbf08-87b2-4bf3-b23d-bd40df7d5eaaR01